Vegetable Fats and Oils

Sabine Krist

Vegetable Fats and Oils

with contributions from
S. Biladt, C. Bulis, G. Buchbauer, B. Ellinger, J. König,
S. Strugger, M. Vala, and M. Wall

Springer

Prof. Dr. Dr. Sabine Krist
Chair of Medicinal Chemistry
Sigmund Freud University of Vienna
Vienna, Austria

Translated by
Dr. Barbara Bauer
Vienna, Austria

ISBN 978-3-030-30313-6 ISBN 978-3-030-30314-3 (eBook)
https://doi.org/10.1007/978-3-030-30314-3

This Springer imprint is published by the registered company Springer Nature Switzerland AG
The registered company address is: Gewerbestrasse 11, 6330 Cham, Switzerland

For Lydia, Patricia, Gregor and Florens

Preface

This encyclopaedia, which scientifically describes 121 fat vegetable oils, waxes and vegetable fats from all continents, was compiled at the Department of Pharmaceutical Chemistry of the University of Vienna and at the Sigmund Freud University Vienna. The oils discussed in this reference book range from well-known and much-used oils such as olive oil, sunflower oil, corn oil and peanut oil to lesser-known oils like kiwi seed oil, tomato seed oil, capsicum seed oil and ootanga oil. Special attention was given to volatile compounds, aroma and the oils' composition with regard to the fatty acid profile and sterol and tocopherol content. To make the description of quality as exact and professional as possible, organoleptic tests were carried out by flavourists, flavour chemists and perfumers on many oils described in this book. This made it possible to describe the characteristics of oils that have not yet been adequately treated in literature. In addition to the characteristic properties of each oil, for example, its appearance, colour, contents, smell and flavour, there is also information on its uses and range of applications in pharmacy, medicine, cosmetics and technology and also, in some cases, on its use as an edible oil. Fat vegetable oils not only have positive effects; some of them also have unwanted side effects. These concomitant phenomena have also been taken into account. In the descriptions of the 121 fat vegetable oils, information on the source plant, its cultivation and the extraction of the oil has been included.

The author would like to thank flavourists Dr. Reiner Gottfried, Isolde Tomann (both from the company ESAROM, Oberrohrbach) and Dr. Reinhild Eberhardt and perfumers Wolfgang Höppner, Volker Hausmann (both from the company Symrise, Vienna) and H. J. Wenzl (from the company Firmenich, Vienna) for the sensory evaluation of the oil samples.

Furthermore, I would like to thank DI Klaus Riediger for his sound advice concerning the "novel food" part of this book.

My special thanks go to my husband for his perpetual support in all aspects of my life.

Vienna, Austria
April 2019

Sabine Krist

Contents

Introduction

1 Definitions

Vegetable Fats

Fats and fat oils consist of triesters of the triad alcohol glycerol with various, mostly even-numbered, unbranched aliphatic monocarboxylic acids (fatty acids). They are neutral compounds. The name recommended by IUPAC[1] for this class of compounds is triacylglycerol. The fatty acids of natural fats are usually unbranched, with an even number of C-atoms; most of them have between 4 and 26 carbon atoms. In vegetable fats, the primary hydroxyl groups on the carbon atoms 1 and 3 of the glycerol are generally esterified with saturated fatty acids, whereas position 2 is occupied by an unsaturated fatty acid. An exception is coconut fat,[2] which is an almost completely saturated fat, like animal fats (Hunnius 2004, p. 574; Ulmer 1996, p. 25).

The fats differ considerably in their consistency. Their melting points depend on the nature of their fatty acids. Fats with a high percentage of unsaturated fatty acids are liquid at room temperature, whereas those with a high percentage of saturated fatty acids are solid, or nearly so. Fat oils that are liquid at room temperature are classified according to their tendency towards autocatalytic oxidation in the presence of oxygen into drying oils (linseed oil and poppyseed oil), semi-drying oils (peanut oil and rapeseed oil) and nondrying oils (olive oil) (Hunnius 2004, p. 574).

The name "fat oils" is the common designation in pharmacy, in order to distinguish them from etheric, volatile oils. In nutritional science, the designation "fat" is usually used for triacylglycerols that are solid at room temperature, whereas "oil" indicates liquid triacylglycerols. In this encyclopaedia, the term "oils" will be used.

[1] IUPAC = International Union of Pure and Applied Chemistry; founded in 1919 to support the worldwide communication of chemists and lay down standardised names and binding recommendations for nomenclature, symbols, etc. (iupac.org/16.01.07).

[2] See p. 232 et seqq.

S. Krist, *Vegetable Fats and Oils*, https://doi.org/10.1007/978-3-030-30314-3_1

Waxes

Waxes are mixtures of monoesters of fatty acids with 16–36 carbon atoms and monovalent higher alcohols, usually of the same length. Waxes are defined by their mechanical and physical properties, as their chemical composition may vary. A substance is called a wax if it is kneadable/solid to brittle at 20 °C, has a coarse to microcrystalline texture, melts at temperatures above 40 °C without decomposing or turning stringy, and if its consistency and solubility largely depend on the temperature (Fleischhacker 2002; Hunnius 2004, p. 1607).

2 Historical Summary

The oldest finds of tools used for the preparation of oil date from the Stone Age. Mills to grind products, the so-called "Jericho mills", consisted of two stones, one with a mould in the form of a pan and the other ball-shaped, between which nuts and almonds were crushed (Matthäus and Münch 2009, p.13).

Since early antiquity, oil was produced from plants, fruit and seeds rich in fat. Oils were not only used for preparing meals, but also in medicine, and especially for massage and skin care. Archaeological finds from Babylonia and Egypt prove that oil was already being extracted from plants 6000 years ago (Roth and Kormann 2000, p. 14).

Olive oil, which is especially rich in valuable components, was the most important source of fat for many centuries. For Persians, Syrians, Hebrews and Romans, the fruit of the olive tree was important as a food, and various methods of extracting the oil were developed early on (Ulmer 1996, p. 11).

The olive harvest began in September and October, before the fruit was fully ripe. The branches of the olive trees were shaken with long poles; in Greece and Italy, large sheets were spread underneath the trees, to facilitate the gathering of the olives. In order to obtain oil of the best quality, the olives were not mashed, but simply gathered in baskets, from which the oil fell in drops into a container (dripping oil) (Roth and Kormann 2000, p. 14). Another method was to place the olives in a cupola-shaped space hewn into rock that had a hole at the bottom, so that the oil drained off. In both methods, the olives were crushed by their own weight (Roth and Kormann 2000, p. 14).

Another method was to pour hot water onto the fruit. The oil then settled on the surface. This method also produced oil of the finest quality, which was used for lamps in holy shrines, for the preparation of sacrifices, and was also poured into bottles and jugs for anointing. The Romans developed this technique further and constructed the oil mill "mola oleara". They put the olives in a shallow vessel and mashed them by rolling a cylinder-shaped stone back and forth. The oil flowed off through a groove and was collected in another vessel. This method is still in use in rural oil mills (Roth and Kormann 2000, p. 14; Ulmer 1996, p. 9).

In antiquity, oil was also extracted using wooden beams. It was probably the Greeks who invented this method. These oil mills were used in Crete and on the

Cyclades around 1500 B.C. The underlying principle is leverage (Roth and Kormann 2000, p. 14). Olives placed in baskets were ground beneath the beams, which in turn were weighted down by stones. The squeezed mixture of oil and water was collected in large underground cisterns and stored until the small suspended particles and foreign substances had settled to the bottom. Afterwards, the lower, watery layer was drained off, and the greenish-yellow oil remained. This oil is called "maiden oil" and is rich in vitamins (Ulmer 1996, p. 9). It was important to remove the watery components completely, as they contain the bitter substance oleuropein, which ruins the olive oil's taste. After this first cold pressing, a second and maybe even third pressing took place by adding hot water, in which the fruit pulp was soaked. The quality of the oil declined, however, with each pressing (Roth and Kormann 2000, p. 14).

After the invention of the screw thread (approx. 50 B.C.), it was possible to generate the high pressure necessary for squeezing oil seeds like flax and poppy. Around this time, Pliny mentioned presses that were moved by means of ropes or straps, which were in turn moved by levers. The importance of oil extraction in antiquity is discernible from excavations in the Mediterranean area, which brought to light whole facilities for oil production, with oil mills, oils presses, oil shops and storerooms. Screw presses of various designs were in use until the eighteenth century (Matthäus and Münch 2009, p. 14, 15).

3 The Production of Vegetable Oils and Fats Today

Vegetable fats and oils are extracted from seeds or fruit of different plants. In vegetable oil production, the primary concern is obtaining the maximum output while preserving the natural quality of the oil. In oil production, a distinction is made between cold-pressed and extracted oils. "Cold pressed" means that the fruit, for example olives, pumpkin seeds or walnuts, is crushed to a pulp under careful mechanical pressure at relatively low temperatures between 40 °C and 60 °C. The result is a high-grade natural oil. Consumers recognise it by labels such as "virgin oil", "cold pressed" or "unrefined" (Löw 2003, p. 79 et seq; Kerschbaum and Schweiger 2001, p. 1; Lawson 1995, p. 49 et seq).

Mechanical Extraction

Crushing or pressing is the most common method. The seeds are reduced to small pieces and then crushed in hydraulic presses or, in the case of more primitive manufacture, in simple wedge presses. Crushing is usually carried out in steps. In general, the seed is first pressed without adding heat. The first pressing provides the most valuable edible oil. After the cold pressing, the seeds are warmed and then pressed a second time. At this pressing, substances end up in the oil that may have a negative effect on its colour, smell or flavour. The advantage of hot pressing is that the oil becomes thinner, the proteins coagulate, and mucilage and gum disintegrate to a greater extent. Hot pressing is especially important for oils and fats that contain components with a high melting point. Viscous oils become thinner by adding heat, and can therefore be separated more easily. A decisive factor is that, with the adding of heat, the yield is about 5–8% higher than when the oil is cold pressed (Roth and Kormann 2000, p. 18).

Machines for the Cold Pressing of Vegetable Oils

Open Presses

Nowadays, open presses are only used for the extraction of olive oil. It is a discontinuous process. The olive oil pulp is spread on frames covered with filter cloths; these are then piled on top of each other. In the middle of the frame is a perforated peg, through which the olive oil flows off. Pressure is created by means of a hydraulic piston (Löw 2003, p. 84).

Closed Discontinuous Presses

Today, closed discontinuous presses are only used for the extraction of oils in small quantities. Among them are filter presses and other presses of similar build. They consist of vessels in which a pestle puts pressure on the fruit or seeds. The oil flows off via a sieve bottom and drains (Löw 2003, p. 84).

Closed Continuous Presses

Closed continuous presses are used to extract natural or cold-pressed vegetable oils. Stamping presses are used to extract poppyseed oil and camelina oil. Screw compressors contain a screw in their horizontal corridor. To intensify the pressure when the material to be pressed passes, and to compensate for pressure loss, the diameter of the winding alley decreases in the direction of conveyance. The screw usually consists of segments of various ascent and form, so that the press can be adapted to different kinds of seeds. As a result of the high pressure, temperatures usually reach about 100 °C; on occasion, they can rise to a maximum of 170 °C. If small quantities of output are acceptable, temperatures can be lowered to about 60 °C (Löw 2003, p. 85).

Extraction

From seeds reduced to small pieces, oil is extracted using solvents. Formerly, carbon disulphide, benzene, carbon tetrachloride, trichloroethylene and fuel were in use as solvents, but as they are poisonous and potentially carcinogenic, they have been substituted with hexane, which is now used almost exclusively. The advantage of extraction is that the yield of the oil is higher than when the seeds are pressed. The disadvantage is that it is difficult to remove the solvent from the oil, and that the oils must be subjected to refining afterwards, which may affect their flavour and quality. However, modern refining procedures clean extracted oils in such a way that the end product's flavour does not differ from that of pressed oils (Roth and Kormann 2000, p. 14).

Cleaning/Refining

As crude oils may contain unwanted substances affecting odour, taste and colouring, they are refined. Refined fats have been cleaned, i.e., neutralised, deodorised, decoloured and steamed. By treating the fats and oils with alkali, the free acids are esterified, and subsequently extracted from the oil by washing. Oils can also be esterified, and thereby neutralised, with glycerin. To remove inconvenient colour, certain oils are decoloured using solid adsorbents such as aluminium silicate or activated carbon. A steam distillation under vacuum removes unwanted flavouring substances. The disadvantage is that by refining, part of the unsaturated fatty acids found in oils are transformed into saturated fatty acids. Associated material, for example vitamin A, vitamin E or lecithin, is removed from the oil in the refining process. On the other hand, refining also removes toxic agents like pesticides, heavy metals and remains of mildew (Roth and Kormann 2000, p. 18).

Generally, refined oils are almost odourless and without aroma. Cold pressed oils, on the other hand, often have a characteristic smell and colour. Refined oils have the advantage that they have a longer shelf life and are more stable and heat-resistant. Yet refining, contrary to pressing, destroys flavouring substances and other valuable ingredients, for example vitamins, carotenoids and mineral compounds. For this reason, natural, cold-pressed oils are considered to be of higher quality (Lawson 1995, p. 49 et seq; Kerschbaum and Schweiger 2001, p. 4).

In addition, it must be noted that the refining of vegetable oils involves high temperatures, which may transform cis configurations into trans configurations. The percentage of trans configurations in refined oils lies between 0.1% and 0.8%, whereas their percentage in cold pressed oils amounts to a maximum of 0.1% (Wagner et al. 2000).

Liquid oils can be transformed into solid fats through hydrogenation of their double bonds, by transesterification and fractionating, or by separation of fractions melting at low temperatures. This process is called fat hardening or elaidinisation and results in hardened fats, which can, for example, be used in margarine production (Hunnius 2004, p. 574). Trans-fatty acids may form in the process; these are unsaturated fatty acids with a trans configuration on the corresponding double bonds (see below).

4 Advantages of Cold-Pressed Oil

Hippocrates (fourth century B.C.) advocated a change of diet to cold-pressed, natural oils for health reasons (Ulmer 1996, p. 7).

Cold pressing techniques do not add heat to the original material in oil production. In this way, the valuable natural associated material is largely preserved. In addition, no organic solvents are needed, and the oil remains chemically untreated. It is not necessary to add synthetic antioxidants, as cold-pressed oils contain many natural antioxidants, which can have a positive health effect (Parker et al. 2003).

Usually, cold-pressed oils are not refined after pressing. In this way, valuable associated material is preserved. Decomposition products that form in oil production when heat is added and subsequently must be removed by refining do not occur in the first place. The oils produced contain many valuable ingredients, such as high-quality fatty acids and associated material like waxes, enzymes, β-carotene, lecithin, vitamin E and various secondary plant substances such as phytosterol, which is considered to have many positive effects. β-Carotene is responsible for the characteristic reddish colour of some cold-pressed oils. Cold-pressed oils have a typical flavour and smell as well as a characteristic colour. Their properties are determined by their source plants and vary depending on the producing area, the time of harvest and the climate.

5 Disadvantages of Cold-Pressed Oil

The disadvantage of cold-pressed oils is that they still contain decomposition products and remains of plant-protective agents, heavy metals and polycyclic hydrocarbons, which have been removed from refined oils. For this reason, the raw material for cold-pressed oils must meet quality standards; there are threshold values for pollutants, so that the oils do not pose a health risk. The majority of cold pressed oils are thermally unstable because of their multiple unsaturated fatty acids and other components, so that harmful decomposition products may form when the oils are heated. For example, if polyunsaturated linolenic acid is heated over a longer period, harmful substances are produced, the so-called HNEs (4-hydroxy-trans-2-nonenals). These HNEs form in decomposition processes and can damage body cells, thus increasing the risk of heart and circulatory diseases, liver diseases and cancer. Vegetable oils with a high percentage of monounsaturated or polyunsaturated fatty acids are not suitable for roasting and frying at high temperatures. When the temperature rises above 200 °C, the unsaturated fatty acids are transformed into saturated fatty acids. If the temperature is increased further, the fatty acids split into glycerin and non-esterified fatty acids, and the glycerin is eventually transformed into the harmful acrolein. The latter stage is reached when the fat starts emitting a pungent smell. For this reason, oil should never be heated so much that smoke rises. As cold-pressed oils become rancid easily, they do not have a long shelf life. They must be stored in a cool, dark place and must be consumed quickly. Due to the various ingredients and possible residues of cold-pressed oils, their use is not recommended for babies in the first year of their life, since their digestive system is not yet fully developed. When preparing baby food, only refined oils should be used, as they still contain valuable fatty acids and relatively high quantities of vitamin E, if it is added again after the cleaning of the oil (Parker et al. 2003).

6 Character of the Selected Vegetable Oils

The character of a fat vegetable oil indicates its colour, smell and flavour. The character provides important clues about the quality of the oil. If the oil is colourless and does not smell, this is an indication that it is refined and thus heat-resistant and suitable for roasting and frying, but it is not likely to contain wholesome ingredients. Aroma is not only one of the most important assessment criteria for the quality of oils, but also for their identity and purity (Krist et al. 2006a, b, c [3]). For this reason, the volatile compounds of the described vegetable oils/fats are mentioned in this encyclopaedia as far as they have been discussed in the relevant literature.

When assessing the character of vegetable oils, both the human senses and instrumental analysis play an important role.

Principles of Sensory Analysis

Sensations originate with the senses of sight, smell, taste, hearing and touch, and also with the senses by which organisms perceive temperature and pain. When characterising a substance, for example food, we experience an oral sensation from all the senses, which is called "flavour". Flavour thus includes aroma, taste and touch. The term "aroma" describes the smell that we notice before or while eating food, whereas the term "off-flavour" indicates a deviation from the usual flavour, caused, for example, by spoilage or by packaging material, which is perceived as unpleasant. The interplay of several factors in the creation of "aroma" and "flavour" is illustrated in Diagram 1 (Blum 1999, p. 1–19; Lawson 1995, p. 318 et seqq; Jellinek 1979, p. 31):

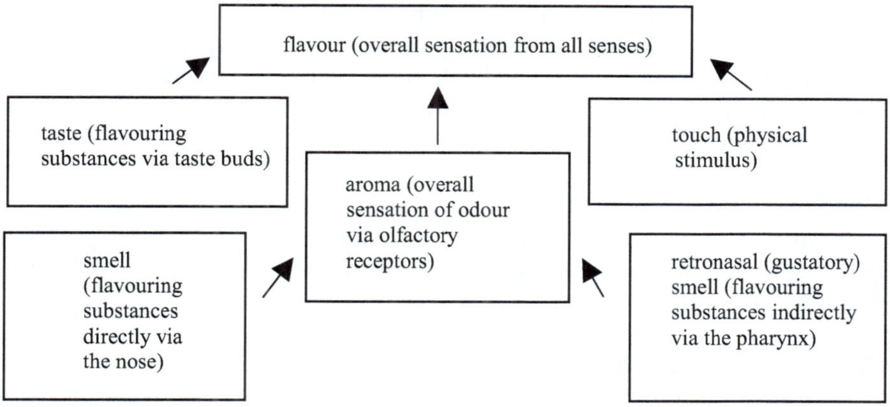

Diagram 1 *Flavour and aroma* (after Blum 1999, p. 19)

Sensory Evaluation of Vegetable Oils – Odour Testing

Many fat vegetable oils discussed in this book have not been described before in the available relevant literature with regard to their aroma components. In order to provide the missing information, odour tests were carried out at the Institute of Pharmaceutical Sciences of the University of Vienna. The oils were arranged according to increasing intensity of smell and then tested by specialists (flavourists, perfumers and flavour chemists). The test persons evaluated the oils' overall olfactory impression. The result was included in the description of the oil and specified in point 3 without bibliographical data.

Instrumental Analysis in the Research on Odorous Substances

The description of the character of fat vegetable oils includes the analysis of volatile compounds. There are often up to several hundred compounds responsible for a complex "aroma". The "aroma" of some vegetable oils has been evaluated in detail, for example that of olive oil or pumpkin seed oil. In research, different methods are used to analyse volatile compounds, which are often responsible for the typical general impression of smell and flavour (Boucek 1997, p. 30; Reiners 1997, p. 3 et seqq).

Aroma Extraction Dilution Analysis (AEDA)

By combining olfactometry and gas chromatography (GC), odorants can be identified from a multitude of volatile compounds, and can thus be distinguished from compounds without aroma. The technique used here is the so-called GC sniffing technique, a combination of machine and human sensory evaluation. Components that do not result in a signal of the detector can be registered by the experimenter, as the human nose is not only selective, but with regard to some substances, also much more sensitive than any chemical or physical detector. In the AEDA, the volatile fraction is gradually diluted by adding a solvent, and analysed by smelling the carrier gas stream. The AEDAs result largely depends on the perceptive faculty of the experimenter. It is a selective method indicating flavouring substances that primarily cause the aroma of food. Further analysis should then concentrate on their identification (Reiners 1997, p. 114 et seq).

Headspace Analysis (HDA)

In this method, successively halved volumes of a gas are analysed using gas chromatography and an odour test. The HDA supplements the AEDA and is especially useful for potent volatile compounds (Reiners 1997, p. 115 et seqq).

SPME-GC-MS Analysis

Modern aroma analysis would be unthinkable without SPME (solid-phase microextraction). It is an adsorption/desorption technique employed to enrich volatile and less volatile compounds directly from gaseous or liquid samples. Its greatest advantage is that solvents are not necessary. After the aromatic substances have been enriched on the SPME fibre, they are analysed using a combination of gas chromatography and mass spectrometry (Krist 2002, p. 47 et seqq). By applying this method, all volatile compounds of a sample are registered, not only the aromatic substances. The resulting data can be used to characterise and identify oils and fats (Krist 2011, p. 9 et seqq).

7 Physical and Chemical Key Figures

These key figures were already used before modern methods of fat analysis (TLC, GC and HPLC) were developed, in order to identify substances, and changes caused by storage, treatment, etc. The following properties are mentioned in this book:

- **density: The density of an oil is the weight ratio of the volume units of the oil in question to water** (Pharmacopoeia Europaea 1976, p. 54).
- **refractive index:** The refractive index of an oil is the ratio of the sine of the angle of incidence of a ray of light in air to the sine of the angle of refraction of the ray in the oil (Pharmacopoeia Europaea 1976, p. 62).
- **iodine number: states how many grammes of i**odine 100 g oil can bind. The iodine number is given in relation to the number of double bonds in the fatty acids and is thus a measure for the proportion of unsaturated fatty acids (Hänsel et al. 1999, p. 236).
- **acid number: states how many milligrammes** KOH are necessary to neutralise the non-esterified fatty acids in 1 g fat oil (Hänsel et al. 1999, p. 236).
- **hydroxyl number: states how many milligrammes** KOH of acetic acid bound in the process of acetylation by 1 g fat are equivalent (Hänsel et al. 1999, p. 236).
- **peroxide number: states how much peroxide** (given in milliequivalents of active oxygen) are contained in 1000 g oil.
- **saponification number:** states how many mg KOH are necessary to neutralise or saponify the acids and esters contained in 1 g oil, and thus indicates the degree of purity (Hänsel et al. 1999, p. 237)
- **unsaponifiable:** the percentage of substances that can be extracted using organic solvents after saponifying the fat oil and that are not volatile at 105 °C. Among these are sterols, tocopherols, hydrocarbon (squalene, amongst others) and triterpene alcohols (Karleskind 1996).
- **melting point:** the temperature at which the oil's aggregate state changes from solid to liquid (Pharmacopoeia Europaea 1976, p. 45).
- **point of solidification:** the temperature at which solidification of the oil begins when the temperature drops (Pharmacopoeia Europaea 1976, p. 42).

8 Shelf Life and Storage

To use fat vegetable oils properly, it is important to know their perishability and storage requirements. Unfavourable conditions in storage can deteriorate the oil's quality and lead to spoilage and rancidity.

Fat Spoilage

Fats, oils and food containing fat are naturally subject to a more or less pronounced, exogenously or endogenously induced spoilage process. This process can be induced either enzymatically by endogenous or microbial phospholipases, lipases or lipoxygenases, or – more importantly – by autoxidative spoilage. The period at the beginning, in which oxidation processes are slow and the product is still marketable, is called the induction phase. Its duration depends on the composition of the oil's fatty acids, the temperature and the proportion of antioxidants. When all antioxidants have been used up, the input of oxygen and, parallel to this, the concentration of primary oxidation products increases exponentially in autoxidative spoilage. A radical chain reaction is triggered, leading to organoleptic changes in the product, a process that is also known as "rancidity". Discolouration due to various polymerisation reactions may also occur (Herr 2003, p. 3).

The mechanism of autoxidation can basically be subdivided into three phases (Herr 2003, p. 4):

- chain propagation,
- chain branching and
- chain termination.

The fatty acid profile influences the fatty acids' stability against oxygen. While saturated fatty acids are relatively stable, unsaturated and especially polyunsaturated fatty acids are more liable to autoxidation. Oxygen and energy in the form of light and warmth are conducive to this process. Traces of metals like iron, copper and manganese also lead to a distinct shortening of the induction phase. Non-esterified fatty acids also have a negative influence on stability (Herr 2003, p. 6 et seq).

Storage

Oil keeps longer when it is stored in a cool (8–12 °C), dark place, but its shelf life can also be lengthened by natural or added antioxidants. Fumigation with inert gases may also suppress the damaging effects of oxygen effectively (Herr 2003, p. 8).

Shelf life of cold pressed oils in appropriate storage nevertheless differs:

Vegetable oil	Shelf life in months	Recommended temperature in storage
Almond oil	10	Room temperature (about 20 °C)
Flax oil	1–2	Refrigerator
Hazelnut oil	6	Room temperature
Hempseed oil	9	Refrigerator
Olive oil	12	Room temperature
Peanut oil	12	Room temperature
Poppyseed oil	9	Cool place
Pumpkinseed oil	12	Refrigerator
Rapeseed oil	12	Cool place
Sunflower oil	9	Room temperature
Thistle oil	9	Room temperature
Walnut oil	6	Room temperature
Wheat-germ oil	24	Room temperature

Antioxidants

Antioxidants are substances that delay the oxidation process and thus counteract rancidity and other negative changes in smell and taste. Antioxidants are subdivided into classes according to their mode of action (Herr 2003, p. 8 et seq):

Food antioxidants			
Primary antioxidants		Secondary antioxidants	
Phenols	Gallates	Oxygen scavengers	Sulphites
	Hydroquinones		Ascorbic acid
	Butylated hydroxyanisoles		Ascorbyl palmitates
	Butylated hydroxytoluenes	Chelating agent	Lecithin
	Tocochromanols		Citric acid

9 Use of Vegetable Fat Oils

Vegetable oils have many uses. Most importantly, they are a food product, and used for the preparation and seasoning of food. Oils also have important external uses, for example in cosmetics and for therapeutic unction. They form part of skincare and hair-care products, and serve as a base for the manufacture of pharmaceutical and cosmetic products. Some oils have a pain-relieving or anti-inflammatory effect, or support cell regeneration; they alleviate sunburn, they help dry or greasy skin to regain its natural balance, and some are effective in treating eczema, for example avocado oil or grapeseed oil. In addition, there are many technical applications, for example in the production of lubricants, paint and varnish. Naturopathy and folk medicine also use vegetable oils. Some oils are even presented as "cure-alls" for almost every human affliction. Whether they are actually effective becomes apparent when the medical usage of different oils is compared (Kerschbaum and Schweiger 2001, p. 46; Ulmer 1996, p.3 et seq; Kircher 2002, p. 84 et seq).

As Food

Many but not all vegetable oils and fats are edible. Cold-pressed oils are used predominately as salad oil. As a basic principle, oils with many polyunsaturated fatty acids should not be heated, as this would promote various processes leading to spoilage, for example oxidation. Some oils have a strong flavour of their own, for example borage seed oil or pumpkinseed oil. It is preferable to use them for seasoning. In the food industry, vegetable oils are used, amongst other things, as raw materials for the production of margarine (Kerschbaum and Schweiger 2001, p. 46; Ulmer 1996, p. 3 et seq; Kircher 2002, p. 84 et seq).

Most importantly, edible fats and oils provide energy for the human organism (about 9 kcal/g fat). They also contain associated material, especially liposoluble vitamins (vitamin A, D, E and K) and carotenoids, and are indispensable for the absorption of these nutrients into the human digestive system. In addition, vegetable oils are a source of vital polyunsaturated fatty acids, which are necessary for synthesising many essential compounds of the immune system, cell membranes and steroid hormones.

As mentioned above, fat contains, independent of its vegetable source, about 9 kcal/g or about 38 kJ/g of energy if all three fatty acids of the triglyceride in β-oxidisation are broken down. From 100 g fat, 51.4 mol ATP are formed, in contrast to 23.5 mol for the breaking down of 100 g glucose, or 20.4 mol for the breaking down of 100 g protein (Löffler et al. 2007). Fats thus have the highest energy density of all foodstuffs and, on the basis of a conventional diet, supply about 40% of the total energy. According to a recommendation of the German Nutrition Society (*Deutsche Gesellschaft für Ernährung*), 30% of the total energy input should derive from fat (DGE et al. 2000). This corresponds to 83 g/d of fat if the average input is set at 2500 kcal/d.

As a result of its lipophilic quality, fat is subject to a relatively complex absorption mechanism. The breaking down of fat to fatty acids and monoglycerides basically happens in the duodenum and the jejunum. A lipase located at the root of the tongue and a gastric lipase also contribute to the breaking down of fat, but only marginally. Bile salts, which are secreted in the duodenum, emulsify the nutritional fats, which are split with the help of peristaltic movements into small droplets, the so-called fat micelles. The surface is thereby enlarged by a factor of 10,000, which facilitates access for lipases from the pancreas. Micelles basically consist of long-chain fatty acids, monoglycerides and bile salts. Via passive diffusion, the fatty acids and other decomposition products reach the mucous cell of the small intestine, where they are again resynthesised to triglycerides, phospholipids and cholesterol esters. The resynthesised products are released into the thoracic duct of the lymphatic system and later into the blood stream in the form of chylomicrons. With the help of lipoproteins of different density (very low density lipoproteins VLDL, low density lipoproteins LDL, high density lipoproteins HDL, and various subforms of these main categories), the lipids are transported to the tissues (Biesalski et al. 1999).

Fats are not only an energy source, but also supply fatty acids and other preliminary substances for the synthesis of phospholipids, which play an important part in the stability, permeability, saltatory conduction and fluidity of the cell membranes of almost all cells. Fatty acids are also pre-stages of eicosanoids (prostaglandins, thromboxanes and leukotrienes), which are important cell mediators at different metabolic levels. Polyunsaturated fatty acids of the omega-6 and omega-3 row are transformed into more or less effective eicosanoids. As the human organism cannot completely synthesise eicosanoids, their preliminary substances must be ingested in the form of linoleic acid and α-linolenic acid. For this reason, these fatty acids are called essential fatty acids. Via the enzyme Δ9-desaturase, the human organism can build a double bond into a saturated fatty acid, for example into stearic acid (C18:0), which results in the corresponding omega-3 fatty acid C18:1, oleic acid. However, animal cells have no enzymes for synthesising omega-3 fatty acids; they only have Δ9-, Δ6- and Δ5-desaturase at their disposal. Only polyunsaturated fatty acids from linoleic acid and linolenic acid from vegetable sources can be synthesised by combining desaturation and chain-extending (elongases). With the help of Δ6-desaturase, by chain-elongation and then by finally adding an additional double bond through the activity of Δ5-desaturase, linoleic acid (C18:2n6) can be synthesised into arachidonic acid (C20:4n6). Other similar reactions synthesise other polyunsaturated fatty acids; only one double bond can, however, be inserted between already existing double bonds and the carboxyl group of the fatty acid (Löffler et al. 2007).

Currently, the recommended supply of essential fatty acids is 2.5% of the energy input for omega-6 fatty acids and 0.5% of the energy input for omega-3 fatty acids. With an energy input of 2500 kcal/d, this corresponds to about 7 g linoleic acid and 1.4 g linolenic acid. The ratio of omega-6 and omega-3 fatty acids is also important; the recommended ratio is 5:1 (DGE et al. 2000).

Evaluation of Vegetable Oils in Nutritional Physiology

The vast majority of vegetable oils and fats, especially vegetable oils that are liquid at room temperature, contain a high amount of fatty acids that are valuable from the point of view of nutritional physiology. As mentioned above, polyunsaturated fatty acids are vital for humans, and they have extensive positive effects on health. The fact that all fats and oils have a high energy content, and that the supply should not exceed 30% of the total energy supply, must, however, be taken into account.

After considering the quantity, it is the quality of fats and oils, in particular, that is important in nutritional physiology. The quality is determined primarily by the content of fatty acids. Not only the absolute quantity of unsaturated fatty acids is relevant, but also the ratio of polyunsaturated fatty acids, monounsaturated fatty acids (mainly oleic acid) and saturated fatty acids in general, as well as the ratio of omega-6 fatty acids and omega-3 fatty acids in particular. The concentration of trans-fatty acids (see below) is also of consequence for evaluating the quality of a fat.

The ratio of the different kinds of fatty acids is expressed with the quotient P:M:S, which indicates the quantity of polyunsaturated fatty acids, monounsaturated fatty acids and saturated fatty acids. According to the current reference values, the ratio should be 0.7:1.3:1. Only a third of the energy supplied in the form of fat should thus come from saturated fatty acids, which corresponds to 10% of the total energy supply. Polyunsaturated fatty acids should contribute about 7% to the total energy input, or, if the amount of saturated fatty acids exceeds the 10% limit, up to 10%. In total and with a total fat supply of 30% of the food energy, the ratio of saturated fatty acids to unsaturated fatty acids should be 1:2 (DGE et al. 2000). Unsaturated fatty acids should originate predominantly from vegetable fats.

The quality of the fats in most vegetable oils meets these criteria; some oils have more unsaturated fatty acids than the ratio stipulates. Vegetable fats that are solid at room temperature contain a high percentage of saturated fatty acids and do not therefore meet the P:M:S quotient. Nevertheless, in terms of nutritional physiology, all fats and oils of vegetable origin are of a higher quality than animal fats, as they do not contain cholesterol.

It is hardly possible to characterise vegetable fats and oils in terms of nutritional physiology in any general way, since much depends on their production and dietary use. For the production of margarine, for example, the vegetable oils are hardened, so that they are easy to spread at room temperature. In this process, unsaturated fatty acids hydrogenate to saturated fatty acids. The advantage of unsaturated fatty acids in nutritional physiology is thus lost, and trans-fatty acids may form under certain conditions (see below). Furthermore, cold-pressed, virgin vegetable oils are not suitable as fats for frying, baking and roasting, as high temperatures destroy the unsaturated fatty acids, so that their advantage in nutritional physiology is again lost. It should also be noted that vegetable oils with a high amount of unsaturated fatty acids have a shorter shelf life than oils that contain a lower amount of unsaturated fatty acids. They must be stored in a cool, dark place, and should be consumed soon after opening.

In general, vegetable oils produced with care are, in terms of nutritional physiology, high-quality fats. Nowadays, about 16% of an adult's energy is consumed in the form of saturated fatty acids, 14% in the form of monounsaturated fatty acids, and only 6% in the form of polyunsaturated fatty acids (Elmadfa et al. 2004). From the perspective of nutritional physiology, it is desirable to consume more vegetable oils, provided that the total consumption of fat does not exceed 30% of the total energy supply. As no single vegetable oil is of particular importance, but rather a variety of vegetable oils is available with different amounts of polyunsaturated fatty acids and associated material, like tocopherols, sterols and phospholipids, a healthy diet should ideally include different kinds of vegetable oils.

Use of Fat Vegetable Oils in Pharmacy and Medicine

This encyclopaedia pays particular attention to the applications of vegetable oils in pharmacy and medicine. Depending on the composition of fatty acids in the different oils and the associated material, there are different therapeutic uses, which are described separately in the discussions of the individual oils.

In Pharmaceutical Preparations

Vegetable oils are frequently part of dermatological formulations, for example ointments, emulsions, lipogels, creams, patches, soaps and liniments. Liniments are either liquid, semi-liquid, gelatinous, or solid but melt at body temperature; they are homogenous mixtures for external use and consist of fat oils, soaps, fats, balms or similar saponifiable or emulsifiable substances. Hardened vegetable oils are also sometimes used, especially hydrogenated peanut oil, cottonseed oil or castor oil. In emulsified form, vegetable oils are absorbed through the skin quite easily, and moisturise it. Salves containing oils are used especially for rough and dry skin, but also for skin scoured by solvents or detergents. Even water leads to a scouring of the skin, especially in the cold winter months or when applied frequently, which makes the use of ointments necessary (Roth and Kormann 2000, p. 29).

Oils can also be used in preparations for the eyes. As oily solutions and suspensions are much more stable towards hydrolysis, they have a longer duration of action and a longer retention period due to their viscosity (deposit effect). Ophthalmic solutions containing vegetable oils, for example *Oleum Amygdalae, Oleum Olivae* or *Oleum Ricini,* must be free from particles, however, to avoid injury and inflammation of the eye. Oleiferous eye drops have the advantage that they can be kept sterile more easily than aqueous eye drops. The disadvantage is that they impair vision, which is why they are used predominantly in the evening/at night (Valenta 2005, p. 70).

Oleiferous nose drops are only used for *Rhinitis sicca*[3]; aqueous nose drops are more common. Preparations for the nose contain fat degradable vegetable oils, as paraffin oils cannot be broken down and thus accumulate in the body over the years (Valenta 2005, p. 86). If tensides are added, oleiferous nose drops come into direct contact with the mucous membrane and immediately take effect. As a result of the viscosity of their oleaginous basis, they do not flow off through the nose too quickly (Hunnius 2004, p. 1300).

Vegetable oils like cocoa butter, palm kernel oil and coconut oil are used for the production of suppositories – this is a classical, magistral formula. They also form the base for vitamin drops (for the liposoluble vitamins A, D, E and K) and are used as a fill material in soft-shelled gelatine capsules (for example rapeseed oil or soya bean oil).

Lipophilic drugs that cannot be taken orally are injected as an oleaginous solution. The oil in these solutions must be subjected to dry heat sterilisation before use. This also demonstrates the fact that the use of oils depends on their thermal resistance. In this case, the oil's acidity must not exceed a value of 0.2 or 0.25. As a ressult of the danger of embolism, the solvents must never be injected into the blood stream or into liquor. Peanut oil, almond oil, corn oil, poppyseed oil, olive oil, castor oil, soya bean oil and sesame oil, in particular, are suitable for injections (Viernstein 2007; Roth and Kormann 2000, p. 29).

Fat vegetable oils can also be used as bath essences. They are not water-soluble and form a thin film on the surface of the bath water, unless surfactants are added. A differentiation is between spreading and non-spreading oil baths. Surfactants are

[3] Rhinitis sicca: a form of chronic coryza with eczema and exudate forming in the lower part of the nose and the inner side of the alae, accompanied by itching and a feeling of dryness (Pschyrembel 1994, p. 1332).

added to spreading oil baths, dispersing the oil in the water. Depending on the quantity of the emulsifying agent, they either look milky (high proportion of emulsifying agents) or clear (very high proportion of emulsifying agents). Typical emulsifying agents for bath essences are ethoxylated or propoxylated fatty alcohols, esters of phosphoric acid, sulphates of ricinoleic acid and alkylether sulphates. In the bath, an oily film forms on the surface of the skin, which protects it from drying out, provided that it is not removed by showering or rubbing afterwards. Non-spreading oils form a thin film on the surface of the water, as they contain no or few emulsifying agents. They moisturise the skin even more effectively, provided that the film is not removed from the skin by mechanical means. For therapeutic use, bath essences without emulsifying agents are preferable. Modern concentrated bath essences made from phosphatidylcholines and oils spontaneously form a mixture of liposomes and nanoparticles if added to water. They have a good skin care effect. Due to their chemical affinity to skin, small quantities are sufficient. Pure oil baths are used for therapeutic purposes, for example to treat dry skin or skin diseases like neurodermatitis (Raab and Kindl 2004, p. 306; Lautenschläger 2004).

Therapeutic Use

Vegetable oils like olive oil, castor oil, peanut oil or linseed oil are mild laxatives if administered orally in a dose of 30–60 ml. If they have not been saponified, they can be used as lubricants; after hydrolysation in the gut lumen, they have a slightly irritating effect on the intestinal mucosa. Fats that release hydroxy fatty acids in hydrolysis are highly anti-absorptive[4] and hydragogue,[5] which leads to increased water excretion and diarrhoea. The Na/K-dependent ATPase is blocked, resulting in an obstruction of the resorption of sodium ions and water. Simultaneously, the influx of electrolyte and water into the gut lumen is increased. A well-known example of an oil with this effect is castor oil. In the small intestine, the actual active component, ricinoleic acid, is released. Glycerides of the ricinoleic acids are also contained in grapeseed oil and African dingil oil. In the case of constipation of stool (*dyschezia*), vegetable oils are inserted rectally to soften solid faeces. Vegetable oils, especially olive oil, are also used to remove scabs and exudates, and to cleanse skin from adherent ointments (Roth and Kormann 2000, p. 29; Hänsel and Sticher 2007, p. 769 et seq; Mutschler et al. 2001, p. 650).

In nutritional medicine, the content of polyunsaturated fatty acids in vegetable oils is the most important factor. They are counted among the essential fatty acids, which cannot be produced by the organism and must therefore be ingested. Vegetable oils mainly supply omega-6-fatty acids, whereas fish supplies omega-3-fatty acids. These unsaturated fatty acids have a cholesterol-lowering effect (Löw 2003, p. 95–96). A study by Wagner et al. showed that ingestion of a vegetable oil with a high percentage of polyunsaturated fatty acids (in this case, corn oil) resulted in a significant reduction of LDL cholesterol. The quantities of VLDL cholesterol, plasma triglycerides and VLDL triglycerides were much lower when the test

[4] Anti-absorptive: Water absorption from the intestines into the lumen is prevented.

[5] Hydragogue: stimulating water absorption (foodnews.ch/06.03.07).

persons were given corn oil than when they were supplied with monounsaturated fatty acids. Corn oil thus has a greater influence on lipometabolism than a diet containing monounsaturated fatty acids. A diet containing the polyunsaturated fatty acids of corn oil can, according to Wagner et al., reduce the risk of coronary cardiac diseases, an effect which is probably also connected to the oil's unsaponifiable substances, for example phytosterol (Wagner et al. 2001). A study by Schwingshackl and Hoffmann shows that a diet with a high proportion of monounsaturated fatty acids also minimises cardiovascular risk factors. The reduction of diastolic blood pressure is particularly noteworthy, since a reduction of blood pressure by 3 mmHg reduces the risk of coronary cardiac diseases by 5–10%, the risk of myocardial infarction by 8–15%, and overall mortality by 5%. The risk of dying from a cardiovascular disease is thus reduced by increasing the supply of unsaturated fatty acids (Schwingshackl and Hoffmann 2011a, b).

Fats and Oils in Cosmetics

Even though they appear to have no spectacular effects, oils and fats are among the most important ingredients in cosmetics. Lipophilic substances are more or less insoluble in water, and are thus used to support the lipid layer of the skin, which forms a barrier against the outside. Lipophilic substances make the skin water-repellent, minimise transepidermal loss of water and protect the skin from drying out. They fill up microscopic dimples and thus help to reduce lines and smooth the skin. Creams containing oils with a low solidification point are easier to apply. Fats and waxes with a higher melting point produce a more solid texture; for this reason, they are mainly used in lipsticks. The interesting effects of the associated material in vegetable fats and fat oils also play a decisive role. Many oils contain liposoluble vitamins and provitamins, for example β-carotene and vitamins A and E. Phytosterols are another important ingredient of vegetable oils. Their structure is similar to that of cholesterol, so that they can replace it in the skin to a large extent, which strengthens the skin's barrier effect. Phytosterols are particularly useful for older skin.

Another important criterion for the selection of oils for cosmetic purposes is the composition of their triglycerides, respectively their fatty acids bound to glycerin. Unsaturated acids make oils liquid and easy to spread. For this reason, vegetable oils containing oleic acid, for example olive oil, are used in skin care. Polyunsaturated vegetable oils release linoleic acid, which counteracts flaky skin because it is, among others, built into ceramide I (a barrier substance). γ-Linolenic acid is anti-inflammatory and soothes itchiness, and is thus used in topical preparations for the treatment of very dry skin and neurodermatitis. The effect of these oils can be heightened if they are administered encapsulated in nanoparticles, which improves penetration. To strengthen the function of human skin, especially of the *stratum corneum*, as a barrier, then vegetable oils definitely take precedence over cosmetic products based on hydrocarbons. Hydrocarbons also form part of the lipid layer of the skin, but paraffin oils foreign to the body and petroleum jelly, applied in high concentration, reduce the regeneration of the skin by forming an impermeable film on its surface. The main reason for adding these substances is that they do not form decomposition products and are stable against the oxygen in air and solar radiation, whereas vegetable oils must be protected by antioxidants such as butylated hydroxyanisole or butylated hydroxytoluene.

As oils and fats are hydrophobic, they must be converted into a stable emulsion before being added to aqueous creams. Today, mainly emulsifying agents are used for this purpose. Membrane-forming components are well suited for cosmetic care products for sensitive skin, as their physical structure corresponds to natural skin conditions. Alternatively, oils can be converted into oleogels, which are used like creams but are anhydrous and thus microbiologically more stable. Oleogels are hydrophobic gels that consist of liquid paraffin and polyethylenes, or of fat oils thickened with the help of colloidal silicic acid, aluminium soap or zinc soap. In cosmetic care products, oils are also used as additives, for example as carrier substances for scents or liposoluble vitamins (Lautenschläger 2004).

10 The Most Important Ingredients of Fat Vegetable Oils

Fatty Acids in Vegetable Oils

Fatty acids are monocarboxylic acids of different lengths with double bonds in different numbers and positions. If they do not have a double bond, they are saturated fatty acids; if they have at least one double bond, they are unsaturated fatty acids. In vegetable oils, the fatty acids have an even number of C-atoms in cis configuration.

Fatty acids are defined by the length of their chains and their number of double bonds (Bauer 1928, p. 53 et seq):

- 12:0 lauric acid
- 14:0 myristic acid
- 16:0 palmitic acid
- 16:1 palmitoleic acid
- 18:0 stearic acid
- 18:1 oleic acid
- 18:2 linoleic acid
- 18:3 α-linolenic acid
- 20:0 arachidic acid
- 20:4 arachidonic acid
- 22:0 behenic acid
- 24:0 lignoceric acid

In contrast to animal fats, fat vegetable oils contain predominantly unsaturated fatty acids, often over 80%. In nutritional physiology, the substitution of vegetable oils for animal fats is recommended, as they supply the human organism with polyunsaturated fatty acids, so-called essential fatty acids. They are components of phospholipids, which are indispensable for the permeability of cell membranes, for the fixation of enzymes on the *crista mitochondrialis* and for ion transport (Kerschbaum and Schweiger 2001, p. 48; Fleischhacker 2002, p. 24).

Oleic Acid

Oleic acid is not an essential fatty acid, but it is nevertheless of great importance since it is found in every fat in various quantities. Chemically, oleic acid is an 18:1, ω-9 fatty acid, which means that it is a monounsaturated fatty acid. Polyunsaturated fatty acids are generally regarded as being more valuable in nutritional physiology, but oleic acid also has cholesterol-lowering properties. Research has shown that a diet rich in monounsaturated fatty acids such as oleic acid lowers the cholesterol level of the blood and strengthens the immune system.

When compared to polyunsaturated fatty acids, oleic acid is more effective in supressing arteriosclerotic developments and reduces the local accumulation of oxidised low-density-lipoprotein (LDL), which is involved in the development of arteriosclerotic plaque. In addition, oleic acid is certain to play an important part in the cell structure of small children and has a positive effect on their learning aptitude (Parker et al. 2003; Fleischhacker 2002, p. 14).

Omega-6 Fatty Acids

Among the ω-6 fatty acids are linoleic acid (18:2, ω-6) and γ-linolenic acid. They are essential fatty acids and are also important as a source material for the synthesis of prostaglandins and leukotrienes. In healthy human organisms, γ-linolenic acid is formed from linoleic acid by the enzyme Δ6-desaturase. An absence of the enzyme results in a lack of γ-linolenic acid and the corresponding metabolites, which are important for the organism. The enzyme's activity is inhibited by saturated fatty acids, alcohol, stress and cholesterol levels that are too high. Inactivity of the enzyme can be circumvented by a selective supply of γ-linolenic acid from vegetable oils (Fleischhacker 2002, p. 25).

Omega-3 Fatty Acids

α-Linolenic acid is an ω-3 fatty acid. It is found in vegetable oils, but only in few foods of animal origin, such as salmon and herring. ω-3 fatty acids have a broad spectrum of activity. Their metabolites are anti-inflammatory, reduce the risk of arteriosclerosis and lower triglyceride and cholesterol levels demonstrably. Walnut oil and rapeseed oil, in particular, are rich in α-linolenic acid (Fleischhacker 2002, p. 25).

Possible Undesirable Components of Vegetable Fats

Trans-fatty Acids

Naturally occurring unsaturated fatty acids normally have a cis configuration, which means that viewed from the double bond, the fatty acid parts lie on the same side. This configuration is part of their physiologic importance, as their spatial arrangement is crucial for the organic substances (for example phospholipids) derived from them. As a consequence of the activity of microorganisms in the gastro-enteric system of ruminants, cis configurations in fatty acids can be transformed into trans configurations, which is why a certain amount of trans-fatty acids can be found in the fat of milk and dairy products. Vegetable fats do not normally contain fatty acids in trans configuration, but during food processing, for example when heating over longer periods of time, cis configurations can be transformed into trans configurations, so that, for example, linoleic acid is changed into elaidic acid (tC18:1). Hydrogenation processes in fat hardening, but also the use of fats with a high proportion of unsaturated fatty acids for frying, can result in the formation of considerable quantities of trans-fatty acids. It is currently assumed that fatty acids with trans configurations are more likely to increase the cholesterol level than unsaturated fatty acids, and are thus atherogenic, even though they have double bonds (Biesalski et al. 1999). As can be shown, trans fats lead to a rise in LDL (low density lipoprotein) cholesterol and a lower HDL (high density lipoprotein) cholesterol level. They thus increase the LDL/HDL ratio and the risk of cardiovascular diseases. Due to the negative effects of trans fats in human organisms, there is global demand for a reduction of trans fats in food to a maximum of 2 g/100 g fat (Wagner et al. 2006; Großgut 2009).

Phthalates

Phthalates are emollients; without them, some synthetic materials would be brittle and thus unfit for use. Phthalic acid esters shift the thermoelastic range towards low temperatures, which causes synthetic materials to have the desired properties at operation temperature. Phthalates are predominantly used as emollients for the widely-used plastic material polyvinyl chloride (PVC). It is known that they are injurious to health and the environment; meanwhile, their use is prohibited in many areas.

Phthalates are not chemically bound to the plastic and can thus be extracted by solvents such as fats and oils. Phthalates that have been classified as toxic to reproduction (phthalate report of the German Federal Environmental Agency 2007) have repeatedly been found in vegetable oils and fats. They enter by way of packing materials, containers, cover seals, conveyors (for example during olive harvest) or directly via unsuitable filling hoses (*Stiftung Warentest*, test on gourmet oils 9/2015). Emollients have been detected, for example, in olive oil, walnut oil, argan oil and sesame oil (www.test.de: test olive oil 2/2017; test gourmet oils 9/2015). When buying vegetable oils, it is thus crucially important to look out for approved quality.

Phthalate report of the German Federal Environmental Agency 2007:

The Most Important Vitamins in Vegetable Oils

Vitamin E (α-tocopherol)

Vitamin E is an umbrella term for all tocopherols and tocotrienols with vitamin E activity. In topical preparations, the esters of vitamin E, acetates, succinates and nicotinates, are used because they are more stable. If they penetrate the skin, esterases split them into their actual active form, unesterified vitamin E. Vitamin E is used as an antioxidant in topical preparations in concentrations between 0.05% and 0.2%. The most important function of vitamin E is its function as a scavenger. The effects can be of a primary or secondary nature. In the first instance, interfering noxae, such as singlet oxygen created by the oxidation of tocopherol, are directly intercepted. In the latter instance, radical chain reactions that have already begun are halted. Radical reactions in the skin are responsible for inflammatory diseases and degenerative processes and, in the worst case, can trigger a carcinoma (Raab and Kindl 2004, p. 260).

The main characteristics of vitamin E in cosmetic preparations are thought to be (Raab and Kindl 2004, p. 260):

- protection against ultraviolet rays by inhibiting erythema, inflammations and oedema;
- preventing ageing of the skin caused by ultraviolet rays by preventing UVA rays from damaging connective tissue;
- preventing the formation of age spots;
- raising the mitotic activity of the *stratum basale*; and
- increasing the capacity of the *stratum corneum* to keep the skin moisturised by regulating the transepidermal water loss and thus improving the relief of the skin surface.

In dermopharmaceutical products, the main effects of vitamin E are (Raab and Kindl 2004, p. 260):

- quicker wound healing;
- controlled scar formation;
- reducing inflammatory processes in the skin; and
- soothing itchiness.

Many activities of vitamin E as an agent are possible because of its molecular structure, which permits excellent penetration into lower dermal layers as a result of its lipophilicity. The molecule's polarity (hydrophilic chromane ring and lipophilic side chain) also allows for direct insertion into intracellular and extracellular membranes. The membranes are thereby stabilised, and physical and chemical interactions maintain the fluidity of the membrane. Another factor is the vitamin's excellent properties as a scavenger, especially in lipophile cell membranes, whereas other scavengers mainly take effect in hydrophile cell compartments.

The vitamin E concentration in cosmetic and topical preparations varies. In cosmetic products, it amounts to 2–10%, and in dermatopharmaceutical products to 10–25%. As the damage that the skin sustains from radical processes was not known until fairly recently, the importance of vitamin E as an active agent is probably even higher than currently believed (Raab and Kindl 2004, p. 260).

All vegetable fats and oils contain vitamin E, but the concentration of the individual vitamers varies. α-Tocopherol, *β*-tocopherol, γ-tocopherol and δ-tocopherol all contain effective compounds of vitamin E in different concentrations, as do the corresponding tocotrienols, albeit in smaller quantities. Apart from the physiological effects of vitamin E, especially of α-tocopherol as a scavenger and stabiliser of cell membranes in the human organism, the various forms of vitamin E also play a fundamental part in oxidation protection of the corresponding fats and oils and are thus ultimately important for the stability of the triglyceride against becoming rancid. In this respect, the various vitamers show different levels of efficacy, which in general terms decreases in the sequence δ-, γ-, β-, α-tocopherol. Different combinations of vitamers differ in effectiveness, however, depending on the composition of fatty acids and external factors of influence.

Although all vitamers are found in vegetable fats and oils, only *γ- a*nd α-tocopherol occur in sufficient quantities to make a difference. Some quantities are specified in the table below.

Due to its molecular structure, vitamin E is able to bind lipoperoxides, which form in vegetable fats and oils in storage over time because oxygen is adsorbed by the double bonds of unsaturated fatty acids. Vitamin E thus stops chain reactions in fat oxidation (Kamal-Eldin and Appelqvist 1996).

Tocopherol and tocotrienol concentration (mg/100 g) of vegetable oils and industrial fats (modified after Schwartz et al. 2008).

Sample/sample number	α-T	α-T3	*β-T*	γ-T	*β-T3*	γ-T3	*δ-T*	*δ-T3*	∑T	∑T3 s	α-T_ eq.
RS ref/10	24	–	tr	39	–	–	0.98	–	63	–	28
RS ref 2/10	22	–	tr	37	–	–	1.9	–	61	–	25
RS ref 3/10	22	–	tr	42	–	–	1.9	–	66	–	26
RS cp/10	19	–	tr	50	–	–	1.5	–	70	–	24
RS bio/8	19	–	tr	51	–	–	1.5	–	71	–	24
OO ref/9	17	tr	0.27	0.89	–	−c	tr	tr	18	tr	17
OO ex virg/10	24	tr	0.30	1.4	–	−c	tr	tr	26	tr	24
Sunflower/10	59	0.11	2.4	1.4	–	tr	0.27	tr	63	tr	60
Corn/2	18	0.94	1.1	44	–	1.3	2.2	0.26	66	2.5	23
Coconut/9	0.20	3.0	tr	0.12	0.17	0.64	0.10	0.32	3.9	1.1	
Sesame/10	7.9	tr	0.41	36	–	0.34	1.2	−c	45	0.3	12
Flax seed/8	1.2	–	tr	52	–	–	0.95	–	54	–	6.4
Flax/6	3.8	–	0.09	72	–	–	1.5	–	78	–	11
Wheat germ/6	192	2.5	65	tr	8.2	–	0.55	0.24	257	11	225
Walnut/1	6.6	tr	−c	39	–	0.17	4.6	tr	50	0.3	11
Frying fat A/1	12	11	−c	6.7	1.2	15	0.15	3.6	18	31	16
Frying fat fat B/1	11	17	0.12	0.10	1.2	19	tr	4.4	12	42	17

Sample/sample number	α-T	α-T3	*β-T*	γ-T	*β-T3*	γ-T3	*δ-T*	*δ-T3*	∑T	∑T3 s	α-T_ eq.
Frying fat C/1	14	4.7	tr	20	0.28	7.2	0.72	1.7	35	14	17
Margarine D/2	13	4.1	0.17	12	–c	5.9	0.34	1.4	25	11	15
Margarine E/1	11	2.9	0.16	14	0.28	4.2	0.66	0.98	26	8.4	13
Margarine F/1	12	0.84	tr	22	1.5	0.63	0.24	34	2.5	15	
Drawing grease G/1	11	5.1	0.24	7.2	0.86	7.5	0.39	1.8	18	15	13
Drawing grease H/2	12	7.9	0.31	7.2	0.97	11	0.28	2.3	20	22	15

α-T α-tocopherol, *α-T3* α-tocotrienol, *β-T* β-tocopherol, *β-T3* β-tocotrienol, *γ-T* γ-tocopherol, *γ-T3* γ-tocotrienol, *δ-T* δ-tocopherol, *δ-T3* δ-tocotrienol, *RS* rapeseed, *ref* refined, *cp* cold pressed, *bio* biological, *OO* olive oil, *ex verg* extra vergine, *Σ-T* sum of tocopherols, *Σ-T3* sum of tocotrienols, *α-T eq.* α-tocopherol equivalents, *tr* trace, *c* coeluting

Vitamin A (Retinol)

Vitamin A appears in different chemical forms with different biological activity levels. We distinguish between retinol, retinal and retinoic acid. The latter is used in the medical treatment of common acne. Retinoic acid should not be used in cosmetics because of the side effects that a therapy may have. Retinoic acid has a strong teratogen effect that continues for about 2 years after the therapy has ended. In addition, it can lead to peeling and reddening of the skin, itching, hair loss and an increase in transaminases in the blood serum. In the skin, metabolic processes transform forms of vitamin A into each other. If retinoids are inactive, they have the form of retinol esters. In this form, they are stored in the skin. Bound to a retinol binding protein, they are transported within the organism and the skin. A deficiency in vitamin A leads to hyperkeratosis. As a result of a thickening of the corneal layer, the skin appears leathery, flaky and wrinkly. These effects are caused by chronic solar radiation. Maximum retinoid absorption lies between 330 nm and 360 nm. For this reason, UVA rays, in particular, lead to a decrease in the amount of retinoid in the skin. Since retinoids play an important part in the differentiation of keratinocytes, the continual renewal process of the skin is severely disrupted. UVA rays also penetrate deeper skin layers. Here, a decrease in vitamin A has a negative effect on the formation of connective tissue fascicles. Fibroblasts become hyperactive and more collagenases are produced, so that normal production of a collagenous network of connective tissue is no longer possible. The skin loses its elasticity and becomes leathery. Treatment of photopathologic skin with vitamin A preparations counteracts this effect. Natural skin ageing looks completely different from skin ageing as a result of excessive exposure to sunlight. In contrast to the leathery skin caused by sunlight, it appears soft and without blemish. The epidermis is significantly thinner and transparent. For this reason, skin that has aged naturally is called xeroderma. The reason behind this ageing process is that with progressing age, both the number of fibroblasts in the dermis and the number of keratinocytes in the epidermis decrease. The capacity of cell division also decreases with age. The activity of enzymes connected to the construction of connective tissue is reduced, and less collagenous connective tissue is produced. Retinoids work against this effect of biological ageing. They stimulate cell growth in the dermis and epidermis. If concentrations of 1% retinol are topically applied, they can increase the growth capacity of fibroblasts by 100%, provided that sufficient penetration over longer periods of time is ensured; the number of keratinocytes rises by 75%. These effects lead to a significant thickening of the epidermis. In cosmetic formulations, retinol is processed in an esterified form as palmitate or acetate. This corresponds to the form in which it is stored in the skin. The esterases of the skin can therefore easily transform retinol esters into their active forms. Retinol esters are more stable than retinol itself. Nevertheless, the preparations must be protected from light and air, and an antioxidant (vitamin E) must be added. Skin care products for the face should be additionally stabilised by filters to prevent the retinol from being inactivated by sunlight. In general, a concentration of 1000–5000 IU vitamin A/g is recommended for cosmetic preparations to treat dry, flaky and ageing skin. One International Unit (IU) corresponds to approx. 0.3 g of vitamin A alcohol or 0.344 g of vitamin A acetate respectively. 1 g of vitamin A amounts to 3.33 million IU. The provitamin A international unit has the same effect as 0.6 g of *β-carotene* (Burger and Wachter 1993, p. 1485; Raab and Kindl 2004, p. 262).

β-Carotene, a provitamin A, is a natural pigment. It belongs to the group of carotenoids, molecules that are constructed from eight isoprene units. Other representatives of this group are lycopene, lutein and zeaxanthin. In medicine, carotenoids are increasingly important as scavengers. They directly intercept free radicals and are thereby oxidised. As carotenoids are lipophilic substances, their site of action is located in the cell membranes.

Due to their intensive colouring, the use of carotenoids in cosmetic products is limited. Oral administration of *β*-carotene in high doses leads to a yellowish-reddish skin colour, especially on the palms of the hands and the soles of the feet. A colouring of the skin may be intended, however, in order to conceal dyschromia[6] (Burger and Wachter 1993, p. 415).

The EU Novel Food Regulation

Since January 1st, 2018, the EU regulation 2015/2283 about Novel Food is in force and applicable. Food is classified as "novel" if it has not been in use for human consumption to a significant degree in the EU before May 15th, 1997, independently of when individual member states joined the EU ("history of safe consumption"), and falls into at least one of the defined categories (Article 3 of the regulation) (www.verbrauchergesundheit.gv.at, consulted 12.4.2019). Food made from plants or their parts is also mentioned in these categories, so that the EU regulation is applicable for vegetable fats and oils.

To determine if there has been "human consumption to a significant degree before May 15th, 1997", the guideline "human consumption to a significant degree" issued by the European Commission is to be consulted (https://ec.europa.eu/food/sites/food/files/safety/docs/novel-food_guidance_human-consumtion_en.pdf). The

[6]Dyschromia is a dermatologically congenital or acquired, localised or general change in skin colour. This pigmentary abnormality is caused by an alteration in the amount of melanin. Hyperpigmentation leads to a brown colouration of skin areas. Depigmentation of certain portions of the skin is called vitiligo (Burger and Wachter 1993, p. 415).

subsuming of the food in question into one of the above named categories or the decision that it falls into none of the categories named is made by the food contractor on his own responsibility. Classifications made up to now and information coordinated between member states about the status of many foodstuffs is available in the Novel Food catalogue of the European Commission (http://ec.europa.eu/food/safety/novel_food/catalogue/search/public/index.cfm).

Approved novel food is listed in this positive list, the so-called Union List, since January 1st, 2018 (regulation (EU) 2017/2470 of the Commission of December 20th, 2017): eur-lex.europa.eu/legal-contenct/DE/TXT/PDF/?uri=CELEX:32017R2470&from=EN. Listed food may be offered for sale by all food contractors, provided that the conditions of use, labelling and specifications are observed (www.bvl.bund.de/DE/01, consulted 12.4.2019). The Union List is continually updated – the current version is published in the Official Journal and can be found on the website of the European Commission (https://ec.europa.eu/food/safety/novel_food/authorisations/union-list-novel-foods_en).

In this list of novel food, the following vegetable fats and oils are currently listed (consulted 12.4.2019):

Algae oil from the micro alga *Ulkenia sp.*
Allanblackia seed oil from *Allanblackia floribunda* and *Allanblackia stuhlmannii*
Argan oil from *Argania spinosa*
Chia oil from *Salvia hispanica*
Coriander seed oil from *Coriandrum sativum*
Oil from the micro alga *Schizochytrium sp.* (ATCC PTA-9695)
Plum kernel oil from *Prunus domestica*
refined oil from *Echium plantagineum*
Sacha Inchi oil from *Plukenetia volubilis*
Seed oil from *Buglossoides arvensis*

The specific food category and maximum contents have to be observed.

Anyone planning to offer novel food for sale that is not yet listed in the Union list of approved novel food has to apply for permission to the European Commission according to Article 10 Novel Food Regulation (EU) 2015/2283 (www.bvl.bund.de/DE/01, consulted 12.4.2019).

If the novel foodstuff has a "history of application as safe food" in a state outside of the European Union, a (simpler) notification according to Article 14 and 15 Novel Food Regulation can be made to the European Commission instead of an application of permission. This is only applicable for food of the categories "microorganisms, fungi, algae", "animals", "plants" and "cell and tissue cultures". In case there are no substantiated objections voiced by the European Commission, the member states or EFSA within 4 months, the Commission enters the food into the list. If there are objections, an entry is only possible via a registration procedure with shortened time-limits. The procedures for notifications of a traditional foodstuff from a non EU-country is regulated in Articles 14 and 15, those for registration in Articles 16–19 of the Novel Food Regulation (EU) 2015/2283. The procedures are specified in the regulation (EU) 2017/2468 (www.bvl.bund.de/DE/01, consulted 12.4.2017).

Applications for permission of a novel foodstuff currently processed and notifications of a traditional foodstuff from a non-EU country are published by the European Commission on the following website, which is updated regularly: ec.europa.eu/food/safety/novel_food/authorisations/summary-applications-and-notifications_en.

Ajowan Oil

Oleum Ajowan

1 Source Plant

Trachyspermum ammi Sprague, Apiaceae *(Umbelliferae),* ajowan, ajowan caraway.

S. Krist, *Vegetable Fats and Oils*, https://doi.org/10.1007/978-3-030-30314-3_2

A

Habitat

Ajowan came to Europe in the early Middle Ages, when the markets in large towns sold it as a fragrant, exotic spice from the Orient. Attempts to cultivate ajowan in Europe met with considerable difficulties due to the unfavourable climate. By the end of the seventeenth century, the attempts were abandoned and the spice was imported from the Middle East instead. Ajowan is a native plant of the region from Egypt to eastern India and Ethiopia, but it is cultivated in India from Punjab to Bengal and south of Deccan, in particular. Ajowan is also grown in Iran, Afghanistan, Pakistan and on the Seychelles. The main harbour of export is Mumbai.

Description

Ajowan is an annual plant, about 90 cm tall, and with a similar appearance to wild parsley. The plant grows upright; it is hairless and somewhat branched. The leaves are few, delicate and pinnate. About 5–15 white flowers form an umbel. The fruit is covered with coarse hair. It is oval, greyish brown and aromatic.

2 Cultivation and Extraction

Cultivation

Ajowan grows in all kinds of soil, especially on loam and clayey loam. It is cultivated in dryland farming as well as under irrigation. In South Asia, it is sown from September to October and gathered about 2 months later, when the umbels turn brown. They are picked, dried on mats and destemmed using the hands or feet, so that the seeds are separated from the umbels.

Extraction of the Oil

Ajowan oil is extracted by pressing the seeds after the volatile compounds have been removed (Narayana et al. 1967).

3 Character

Colour

Oil from raw seeds: colourless (Nagalakshmi et al. 2000).

Oil from roasted seeds: light yellow (Nagalakshmi et al. 2000) to brown.

Odour

Reminiscent of thyme or camphor, herbal.

Volatile Compounds

Concentration acc. to GC-MS analysis of ajowan oil (Nagalakshmi et al. 2000):

α-Pinene 1.48%	Menth-2-en-1-ol 0.13%
β-Pinene 5.45%	Linalool 0.07%
β-Myrcene 1.40%	Terpinen-4-ol 0.12%
α-Terpinol 0.09%	α-Terpineol 0.12%
ρ-Cymene 19.47%	Cuminaldehyde in traces
Limonene 0.48%	Thymol 39.36%
γ-Terpinene 30.97%	Cuminol in traces
ρ-Cymenene 0.06%	Carvacrol in traces
Terpinolene in traces	

Flavour

Spicy, sweet and herbaceous.

Pungent, acrid taste.

4 Ingredients

Dried ajowan fruit contain up to % etheric oil, of which thymol is the main component.

Physical Key Figures of Ajowan Oil

refractive index: n_D^{25} 1.4951–1.4930
density: d_{25}^{25} 0.9205–0.9137
saponification number: 182
iodine number: 100
acid number: 0.36–0.39
unsaponifiable: 2.3%
point of solidification: −4 °C
(Nagalakshmi et al. 2000; Roth and Kormann 2000, p. 157)

Shelf Life

Ajowan oil has a shelf life of about 1 year.

5 Application

In Pharmacy and Medicine

In Treatments for Fungi, Bacteria and Worms

Due to its high amount of thymol, ajowan oil has a strong antibacterial and fungicidal effect. Ajowan oil is used in treatment for ascaris and hookworms (Neetu and Meenakshi 2003).

To Treat Indigestion

Ajowan oil has a strong digestive, antispasmodic effect and is therefore employed to treat colic, indigestion, diarrhoea and flatulence (Nagalakshmi et al. 2000).

To Treat Colds

Because it is antiseptic, ajowan oil is used to treat colds, coughs, influenza and sore throats (Umanskii and Krutik 1945).

To Treat Arthritis and Rheumatism

Ajowan oil is used as a liniment to improve mobility and alleviate pain.

To Treat Diseases of the Skin

Since it contains furanocoumarins and psoralenes, ajowan oil has been used to treat vitiligo for centuries. Vitiligo is a loss of pigment in the skin that causes sharply bounded white spots of irregular size to appear, especially on the face, neck and hands. Research indicates that ajowan oil is also suitable for treating psoriasis (Ivie 1978; Roche Lexikon Medizin 1993, third edition, p. 1745).

In Folk Medicine

In Iran and Egypt, toothache, earache, rheumatism and kidney stones are treated with ajowan oil. It is also thought to help in the treatment of female infertility, menstrual complaints and postnatal complications, and is thought to be an aphrodisiac. These applications have not yet been clinically proven.

In Industry

As a result of its high thymol content, ajowan oil is used in the technical production of thymol (Nagalakshmi et al. 2000).

In Cosmetics

As a result of its strong antiseptic effect, ajowan oil is added to mouth-wash and toothpaste. It is also used in the perfume industry because of its characteristic smell.

6 Possible Unwanted Side Effects

Ajowan oil may cause allergic and especially photosensitive reactions in babies and children. The oil should be avoided during pregnancy because it can cause contractions of the uterus. As it promotes digestion, it is not suitable for persons producing too much gastric acid.

Photosensitive effects were proven in a study on sheep that were fed ajowan seeds (up to 8 g/kg body weight). When the sheep were subjected to solar light, they suffered from corneal opacity, conjunctivitis, photophobia and oedema on the nose and ears (Witzel et al. 1978).

In an in vitro study on human sperm, ajowan oil was proven to have a spermicidal effect (Buch et al. 1988).

Algae Oil

synonym: Single cell oil (E)

1 Source Plant

Many species can be used for oil extraction, for example *Chlorella, Spirulina, Nitzschia, Ochromonas, and many more*.

S. Krist, *Vegetable Fats and Oils*, https://doi.org/10.1007/978-3-030-30314-3_3

A

Habitat

Most algae live in water. They can be classified according to the pigments of their plastids, the reserve substances they produce, morphological, cytological and developmental characteristics or the chemical composition of their cell walls. Algae exist in many forms, from simple unicellular organisms, thread-shaped, branched or unbranched forms to highly organised algae with a length of several metres. They grow nearly everywhere and are either afloat or sessile. Brown algae, red algae, Dinophyceae and Chrysophyceae are mostly found in salt water, green algae and Xanthophyceae prefer fresh water or land.

Spirulina sp. is a good example. *Spirulina* are cyanobacteria that live in salt water in tropical and subtropical regions. They are thread-shaped and about 250 μm long. As they contain blue and green pigments (chlorophyll and phycocyanin), they look like bluish-green threads under a microscope. Spirulina are ideal phytogenetic raw material because they reproduce quickly and have a tendency to build colonies. There are 35 wild species of spirulina, which grow in salty lakes around the world: in Lake Chad (Chad, Africa), Lake Texcoco (Mexico, North America) and Lake Turkana (Kenya, Africa) (Simopoulos 1995, p. 32 et seq).

Description

Traditionally, algae (*phycophyta*) are considered to be plants and are organised into 12–14 classes. In more recent systematics, they are regarded as an organisation type of plants and subdivided into nine departments. They are photoautotroph thallophytes of different colour, nucleated, and consist of one or more cells. Cyanobacteria, traditionally known as blue algae, perform photosynthesis, but they are prokaryotically organised. The plastids of almost all algae contain chlorophyll a, as well as other chlorophyll components. Reproduction is either vegetative or sexual. Vegetatively, algae reproduce either by bisection (schizotomy), production of several daughter cells (schizogony) or disintegration of multi-cellular species into parts containing fewer cells. Alternate vegetative and sexual reproduction is also possible. Organs producing spores are always unicellular, as are gametangia (spermatogonia). Female gametangia are termed oogonia; most of these propagation cells are flagellate.

2 Cultivation and Extraction

Cultivation

Algae are grown for scientific as well as for economic reasons. In Asia, in particular, they are used as food, for example red algae (nori) and brown algae. Gelling agents and mucilages are produced from both. Agar-agar is made from red algae, and alginic acid and seaweed meal from brown algae.

Several hundred tons of spirulina are produced for the health food market. To produce algae on an industrial scale, solar light, CO_2, mineral salts and water are required. Ideal conditions are anywhere along the 35th degree of latitude north and south of the equator, a region with high solar radiation, marginal differences in temperature, moderate precipitation and available CO_2 (Simopoulos 1995, p. 33 et seq).

Extraction of the Oil

Since there are many different algae species, there are different kinds of oil and production modes, for example extracts of blue algae prepared with a fat oil, like soya bean oil, or brown algae macerate in soya bean oil.

3 Character

Colour

Light green oil, liquid at room temperature.

Extract from brown algae, basis soya bean oil: yellowish-green oil.

Odour

Soapy, fresh, smells of the sea.

4 Ingredients

Fat from algae contains considerable quantities of fatty acids with an uneven number of C-atoms, predominantly palmitic acid and oleic acid, but also great quanities of linoleic acid and linolenic acid. The fat content depends on the species and varies from 10% to 85% (Löw 2003, p. 78).

The following table lists the fat content of different species of algae:

Species (algae)	Content in % acc. to Karleskind (1996), p. 259
Chlorella ellipsoida	85
Chlorella pyrenoidosa	85
Chlorella sorohiniana	39
Chlorella vulgaris	28
Ochromionas danica	53
Thalassiosira fluviatilis	50
Euglena gracilis	23
Monodus	32
Navicula pelliculosa	32
Nitzschia closterium	23

A

Species (algae)	Content in % acc. to Karleskind (1996), p. 259
Nitzschia palea	40
Oocystis polymorpha	18
Scenedesmus quadricuada	38
Tribonema aequale	23
Uronema gigas	21

Composition of Fatty Acids in Extracts of Algae

(content in %)

Fatty acids	(DHASCO[1] oil acc. to Arterburn et al. 2000)	CO®algae oil 1 acc. to Frankel et al. (2002)	Algae oil 2 acc. to Frankel et al. (2002)	Algae oil 3 acc. to Frankel et al. (2002)
Capric acid	0.6	–	–	–
Lauric acid	4.5	5.0	5.1	4.4
Myristic acid	15.4	17.8	18.4	15.6
C14:1 n-5	0.2	–	–	–
Palmitic acid	11.6	15.0	15.6	12.8
Palmitoleic acid	2.3	1.4	1.5	1.4
Stearic acid	0.3	–	–	0.9
Oleic acid	11.3	8.4	8.7	21.6
Linoleic acid	0.8	–	–	0.8
γ-Linolenic acid	–	–	–	–
α-Linolenic acid	0.2	–	–	–
Arachidic acid	0.1	–	–	–
Eicosenoic acid	0.1	–	–	–
Behenic acid	0.1	–	–	–
C22:6 n-3	51.7	52.4	50.8	42.4

Composition of fatty acids in three different oils of the company Martek Bioscience Corp., Columbia, MD (Frankel et al. 2002)

[1]DHASCO? docosahexae-noic acid single cell oil; a triglyceride extracted from algae containing 40-50% DHA (docosahexaenoic acid).

Composition of Tocopherols (Content Given in ppm)

Tocopherols	Algae oil 1 acc. to Frankel et al. (2002)	Algae oil 2 acc. to Frankel et al. (2002)	Algae oil 3 acc. to Frankel et al. (2002)
α-Tocopherol	836	420	1729
$\beta + \gamma$-tocopherol	–	–	208

Composition of tocopherols in three algae oils of the company Martek Bioscience Corp., Columbia, MD (Frankel et al. 2002*).*

Algae oil contains mainly α-tocopherol (420–1729 ppm) and carotenoids (577–2823 ppm) (Frankel et al. 2002).

Characteristics of the Oil

EPA (eicosapentaenoic acid) can be extracted from algae in three steps, with a yield of 70%. First, the algae are simultaneously extracted and esterified. In a second step, the raw extract is cleaned using chromatography, and finally the pigments are removed, also by chromatography. The EPA thereby gained is of a better quality than comparable fish oil (Belarbi et al. 2000).

5 Application

In Pharmacy and Medicine Animal Experiment as Tumour Therapy

A study on nude mice shows that a diet with 8% corn oil in combination with 16% golden algae oil (corresponds to a diet rich in fat containing an ω-3 fatty acid: docosahexaenoic acid) significantly obstructs the growth of carcinoma of the human colon, in comparison to diets with (a) 8% corn oil, (b) 24% corn oil and (c) 8% corn oil in combination with 16% menhaden oil. Golden algae oil reduces the activity of various genes associated with angiogenesis, for example VEGF (vascular endothelial growth factor) and an initial stage of the insulin receptor IRP (insulin receptor precursor). This leads to a reduction of angiogenesis and tumourigenesis. In addition, the expression of CRAF-1 (CD40 receptor associated factor) in the tumour decreases. This protein can interact with a region similar to TNF-α (tumour necrosis factor alpha). Additional research is necessary, however, to verify the data (Kato et al. 2002).

In Formulations for Babies

According to recommendations of ESPGAN (European Society for Paediatric Gastroenterology, Hepatology and Nutrition) and FAO/WHO (Food and Agriculture Organisation of the United Nations/World Health Organization), DHA (docosahexaenoic acid) and AA (arachidonic acid), in the same concentrations as in breast-milk, should be added to formulations for babies. DHA from vegetable sources can be extracted from algae oil, AA from oil made from fungi. A combination of the two that simulated their content in breast-milk was tested on rats, both in utero during gravidity as well as on the young. Diets with 18 g DHA + AA/kg, 60 g DHA + AA/kg and 120 g DHA + AA/kg were tried. This study aimed at examining whether DHA and AA from algae and fungi are a suitable addition to baby food.

The results of the tests on rats showed that in the applied doses, these oleic acids have no adverse effect (Burns et al. 1999).

A study on premature babies with a birth weight of 846–1560 g who were fed premature infant formula and subsequently a formula to which DHA (docosahexaenoic acid) from algae oil and AA (arachidonic acid) from fungi had been added for 57 weeks, showed that no adverse effects on growth were detectable if 0.33% DHA from algae oil and 0.60% AA from oil extracted from fungi were added to the diet. The babies put on more weight than the babies in the control group; this increase in weight was even comparable to that of babies fed with breast milk. As LCPUFAs (long chain polyunsaturated fatty acids) are decisive for the brain development in early periods of life as well as for the growth of premature babies in the first weeks, it is recommendable to add DHA and AA to premature infant formulae, in the same amounts as contained in breast milk (Diersen-Schade et al. 1998).

In the Pharmaceutical Industry

PUFAs (polyunsaturated fatty acids), especially of the ω-3 kind, have an important function in biological membranes, as the initial stages of several lipid regulators for cell metabolism, and as initial stages of eicosanoids such as prostaglandins, prostacyclins, thromboxanes and leukotrienes. Malfunctions in eicosanoids from PUFAs may result in cardiac and coronary diseases, respiratory diseases, gastrointestinal diseases, diseases of the kidneys, skin and immune system as well as disorders of the brain, underdeveloped eyes, and cancer. The food industry's and the pharmaceutical industry's increasing demand can be met by producing PUFAs from microalgae, for example from *Mortierella alpina* ATCC 32222 (52.7% PUFAs) or *Monodus subterraneus,* one of the few fresh-water algae that have a high content of EPA (eicosapentaenoic acid) of up to 34% (Chen and Jiang 2001, p. 1 et seqq).

In Cosmetics

Extracts from algae are used in skin care products, as they are rich in protein, vitamins, allantoin, mineral salts, vitamin B_1 and vitamin B_{12}. They stimulate blood circulation and have a smoothing effect on the subcorneous layer. The active substances in algae regulate the internal pressure of the skin cells, the oil has a moisturising and smoothing effect. It is used in 1–3% creams, body lotions, baby care products, and many other products.

In the Food Industry

Oil of the microalga *Schizochytrium sp. is rich in* DHA (docosahexaenoic acid) and is added to dairy products, cereals and spreads, and is also used in dietary supplements, in medical diets to reduce weight, and for other purposes.

As Diesel Oil

After the lipids of micro algae have been extracted and esterified, the resulting fatty acid methyl ester can be used as an alternative to diesel oil. For esterification, the catalyst used, its concentration, the reaction rate, the temperature and the quality of the lipids are decisive. A study tested various solvents, of which 1-butanol proved to be the most effective with 90%, whereas the largest yield of fatty acid methyl esters was achieved by using 0.6 N hydrochloric acid in methanol at 70 °C and 0.1 h (Nagle and Lemke 1990).

Almond Oil

Oleum Amygdalae synonyms: Mandelöl (D); huile d'amande (F); bitter: almond oil (E); huile d'amande amere (F); sweet: sweet almond oil (E); huile d'amande douce (F)

S. Krist, *Vegetable Fats and Oils*, https://doi.org/10.1007/978-3-030-30314-3_4

1 Source Plant

Prunus dulcis (Mill.) D.A. Webb *(Rosaceae),* almond, *Prunus dulcis var dulcus* – sweet almond, *Prunus dulcis var amara* – bitter almond

Habitat

The origins of the almond probably lie in subtropical China and the Middle East. It has been cultivated for its seeds in many warm, temperate zones for millennia. Today, it is found especially in the Orient, in the Mediterranean area, Southern Europe, California, southern Australia and South Africa. The almond prefers warm, airy, slightly calciferous soil in sunny and warm locations. It survives cold winters with a minimum temperature of –20 °C. Changing temperatures can, however, be a problem (Hager 1978, volume 6a, p. 938; Roth and Kormann 2000, p. 136).

Description

Almonds are trees or shrubs of medium height with relatively weak roots. The branches are a reddish colour; the wild form also has thorns, which the cultivars lack. The leaves have a glandular stem 1.2–2.5 cm long and a hairless, rough blade that is a shiny dark green colour; its edges are glandular, sharply serrate or crenated. There are usually two flowers on very short stems that sometimes surmount the bud scales; they appear before the leaves. The fruit are oval-lanceolate and slightly compressed, up to 4.6 cm long and about 3.5 cm thick, greyish green and covered with felty, velvety hair. The seeds are cinnamon brown, flattened, and about 2 cm long and 1.3–1.5 cm broad (Hager 1978, volume 6a, p. 938; Löw 2003, p. 59).

2 Cultivation and Extraction

Cultivation

Today, almonds are mainly grown in California and Spain. In the large growing areas of California, authentic sweet almonds are cultivated. Spanish almonds have a thicker shell than Californian almonds. In the Mediterranean area, almond trees bear sweet as well as bitter almonds. Each harvest of sweet almonds therefore contains a natural amount of bitter almonds. As bitter almonds contain hydrocyanic acid, the amount may not exceed 5% (Hunnius 1998, p. 1136; Hager 1978, volume 6a, p. 938).

Extraction of the Oil

According to Ph.Eur. 2002, almond oil can be obtained from the ripe seeds of *Prunus amygdalus var. amare* and *var. dulcis*, which means that the oil may originate from sweet almonds, bitter almonds, or from a mixture of both. It is cold pressed and sometimes then refined. After the impurities and shells have been removed, the dry almonds are pressed. The oil is left standing for several days so that the mucilage settles to the bottom and can be separated using filter presses. The press residue is sold as almond bran and still contains 10% fat oil. One ton of peeled almonds results in about 400 litres almond oil. Ph.Eur. 2002 lists monographs of two different qualities of almond oil (Roth and Kormann 2000, p. 136; Hänsel et al. 1999, p. 254; Hager 1978, volume 7b, p. 193; Hunnius 1998, p. 1136):

- natural, cold-pressed almond oil (*Amygdalae oleum virginum*)
- refined almond oil (*Amygdalae oleum raffinatum*), an oil made odourless, to which a suitable antioxidant may be added. The nature and amount of the addition must be declared.

3 Character

Colour

Light yellow (Roth and Kormann 2000, p. 136; Hunnius 1998, p. 1136).
Slightly yellow (Hager 1978, volume 7b, p. 193).
Varying from yellow to colourless (Kerschbaum and Schweiger 2001, p. 20).

Odour

Mildly nutty, rounded, slight touch of vanilla.
Odourless (Roth and Kormann 2000, p. 136; Hunnius 1998, p. 1136).
Nearly odourless (Hager 1978, volume 7b, p. 193).

Volatile Compounds

Analysis of volatile compounds of almond oil by neams of liquid extraction, and liquid and gas chromatography (Del Mar Caja et al. 2000):

benzaldehyde
nonanal
decadienal
(E,E)-2,4-decadienal

A

Flavour

Mild, oily (Roth and Kormann 2000, p. 136).

Mild (Hager 1978, volume 7b, p. 193).

Decidedly reminiscent of almonds (Kerschbaum and Schweiger 2001, p. 20).

4 Ingredients

Composition of Fatty Acids

Fatty acids	Content in % acc. to Hager (1978), volume 7b, p. 193	Content in % acc. to Roth and Kormann (2000), p. 136	Content in % acc. to Kerschbaum and Schweiger (2001), p. 19
Palmitic acid	3–5	3	6.7
Linoleic acid	17–20	19.9	21.6
Palmitoleic acid	–	–	0.4
Oleic acid	77	77	66.4
Stearic acid	–	3	1.7
Arachidic acid	–	–	0.1
Vaccenic acid	–	–	2.7
Myristic acid	1.2	–	–

Composition of Triglycerides

Triglycerides	Content in % acc. to Karleskind (1996), p. 161
LLL	1–4
OLL	12–20
PLL	1–4
LOO	21–28
POL	7–11
OOO	23–40
POO	7–10
PPO	<0.4
SOO	1–5

Demanded Composition of the Sterol Fraction of Almond Oil

(Ph.Eur. 2002)

Cholesterol:	0.7% maximum
Campesterol:	4% maximum
Stigmasterol:	3% maximum
β-Sitosterol:	73.0–87.0%
Δ5-Avenasterol:	10% minimum
Δ7-Avenasterol:	3% maximum
Δ7-Stigmasterol:	3% maximum
Brassicasterol:	0.3% maximum

Composition of Sterols

(concentration in mg/100 g and content in %)

Sterols	Almond oil acc. to Firestone (1999), p. 7	Almond oil acc. to Karleskind (1996), p. 162	Refined almond oil acc. to Carstensen (2001), p. 42	Cold-pressed almond oil acc. to Carstensen (2001), p. 42
Total content	266	120–400	165.7	324.3
Cholesterol	–	<1	–	0.9
Brassicasterol	–	–	–	0.2
Campesterol	2–4	2–3	1.5	1.9
Stigmasterol	1–2	0–1	0.9	0.4
β-Sitosterol	80	74–87	84.9	84.3
Δ5-Avenasterol	10–12	7–20	5.2	9.1
Δ7-Stigmasterol	1–2	0–5	1.1	1.5
Δ7-Avenasterol	1–2	0–4	–	1.8
Fucosterol	–	1–2	–	–

Composition of Tocopherols

(concentration in mg/100 g and content in %)

Tocopherols	Almond oil acc. to Karleskind (1996), p. 162	Refined almond oil acc. to Carstensen (2001), p. 44	Cold-pressed almond oil acc. to Carstensen (2001), p. 44
Total content	25–56	44.6	63.4
α-Tocopherol	92–99	96.7	92.4
β-Tocopherol	<1	–	1.6
γ-Tocopherol	1–7	3.3	3.1
Δ-Tocopherol	–	–	3.0

Other Ingredients

(Kerschbaum and Schweiger 2001, p. 20; Ulmer 1996, p. 83)

Squalene
Glutamine
Calcium
Iron
Vitamin A
Vitamin B
Magnesium
Sulphur
Sodium
Phosphorus
Potassium
Sterols

A

Physical Key Figures of Almond Oil

refractive index: n_D^{25} 1.4702–1.4715
density: d_4^{15} 0.914–0.920
saponification number: 183–207
acid number: not over 8
iodine number: 85–106
unsaponifiable: max 1.5%
point of solidification: −10 to −21 °C
(Roth and Kormann 2000, p. 136)

Shelf Life

As almond oil becomes rancid easily, it should be stored in a cool, dark place and, if possible, in completely filled containers. Under these conditions, it has a shelf life of up to 12 months (Roth and Kormann 2000, p. 136; Hager 1978, volume 7b, p. 193).

Characteristics of the Oil

Almond oil remains clear at a temperature of −10 °C and only solidifies when cooled to almost −20 °C (Hager 1978, volume 7b, p. 193; Hunnius 1998, p. 1136).

The properties and uses of almond oil are very similar to those of apricot oil. Almond oil contains α-tocopherol as well as γ-tocopherol, whereas apricot oil only contains γ-tocopherol. This is also important for differentiating between marzipan and persipan. Marzipan is made from almonds, whereas persipan, which is similar to, slthough slightly stronger than marzipan in taste, is made from apricot kernels and is therefore a cheap substitute for marzipan (Kerschbaum and Schweiger 2001, p. 36).

Sweet almond oil is used externally, for example in ear drops, as well as internally as a laxative. Bitter almond oil may only be applied externally and may only be added to food in very small amounts (Löw 2003, p. 106).

5 Application

In Pharmacy and Medicine

Almond oil is a solvent and a dispersant for injections of oily formulations of lipophilic medicines that cannot be administered perorally. It is not, however, used as often as castor oil (Hänsel et al. 1999 p. 254; Hager volume 7b p. 193).

Due to its soothing quality, almond oil is employed as an excipient in oily eye drops. Since its oleaginous solutions are viscous, the active ingredient solution does not flow off as quickly as an aqueous solution, which ensures the desired deposit effect. In additions, oleaginous solutions do not constitute nutrient media for microorganisms, and it is therefore not necessary to add preservatives. The disadvantage is that oleaginous solutions cause short-term impairment of vision (Hänsel et al. 1999, p. 254; Hunnius 1978, p. 1136).

To soothe earache, the oil is warmed and instilled into the ear (Roth and Kormann 2000, p. 136).

Almond oil is also applied to remove *crusta lactea* of babies.

In Iran, almond oil is a common laxative (Filsoof et al. 1976).

As Food

Almond oil is a good salad oil. It is suitable for frying, but is actually too valuable for this purpose because of its high amount of unsaturated fatty acids. Due to its almond aroma, it is also suitable for sweets (Roth and Kormann 2000, p. 136; Kircher 2002, p. 79; Löw 2003, p. 106; Kerschbaum and Schweiger 2001, p. 20).

In Cosmetics

Almond oil is one of the most valuable skin care oils, and has been used for this purpose for centuries. It is a kind of "all-rounder", used in the production of (Ulmer 1996, p. 83; Löw 2003, p. 128):

- skin oil;
- massage oil;
- oil baths;
- lip balm, ointments, cold creams;
- baby-care products;
- face and oil packs; and
- hair-care products.

As a topical product, it penetrates the skin easily and moisturises it, which has a soothing effect, especially on raw, flaky and dry skin. In this respect, almond oil has the following properties (Roth and Kormann 2000, p. 136; Hänsel et al. 1999, p. 248; Kerschbaum and Schweiger 2001, p. 45; Löw 2003, p. 128; Dehelean et al. 2003):

- soothing;
- pain-relieving;
- circulation-enhancing;
- moisturising; and
- it is a natural filter.

A special feature in care cosmetics is almond milk, a liquid oil-in-water emulsion that cleanses the skin (Roth and Kormann 2000, p. 136).

As *Oleum Amygdalae* becomes rancid easily, it is often substituted with synthetic oils (Roth and Kormann 2000, p. 136; Hänsel et al. 1999, p. 248).

Even residual oil is used: Almond powder and almond bran are important ingredients of creams and ointments (Roth and Kormann 2000, p. 136).

Technical Uses

Almond oil is a lubricant in fine mechanics for clocks, instruments and arms (Roth and Kormann 2000, p. 136; Löw 2003, p. 128).

In the Food Industry

Almond oil is used for the production of marzipan, but only rarely, as it is expensive and does not keep long (Roth and Kormann 2000, p. 136).

According to a single case study, the oil had a positive effect on the treatment of colon cancer of rats (Davis and Iwahashi 2001).

In Folk Medicine

Since almond oil has been used for millennia, its applications in folk medicine are manifold (Ulmer 1996, p. 83; Kircher 2002, p. 77; Löw 2003, p. 106). It is used to treat:

- mucous obstruction of the bronchial tubes;
- pneumonia;
- gastric disorders;
- gripes;
- complaints of the bladder and kidneys (external application);
- stomach ache and colic of children;
- loss of appetite;
- furuncles;
- hair loss and dandruff;
- skin irritations; and
- diaper rash and bedsores.

Amaranth Oil

synonyms: Amaranthol (D); Huile d'Amarante (F)

1 Source Plant

Amaranthus caudatus (Amaranthaceae), amaranth.

S. Krist, *Vegetable Fats and Oils*, https://doi.org/10.1007/978-3-030-30314-3_5

Habitat

Today, amaranth is an ornamental plant often cultivated in kitchen gardens. It is also among the oldest cultivated plants in the world and was one of the first food plants grown in Native American agriculture. Amaranth is a native pant of America, especially Central America and the South American Andes. Cultivation in these areas has, however, significantly decreased (Danert 1993, p. 536–537).

Description

There are more than 60 varieties of amaranth, but only the following three produce seeds: *Amaranthus hypchondriacus*, *Amranthus cruentus* and *Amranthus caudatus* (Seibel 2005, p. 57).

Amaranth is an annual plant and can grow up to 2 m high. It has large leaves and produces terminal, branched, spiked, clustered, usually long, partly pendent spadices. Amaranth is a monoecious plant; either the male flower is located at the base and the female flower at the top of the spadices, or there are both male and female racemose inflorescences. The flowers are a greenish colour and consist of a perianth in 3–5 parts, 2–5 stamina and usually three carpels. The superior ovary ripens to a circumscissile capsule, more rarely to a nut. The capsule contains a round, brownish-black seed about 1–2 mm thick (Lieberei and Reisdorff 2007, p. 90).

Amaranth seeds are similar to cereal both in content and use, but as amaranth is not a grass, it is classified as a pseudocereal. Not only the seeds, but also the leaves can be used, the latter mostly as vegetables (Aufhammer 1998, p. 160).

2 Cultivation and Extraction

Cultivation

After Latin America had been conquered by Spain, agriculture gained importance in the region of the former Aztec realm. Documents listing tributes contain information on what and how much each province had to grow and deliver to the sovereign. Nearly all the provinces had to grow amaranth.

The Aztecs used amaranth in religious rites. The Spanish regarded this as contempt for Christian customs and as satanism, which is why they prohibited the use of amaranth, and as a result its cultivation decreased rapidly. Only in the twentieth century was it cultivated in Central and South America again (Danert 1993, p. 538).

The colour of the amaranth corn varies from white to yellow, brown and black. As golden-yellow corn is in demand, varieties that produce corn of this colour are cultivated most frequently.

The cultivated areas where pseudocereals are grown are small compared to the areas where genuine cereals are grown. They also play a subordinate role as far as plant breeding is concerned, and therefore still have many of the unwanted side effects of wild plants (Aufhammer 2003, p. 31–32).

Global cultivation area and amaranth corn production in the year 1998 (Aufhammer 2003, p. 32)			
1998	Cultivation area (ha 10^3)	Production (t 10^3)	Crop (dt ha^{-1})
Global	300.0	150.0	7.550.561

Extraction of the Oil

The oil can be extracted via pressing or chemical extraction.

Pressing results in a higher yield of squalene and tocopherols (Martirosyan et al. 2007).

To extract the oil chemically, the seeds are first ground in an abrasion mill. The purpose of this is to remove the endosperm, as most of the oil is contained in the seedling. The parts of the seed can be separated by screening (Lyon and Becker 1987).

If the corn is finely ground, greater amounts of oil are extracted. Only small amounts of oil can be extracted from unground seeds, as the solvent can scarcely penetrate the aril and reach the embryo to extract the lipids (Han-Ping He et al. 2002).

Too small brans can also pose a problem: They are more susceptible to enzymatic degradation, and the solvent may wash them away. To prevent this, the brans are pressed. The optimal corn size was found to be 0.71–0.89 mm (Sun et al. 1995).

The oil is extracted in a Soxhlet extractor; the yield depends on velocity of flow, pre-treatment, temperature and pressure. In order to achieve a good yield, the pressure is set at 300 bar, and the temperature at 70 °C (Westerman et al. 2006).

If the oil is to be used in natural cosmetics, extraction with hexane is not allowed. In these cases, the oil is extracted by adding supercritical carbon dioxide. Further treatment is not necessary, as this process takes place under mild conditions and exclusion of oxygen. The remnants of solvents are removed, so that further filtrations can be omitted (Quirin 2009).

3 Character

Colour

Yellow (Lyon and Becker 1987).

Odour

Characteristic, fusty especially when extracted using CO_2. Pressed oil smells unobtrusive (olionatura.de).

A

Flavour

Amaranth oil is an edible oil of acceptable taste (www.worldlingo.com/ma/enwiki/de/Amaranthoil).

4 Ingredients

Composition of Fatty Acids

Fatty acids	Content in % acc. to Leon-Camacho et al. (2001)	Content in % acc. to Martirosyan et al. (2007)	Content in % acc. to Gamel et al. (2007)
Total content of fat in the seed	6.34	6–9	7.1
Palmitic acid	20	19	–
Stearic acid	4	3.4	–
Oleic acid	33.3	34	25
Linoleic acid	38.2	33	50
Linolenic acid	1	–	1

The ratio of saturated fatty acids and unsaturated fatty acids is 1:3 (Martirosyan et al. 2007).

Composition of Triglycerides

Triglycerides	Content in % acc. to Leon-Camacho et al. (2001)
LLL	4.0
LnLO	0.6
LnLP	0.5
OLL	12.1
PLL	13.8
OLO	11.8
PLO + SLL	20.0
PPL	7.5
OOO	7.9
POO+SOL	12.5
PLS	2.1
POP	3.8
SOO	2.2
SOS	1.3

Composition of Sterols

Sterols	Content in % acc. to Leon-Camacho et al. (2001)	Content in mg/100 g acc. to Pina-Rodriguez and Akoh (2009)
Total content	24.6×10^3 ppm	846
24-Methylene-cholesterol	0.3	–
Campesterol	1.6	20.7
Stigmasterol	0.9	306.6

Sterols	Content in % acc. to Leon-Camacho et al. (2001)	Content in mg/100 g acc. to Pina-Rodriguez and Akoh (2009)
Δ7-Campesterol	24.8	–
Clerosterol	42.0	–
Beta-sitosterol	1.3	497.7
Δ5-Avenasterol	2.0	–
Δ7-Stigmasterol	15.2	–
Δ7-Avenasterol	11.9	–

The majority of sterols in amaranth oil are esterified. This is a peculiarity because in nearly all vegetable oils, the percentage of free sterols is higher. Another characteristic is the Δ7 structure of more than 50% of the sterols; the other half consists of clerosterol (42%).

In contrast to other vegetable oils, amaranth oil contains a remarkably high amount of sterols, which increases its potential in pharmaceutical use, for example in lowering the cholesterol level. There are indications of anti-inflammatory and analgesic effects. Clerosterol has antibacterial effects. Stigmasterol and sitosterol are used as semi-synthetic reagents in the production of steroid drugs (Leon-Camacho et al. 2001).

Composition of Tocopherols

Tocopherols	Content (ppm) acc. to Leon-Camacho et al. (2001)	Amaranth oil content (mg/100 g) acc. to Pina-Rodriguez and Akoh (2009)
Total content	–	65.3
α-Tocopherol	248	13.3
β-Tocopherol	546	26.8
γ-Tocopherol	–	5.7
δ-Tocopherol	8	19.5

α-Tocopherol is antioxidative, but β-tocopherol is not. As a result of the higher content of β-tocopherol, these compounds cannot protect amaranth oil against autoxidation (Leon-Camacho et al. 2001).

Other Ingredients: (Leon-Camacho et al. 2001)

Phytosterols	
Vitamin E	
Vitamin C	
Squalene	6–8%

Physical Key Figures of Amaranth Oil: (Singhal and Kulkarni 1990)

refractive index: n_D^{20}	1.470
relative density:	9.8155
saponification number:	217

iodine number:	100.0
colour value:	3 R + 2.8 Y
viscosity:	40.1
peroxide number:	5.1

Nutritional Value of Amaranth Oil per 100 g: (Martirosyan et al. 2007)

large calories: 711 kcal
kilojoule: 87 kJ

Shelf Life

Amaranth oil has a shelf life of about 9 months. (oilnatura.de).

5 Application

In Pharmacy and Medicine

As a result of its high content of squalene and polyunsaturated fatty acids, amaranth oil lowers the cholesterol level. In a study by Martyrosan et al., the effect of amaranth oil was tested on chickens. The result showed the oil's stabilising effect on cell membranes and its modulating effect on their fluidity. This is of interest because damage to the cell membrane disrupts the sodium/potassium exchange, which leads to an increase in blood pressure. Amaranth oil therefore reduces the overall risk of cardiac and circulatory diseases. It should not be expected that a therapy with amaranth oil will lead to weight loss, but positive effects have been observed in cases of breathlessness during physical activity, headaches, fatigue and leg oedema. A normalisation in the rhythm of the heart has been observed in the electric diagram, and symptoms of cardiac and circulatory diseases have improved. As Martirosyan et al. have ascertained, a therapy of 18 ml amaranth oil per day can reduce the total cholesterol level, triglycerides, LDL and VLDL, and consequently also the systolic and diastolic blood pressure as well as the heart rate (Martirosyan et al. 2007).

In Cosmetics

Amaranth oil is used in cosmetics as a result of its high squalene content. Squalene is absorbed quickly by the skin and is a natural moisturiser. The oil's considerable amount of phytosterols, natural tocopherols and tocotrienols is also important in this respect. Together with squalene, they act as antioxidants. Amaranth oil has a soothing, healing and anti-inflammatory effect and is used in the treatment of dry and sensitive skin. It counteracts skin ageing, smooths the skin and has a soothing effect (Quirin 2009).

As Food

A recent study by Pina-Rodriguez and Akoh discusses the use of amaranth oil in breast milk substitutes. For this purpose, amaranth oil is modified by enzymatic ester interchange to achieve a lipid component called DCAO (DHA-containing customised amaranth oil), which corresponds to breast milk. The modification is performed in two steps. First, the content of palmitic acid is increased. Docosahexaenoic acid, which is part of natural breast milk, is added. According to Pina-Rodriguez and Akoh, this process results in an ideal structural lipid that corresponds to natural breast milk. The study entitled "Synthesis and characterization of a structured lipid from amaranth oil as a partial fat substitute in milk-based infant formula" by Pina-Rodriguez and Akoh approves application by characterising the modified structural lipid physically and chemically. Further research on application and stability is necessary, however; the stability of the adapted amaranth oil containing DHA could be improved by adding antioxidants (Pina-Rodriguez and Akoh 2009).

Since amaranth oil has a similar composition to corn oil and cotton seed oil, it can be used as an edible oil (Lyon and Becker 1987).

Andiroba Oil

synonyms: Bastard mahogany oil, Carapa oil (E)

1 Source Plant

Carapa guianensis Aubl., *(Meliaceae), bastard mahogany.*

Habitat

Bastard mahogany is common along the Atlantic Cast of Central America, in Costa Rica, Colombia, Brazil, Ecuador, Cuba, and Trinidad and Tobago (Noldt 2000, p. 9).

Description

The tree is about 40 m tall, with oblong, green leaves and small, yellow, red or greenish flowers that smell disagreeable. The fruit grow in racemes and resemble chestnuts. They contain a seed rich in fat (approx. 63%), which is hard and a light-pink colour. A tree may bear up to 200 kg fruit.

2 Cultivation and Extraction

Cultivation

Bastard mahogany grows mainly in the Amazon basin, Pará and Guyanas (Brazil), especially on river banks. In order to save natural resources and support the

S. Krist, *Vegetable Fats and Oils*, https://doi.org/10.1007/978-3-030-30314-3_6

sustenance of the native population, areas that have been slashed and burned are reforested with bastard mahogany.

Extraction of the Oil

One tree produces seeds for the extraction of about 7 l of oil. The seeds are collected from the ground, and the oil is cold pressed, either manually by the native population, or industrially by the Brazilian soap industry.

3 Character

Colour

Milky-white to yellow oil, liquid at room temperature.

Odour

Sweet, reminiscient of cinnamon, balmy, biscuit flavour, pleasantly nutty.

Flavour

Bitter taste.

4 Ingredients

Composition of Fatty Acids

Fatty acids	Content in %, acc. to greencottage.com/ 28.07.04	Content in %, acc. to Abreu et al. (2004)	Content in %, acc. to augustus-oils.ltd.uk/ 2005
Palmitic acid	28	27	20.90
Palmitoleic acid	1	1 (n-7)	1.00
Stearic acid	8.1	7	12.10
Oleic acid	50.5	49 (n-9)	51.20
Linoleic acid	9	16	13.00
α-Linolenic acid	0.3	–	0.50
Arachidic acid	1.2	–	1.30

Other Ingredients

The oil's bitter-tasting components are terpenes, so-called meliacins, for example gedunin, but also limonoids like andirobin and epoxyazadiradione.

Physical Key Figures of Andiroba Oil

refractive index: n_D^{20} 1.463–1.4710
unsaponifiable: approx. 5%
acid number: *<100*

5 Application

In Pharmacy and Medicine

Gedunin, a type of meliacin, has antiparasitic properties and is comparable to quinine in its effect against malaria. Its anti-inflammatory properties as well as its effect as an insect repellent are ascribed to limonoids, especially andirobin. Another limonoid, epoxyazadiradione, is in vitro effective in treating tumours. This was tested on neuroblastoma and osteosarcoma cancer cells. In topical application, andiroba oil chemically inhibits the formation of cellulitis by blocking enzymes, but it does not affect already existing cellulitis.

As Insect Repellent

A comparative study tested the topical effectiveness of andiroba oil as an insect repellent against healthy females of *Aedes sp.*, a species of mosquito that can spread dengue fever. It was compared to 50% DEET (N,N-Diethyl-3-methylbenzamide), which is currently the most effective repellent, but may have topical and systemic side effects. The forearms of the four test persons were covered with 100% andiroba oil and 50% DEET. The time until the first and third mosquito bite was measured, at first without protection, then with the products applied. Without protection, 17.5 s passed on average until the first bite and 40.0 s until the third bite. With 100% andiroba oil, the bites were registered after 56.0 s and 142.5 s; in comparison, no bites were registered after 3600 s when DEET was used. The oil is thus only a weak protection against mosquitoes, but the advantage is that it has no side effects (Miot et al. 2004).

A

In Cosmetics

Andiroba oil is used in body care products, such as body lotions and creams, including for irritated skin, since it is anti-inflammatory. It is applied as skin oil, massage oil and in emulsions for sensitive skin or for combination skin with a tendency towards impurities. As an analgesic massage oil, it is useful in cases of contusions or swelling because it stimulates blood circulation. It also forms part of cosmetic products to treat cellulitis and of insect repellents applied before and after an insect bite.

In Folk Medicine

In Marajó (eastern Amazon basin), the oil is used to treat arthritis, adenopharyngitis, insect bites, cuts, inflammation, contusions, bruises, skin diseases, diarrhoea and uterine cancer (Hammer and Johns 1993).

6 Possible Unwanted Side Effects

The oil may cause skin irritation.

Apple Seed Oil

synonym: Apfelkernöl (D)

1 Source Plant

Malus domestica (Rosaceae), Apple.

S. Krist, *Vegetable Fats and Oils*, https://doi.org/10.1007/978-3-030-30314-3_7

Habitat

The apple tree has its origins in Central Asia, where its domestication began more than 8000 years ago. Apples were smaller and sweeter at that time. Later, the first cultivars were introduced to western regions. In the ninth century B.C., apples were cultivated in Greece, and in Italy soon after; with the Romans, they spread to Central Europe. Systematic cross-breeding began as late as the nineteenth century. Today, there are about 30,000 kinds of apple (Lieberei and Reisdorff 2007, p. 197).

Description

Apple trees blossom from March to May. The harvest takes place from June to October, and tree care from November to February (Hachfeld et al. 1999, p. 130).

The flowers have five white petals, which are sometimes a delicate shade of pink on the underside; they are arranged in terminal racemes, with short stems. The five carpels are surrounded by the receptacle, so that only part of the style protrudes. Each carpel has two ovules, but often only one of them develops further. After pollination, the receptacle becomes a fleshy fruit (Lieberei and Reisdorff 2007, p. 197).

The five carpels form the apple core, which – depending on the cultivar – contains about ten seeds. A brown, leathery seed coat surrounds the soft, white albumen and the seedling. The pips are flat and oval, tapering at the end where the seedling is located (Hackbarth 1944, p. 341).

2 Cultivation and Extraction

Cultivation

The apple tree is one of the world's most common fruit trees. Production has been growing continuously since the Second World War. In 1980, the growth rate stagnated, but rapidly increased again in 1990, due to increased appletree cultivation in China. At the beginning of the 1990s, Chinese apple production amounted to about four million tons, which corresponds to 10.7% of the global apple production. In 2000, this percentage further increased to 36.7% (Ferree and Warrington 2003, p. 15–16).

From 1948 to 1950, France was the main producer of apples, followed by the USA, Germany, the former USSR, Italy and the United Kingdom. These six countries had a share of 65% of the global apple production. In 1996, China succeeded France as the main producer, and in 2000, Chinese apple production increased to 22.89 million tons (Jackson 2003, p. 8–9).

In 2005, 72 million tons of apples were produced worldwide, with China remaining the number one producer with 22 million tons. Of the apples produced in China, three million tons are used to manufacture juice, with 0.9 million tons of pressed fruit pulp and 12,000 tons of pips as a by-product. These large amounts make it easy to understand why interest in a possible use for apple pips is increasing (Yu et al. 2007).

Extraction of the Oil

To extract the oil, the moistened pips are dried until their weight is constant. They are then ground and the oil either extracted using a Soxhlet extractor, with n-hexane as a solvent (Yu et al. 2007), or cold pressed.

3 Character

Colour

Light-yellow.

Odour

Aromatic.

Flavour

Weak.

4 Ingredients

Composition of Fatty Acids

Fatty acids	Content in g/100 g acc. to Yukui et al. (2009)	Content in % acc. to Lu and Foo (1998)	Content in % acc. to Yu et al. (2007)	Content in % acc. to Morice et al. (1971)
Myristic acid	–	–	–	–
Palmitic acid	5.606	10.49	7.1	4.8–8.5
Palmitoleic acid	0.060	–	–	0.1–0.5
Stearic acid	1.466	4.33	2.4	1.0–2.5
Oleic acid	26.473	4.12	39.7	24.4–42.1
Linoleic acid	43.031	51.15	49.6	48.2–64.1
Linolenic acid	0.600	5.60	–	0.1–0.8

A

Fatty acids	Content in g/100 g acc. to Yukui et al. (2009)	Content in % acc. to Lu and Foo (1998)	Content in % acc. to Yu et al. (2007)	Content in % acc. to Morice et al. (1971)
Arachidic acid	1.311	–	–	0.5–1.7
Behenic acid	0.270	–	–	0.1–0.3
Lignoceric acid	0.089	–	–	–

Morice et al. compared different cultivars with regard to their content of fatty acids and physical key figures. The study showed that *Granny Smith* contains greater amounts of palmitic acid and linoleic acid than *Dougherty* and *Sturmer*, but less oleic acid. The differences between *Dougherty* and *Sturmer* were not significant. With regard to physical key figures, it became apparent that *Granny Smith* has a higher iodine number. Apart from this, no significant differences could be detected (Morice et al. 1971).

Composition of Triglycerides

Triglycerides	Content in atomic percent acc. to Deineka and Deineka (2004)
$LnL_2 + Ln_2O$	0.5
$L_3 + LnLO$	16.7
$LnLP + Ln_2S$	0.9
$L_2O + LnO_2$	27.0
$L_2P + LnLS + LnOP$	9.0
LO_2	19.2
$L_2S + LOP + LnOS$	10.7
$LP_2 + LnPS$	1.6
O_3	7.2
$LOS + O_2P$	4.2
$LPS + OP_2$	0.8
O_2S	2.1

Tocopherol Content

The oil contains 165 mg vitamin E per 100 g (Yu et al. 2007).

Physical Codes of Apple Seed Oil

	Yu et al. (2007)	Hackbarth (1944)
Refractive index	1.4734 (20 °C)	1.4713 (21 °C)
Saponification number	186.5	189.5
Iodine number	1.293 (mg/g)	119.8
Acid number	–	–
Unsaponifiable	1.675%	1.2
Relative density	0.9219 (20 °C)	0.9016 (20 °C)

5 Application

In Pharmacy and Medicine

Apple seed oil only contains about 9.5% saturated fatty acids, but a high amount (up to 39.7%) of oleic acid, a monounsaturated fatty acid, and up to 49.6% of the essential fatty acid linoleic acid. It thus has a positive effect on the heart, reduces the risk of heart diseases and contributes to lower LDL and cholesterol levels (Yu et al. 2007).

Other Uses

Due to its relatively high iodine number, apple seed oil is part of the group of semi-drying oils and is used in the production of alkyd resins, shoe polish and varnish (Yu et al. 2007).

Apricot Oil

Oleum Pruni armeniacae synonyms: Marillenkernöl, Aprikosenkernöl (D); apricot kernel oil (E)

1 Source Plant

Prunus armeniaca L. (Rosaceae), apricot.

S. Krist, *Vegetable Fats and Oils*, https://doi.org/10.1007/978-3-030-30314-3_8

A

Habitat

The origins of the apricot are presumed to lie in Manchuria. From there, it spread west, first to Uzbekistan. The Greeks and Romans imported the cultivation of apricots to the northern Mediterranean area, and the Arabs to the southern Mediterranean area. The fruit also spread from Central Asia along both sides of the Black Sea to Turkey and Eastern Europe. In the eighteenth century, apricots were imported to the USA, especially to Washington and California, to Canada, South Africa, Australia and New Zealand (Wurm 2002, p. 11).

Description

The apricot is a tree that grows to a height of 3–4 m. It has hairless, roundish, shiny branches of a green to red colour. The leaf stems are 2–3 cm long and usually covered with several large glands. The leaf blade is about 10 cm long and up to 7 cm broad, roundish to oval, pointed, heart-shaped on the base, crenate to serrate at the edges, usually hairless on both sides, and slightly shiny on the upper side. The flower stems are very short and downy. The flowers are either solitary or arranged in pairs, and appear before the leaves. The calyx is reddish brown, with a 7–8 mm long hypanthium with velvety hair on the base, and blunt, hairless or downy tips bending downwards. The petals are roundish to obovate, about 1–1.5 cm long, at first light pink, then white. The fruit is roundish, more rarely elongated, 4–8 cm long, with a longitudinal furrow down one side, and is umbilicate at the base. It is covered with velvety hair and is a light-yellow to orange-red colour; on the sunny side it is often dappled crimson. The stone is smooth, oval, round-bodied, and up to 3 cm in length. On one side, it is triangular, on the other rounded. It can be easily separated from the yellow-orange fruit pulp. The seed tastes either bitter or sweetish (Hager 1978, volume 6a, p. 941 et seq).

2 Cultivation and Extraction

Cultivation

Today, apricots are mainly cultivated in Europe (Austria, Hungary, Italy, etc.), in the Middle East, the USA and Australia. Global production amounts to more than 2.3 million tons of fruit, of which about a third is from Europe (including Ukraine and Russia) and another third from the Middle East. The largest producer is Turkey with a share of 20% of the global production. The main part of the harvest is processed further to produce dried apricots in a practically monopoly position; this kind of processing is of importance in the entire Middle East. In Europe, Spain, France and Italy produce particularly large quantities of apricots. Here, fresh apricots are more important, and the proportion of industrially processed products amounts to only 25% of the total yield (Wurm 2002, p. 11).

Extraction of the Oil

Apricot oil is obtained from the kernels of ripe apricots. They are cold pressed. If the oil is not used for flavouring purposes, it is usually refined.

3 Character

Colour

Light to medium yellow.

Odour

Unrefined: like marzipan, bitter almonds, benzaldehyde, hydrocyanic acid.

Refined: slightly nutty, like olives, weakly fruity – sourish, delicate smell of marzipan, characteristically of apricots.

Volatile Compounds

Almost no results of analyses of the volatile compounds of apricot oil are available in literature. Some volatile compounds were described by Compton and Stout using headspace/GC-FTIR-analysis (Compton and Stout 1992):

H_20	Ethyl caproate	Linalyl acetate
Ethyl acetate	α-limonene	α-phellandrene
Acetous acid	Geranial (citral A)	Diethyl oxalate
Propyl acetate	Ethyl caprylate	2-ethyl-4-methyl acetate
1,2 propandiol	Valeric acid	Hexadecanoic acid pentyl ester
Butyric acid ethyl ester	Anethole	2-carboxy-4-octyl-butyrolactone
3-Methyl butyric acid ethyl ester	Triacetin	Benzaldehyde
1-Pentanol	Ethyl caprate	

Flavour

The weak marzipan fragrance and taste indicate the close affinity between apricot oil and almond oil.

Slight, mild taste of marzipan (Kerschbaum and Schweiger 2001, p. 20), characteristically of apricots.

A

4 Ingredients

Composition of Fatty Acids

Fatty acids	Content in % acc. to Nature certificate of analysis/28.08.03	Content in % acc. to Kerschbaum and Schweiger (2001), p. 19
Palmitic acid	5.9	5.4
Palmitoleic acid	0.6	0.7
Stearic acid	1.6	0.8
Oleic acid	66.1	66.4
Linoleic acid	25.3	21.6
Vaccenic acid	–	3.2
Eicosenoic acid	–	0.1

Composition of Sterols

(Firestone 1999, p. 9)

Sterols	Amount (%) of the total content of sterols
Cholesterol	0.6
Campesterol	6
Stigmasterol	1–2
β-Sitosterol	88–89
Δ5-Avenasterol	3–4

Composition of Tocopherols

Tocopherols	Content (mg/kg) acc. to Firestone (1999), p. 9	Content (mg/100 g) acc. to Kerschbaum and Schweiger (2001), p. 36
Total content of tocopherols	200–840	–
α-Tocopherol	10–22	–
γ-Tocopherol	170–794	34.2
δ-Tocopherol	20–24	–

Other Ingredients

(Hager 1978, volume 6a, p. 942; Kerschbaum and Schweiger 2001, p. 43)

Vitamin A
Vitamin B
Vitamin B_{15}
Vitamin E
Carotene
Niacin
Folic acid
Pangamic acid

Flavonoids
Squalene 0.02%
β-Sitosterol
Campestrin
Emulsin

The composition of apricot oil is similar to that of almond oil. There is, however, a significant difference concerning the ingredient tocopherol. Almond oil contains both α- and γ-tocopherol, whereas apricot oil contains only γ-tocopherol. This is important in order to distinguish marzipan from persipan (Kerschbaum and Schweiger 2001, p. 36).

Physical Key Figures of Apricot Oil

refractive index: n_D^{25} 1.469–1.473
density: 0.1914–0.920
saponification number: 188–197
iodine number: 96–109
unsaponifiable: 0.4–1.3%
point of solidification: −4 to −21 °C
(Roth and Kormann 2000, p. 160)

Shelf Life

Apricot oil has a shelf life of about 1 year.

5 Application

In Pharmacy and Medicine

For the Treatment of Skin Diseases

Apricot oil is part of formulations for the treatment of skin diseases that result in skin dehydration.

To Treat Migraines

Apricot oil contains pangamic acid, which is thought to stimulate oxygen supply in the cells. It therefore alleviates the consequences of a lack of oxygen, such as fatigue and listlessness. How effective a treatment for migraines is with apricot oil alone must still be scientifically examined.

In Cases of Anginal Complaints and Cirrhosis of the Liver

Since Oleum pruni armenicae contains pangamic acid, it is used in cases of angina pectoris and liver damage (Hunnius 1998, p. 1461).

In Cosmetics

Apricot oil is rich in oleic acid and linoleic acid and serves as a base of skincare products (ointments, creams). It is mild, not too greasy, disseminates well and penetrates the skin quickly. It is especially suitable for ageing and sensitive skin (Hager 1978, p. 942; Löw 2003, p. 117).

Apricot oil is thought to have the following effects (Löw 2003, p. 117):

- stabilising tissue;
- activating skin metabolism;
- supporting cell regeneration;
- retaining moisture;
- soothing irritated skin;
- nurturing flaky and chapped skin;
- antibacterial effect; and
- refreshing pale complexions.

Apricot oil is therefore used in:

- anticellulite products;
- creams for the treatment of chapped calluses of the feet;
- anti-dandruff shampoo;
- products to treat premature symptoms of old age (oil to treat eye wrinkles);
- aftersun lotions;
- products for the treatment of sensitive and inflamed skin;
- baby-care products; and
- massage oil.

As a Mild Laxative

In Iran, apricot oil is sometimes used as a laxative instead of expensive almond oil (Filsoof et al. 1976).

In the Sweets Industry

Similarly to almond oil, apricot oil is used in the sweets industry because of its taste of marzipan, but only rarely. It has been tested with a view to economising the production of cakes. As it is non-toxic (the oil does not contain amygdalin) and does not change the products' taste, colour or consistency, it provides a good future alternative (El-Aal et al. 1986).

As Food

Apricot oil is only used rarely as an edible oil, as it becomes rancid quickly. Due to its short shelf life, only small amounts are commercially available.

Argan Oil

synonyms: Arganöl, Marokkanisches Olivenöl, Ardjanöl (D); huile d'argan (F)

1 Source Plant

Argania spinosa (L.) Skeels *(Sapotaceae)*, argania, argan.

S. Krist, *Vegetable Fats and Oils*, https://doi.org/10.1007/978-3-030-30314-3_9

Habitat

Argan is a slow-growing, thorny bush or tree native to south-western Morocco between Essaouira and Agadir. It grows wild in a region of about 740,000 hectares (Charrouf and Guillaume 1999). In 1999, UNESCO added this area to its list of biosphere reserves.

Description

Argan grows 7–10 m high and can reach a circumference of up to 15 m and an age of 125–150 years. The trunk is usually gnarled and crooked. The tree starts bearing fruit when it is 5 years old. The fruit is round or oval, green and fleshy like an olive, but larger. It contains a nut with a very hard shell. When ripe, the fruit is bright yellow. Each fruit contains 1–3 oval, smooth brown seeds about 2 cm long. The seed contains a light brown oil, which makes up about 50% of its content. Within the seed coat, there is a white kernel rich in oil (Axtell and Fairman 1992, p. 11; Charrouf and Guillaume 1999).

2 Cultivation and Extraction

Cultivation

In southwestern Morocco, argan has been commercially used for decades because it survives dry conditions. It protects the soil from erosion and thus prevents the desert from spreading. In recent years, however, the tree population has been continually decreasing. The reason for this is that the trees are increasingly used for their wood and as forage (Charrouf and Guillaume 1999), in connection with the fact that there are no plantations, but only a population of wild growing trees (Axtell and Fairman 1992, p. 11).

Apart from Morocco, the tree is also cultivated in two different regions of the Negev desert in Israel (Yaghmur et al. 2001).

Extraction of the Oil

For oil production, only fruit that have fallen to the ground are used; they are gathered from June to September. Harvesting would otherwise be difficult, as the wood is too brittle for the tree to be shaken, and the thorns do not allow the fruit to be plucked by hand.

Depending on the extraction method, the seeds contain 30–55% fat oil.

Hand-pressed Argan Oil

According to the traditional method, the oil is extracted manually. The ripe fruit pulp and the shell are removed, and the nuts are broken using stones. The seeds are air-dried in clay pots and then roasted slightly at low temperatures. The roasted seeds are cooled and ground to a coarse dough. To extract the oil, the dough is mixed manually with water and kneaded until it is solid. The brown emulsion contained in the dough is decanted. After several minutes, a clear oil with a hazelnut flavour is obtained. The remaining press cake is dark brown to black and still contains 10% of the oil; it is used as forage. This is a slow method, which only produces about 1 l of oil in 10 hours. It only extracts about 30% of the oil, which has a short shelf life because water has been added (Charrouf and Guillaume 1999).

Extraction with Mechanical Presses

Using this method, the seeds can be directly pressed, without adding water. The oil is extracted in the same way as described above. Up to 45% of the oil is extracted, 1 l in about half an hour. The oil contains less acid than hand-pressed oil (Charrouf and Guillaume 1999).

Extraction with Solvents

For industrial purposes and the use in laboratories, the oil can be extracted using any volatile, lipophilic solvent. After the solvent has volatilised, 50–55% of the oil remains. This oil has disadvantageous organoleptic properties, however, and is only used in the cosmetics industry (Charrouf and Guillaume 1999).

The so-called "enriched argan oil" is also only used in the cosmetics industry. It is produced by short path distillation of oil that has been extracted with solvents under reduced pressure and at a temperature of 270 °C. The unsaponifiable portion is three times larger than that of pressed oil (Charrouf and Guillaume 1999).

3 Character

Colour

Yellow to light orange oil, liquid at room temperature (Seidemann 1998).

Odour

Pungent, sweet, sour, disagreeable, similar to olive oil.

Flavour

Reminiscent of hazelnuts (Charrouf and Guillaume 1999). Roasted argan oil: nutty taste.

A

4 Ingredients

99% of the oil consist of glycerides; the largest part of these are triglycerides with a percentage of 95%. Unsaturated fatty acids, mainly oleic acid and linoleic acid, constitute the greater part of the oil (Charrouf and Guillaume 1999).

Composition of Fatty Acids

(sample 1: argan oil from a pharmacy in Marrakesh; sample 2: argan oil from a market in Agadir)

Fatty acids	Content in % acc. to Charrouf and Guillaume (1999)	Content in % acc. to Seidemann (1998), sample 1	Content in % acc. to Seidemann (1998), sample 2	Content in % acc. to Firestone (1999), p. 10	Content in % acc. to Khallouki et al. (2003)
Myristic acid	0–0.2	0.2	0.1	0.1–0.2	–
Pentadecanoic acid	0–0.1	–	–	–	–
Palmitic acid	13.5–13.9	12.5	11.8	14–16	15.6–16.5
Palmitoleic acid	0–0.2	0.2	0.1	0.1	–
Margaric acid	0–0.1	–	–	–	–
Stearic acid	5.6	4.5	3.6	–	3.7–8.5
Oleic acid	45.2–46.9	42.7	22.9	43–50	41.2–45.0
Linoleic acid	31.6–34.6	34.8	51.4	30–34	35.0–37.9
α-Linolenic acid	0–0.1	3.5	8.8	0.1–0.3	–
C 19:0	0–0.1	–	–	–	–
Arachidic acid	0–0.4	0.4	0.3	0.2–0.4	–
Timnodonic acid	–	0.2	0.3	–	–
Eicosenoic acid	0–0.5	0.5	0.3	–	–
Behenic acid	0–0.1	–	–	–	–

Composition of Triglycerides

The greater part of triglycerides consists of three oleic acid radicals (OOO), two linoleic acid radicals and one oleic acid radical (LLO), one radical each of palmitic acid, oleic acid and linoleic acid (POL), or one palmitic acid radical and two oleic acid radicals (POO). Stereospecific analysis shows that saturated fatty acids like palmitic acid and stearic acid are esterifed at the glycerol in the outer positions (Sn-1 and Sn-3), but linoleic acid mainly at the secondary alcohol (Sn-1) of the glycerol. Oleic acid can take up all positions. The following table shows the distribution of fatty acids to the three positions of glycerol (Charrouf and Guillaume 1999):

Fatty acids	Sn-1 (in %)	Sn-2 (in %)	Sn-3 (in %)
Palmitic acid	54.0	9.4	36.6
Stearic acid	19.4	1.7	78.9
Oleic acid	33.3	39.7	27.0
Linoleic acid	29.5	40.0	30.5

Composition of Sterols

Sterols	Content in % acc. to Firestone (1999), p. 10	Content in % acc. to Charrouf and Guillaume (1999)
Δ7-Stigmasterol	48	–
Δ7-Avenasterol	4	–
Spinasterol	–	44
Schottenol	–	48
Stigmasta-8,22-dien-3β-ol	4	4
5α-Stigmasta-7,22-dien-3β-ol	44	4

Composition of Tocopherols

(concentration in mg/kg and content in %)

Tocopherols	Argan oil acc. to Firestone (1999), p. 10	Argan oil acc. to Charrouf and Guillaume (1999)
Total content	167–635	620
α-Tocopherol	–	69
β-Tocopherol	–	16
γ-Tocopherol	–	13
δ-Tocopherol	–	2

Other Ingredients

(Charrouf and Guillaume 1999; Khallouki et al. 2003)

Butyrospermol: 18.1%
Tirucallol: 27.9%
β-Amyrin: 27.3%
Lupeol: 7.1%
24-Methylen-cycloartanol: 4.5%
Citrostadienol: 3.9
Cycloeucalenol: <5%
Vanillic acid: 6–123 g/kg
Syringic acid: 68 g/kg
Ferulic acid: 3147–3470 g/kg
Tyrosol: 3–52 g/kg
Carotene: 37%
Quercetins
Myricetin

A

Physical Key Figures of Argan Oil

density: 0.9 g/ml
refractive index: 1.468
saponification number: 195.2
iodine number: 98.1
unsaponifiable: 1.0
acid number: 1
(Charrouf and Guillaume 1999)

5 Application

In Pharmacy and Medicine

Effects on the Cholesterol Level

A study on 96 test persons showed that an average daily consumption of 15 g native, cold pressed argan oil reduces the plasma LDL cholesterol level considerably in comparison to that of test persons who did not take the oil. In addition, the vitamin E plasma level is higher and the lipoperoxide level lower, which suggests that the oil has antioxidant activity. In vitro, the antioxidant effect on LDL oxidation was proven. These results show that argan oil has a positive effect on the lipid metabolism and thus lowers the risk of cardiovascular diseases (Drissi et al. 2004).

As an Antioxidant

The predominating tocopherol in argan oil is γ-tocopherol, with a share of 465–504 mg/kg oil. γ-Tocopherol hems the activity of cyclooxygenase, and as a dietary supplement protects against against inflammation caused by reactive oxygen species. The oil also contains a relatively high amount of squalene (313 mg/100 g oil), which is thought to protect from skin cancer. The sterols spinasterol (40%) and schottenol (48%) are not common in vegetable oils, but they form the main part of the sterols in argan oil. In animal experiments with mice, spinasterol caused a reduction in the plasma and cholesterol levels of the liver and was found to have antitumour potential. Schottenol has anticarcinogenic and cytotoxic potential (Khallouki et al. 2003).

Antihypertensive Effects

Argan oil contains a high percentage of monounsaturated and polyunsaturated fatty acids (MUFAs and PUFAs) and only a low percentage of saturated fatty acids. It also contains a high proportion of the antioxidant α-tocopherol, which in an experimental model reduced blood pressure. During a study, SHR (spontaneously hypertensive rats) and WKY (Wistar-Kyoto rats with normal blood pressure) were fed argan oil over 7 weeks, and the effects of this diet on blood pressure and endothelial dysfunction were registered. The SHR reacted to the daily administration of argan oil with a continuous reduction of the average blood pressure, from the fifth week onwards significantly so. No change was observed in the WKY group. Although the blood pressure of the SHR was lower, their heart rate stayed the same, as did their body weight. The endothelial function was measured by the carbachol-induced degree of relaxation on arteries that had been previously narrowed by phenylephrine. After 7 weeks, the endothelium-dependent relaxation of the aortic arterial rings significantly increased in the SHR group that was administered argan oil, in comparison to the control group. The result shows that due to its high amount of linoleic acid, argan oil plays a role in the regulation of blood pressure, and that the amount of α-tocopherol has an influence on antihypertensive effects (Berrougui et al. 2004).

In Cosmetics

Argan oil increases antioxidant cell activity. This effect is ascribed to its vitamin E content. Application on the skin reconstructs the water-lipid-layer, increases the amount of nutrients in the skin cells, stimulates intercellular oxidation, and neutralises free radicals; the oil thus protects the connective tissue. For this reason, argan oil is increasingly used in the cosmetics industry for skin protection products (Charrouf and Guillaume 1999, p. 7 et seqq). Argan oil protects the skin from drying out; it has a lifting, smoothing effect and makes the skin supple. Its protective, smoothing and moisturising properties have a positive effect on juvenile acne, skin allergies and ageing skin.

As Food

Argan oil is an important edible oil and is used as a cooking oil in Morocco (Axtell and Fairman 1992, p. 11). In Europe, argan oil is considered a gourmet edible oil for sophisticated gastronomy, on the level of caviar and truffles.

In order to examine the usefulness of argan oil as a fat for frying, 2 l of the oil were heated to 180 °C for 24 hours. To simulate controlled frying, 2 ml water were added every 4 hours. The experiment showed that argan oil has a similar or even better quality than cotton seed oil and olive oil. The experiment examined the composition of fatty acids, the peroxide number, acid number, induction period, colour index, viscosity, the content of conjugated double bonds and the contact angle between the oils and a potatoe camouflage in dependence of the temperature (Yaghmur et al. 2001).

Artichoke Seed Oil

synonyms: Artischockenöl (D); huile de Cynara (F), Olio di semi di carciofo (I), Alcachofa aceite (ESP), enginar tohumu yagi (TUR)

1 Source Plant

Cynara cardunculus L. (Asteraceae), artichoke.

S. Krist, *Vegetable Fats and Oils*, https://doi.org/10.1007/978-3-030-30314-3_10

Habitat

The artichoke originates in the Mediterranean area. In Ancient Egypt, it was used on sacrificial altars and features on pictures showing fruit trays. Evidence that artichokes were used in Greece dates from the eighth century B.C. For the Romans, artichokes were a luxury good, which in the first century A.D. resulted in plans to cultivate artichokes. When the Roman Empire fell apart, the uses of the artichoke were forgotten and only rediscovered in the fifteenth century (Baier 2003, p. 9).

Even today, artichokes are mainly found in the Mediterranean area. 85% of the global production are located here, and Italy is the most important producer with a share of over 50% in the total production (Halter 2003, p. 17).

Description

Artichokes have a phyllotactic rosette consisting of large, pinnatisect leaves. The long inflorescences emerge from the axil. They are closed at first and open later to show purple flowers. The bracts are broad and oval, surrounding the receptacle like roof tiles. They are fleshy at the base (Lieberei and Reisdorff 2007, p. 279).

The artichoke is a perennial shrub. For the plant to produce flowers, days need to be at least 10.5 hours long (Halter 2003, p. 17).

The fruit is a brown achene 3.5–7 mm long and 2–5 mm broad. The root functions as a storage organ and can reach to a depth of 90–120 cm (Baier 2003, p. 12–13).

2 Cultivation and Extraction

Cultivation

Domestication probably began in Spain or Tunisia, with two aims: An increase in the size of inflorescences led to the cultivated artichoke, and an increase in the size of leaf stalks led to the cultivated cardoon (Baier 2003, p. 14).

The ideal temperature for growing artichokes is 12–14 °C by night and 20–22 °C by day. Artichokes also grow at temperatures between 0 and 35 °C. If the temperature is too high or too low, the artichoke enters a period of dormancy. This usually happens in the summer months. With increased precipitation in autumn, it starts to form buds and new rosettes. The harvest takes place between September and June; afterwards, the plants are cut and chopped. There is another phase of dormancy until the next cycle starts with the sprouting of buds underneath the soil. In the Mediterranean area, the artichoke is a typical autumn/winter vegetable as well as a national dish (Halter 2003, p. 18).

All parts of the plant contain bitter substances, which stimulate the production of bile. The artichoke is used to treat gall bladder problems, indigestion and nausea. It also has cholesterol-lowering properties and protects the liver (Chevallier 2001, p. 198).

In 2007, the global crop amounted to an average of 12 t/hectare; 1.3 million tons of artichokes were produced (Lieberei and Reisdorff 2007, p. 279).

The most important producer countries of 2005 in 1000 t (Lieberei and Reisdorff 2007, p. 279)	
Italy	470
Spain	189
Argentina	88
Egypt	70
Peru	68
China	55
Morocco	53
France	50
USA	38
Algeria	37

Extraction of the Oil

Extraction with solvents is the most common method. The seeds are gathered at the end of the vegetative cycle, ground finely and dried. The oil is then extracted with a Soxhlet extractor by using ethyl ether (Micelli and De Leo 1996).

3 Character

Colour

Light-yellow (Micelli and De Leo 1996).

Odour

Pleasant (Sharaf-Eldin 2002).

Flavour

Good (Sharaf-Eldin 2002).

A

4 Ingredients

Composition of Fatty Acids

Fatty acids	Content in % acc. to Curt et al. (2002)	Content in g 100 g^{-1} acc. to Maccarone et al. (1999)	Content in % acc. to Micelli and De Leo (1996)	Content in % acc. to Hassanein et al. (2011)
Myristic acid	0.11	–	–	0.1
Myristoleic acid	0.01	–	–	–
Palmitic acid	10.66	10.88–11.03	13.9	11.3
Palmitoleic acid	0.16	–	<0.1	0.1
Margaric acid	0.05	–	–	–
C17:1	0.01	–	–	–
Stearic acid	3.71	3.25–3.60	3.4	3.2
Oleic acid	24.99	28.62–29.92	30.0	30.2
Linoleic acid	59.74	53.63–54.42	51.2	53.3
Arachidic acid	0.36	–	0.3	0.4
C 20:1	0.13	–	–	0.2
Behenic acid	0.03	–	<0.1	–

Composition of Triglycerides

Triglycerides	Content in % acc. to Hassanein et al. (2011)
LLL	19.2
XLO	0.3
LLO	19.1
LLP	11.8
LOO	12.5
LOP	13.0
LPP	2.3
OOO	6.4
LOS	2.6
OOP	5.8
LSP	1.1
POP	1.6
OOS	2.1
LSS	0.3
POS	0.9
SPP	0.1
SOS	0.1

Composition of Sterols

Sterols	Content in mg 100 g^{-1} acc. to Maccarone et al. (1999)	Content in % acc. to Micelli and De Leo (1996)	Content in % acc. to Hassanein et al. (2011)
Cholesterol	–	0.2	–
Campesterol	44.78	12.8	9.5
Stigmasterol	76.25	16.2	22.7
β-Sitosterol	181.43	45.6	33.1
Spinasterol	–	–	2.2
Isofucosterol	–	–	0.4
Δ7-Stigmasterol	183.97	18.4	25.7
Δ7-Avenasterol	–	6.6	6.4

The total content of sterols in artichoke seed oil amounts to 580 mg/100 g (Hassanein et al. 2011).

Composition of Tocopherols

Tocopherols	Content in mg 100 g^{-1} acc. to Maccarone et al. (1999)	Content in % acc. to Hassanein et al. (2011)
Total content	–	311 ppm
α-Tocopherol	54.6	96.5
β-Tocopherol	–	–
γ-Tocopherol	–	2.5
δ-Tocopherol	–	1.0

Physical Key Figures of Artichoke Seed Oil

refractive index: n_D^{20}	1.4682
saponification number:	182.9 (mg/g)
acid number:	22.4 (mg/g)
iodine number:	108.4 (g/100 g)
unsaponifiable:	0.9%
molecular weight:	306.2
(Micelli and De Leo 1996)	
peroxide number:	4.77 (mEq O2/kg)
density:	0.924
calorific value:	32.99 (MJ/kg)
viscosity 40°:	31.3 (mm/s)
cetane number:	51.4
flash point:	350 (°C)
(Fernández and Curt 2004)	

5 Application

In Pharmacy and Medicine

Artichoke seed oil contains many essential fatty acids, linoleic acid and linolenic acid in particular. These cannot be synthesised by the organism itself and play a part in the production of cells and the development of the functions of the brain and nervous system. Other important ingredients are phytosterols, especially β-sitosterol, which lowers blood pressure and the serum cholesterol level. The risk of chronic heart diseases is thus reduced. The phytosterols in artichoke seed oil are natural antioxidants and protect the organism from free radicals (Hassanein et al. 2011).

Technical Uses

Artichoke seed oil is used in the production of biodiesel. In Europe, biodiesel is predominantly made from rapeseed oil. This, however, is not profitable in the Mediterranean area, as the yield is not high enough in dry climates. These regions prefer sunflower oil as a base for biodiesel, but this is expensive. The aim is thus to find raw material that is cost-efficient as well as easy to produce in dry conditions. Artichoke seed oil is cheaper than sunflower oil. Costs can be minimised because two substances can be produced at once: on the one hand lignocellulose, which is used as green fuel for electrical power generation, and on the other hand seeds for the extraction of the oil, which is in turn a raw material for biodiesel (Fernández and Curt 2004).

As Food

Artichoke oil contains 91.4% essential polyunsaturated fatty acids and has a pleasant taste. It is a high-quality edible oil (Sharaf-Eldin 2002).

Other Applications

Artichoke seed oil is used for the production of soap, shampoo, alkyd resin and shoe polish (Micelli and De Leo 1996).

Avocado Oil

Oleum Perseae synonyms: Avocadoöl (D); avocado pear oil (E); huile d'avocat (F) – ill. 5

1 Source Plant

Persea americana Mill. (Lauraceae), avocado tree.

S. Krist, *Vegetable Fats and Oils*, https://doi.org/10.1007/978-3-030-30314-3_11

Habitat

The genus *Persea* consists of 135 species, which are common on the Atlantic side of North America as well as in Chile, India, Malaysia and Japan. Avocado trees grow in all kinds of soil, although they prefer eutrophic soil. The water supply may pose a problem, however, as on the one hand, the plants need considerable amounts of water, but on the other hand may develop root rot if the soil is saturated. A regular water supply is ideal, which is why avocado trees are also cultivated under irrigation. In addition, avocado trees need plenty of light (about 2000 hours of sunshine annually) and wind protection, as their branches break easily.

Description

The avocado tree is 4–20 m tall; its trunk is about 60 cm thick. Trees that are allowed to grow freely usually grow upright with a significant apical dominance. The bark of the trunk is rather thin, fairly smooth, and ash-coloured.

The leaves are evergreen, light green at first and, in sunny locations, later change to a shiny dark-green colour. The flowers of *Persea americana* are small, 5–8 mm long, and yellowish to greenish in colour. The inflorescence is panicle-shaped, either axial or terminal. There are two types of flower, one plant having only one type. Both types are only ready to receive pollen at a particular time of day. For purposes of cultivation, this means that both types of plant must be grown, as pollination is otherwise (at least theoretically) impossible. The fruit can be the size of a fist, but also as large as a gherkin; it is oval to pear-shaped, and measures up to 18 cm in length. It is a berry varying in colour from dark green to brownish-red. It is easy to peel. The fruit pulp is soft and white to creamy yellow. The seed can be roundish, conical or elongated, and makes up 8–24% of the fruit's weight (Hager 1978, volume 6a, p. 533).

2 Cultivation and Extraction

Cultivation

Avocados are mainly grown in tropical and subtropical regions, especially in America and Asia. The USA and Mexico are the main producers. Exports on a larger scale have only been taking place since the Second World War. The export market is essentially limited to the USA, European countries and several countries in East Asia. Avocados sold in Europe usually come from Israel, South Africa or Spain. There are about 200 cultivars, but only two, "Fuerte" and "Hass", dominate the European market (Salunkhe et al. 1992, p. 511).

Extraction of the Oil

Avocado oil is cold pressed; only the fruit pulp of overripe fruit that is no longer in a saleable condition is used. It can also be extracted by centrifugalising the fruit pulp that has been mashed to puree or by aggrading it with neutral material, usually kieselguhr, and then distilling it with petroleum ether (Roth and Kormann 2000, p. 117).

3 Character

Colour

Dark yellow to brown, greenish (Roth and Kormann 2000, p. 117).

The "raw" oil is green and cloudy. Chemical treatment removes colour material and turbid material, so that the result is a light-yellow oil.

Odour

Avocado oil: green, fruity, mellow, slightly fusty, acetous. Avocado oil green: fruity, like vinegar, nutty, woody.

Volatile Compounds

Analysis of the volatile compounds in avocado oil using microwave and solvent extraction (Ortiz Moreno et al. 2003):

Hexanal	2,6-dimethylundecane	Cyclotetradecane
1,4-Dimethylbenzene	Trans,trans-2,4-decadienal	Germacrene D
1,2-Dimethylbenzene	Tridecane	Pentadecane
Heptanal	2,4-decadienal	(E,E)-α-farnesen
1,2,4-Trimethylbenzene	α-Cubebene	β-Bisabolene
Decane	2-docen-1-al	Caryophyllene oxide
Octanal	α-Copaen	α-Bergamotene
4-Ethyl-1,2-dimethylbenzene	Tetradecane	α-Humolene
Undecane	β-Caryophyllene	Dodecane
Nonanal	Trans-α-bergamotene	Decylaldehyde
2-Hydroxymethylester-benzoic acid	2-ethylhexylester-2-propanoic acid	2-methyl-3-hydroxy-2,4,4-trimethylpentylester-propanoic acid
2-Methyl-1-(1,1-dimethyl)-2-methyl-1,3-propamedylester-propanoic acid		

Flavour

Pleasant, reminiscent of hazelnuts (Roth and Kormann 2000, p. 117). Neutral (Kerschbaum and Schweiger 2001, p. 22).

4 Ingredients

Composition of Fatty Acids

The composition of fatty acids varies depending on cultivation area and cultivar:

Fatty acids	Content in % acc. to Nature certificate of analysis/28.08.03	Content in % acc. to Hager (1978), volume 7b, p. 164	Content in % acc. to Kerschbaum and Schweiger (2001), p. 21
Palmitic acid	16.2	18.7–30.8	20.4
Linoleic acid	13.5	39.9–58.5	12
Palmitoleic acid	5.6	–	9.8
Oleic acid	62.8	45	49.9
Linolenic acid	0.7	0.4–3.3	0.4
Myristic acid	–	–	–
Stearic acid	1.1	0.2–0.5	0.4
Arachidonic acid	–	–	0.1
Vaccenic acid	–	–	6.8

Composition of Triglycerides

Triglycerides	Content in % acc. to Karleskind (1996), p. 244
OLL	5–8
PLL + PoOL	2–7
OOL	10–18
POL + PoOO	12–27
PPL + PPO	0–3
OOO	12–30
POO	19–25
PPO	3–11
PPoO	<0.2

Composition of Sterols

(concentration in mg/100 g and content in %)

Sterols	Avocado oil acc. to Firestone (1999), p. 10 et seq	Raw avocado oil acc. to Carstensen (2001), p. 42	Refined avocado oil acc. to Carstensen (2001), p. 42	Avocado oil acc. to Karleskind (1996), p. 245
Total content	404	455.7	402.8	350–560
Cholesterol	0–0.2	1.0	0.3	1–2

Sterols	Avocado oil acc. to Firestone (1999), p. 10 et seq	Raw avocado oil acc. to Carstensen (2001), p. 42	Refined avocado oil acc. to Carstensen (2001), p. 42	Avocado oil acc. to Karleskind (1996), p. 245
Brassicasterol	2	–	–	–
Campesterol	6–8	4.8	5.6	5–12
Stigmasterol	0–2	0.8	0.5	1–10
β-Sitosterol	89–92	86.8	88.3	60–92
Δ5-Avenasterol	3	2.1	3.5	2–10
Δ7-Stigmasterol	–	–	–	1–2
Δ7-Avenasterol	0.2	1.9	0.5	1–2
Cholestanol	–	–	–	–
Campestanol	–	–	–	–

Composition of Tocopherols

(concentration in mg/100 g and content in %)

Tocopherols	Avocado oil acc. to Firestone (1999), p. 11	Raw avocado oil acc. to Carstensen (2001), p. 44	Refined avocado oil acc. to Carstensen (2001), p. 44	Avocado oil acc. to Karleskind (1996), p. 245
Tocopherol content	8.3–10	12.1	10.5	8–15
α-Tocopherol	64–100	79.9	84.9	77–100
β-Tocopherol	–	–	–	–
γ-Tocopherol	0–19	20.1	15.1	0–23
δ-Tocopherol	–	–	–	–

Other Ingredients

(Kerschbaum and Schweiger 2001, p. 43; Hager 1978 volume 7b, p. 164)

Vitamin A
Vitamin B
Vitamin D
Vitamin E
Lecithin
Phytosterols
Squalene
Biotin
Hydroxy-α-carotene
α-Carotene
β-Carotene
Lutein

Physical Key Figures of Avocado Oil

refractive index: n_D^{20} 1.470–1.472
density: d_{25}^{25} 0.908–0.921

saponification number: 170–197
iodine number: 65–95
unsaponifiable: 1–2%
(Roth and Kormann 2000, p. 117)

Shelf Life

Avocado oil has a shelf life of up to 12 months if stored in the refrigerator.

Characteristics of the Oil

Like in macadamia oil, the amount of palmitoleic acid in avocado oil is far higher than is usual for vegetable oils. In addition, both avocado oil and macadamia oil have a natural sun protection factor (Kerschbaum and Schweiger 2001, p. 20).

5 Application

In Cosmetics

Avocado oil is well suited as a basis for ointments, especially for cosmetic creams, oily lotions and emulsions. Skin care products containing avocado oil have the following characteristics (Hunnius 1998, p. 167):

- good spreading ability as a result of the oil's phytosterol content
- hydrating effect
- good penetration of the stratum corneum
- natural sun protection factor

Avocado oil moisturises the skin effectively and soothes raw, flaky and dry skin, in particular. It is thus suitable for the treatment of skin diseases like neurodermatitis and psoriasis, and for complaints concerning the connective tissue. Avocado oil is also effective in treating skin impurities; it improves skin texture and promotes scarring (Kerschbaum and Schweiger 2001, p. 117; Hänsel et al. 1999, p. 248; Löw 2003, p. 107 et seq).

Technical Uses

Avocado oil is used in the soap industry. Due to its emollient effect, it is ideal for super-fatted soaps. It also produces a fine, creamy foam.

In Medicine

Applied externally, avocado oil alleviates pain and improves cell regeneration. Preparations with avocado oil are recommended in cases of parasitically induced skin damage and eczema. The oil can also be used to treat high cholesterol levels. An Australian study confirms the cholesterol-lowering effect of avocado oil. According to the independent Australian organisation "Heart Foundation", avocados are "heart-healthy food" (Kerschbaum and Schweiger 2001, p. 45; Whiley et al. 2002, p. 11).

As Food

In small quantities, avocado oil is used as an edible oil and to produce mayonnaise. It is suitable as a salad oil, for stewing and frying. The press cake is used as forage (Roth and Kormann 2000, p. 117; Löw 2003, p. 107; Kerschbaum and Schweiger 2001, p. 45).

Babassu Oil

Oleum Babassu

1 Source Plant

Orbignya martiana Barb. Rodr. *(Arecaceae), syn. Attalea speciosa* Mart., babassu palm.

Habitat

Wild babassu palms are numerous in the Brazilian jungle. They grow in an area of about 13 million hectares, especially in Maranhão and Piani. In Europe, babassu palms have only been known to supply oil since the beginning of the nineteenth century. Babassu fruit were first exported to England in 1867, but their value was not recognised until the beginning of the last century (Esdorn 1981, p. 49).

Description

The babassu palm is the largest palm of Brazil. Its smooth stem grows up to 20 m tall and rises steeply. At the top, there are about 20 pinnate palm fronds, which can reach a length of 6 m. The monoecious flowers sit on long, drooping spadices between the fronds. Each palm carries about 200–400 fruit bunches about 1–2 m in length. They are densely covered with fruit of varying size and form. In general, the fruit are 10–12 cm long and about 6 cm thick. The mesocarp is very hard and encloses 3–5 seeds of about 4 cm length. The thin, brown seed coat is mottled. The interior of the pit is white and contains the oil (Esdorn 1981, p. 49).

S. Krist, *Vegetable Fats and Oils*, https://doi.org/10.1007/978-3-030-30314-3_12

B

2 Cultivation and Extraction

Cultivation

Since there is an abundance of wild growing plants, cultivation is still in its early stages. Babassu palms need medium humidity, relatively solid ground and plenty of light. Propagation is possible via seeds, which are pre-grown in a seedbed. The palm grows slowly and bears fruit from its 10th year onwards, but then continues to produce fruit for nearly 200 years. The ripe fruit are gathered from the ground. As long as they are fresh, they are easy to split, but after several days the pericarp becomes as hard as a stone and can then only be split with an axe or cracked mechanically (Esdorn 1981, p. 50).

Extraction of the Oil

There are two methods: In the hot water flotation method, the seeds are lightly roasted and then ground in a mortar. Subsequently, water is added and the mixture is brought to the boil. The fat is released, rises to the surface, and is decanted. The other method is cold pressing the ground seeds; usually the oil is refined afterwards (Axtell 1994, p. 19).

3 Character

Colour

Pale (Roth and Kormann 2000, p. 118).

Colourless (Axtell 1994, p. 23).

At room temperature, babassu oil resembles a white paste (Franke 1989, p. 181).

Golden yellow (Jackson and Longenecker 1944).

Solidified, the oil is pale yellow or white (Dietrichs and Knorr 1920). In a melted state, it is almost as clear as water (Dietrichs and Knorr 1920).

Odour

Nutty (Roth and Kormann 2000, p. 118).

Reminiscent of coconut (Jackson and Longenecker 1944).

Flavour

Mild taste reminiscent of coconut (Lude 1934).

4 Ingredients

Composition of Fatty Acids

Fatty acids	Content in % acc. to Roth and Kormann (2000), p. 118	Content in % acc. to Axtell (1994), p. 19	Content in % acc. to na-turseife.com/13.07.06
Lauric acid	44–45	44–46	50
Myristic acid	15–17	15–20	20
Oleic acid	12–16	12–18	12
Palmitic acid	5–9	6–8	11
Capric acid	6–8	6–7	–
Caprylic acid	4–6	4–6	–
Stearic acid	2.5–5.5	3–5	3.5
Linoleic acid	1–3	1.4–2.8	–
Caproic acid	0.1	–	–

Physical Key Figures of Babassu Oil

refractive index: n_D^{40} 1.4494–1.4503
density: d_{15}^{15} 0.903–0.924
melting point: 22–26 °C
point of solidification: 21–25 °C
saponification number: 241–253
iodine number: 10–17
unsaponifiable: 1%
(Roth and Kormann 2000, p. 118)

Shelf Life

As babassu oil consists mainly of glycerides of saturated fatty acids, it has a shelf life of about 2 years.

5 Application

In Pharmacy and Medicine

Production of Lipase, Protease and Amylase by Penicillium Restrictum

By-products of the production of babassu oil serve as base material for the culture medium of *penicillium restrictum*. Peptones, olive oil and Tween-80 are also added to the culture medium. Lipase, protease and amylase are digestive enzymes. They are administered in cases of exocrine pancreatic insufficiency, which may develop as a consequence of chronic pancreatitis, pancreas excision or mucoviscidosis (Palma et al. 2000; Mutschler et al. 2001, 8th edition, p. 633).

To Treat Skin Diseases

As a result of its high proportion of lauric acid and myristic acid, babassu oil forms a protective, soothing film on the skin and, for this reason, is applied to treat eczema and itching, dry and inflamed skin.

In Cosmetics

As Massage Oil

Babassu oil contains many vitamins and mineral nutrients; for this reason, it is used as a massage oil. It is quickly absorbed and does not leave a fatty film on the skin, but rather leaves it feeling comfortable and velvety soft.

In Skincare and Hair-Care Products

Babassu oil is a rich oil and well suited for sensitive and dry skin that needs moisture, but no additional fat. The oil is added to various lotions and balms, but also to facial creams, body lotions and lip balms. As it melts at skin temperature, it has a cooling effect. Babassu oil has been used for over a hundred years to produce soap that is hard and foamy. As a result of the fluffy, abundant foam, babassu oil is ideal for shampoo soaps, especially those for damaged and/or dry hair.

As Food

Babassu oil is an edible oil and is used in greater quantities to produce margarine (Roth and Kormann 2000, p. 118). Since it has a mild taste and a pleasant smell, it is used for baking cakes, rolls and cookies (Axtell 1994, p. 51).

Balanos Oil

synonyms: Balanitesöl (D); desert date oil, soap berry oil, Lalobe kernel oil (E)

1 Source Plant

Balanites aegyptica Delile *(Zygophyllaceae)*, zachum oil tree.

Habitat

The zachum oil tree is common in the dry regions of tropical Africa. It is a plant of the savannah from Mauritania to Kenya, from Egypt to Zambia, and is also found in the Sahel (Neuwinger 1998, p. 901).

Description

The tree is between 6 and 15 m tall, with a rounded or flattened crown. The branches are dark green, pendant, either smooth or covered with hairs. Above the axils and running spiral along the branches are strong, woody thorns of about 8–12 cm length. The leaves are alternate and characteristically consist of two small, stemmed leaflets standing apart. These are elliptical, obovate or roundly rhomboid, leathery and short stemmed. The flowers are a greenish-white or greenish-yellow colour and grow in large, upright, hairy clusters about 2.5 cm in diameter. The unripe fruit is green, while the ripe fruit is yellow, smooth, fleshy, and oily; they resemble dates. They are stone fruit 2.5–5 cm long, elliptical, with a thin, hard shell. The fruit pulp is sticky, refreshingly bittersweet, and a brownish-yellow colour. The fruit usually contains only one seed (Neuwinger 1998, p. 901).

S. Krist, *Vegetable Fats and Oils*, https://doi.org/10.1007/978-3-030-30314-3_13

B

2 Cultivation and Extraction

Cultivation

The zachum oil tree is well suited to dry regions and is common in the African savannah. The main cultivation area is Sudan, but the tree is also cultivated in Nigeria and the Arabian peninsula (Axtell and Fairman 1992, p. 25 et seq).

Extraction of the Oil

To extract the oil, the very hard seeds are crushed and then pressed (Axtell and Fairman 1992, p. 25 et seq).

3 Character

Colour

Golden yellow, very stable oil (Axtell and Fairman 1992, p. 25 et seq).

4 Ingredients

Fat is one of the main components of the seed, making up 49% (Mohamed 1998, p. 69). Oleic acid and linoleic acid are the main ingredients, followed by palmitic acid and stearic acid. The unsaponifiable portion contains β-sitosterol (Jain and Banerjee 1988).

Composition of Fatty Acids

Fatty acids	Content in mg/g oil at 50 °C pressing temperature acc. to Mohamed (1998), p. 69
Palmitic acid	95.6
Stearic acid	94.5
Oleic acid	287.6
Linoleic acid	263.9
α-Linolenic acid	12.6

Physical Key Figures of Balanos Oil

density: d_{25}^{25} 0.9156
refractive index: n_D^{40} 1.4643
saponification number: 172.7

iodine number: 56.5
unsaponifiable: 0.7
acid number: 0.58
(Hussain et al. 1949; Jain and Banerjee 1988)

5 Application

As Food

Balanos oil is an edible oil (Axtell and Fairman 1992, p. 25).

As Fuel

Since the viscosity of balanos oil is about 10–20 times higher than that of diesel oil, it is regarded as a potential source of fuel in Kenya (Neuwinger 1998, p. 904).

In Folk Medicine

In folk medicine, the oil is used to treat sore throats, colic, epilepsy and toothache. Since it has an analgesic effect, it is applied as a shampoo for stressed hair to soothe the scalp.

As Insecticide

The oil is not toxic for humans, but it is effective as an insect repellent against *Periplaneta americana, with a comparable activity to the insecticide B*aygon. In addition, it is a good natural source of the essential fatty acid linoleic acid (Jain and Banerjee 1988).

Baobab Oil

Oleum Baobab synonyms: Affenbrotbaumöl (D); Kuka oil (E)

1 Source Plant

Adansonia digitata L. *(Bombacaceae)*, baobab.

S. Krist, *Vegetable Fats and Oils*, https://doi.org/10.1007/978-3-030-30314-3_14

B

Habitat

The baobab is a tree that is part of the natural vegetation in northern Nigeria. It originates from Africa and Saudi Arabia; sailors transported it to Zanzibar, Madagascar and Mauritius. As a consequence of commercial relations with India, the baobab was imported to India, Malaysia and Indonesia. Today, it is also found in the Caribbean and on the Guyana coast. However, it remains the characteristic tree of the dry tree-savannahs of Africa. In the Nigerian Sahara, in particular, its leaves, fruit pulp and seeds are used. During droughts, the fruit pulp is crushed to a mash and added to porridge, the finely ground leaves are used to thicken soups, and the ground fruit is used as flour (Lockett et al. 2000).

Description

The baobab is a tropical tree. It is a succulent plant that is able to store large quantities of water. Its trunk is relatively short, but extremely thick – the reason for this is its capacity to store enormous amounts of water, namely up to 3700 l. The shape of the trunk can vary from cylindrical or bottle-shaped to pointed. When the tree is full-grown, it has a diameter of 8–10 m. The bark is smooth, silvery grey, red or purple, and contains a yellow or green photosynthesising layer, which consists of thick, hard longitudinal fibres. Young trees have a tap-root, but with increasing age, a system of lateral roots develops, which reaches 1.8 m deep and is wider than the tree is high. The tree grows very fast, up to a height of 20 m, and lives for an exceptionally long time: in the view of some botanists, between 1000 and 3000 years. It sheds its hand-shaped, dark-green leaves during dry spells. The flowers develop 4 weeks after the leaves. The flower is large, waxy-white, and consists of five petals and 720–1600 stamina. Within 8 months after pollination, capsules form on long stalks. The fruit hanging on the stalks have a length of 25–40 cm. As they ripen, their colour changes from green to yellow and finally to greyish-brown. The ripe fruit contains a floury white fruit pulp and many seeds (Shukla et al. 2001).

2 Cultivation and Extraction

Cultivation

The baobab is common in the hot, dry regions of tropical Africa. In Sudan, it grows in sandy soil along seasonal watercourses, called khors, in short-grass savannahs. Cultivation is limited to areas with not more than 1 day of frost per year. The tree has an extensive root system, which enables it to store large amounts of water. This ability, as well as its fire resistance, ensures its survival in dry climates. It is inured to variations of the soil's pH value and tolerates shallow, lateritic soil. It can also be found on hills, in calcareous soil and in places where water collects or flows off. In Sudan, the tree flowers between May and July, and bears fruit from August to

October. It propagates via seeds, but propagation is also possible vegetatively via root cuttings, grafting or pullulation (Gebauer et al. 2002).

Extraction of the Oil

To extract the oil, the dried fruit are reduced to small pieces using a pestle, to separate the seeds from the fruit pulp. The seeds are then washed, dried in the sun, and finally mechanically ground. The oil is extracted by adding solvents (Eteshola and Oraedu 1996). The seeds can also be cold pressed.

3 Character

Colour

Clear, viscous, golden-yellow oil (Shukla et al. 2001).

Light-yellow oil (Odetokun 1996).

Odour

Nutty. Smells slightly of nuts.

Flavour

Nutty taste.

4 Ingredients

The seeds of *Adansonia digitata* contain 15.12% protein, 11.56% fat, 17.84% carbohydrates and 49.72% fibres (Lockett et al. 2000).

Composition of Fatty Acids

Fatty acids	Content in % acc. to Firestone (1999), p. 12	Content in % acc. to Ezeagu et al. (1998)	Content in % acc. to Andrianaivo-Rafehivola and Gaydou (1994)
Lauric acid	0–0.3	–	–
Myristic acid	0.3–1.5	0.19	4.6
Pentadecanoic acid	–	–	2.4
Palmitic acid	25–46	15.50	37.3
Palmitoleic acid	0.3–1.7	0.20 (n-7)	0.3 (n-7)

Fatty acids	Content in % acc. to Firestone (1999), p. 12	Content in % acc. to Ezeagu et al. (1998)	Content in % acc. to Andrianaivo-Rafehivola and Gaydou (1994)
Margaric acid	–	–	0.3
Stearic acid	0–4	3.12	4.2
Oleic acid	21–59	24.69 (n-9)	19.7 (n-9); 1.7 (n-7)
Linoleic acid	12–29	19.11 (n-6)	13.5 (n-6)
α-Linolenic acid	0–8	0.39 (n-6)	0.1 (n-3)
Arachidic acid	0.5–1.0	0.74	0.7
Eicosenoic acid	0–3.6	0.19	0.1 (n-9)
Behenic acid	–	0.36	0.1
Lignoceric acid	–	0.31	–
Malvalic acid	1–7	–	6.3
Sterculic acid	1–8	–	6.5
Dihydrosterculic acid	2–5	–	1.8

Composition of Sterols

Sterols	Content in % acc. to Firestone (1999), p. 12	Content in % acc. to Gaydou et al. (1979)
Cholesterol	2	1.9
Campesterol	6	6.3
Stigmasterol	1–2	2.0
β-Sitosterol	75	81
Δ5-Avenasterol	0.5	3.4
Δ7-Stigmasterol	0.6	4.8
Δ7-Avenasterol	12	0.6

Physical Key Figures of Baobab Oil

density: d_{15}^{15} 0.914
saponification number: 190–192
iodine number: 76–78
point of solidification: −3 to +34 °C
peroxide number: 5.14 ± 0.12
acid number: 7.79 0.33
(Roth and Kormann 2000, p. 157; Odetokun 1996)

Shelf Life

Baobab oil has a shelf life of about 2 years.

5 Application

In Folk Medicine

In folk medicine, baobab oil is used as an antipyretic drug in cases of febrile diseases and malaria. The oil is also applied to treat inflammation of the gums, and to relieve toothache. In North Africa, in particular, the native peopole use it as an antidote against poisoning by species of *Strophanthus*. Since they kill animals with an arrow poison made from *Strophanthus*, they neutralise the poison before eating the meat by cleansing the shot wound with baobab oil.

In Cosmetics

Baobab oil is mainly used in skincare and hair-care products. It is absorbed through the skin easily, increases the skin's elasticity and, for this reason, is added to a variety of creams and lotions. As a result of its high content of vitamin A, D and E, baobab oil is used as a massage oil and as a bath essence for dry skin. Baobab oil supports the regeneration of skin cells and has non-comedogenic properties, and it is therefore applied in the treatment of blackheads, psoriasis and eczema. When used as a shampoo, it moisturises and helps dry, brittle hair to regain its elasticity.

As Food

Baobab oil contains considerable amounts of linoleic acid, namely 12–29%. Linoleic acid is an ω-6 dooble-unsaturated essential fatty acid, necessary for the growth and regeneration of cell tissue. The oil is also ideal as a source of unsaturated fatty acids because it contains oleic acid, but since it contains a high amount of cyclopropane fatty acids, only refined baobab oil should be used (Eteshola and Oraedu 1996).

6 Possible Unwanted Side-Effects

Baobab oil contains 2–17% cyclopropane fatty acids (CPFAs), for example malvalic acid and sterculic acid. In animals that have consumed the oil, toxicological effects have been observed, for example retarded growth in mammals. A study on rats that were fed baobab oil containing CPFAs showed considerably retarded growth. After 30 days, the rats weighed 50% less than rats fed on a mixture of palm oil and sunflower oil. If rats were given heated baobab oil, this improved their growth, but they were still behind those that received the mixed oil (Andrianaivo-Rafehivola and Gaydou 1994).

Another study on rats that were given fresh baobab oil showed that more than half of the cyclopropane fatty acids contained in the diet (up to 8%) ended up in the perirenal adipose lipids of the tissues. If rats were given heated baobab oil, only traces of these fatty acids were found. In addition, a decrease in monounsaturated fatty acids and an increase in saturated fatty acids was observed in the perirenal adipose lipids of the tissues, triglycerides of the liver, cholesterol esters in the liver and phospholipids of the suprarenal glands of the rats receiving fresh oil. A diet based on the heated oil, however, had a similar effect to that of the control group (which was given a mixed oil: palm oil 93.5%, sunflower oil 6.5%). Obstruction of the biosynthesis of ω-6 polyunsaturated fatty acids caused by cyclopropane fatty acids leads to a decrease in the amount of arachidic acid in the phospholipids of the tissues. This can influence the quality of the membranes in liver, kidneys and suprarenal glands as well as the eicosanoid synthesis. This is a possible explanation for the skin alterations that had been observed in the rats. In addition, an obstruction of the biosynthesis of docosapentaenoic acid can result in dysfunction of the suprarenal glands and thus in a disrupted biosynthesis of corticoids (Cao et al. 1996).

Basil Oil

synonyms: Basilikumöl (D)

1 Source Plant

Ocimum basilicum L. (Lamiaceae), basil.

S. Krist, *Vegetable Fats and Oils*, https://doi.org/10.1007/978-3-030-30314-3_15

B

Habitat

The origins of basil are unknown. It is presumed to have originated in Africa, since all plants closely related to it are of African origin, but Asia and India are also possible areas of origin (Small 2006, p. 591).

Description

Ocimum basilicum is an annual plant, growing up to a height of 60 cm. The roots are thin and branched. The leaves have different oval and elongated forms, and the flowers sit in false whorls on the top of the stem. Basil blooms in June and July. The seeds are either oval or nut-shaped and remain germinable for 4–5 years (Dachler and Pelzmann 1999, p. 141).

The flowers usually open on the lower false whorls first, and later on the ones at the top. Ripe seeds are thus found in the bottom part first. Due to this periodical ripening and the delayed harvest, the much-used herb contains about 20–25% seeds, which are separated from the plant as a by-product and then processed further. The other possibility is to wait until all the seeds are ripe, in which case the herb is a by-product (Domokos and Perédi 1993).

2 Cultivation and Extraction

Cultivation

Basil has been cultivated in Asia for about 3000 years. Presumably, it was already known in Egypt at the time of the pharaohs. In antiquity, the Greeks cultivated it; from Greece, it spread to Rome and Southern Europe. The English used it from the sixteenth century onwards (Small 2006, p. 592).

As basil is frost-susceptible and sensitive to low temperatures, the seeds germinate best at temperatures between 13 and 25 °C. For growth, basil needs temperatures of 5–31 °C, the optimal temperature being 19 °C. Basil requires long days and plenty of sunshine for optimal growth (15 hours of sunlight or more), as well as considerable amounts of water (Small 2006, p. 595).

Extraction of the Oil

The oil is cold pressed (Domokos and Perédi 1993).

Another method is to wash the ripe seeds, grind them finely and extract them with a solvent in a Butt-type apparatus (Angers et al. 1996a).

3 Character

Colour

Yellow (Domokos and Perédi 1993).

Odour

Pleasant (Domokos and Perédi 1993).

Flavour

Characteristic, aromatic (Domokos and Perédi 1993).

4 Ingredients

Composition of Fatty Acids

Fatty acid	Content in % acc. to Angers et al. (1996b)	Content in % acc. to Domokos and Perédi (1993)	Content in % acc. to Earle et al. (1960)	Content in % acc. to Matthaus et al. (2003)
Palmitic acid	6.8–8.8	6–9	–	7.33
Palmitoleic acid	0.2–0.3	–	–	0.12
Stearic acid	2.0–2.8	2–3	–	2.6
Oleic acid	8.7–11.6	9–15	15	7.43
Linoleic acid	18.3–21.7	17–25	22	24.89
Linolenic acid	57.4–62.5	50–63	50	54.58
Arachidic acid	0.2	–	–	0.16
Behenic acid	–	–	–	0.04

Composition of Triglycerides

The oil contains 96% triglycerides, but only 2% mono- and diglycerides each. 70–84% of the triglycerides have 54 C atoms, 15–27% 52 C atoms and 1–3% 50 C atoms (Angers et al. 1996b).

B

Composition of Tocopherols

Tocopherols	Content in mg/kg oil acc. to Matthaus et al. (2003)
Total content	928
α-Tocopherol	52
β-Tocopherol	–
γ-Tocopherol	828
δ-Tocopherol	47

Physical Key Figures

	Angers et al. (1996b)	Domokos and Perédi (1993)	Earle et al. (1960)
Refractive index	1.479–1.481	1.480 (20 °C)	1.4742
Density	–	0.929 (20 °C)	–
Saponification number	199–200	–	188
Iodine number	184–198	–	191
Peroxide number	–	1.5	–
Unsaponifiable	–	0.6	–

5 Application

In Pharmacy and Medicine

As a result of its high proportion of omega-3 fatty acids, basil oil is used in nutritional therapy (Domokos and Perédi 1993).

In Cosmetics

Basil oil softens the skin and is easy to spread. It moisturises the skin and reduces transepidermal loss of water. For these reasons, it is often added to cosmetic products. In addition, its neutral smell can be easily masked (Domokos and Perédi 1993).

As Food

As a result of its pleasant smell and taste, basil oil is a good edible oil (Domokos and Perédi 1993).

Other Uses

The oil is used in the mineral industry, in the petroleum industry and for the production of varnish, printing ink and soap (Angers et al. 1996b).

Beechnut Oil

Oleum Fagi seminis- or Oleum Fagi sylvaticae synonyms: Bucheckernöl (D); huile de faîne (F)

1 Source Plant

Fagus sylvatica L. *(Fagaceae)*, common beech.

S. Krist, *Vegetable Fats and Oils*, https://doi.org/10.1007/978-3-030-30314-3_16

B

Habitat

Common beeches are characteristic of oceanic climates. They are native trees of Europe, the Caucasus countries and northern Iran, but they are also common in North America, on the Eurasian continent and from Malaysia to Australia and New Zealand.

The common beech grows in plains and low mountain ranges (up to 1700 m), forming large, pure stocks. It needs friable, calcareous soil to thrive, and is therefore more common in regions with such soil.

Description

The common beech is an upright tree up to 30 m tall; it can live to an age of 120 years. The trunk is straight and cylindrically round. The bark is smooth and silvery grey, especially if the tree is young, and often covered with crustose lichen. The branches are strong, reaching upwards, and branch out horizontally and downwards. The leaves develop from pointed buds on long and short shoots. They are nearly entire, light green and slightly paler on the underside, shiny, oval, pointed, with straight lateral nerves and a slightly corrugated edge with silky hair. The flowers sit in a brown sheath with soft prickles, which later develops into a lignifying, quadrifid capsule usually containing two fruit. The fruit is one-seeded, triangular, reddish brown and shiny. It is called beechnut; the oil makes up 50% of the fruit.

2 Cultivation and Extraction

Cultivation

The fruit used for oil extraction do not originate from organic farming, but are rather hand-picked and gathered in the woods. As the common beech only bears fruit after several years, cultivation is not economically efficient.

Extraction of the Oil

To extract 1 l of cold pressed beechnut oil, an average of 7 kg fresh, germinable beechnuts is required. After the process of cleaning and drying, only about 4 kg remain. When the nuts are handpicked, it is important to make sure that no mouldy beechnuts enter the product. The triangular nuts with their shells are then either cold pressed or pressed by adding heat. Due to irregular fructification, the oil has no commercial relevance (Hager 1978, volume 4, p. 917; Roth and Kormann 2000, p. 120).

3 Character

Colour

Light to deep yellow.

Odour

Nutty, with a touch of wood and mushrooms.

Nearly odourless (Roth and Kormann 2000, p. 120).

Flavour

Pleasant (Roth and Kormann 2000, p. 120).

Nutty, very slightly bitter.

4 Ingredients

Composition of Fatty Acids

Fatty acids	Content in % acc. to Hager (1978), volume 7b, p. 164	Content in % acc. to Roth and Kormann (2000), p. 120
Palmitic acid	5	5.2
Linolenic acid	0.4	0.4
Stearic acid	2.5	3.7
Oleic acid	77	81
Linoleic acid	9	9.7

Physical Key Figures of Beechnut Oil

refractive index: n_D^{15} 1.4729–1.4752
density: d_4^{15} 0.9220–0.9225
saponification number: 181–196
iodine number: 101–111
unsaponifiable: 0.5–1%
point of solidification: −17.5 to −17 °C
(Roth and Kormann 2000, p. 120)

Shelf Life

Beechnut oil keeps for several years without becoming rancid.

B

5 Application

Beechnut oil is an excellent edible oil, but as beeches do not bear fruit regularly, the oil has no particular commercial relevance. Since it contains linoleic acid, it is not suitable for frying. Preferably, it should be used as a salad oil. During the Second World War, it gained importance because it presented a practical alternative method of self-sufficiency in times of food shortage (Hager 1978, volume 7b p. 165). Beechnut oil is also used for producing margarine and soap (Roth and Kormann 2000, p. 120).

6 Possible Unwanted Side Effects

Poisoning with beechnuts leading to gastrointestinal complaints has been reported. The poisonousness of beechnuts seems, however, to vary in intensity. Usually, it only results in local irritation such as gastric discomfort, eructation, nausea and vomiting. Domestic animals (especially horses and calves, but not pigs) were observed to show symptoms such as staggering, shivering, dyspnoea, paresis of the lower extremities, convulsions, etc. Guinea pigs that were administered large doses had cramps similar to those observed after administration of strychnine, as well as dyspnoea; one section regularly developed defects of the epithelium that looked like stomach ulcers. These symptoms are probably due to oxalic acid. Unlike the beechnuts themselves, the oil is completely non-poisonous (Hager 1978, volume 4, p. 917).

Ben Oil

Oleum Moringae synonyms: Moringa oil (E); Behenöl (D)

1 Source Plant

Moringa oleifera Lam., *Moringa peregrina* (Forssk.) Fiori, *Moringa concanensis* Nimmo *(Moringaceae)*, horseradish tree, ben oil tree, drumstick tree.

S. Krist, *Vegetable Fats and Oils*, https://doi.org/10.1007/978-3-030-30314-3_17

Habitat and Description

The genus Moringa Adans. consists of 16 species; the following three are important for oil extraction.

B

Moringa oleifera Lam

This species is a small to medium-sized tree with slender branches and usually incompletely pinnate, alternately arranged leaves. Their leaflets are 30–60 mm long. The white or yellow flowers grow in large, puberulent panicles and smell of honey. The fruit is a pendant capsule about 22–45 cm in size; it is triangular with nine ribs. The seed contained in it is also triangular, light brown to blackish brown, and has three paper-thin, whitish wings; the seed coat makes up 30% of the seed. The tree is native to north-western and eastern India (Hager 1993, p. 851 et seqq).

Moringa peregrina (Forssk.) Fiori

This is a tree resembling the common broom, 4–10 m tall and sparsely covered with bipinnate leaves. The pinkish white or yellow flowers are arranged in panicles, which can reach a length of 30 cm. The fruit is a 10–30 cm long, corrugated, pendant brown capsule with one compartment that contains seeds arranged in a row with spongy proliferations in between. The seeds fare 1 cm and have no wings. The seed coat makes up 50% of the seed. This tree is a native of the Arabian and African deserts (Hager 1993, p. 851 et seqq).

Moringa concanensis Nimmo

Depending on the location, this is either a small or a large tree with bipinnate leaves, the leaflets being 15–30 mm long. The flowers have either a red stripe or are reddish at the base. The tree grows in India (Hager 1993, p. 851 et seqq).

2 Cultivation and Extraction

Cultivation

The tree originates in the Himalayan region of northwestern India, but Moringa is cultivated in many regions, for example in the tropics, India, Saudi Arabia, Egypt, Sudan, West Africa (Senegal, Nigeria), southern North America (Florida), the West Indies, Central America, Peru, Indonesia and the Philippines. Propagation is relatively simple via hardwood cuttings. The tree grows in hot, semi-arid climates with medium amounts of rainfall (between 250 and 1500 mm/year), but also in regions with high air humidity and an annual precipitation of up to 3000 mm. It prefers well-drained, loamy, sandy soil, but also grows in heavy loam. It can cope with light frost (Hager 1993, p. 851 et seqq).

Extraction of the Oil

To extract the oil, the seed coat is removed, and the seeds are crushed and subsequently pressed. The product is then heated to 70 °C, mixed with 18 vol% of water, cooled and centrifugalised to remove the gum. It is then dried and filtered (Anwar and Bhanger 2003, p. 6559). Another method is to extract the oil using n-hexane, chloroform/methanol or other lipophilic solvents (Lalas et al. 2003, p. 23 et seqq).

3 Character

Colour

Light-yellow to light-brown oil (Roth and Kormann 2000, p. 119). Clear, light-yellow oil (Tsaknis et al. 1999).

Odour

"Characteristic" (Tsaknis et al. 1999).

Flavour

Weak, pleasant taste (Roth and Kormann 2000, p. 119).

4 Ingredients

Composition of Fatty Acids

Composition of Fatty Acids in % Acc. To Le Poole (1996)

Fatty acids	Ben oil of M. peregrina	Ben oil of M. oleifera	Ben oil of M. concanensis
Caprylic acid	–	–	–
Myristic acid	–	–	–
Palmitic acid	9.3	3.1	3.5
Palmitoleic acid	2.4	–	–
Margaric acid	–	–	–
Stearic acid	3.5	8.0	8.2
Oleic acid	78.0	71.0	72.8
Linoleic acid	0.6	0.1	0.7
α-Linolenic acid	1.6	–	–
Arachidic acid	1.8	7.8	7.2
Eicosanoic acid	–	–	–
Behenic acid	2.6	3.5	3.0
Lignoceric acid	–	5.8	4.6

B

Composition of Fatty Acids in %

Acc. To Anwar and Bhanger (2003) and Tsaknis et al. (1999)

Fatty acids	M. oleifera seed oil from the temperate regions of Pakistan acc. to Anwar and Bhanger (2003)	M. oleifera var. Mbololo seed oil from Kenya acc. to Tsaknis et al. (1999)		
	H	CP	H	CM
Caprylic acid	–	0.03	0.03	0.02
Myristic acid	–	0.11	0.11	0.11
Palmitic acid	6.50 ± 0.40	5.73	6.04	5.81
Palmitoleic acid	1.00 ± 0.08	–	–	–
Palmitoleic acid cis *Ω*-9	–	0.10	0.11	0.10
Palmitoleic acid cis *Ω*-1	–	1.32	1.46	1.44
Margaric acid	–	0.09	0.09	0.09
Stearic acid	5.67 ± 0.39	3.83	4.14	4.00
Oleic acid	76.00 ± 1.23	75.39	73.60	73.91
Linoleic acid	1.29 ± 0.10	0.72	0.73	0.71
α-Linolenic acid	–	0.20	0.22	0.20
Arachidic acid	3.00 ± 0.20	2.52	2.76	2.70
Eicosanoic acid	1.20 ± 0.06	2.54	2.40	2.46
Behenic acid	5.00 ± 0.33	5.83	6.73	6.38
Cetoleic acid	–	0.15	0.14	0.14
Lignoceric acid	–	–	–	–
Cerotic acid	–	0.96	1.08	1.06
Saturated fatty acids in total	–	19.1	20.98	20.17

Oil Content of Ben Oil

(CP = cold pressed, H = extraction with n-hexane, CM = extraction with chloroform/methanol 50:50)

	Content in % in M. oleifera seed oil from temperate regions of Pakistan acc. to Anwar and Bhanger (2003)	Content in % in M. oleifera var. Mbololo seed oil from Kenya acc. to Tsaknis et al. (1999)			Content in % in M. peregrina acc. to Le Poole (1996)	Content in % in M. oleifera acc. to Le Poole (1996)	Content in % in M. concanensis acc. to Le Poole (1996)
	Extraction with n-hexane	CP	H	CM			
Oil content	38.00–42.00	25.8 ± 2.6	35.7 ± 2.4	31.2 ± 2.0	53.4	34.4	35.0

Composition of Sterols

Sterols	Content in % in M. oleifera seed oil from temperate regions of Pakistan acc. to Anwar and Bhanger (2003)	Content in % in M. oleifera var. Mbololo seed oil from Kenya acc. to Tsaknis et al. (1999)		
	H	CP	H	CM
Cholesterol	–	0.13	0.13	0.12
Brassicasterol	–	–	0.06	0.06
24-Methylencholesterol	1.49 ± 0.12	0.85	0.88	0.98
Campesterol	16.00 ± 1.03	14.03	15.13	14.12
Campestanol	–	–	0.35	0.35
Δ7-Campestanol	0.50 ± 0.07	–	–	–
Stigmasterol	19.00 ± 1.00	17.27	16.87	16.78
Ergostadienol	–	–	0.39	0.28
Clerosterol	1.95 ± 0.13	0.95	2.52	0.84
Stigmastanol	1.00 ± 0.07	1.05	0.86	0.80
β-Sitosterol	46.65 ± 1.50	49.19	50.07	50.00
Δ7-Avenasterol	0.96 ± 0.06	0.94	1.11	1.04
Δ5-Avenasterol	10.70 ± 0.93	12.79	8.84	11.41
28-Isoavenasterol	0.50 ± 0.06	1.01	1.40	1.14
Δ7,14-Stigmastanol	–	0.83	0.44	0.52

Composition of Tocopherols

(concentration in mg/kg oil)

Tocopherols	M. oleifera seed oil from temperate regions of Pakistan acc. to Anwar and Bhanger (2003)		M. oleifera var. Mbololo seed oil from Kenya, gum not removed, acc. to Tsaknis et al. (1999)		
	Gum not removed	Gum removed	CP	H	CM
α-Tocopherol	134.42 ± 10.00	110.0 ± 2.60	101.46	98.82	105.02
γ-Tocopherol	93.70 ± 7.00	81.63 ± 2.83	39.54	27.90	33.45
δ-Tocopherol	48.0 ± 4.65	41.00 ± 1.96	75.67	71.16	77.60

Physical Key Figures of Ben Oil

Characteristics	M. oleifera seed oil from temperate regions of Pakistan acc. to Anwar and Bhanger (2003)	M. oleifera var. Mbololo seed oil from Kenya acc. to Tsaknis et al. (1999)		
	H	CP	H	CM
Density in mg/ml at 24 °C	0.9036–0.9080	0.9037	0.8809	0.9182

Characteristics	M. oleifera seed oil from temperate regions of Pakistan acc. to Anwar and Bhanger (2003)	M. oleifera var. Mbololo seed oil from Kenya acc. to Tsaknis et al. (1999)		
	H	CP	H	CM
Refractive index n_D 40 °C	1.459–1.4625	1.4591	1.4549	1.4581
Saponification number	180.6–190.5	179.80	178.11	176.23
Iodine number in g of 1/100 g oil	68.00–71.80	66.81	66.83	66.66
Unsaponifiable in %	0.70–1.10	–	–	–
Acid number in % as oleic acid	0.27–0.48	1.01	0.85	0.98
Melting point in °C	196–203	201	198	202
Colour				
Red components	0.95–1.10	1.9	0	3.3
Yellow components	20.00–35.30	30	40	72
Viscosity in mPas	–	103	57	66

Other Ingredients

Calcium
Magnesium
Potassium
Iron
Vitamin C

5 Application

In Pharmacy and Medicine

Esters derived from ben oil via alkali catalysis make cosmetic and pharmaceutical products more stable and prolong their shelf life. They also improve application and covering power (Kleiman et al. 2004).

Calculated esterification creates a product with a melting point of 39 °C, accidental esterification of the hydrogenated oil results in products with a melting point of either 47 °C or 44 °C. Fats with high melting points are suitable as a coating for time-release drugs (Sengupta et al. 1974).

In Folk Medicine

In folk medicine, ben oil is used as a liniment for the treatment of rheumatism and skin diseases (Hager 1993, p. 853).

In Cosmetics

In cosmetics, ben oil is used for producing ointments and creams, as well as for enfleurage in the production of essential oils. Besides being highly stable, ben oil also has scent-fixing properties that make it an ideal ingredient for cosmetic products. The oil of *M. oleifera* also contains a chemical component useful for water treatment (Le Poole 1996).

As Antioxidant

A study examined the suitability of ben oil as a natural antioxidant. After extracting the oil with a 50:50 mixture of chloroform/methanol, the gum was removed by heating the oil to 75 °C, adding 20% of boiling water and mixing the liquid for 10 minutes. The mixture of oil and water was then cooled down to 40 °C and centrifugalised for 10 minutes. The gum was collected and freeze-dried under vacuum at −10 °C until a fine powder formed. The powder was stored at −4 °C in sealed bottles for a maximum of 3 days. To test the antioxidant activity, the freeze-dried gum was added to the oil from which it had been separated. The induction period proved to be only slightly shorter than that of oil from which the gum had not been removed. The reason for this shortening of the induction period is that other antioxidant substances, like tocopherols, were partly destroyed during the removal of the gum. When 200 ppm of the freeze-dried powder was added to sunflower oil, the induction period at 90 °C rose from 8.65 h without gum to 12.99 h with the gum. The pulverised gum was subsequently fractionated and the individual fractions tested for their antioxidant properties. The fraction with the highest activity was myricetin, a 3,5,7,3′,4′,5′-hexahydroxyflavone. It was examined using HNMR and ^{13}CNMR spectroscopy, mass spectrum, melting point and UV-spectrum. The gum contains 330.8 ppm myricetin; in the oil, the substance amounts to 29.11 ppm. It has a significantly higher antioxidant activity than the synthetic antioxidant BHT (butylhydroxytoluene). The use of ben oil as a natural resource for antioxidants should therefore be taken into consideration (Lalas and Tsaknis 2002).

As Food

Ben oil is an edible oil in India and Africa, in particular (Hager 1993, p. 853).

Other Uses

In industry and technology, ben oil is used as a machine oil for delicate instruments, for example clocks (Hager 1993, p. 853).

Tests for the suitability of vegetable oil as a substitute for diesel fuel showed that ben oil has a total content of fatty acids of 22.3%, and its methyl esters have a cetane index of 60.4. The oil would thus be a suitable substitute (Kalayasiri et al. 1996, p. 471 et seq).

Blackberry Seed Oil

synonyms: Brombeerkernöl (D)

1 Source Plant

Rubus fruticosus L., *Rubus sp. (Rosaceae)*, blackberry.

S. Krist, *Vegetable Fats and Oils*, https://doi.org/10.1007/978-3-030-30314-3_18

B

Habitat

The blackberry originates in North America and Eurasia. Today, it is common throughout the temperate zone (Lehari 2001, p. 8).

Blackberries grow predominantly at the edge of forests, in clearings, or in bushes and hedges. Slightly shady positions and wet soils are ideal (Mayer and Nerger 2000, p. 222).

Description

Rubus fruticosus is a collective name for many different species. In Central Europe alone, there are about 70 different species, both cultivars and wild growing plants. They differ in the colour of their fruit, which are usually black, blackish red or blue (Lieberei and Reisdorff 2007, p. 202).

The blackberry is an aggregate fruit, consisting of many connate, fleshy stone fruits ripening from August to October. The shoots are often covered with brownish-red prickles. The pinnate, oval leaves have a serrated edge and woolly white hair on the underside. The plant is in flower from June to August (Mayer and Nerger 2000, p. 222).

The petals are either white or pink and develop on the biennial side shoots. Blackberry bushes can also propagate vegetatively (Lieberei and Reisdorff 2007, p. 202).

2 Cultivation and Extraction

Cultivation

The blackberry has been cultivated since the nineteenth century. In the early 1950s, the first blackberry without thorns was bred (Lehari 2001, p. 8).

In 2005, 154,603 tons of blackberries were produced worldwide. The largest cultivated area is in Europe, but North America is the main producer with 65,154 tons/year. Sixty-five percent of the blackberries cultivated in the USA are grown in Oregon; the most common cultivars are *Marion* (61%), *Boysen* (15%), *Thornless Evergreen* (11%) and *Silvan* (7%). Globally, the cultivated area increased by 40% from 1995 to 2005. The main areas of growth are located in Mexico, the USA, China and Costa Rica (Strik et al. 2006).

2005	Production (tons)	Area (acres)
World	154,603	49,507
Europe	47,386	19,007
North America	65,154	17,690
Central America	1752	4053
South America	7031	3946
Asia	29,038	3830
Oceania	4022	734
Africa	220	247

Strik et al. (2006)

Recently, a large production increase in the genus *Rubus* was registered in Korea. In 2004, about 1840 tons were produced, of which 950 tons were used for the production of wine. The pomace accumulating in wine production contains the fruit pulp and seeds. It is used as either forage or manure. The seeds that it contains can also be used for oil production. Blackberry seed oil is thus produced from a by-product; for this reason, it is interesting from an economic point of view (Oh et al. 2007).

Extraction of the Oil

If blackberry seeds are taken from fresh fruit, they are squeezed by hand under cold tap water and then left to soak for 24 hours. Subsequently, the remaining fruit pulp is removed by gently rubbing the seeds between the palms of the hands. The fruit pulp remains suspended in water, whereas the seeds sink to the ground. By decanting the fruit pulp and water, the seeds are separated. They are then washed, dried and ground; the oil is extracted with solvents (Oh et al. 2007).

Blackberry seeds can also be cold pressed (van Hoed et al. 2009).

3 Character

Colour

Blackberry seed oil is dark green to yellow because it contains chlorophyll. When light falls on it, it is a fluorescent red colour (Krzizan 1908).

Odour

The smell is typical (Krzizan 1908).

4 Ingredients

Composition of Fatty Acids

Fatty acids	Content in g/100 g acc. to van Hoed et al. (2009)	Content in % acc. to Oh et al. (2007)	Content in % in blackberry "Marion" acc. to Bushman et al. (2004)	Content in % in blackberry "Evergreen" acc. to Bushman et al. (2004)
Lauric acid	0.04	–	–	–
Myristic acid	0.05	–	–	–
Palmitic acid	3.71	3.6–4.2	3.4	4.5
Stearic acid	2.18	1.6–2.8	2.1	3.3
Oleic acid	14.72	7.5–15.4	15.1	17.3

Fatty acids	Content in g/100 g acc. to van Hoed et al. (2009)	Content in % acc. to Oh et al. (2007)	Content in % in blackberry "Marion" acc. to Bushman et al. (2004)	Content in % in blackberry "Evergreen" acc. to Bushman et al. (2004)
Vaccenic acid	–	0.4–0.7	0.7	0.7
Linoleic acid	61.22	60.8–69.8	62.7	53.1
Linolenic acid	17.60	12.5–15.3	15.2	19.9
Arachidic acid	0.47	1.0–1.1	0.5	0.8
20:1	–	0.3–0.5	0.3	0.3
20:2	–	0.1–0.2	–	–
20:3	–	0.1	–	–
Behenic acid		0.1–0.2	–	–

According to van Hoed et al., blackberry seed oil contains 6.45 g/100 g saturated fatty acids, 14.81 g/100 g monounsaturated fatty acids and 78.74 g/100 g polyunsaturated fatty acids (van Hoed et al. 2009).

Composition of Sterols

Sterols	Acc. to van Hoed et al. (2009) in %
Cholesterol	0–0.5
Brassicasterol	0–0.1
Campesterol	0–4.0
Stigmasterol	0–4.0
β-Sitosterol	75–80
Δ5-Avenasterol	4–14
Δ7-Stigmasterol	0–0.5
Δ7-Avenasterol	–

Total content of desmethylsterol: 403.7 mg/100 g (van Hoed et al. 2009).

Composition of Tocopherols

Tocopherols	Content in mg/kg acc. to van Hoed et al. (2009)	Content in mg/100 g seeds of blackberry "Marion" acc. to Bushman et al. (2004)	Content in mg/100 g seeds of blackberry "Evergreen" acc. to Bushman et al. (2004)
Total content	1388.7	10.2	23.3
α-Tocopherol	25.4	1.6	2.0
β-Tocopherol	–	–	–
γ-Tocopherol	1311.7	8.5	20.7
γ-Tocotrienol	20.0	–	–
δ-Tocopherol	31.7	–	–

Other Ingredients

Blackberry seed oil contains 17.0 mg/100 g squalene (van Hoed et al. 2009).

Physical Key Figures

specific weight at 15 °C:	0.9256
saponification number:	189.5
iodine number:	147.8
Reichert value:	0.0
Hehner number:	96.3
acid number:	2.03
(Krzizan 1908)	

5 Application

In Pharmacy and Medicine

Since blackberry oil has a high proportion of polyunsaturated fatty acids, it protects the cardiovascular system. It also supposedly offers protection against autoimmune diseases (van Hoed et al. 2009).

Linoleic acid and linolenic acid, of which blackberry seed oil contains relatively high proportions, have a good anti-inflammatory effect (Oh et al. 2007). Additionally, α-linolenic acid and its derivates have positive effects on the brain and the retina (Bushman et al. 2004).

In cosmetics

Blackberry seed oil is also used in cosmetics. It is added to hand and body lotions and shampoo (van Hoed et al. 2009).

Black Cumin Seed Oil

Oleum Nigellae sativae synonyms: Schwarzkümmelöl (D); huile de nigelle (F)

1 Source Plant

Nigella sativa L. (Ranunculaceae), black cumin.

S. Krist, *Vegetable Fats and Oils*, https://doi.org/10.1007/978-3-030-30314-3_19

Habitat

Nigella has its origins in Southern Europe, the Mediterranean area, North Africa and Asia Minor, where the seeds have been used as a spice and in medicine for a long time. In Central Europe, the plant is cultivated on fields, and grows wild on unploughed strips and meadows (Fischer and Krug 1984, p. 228). Today, nigella is mainly cultivated in Egypt, India, the CIS and Turkey.

B

Description

Nigella is an annual plant with a branched stem (Roth and Kormann 2000, p. 80). It grows to a height of 30 cm and is slightly hairy. The white, terminal flowers appear from June to September; the edges of the petals have a slight greenish or bluish hue. The vesicular fruit capsules contain many black seeds (Ulmer 1996, p. 71). They are tiny and triangular. At first, they taste bitter, and then spicy like pepper. Their smell is reminiscent of nutmeg (Roth and Kormann 2000, p. 80). The seeds contain 0.5–1.5% essential oil and up to 40% fat oil, in addition to protein, sugar, gum, saponins and melanthin (up to 1.5%), bitterns, nigellin, tanning agents, nigellon and thymoquinone (Fischer and Krug 1984, p. 229).

2 Cultivation and Extraction

Cultivation

Nigella is sown in late spring. The seedlings are thinned out until they are at a distance of 30 cm from each other. The ripe seed capsules are harvested before they open, and the seeds removed from the dried capsules (Roth and Kormann 2000, p. 80).

Extraction of the Oil

The oil is obtained by cold pressing at high pressure and by adding natural carbon dioxide. Careful processing is important to retain the essential fatty acids (Roth and Kormann 2000, p. 149). Black cumin seed oil can also be extracted with either hexane or a mixture of chloroform and methanol acting as a solvent (Ramadan and Mörsel 2002a, b).

3 Character

Colour

Reddish (Roth and Kormann 2000, p. 149).

Odour

Peppery, herbaceous, spicy, like "fresh paint", typical touch of caraway, woody.

Flavour

Slightly bitter touch of caraway.

4 Ingredients

Composition of Fatty Acids

Fatty acids	Content (%) acc. to Roth and Kormann (2000), p. 149
Myristic acid	0.1–0.4
Palmitic acid	12–13
Stearic acid	2.4–4
Oleic acid	18–25
Linoleic acid	50–60
α-Linolenic acid	0.1–1
Arachidic acid	0.1–0.4
Arachidonic acid	0.1–0.4

Composition of Triglycerides of Black Cumin Seed Oil

Triglyceroles	Equivalent carbon number	Origin: Egypt commercial Content in % acc. to Hassanein et al. (2011)	Origin: Iran Content in % acc. to Khoddami (2011)	Origin: Egypt traditional Content in % acc. to Stübiger (2019)	Origin: Austria Content in % acc. to Stübiger (2019)
LnLL	40	0.4	–	–	–
LLL LnLO LnLnS	42	18.5*	19.2–20.6	21.3*	24.2*
LnLP	42	0.2	–	–	–
LLO	44	16.1	16.0–16.9	16.9	14.8
LLP	44	16.5	12.4–18.5	22.2	24.2
LnPP	44	0.2	–	1.4	1.0
LL20:2 LnO20:2 LnL20:1	44	–	–	3.2*	2.1*
LOO	46	9.6	9.0–9.4	13.5	13.5

B

Triglyceroles	Equivalent carbon number	Origin: Egypt commercial Content in % acc. to Hassanein et al. (2011)	Origin: Iran Content in % acc. to Khoddami (2011)	Origin: Egypt traditional Content in % acc. to Stübiger (2019)	Origin: Austria Content in % acc. to Stübiger (2019)
LOP	46	13.4	10.7–13.9	9.1	8.1
LPP	46	2.2	–	3.3	2.8
LO20:2 LnO20:1 LL20:1	46	–	–	1.5*	2.2*
OOO LOS	48	6.3*	–	2.0*	2.1*
OOP LSP	48	4.2*	1.3–1.8	2.9*	2.0*
POP	48	1.2	1.5–4.2	0.7	0.4
PPP	48	0.2	–	–	–
OO20:2 LO20:1	48	–	–	1.5*	0.5*
OOS	50	1.1	–	–	–
LSS	50	0.3	–	–	–
POS	50	0.7	–	0.7	0.2
SS20:2 OS20:1	52	–	–	0.2*	0.1*

Fatty acids: *Ln* Linolenic acid, *L* Linoleic acid, *O* Oleic acid, *S* Stearic acid, *P* Palmitic acid, *20:2* Eicosadienoic acid, *20:1* Eicosenoic acid
*Sum parameters

Composition of Tocopherols

(concentration and content in µg/g)

Tocopherols	Black cumin seed oil (extraction with hexane) acc. to Ramadan and Mörsel (2002a, b)	Black cumin seed oil (extraction with chloroform/methanol) acc. to Ramadan and Mörsel (2002a, b)
Total content	597	522
α-Tocopherol	284	258
β-Tocopherol	40	30
γ-Tocopherol	225	203
Δ-Tocopherol	48	31

Other Ingredients

(Ramadan and Mörsel, 2002a, b)

β-Carotene
Vitamin K1
Amino acids

Physical Key Figures of Black Cumin Seed Oil

refractive index:	1.4649–1.473
density: 15 °C	0.925
20 °C	0.919
iodine number:	107–123
saponification number:	192–207
acid number:	7.78
point of solidification:	<0 °C
unsaponifiable:	approx. 0.5%
(Roth and Kormann 2000, p. 149)	

5 Application

In Medicine and Folk Medicine

The healing properties of nigella were already known in Ancient Egypt. The seeds were not only used as a spice, but also by the personal physicians of the pharaoh, who used them to treat digestive disorders, headaches, toothaches and inflammation (Schleicher and Saleh 2007, p. 8–9). In folk medicine, black cumin seed oil is practically a "universal medicine" for everything from acne and high blood pressure to pains in the joints (Schleicher and Saleh 2007, p. 55–100).

Medical studies proved the positive effect of black cumin seed oil in cases of rheumatoid arthritis and allergic rhinitis (Gheita and Kenawy 2011; Nikakhlagh et al. 2011).

Most medically effective properties can probably be traced back to one of the main volatile components of black cumin seed oil, thymoquinone. According to Woo et al., thymoquinone acts against inflammation and cancer on several regulatory levels (Woo et al. 2012). The anti-inflammatory and anticarcinogenic effect (against pancreatic cancer, pulmonary cancer and breast cancer, among others (Banerjee et al. 2009; Chehl et al. 2009; Jafri et al. 2010; Woo et al. 2011)) has so far been partly proven only in tests on animals and *in vitro* examinations (Frohne 2002, p. 381). Several studies established the antibacterial and fungicidal effect of thymoquinone (Chaieb et al. 2011; Halawani 2009; Al-Jabre et al. 2003).

In Veterinary Medicine

Black cumin seed oil is applied internally and externally. The largest field of application is horse farming. The oil is applied internally against bronchial asthma, respiratory diseases and allergic reactions; it has antiparasitic properties and stabilises the immune system. Applied externally, it has a grooming effect on the coat and supports the healing process of small wounds and injuries. Dairy cows have been successfully treated with black cumin seed oil in cases of mastitis; the oil is repeatedly rubbed onto the affected areas of the udder. Fowl susceptible to infections showed an increase in immunity after black cumin seed oil had been mixed into their drinking water.

As Food

The spicy, high-quality oil can be used for cooking and as a salad oil (Roth and Kormann 2000, p. 149).

Blackcurrant/Redcurrant Seed Oil

synonyms: Johannisbeersamenöl (D); blackcurrant pip oil (E); huile de pépins de cassis (F)

S. Krist, *Vegetable Fats and Oils*, https://doi.org/10.1007/978-3-030-30314-3_20

1 Source Plants

Ribes nigrum L., *Ribes rubrum* L. *(Grossulariaceae)*, blackcurrant, redcurrant.

Habitat

Blackcurrants are native plants in European and Asian woodlands up to Manchuria and the Himalayas, and are cultivated in Central Europe (Hunnius 1998, p. 1188).

Redcurrants are native plants of Western Europe. In France, Belgium and England, in particular, as well as in Asia and North America, they grow in wet ash and alder forests, along ditches and in bushes (Hager 1994, p. 466 et seqq)

Description

The blackcurrant is a summergreen shrub up to 2 m tall, with three- to five-lobed, double serrated, heart-shaped leaves. The greenish-white flowers are arranged in pendant racemes. The black fruit are round berries, dotted with glands; they have a peculiar smell reminiscent of bugs (Hager 1994, p. 466 et seqq).

The redcurrant is a shrub 1–2 m tall, also with three- to five-lobed leaves. The flowers are greenish in colour and form pendant, multiflorous racemes. The fruit are red berries; cultivars may also produce pink or white fruit. They taste sour (Hager 1994, p. 466 et seqq).

2 Cultivation and Extraction

Cultivation

Species of *Ribes* are cultivated throughout Europe and in many countries outside Europe. The berries are harvested when they are ripe (Hager 1994, p. 466 et seqq).

Extraction of the Oil

The seeds are either cold pressed (Löw 2003, p. 125) or extracted with hexane (ratio 1:3). The seeds are washed beforehand using alcohol at 70 °C and subsequently dried at 90 °C (Karleskind 1996, p. 176).

3 Character

Colour

Clear, yellow oil, liquid at room temperature.

Odour

Fatty, green, floral, fruity, waxy, smoky.
Blackcurrant: fruity, fatty, touch of blackcurrant.

Flavour

Very fruity (Löw 2003, p. 125).

4 Ingredients

Composition of Fatty Acids

Blackcurrant

The oil of *Ribes nigrum* is among the natural oils with the highest amount of γ-linolenic acid (Hager 1994, p. 471).

Fatty acids	Content in % acc. to Rossell and Pritchard (1991)	Content in % acc. to Löw (2003)	Content in % acc. to Barre (2001)	Content in % acc. to Traitler et al. (1988)	Content in % acc. to Firestone (1999), p. 15
Myristic acid	0.1	–	–	–	0.1
Palmitic acid	6–7	–	7	6–8	6–8
Palmitoleic acid	–	–	–	–	0–0.2
Stearic acid	1–2	–	1	1–2	1–2
Oleic acid	9–10	11	11	9–13	9–13
Linoleic acid	45–50	47	47	44–51	45–50
α-Linolenic acid	12–14	11	12	12–14	12–15
ϒ-Linolenic acid	15–19	17	–	15–20	14–20
Stearidonic acid	2.5–4	–	–	2–4	2–4
Arachidic acid	0.5	–	–	–	0.2

Fatty acids	Content in % acc. to Rossell and Pritchard (1991)	Content in % acc. to Löw (2003)	Content in % acc. to Barre (2001)	Content in % acc. to Traitler et al. (1988)	Content in % acc. to Firestone (1999), p. 15
Eicosanoic acid	0.5	–	–	–	0.9–1.0
Behenic acid	–	–	–	–	0.1
Lignoceric acid	–	–	–	–	0.1
Total oil content	–	28–32	–	–	–

Redcurrant

Fatty acids	Content in % acc. to Traitler et al. (1988)	Content in % acc. to Löw (2003)
Palmitic acid	4–5	–
Stearic acid	1–2	–
Oleic acid	14–16	–
Linoleic acid	41–42	42
α-Linolenic acid	29–31	30
Γ-Linolenic acid	4–5	6
Stearidonic acid	2–4	–
Total oil content	–	18–26

Composition of Triglycerides

(all content in %)

Via GC

Chain lengths of the triglycerides	Blackcurrant seed oil acc. to Karleskind (1996), p. 177
C48	–
C50	0.6
C52	19.3
C54	77.0
C56	3.1

Via HPLC

Triglycerides	Blackcurrant seed oil acc. to Karleskind (1996), p. 177	Triglycerides	Blackcurrant seed oil acc. to Karleskind (1996), p. 177
LnLnγLn	Together 5.5	OLLn	3.7
LnγLnyLn		OLγLn	5.1
γLnyLnγLn		PLLn	2.4
S4LnL		PLγLn	3.2
S4γLnL		OLL	6.8
S4S4P		PLL	6.3
LnγLnL	7.9	POLn	Together 2.1
LLLn	10.2	PPγLn	
OLnLn	Together 6.0	OOL	0

Triglycerides	Blackcurrant seed oil acc. to Karleskind (1996), p. 177	Triglycerides	Blackcurrant seed oil acc. to Karleskind (1996), p. 177
OγLnγLn		SLL	1.7
OLS4		LLL	12.9
PLnγLn			

Distribution of Fatty Acids at the Positions 1,2,3 of the Triglycerides

(Karleskind 1996, p. 177)

Fatty acids	Sn-1 (in %)	Sn-2 (in %)	Sn-3 (in %)
Palmitic acid	14.2	2.0	4.4
Stearic acid	4.9	1.5	–
Oleic acid	12.6	14.1	5.1
Linoleic acid	42.7	53.1	43.8
α-Linolenic acid	17.2	8.1	14.1
Γ-Linolenic acid	4.1	17.4	25.8
Stearidonic acid	0.7	2.6	–

Composition of Sterols

Sterols	Content in % acc. to Firestone (1999), p. 15	Content in % acc. to Karleskind (1996), p. 178
Cholesterol	0.2–0.7	0.7
Campesterol	7.2–10.4	7.2
Stigmasterol	0.5–1.0	0.2
β-Sitosterol	70–85	85.4
Δ5-Avenasterol	2–3	3.0
Δ7-Stigmasterol	0.4–4.5	0.4
Δ7-Avenasterol	0.4–2	0.4

Composition of Tocopherols

Tocopherols	Content in mg/kg acc. to Firestone (1999), p. 15
Total content	1043
α-Tocopherol	320
β-Tocopherol	8
Γ-Tocopherol	647
Δ-Tocopherol	68

Physical Key Figures of Blackcurrant Seed Oil

refractive index: n_D^{20} 1.474–1.477
saponification number: 195–197
iodine number: 160–175
unsaponifiable: 1.8–2.3%
(Roth and Kormann 2000, p. 160)

Shelf Life

Blackcurrant seed oil should be kept in a cool (10–20 °C), dark, dry place. In unopened packages, it has a shelf life of 18 months.

B

5 Application

In Pharmacy and Medicine

Against Rheumatoid Arthritis

In a study on patients suffering from rheumatoid arthritis, interleukin-1β, interleukin-6 and PGE_2 (pro-inflammatory cytokines) decreased significantly when the oil of *Ribes nigrum* was administered in a dose of 525 mg/d in capsules over a period of 6 weeks. Other studies reported a decrease in the plasma triglyceride level, cholesterol and LDL cholesterol levels, and an increase in HDL cholesterol in hyperlipidaemia patients of the type IIa and IIb after ingesting blackcurrant seed oil with 450 mg γ-linolenic acid per day over a period of 12 weeks (Barre 2001).

Benefits of γ-Linolenic Acid

γ-Linolenic acid is an essential fatty acid for patients with a deficiency in Δ-6-linolenic acid desaturase as well as an essential intermediate stage for the biosynthesis of prostaglandin, prostacyclin and thromboxane. A diet with γ-linolenic acid has anti-inflammatory and antiproliferative effects, stimulates the β-oxidation of fatty acids in the liver and has a cytotoxic effect on superficial bladder cancer. A study showed that blackcurrant seed oil contains 19% γ-linolenic acid, and is thus among the oils with the highest γ-linolenic acid content, the others being evening primrose oil (7–10%) and borage seed oil (17–25%) (Goffman and Galletti 2001).

Another study on rabbits fed with blackcurrant seed oil showed that the oil has an antithrombotic effect on the vascular walls, which is probably the result of selective absorption and ensuing metabolism of linoleic acid and γ-linolenic acid. As a consequence, the synthesis of 13-hydroxyoctadecadienoic acid is stimulated and the level of arachidonic acid decreases. The antithrombotic effect is further increased by the presence of γ-linolenic acid, compared to the administration of linoleic acid alone. The reasons for this phenomenon are still unclear (Bertomeu et al. 1990).

Benefits of Tocopherols

The oil is an excellent source of vitamin E. Tocopherols are natural antioxidants with biological activity. Their main biochemical function is to protect polyunsaturated fatty acids from peroxidation. They also seem to lower the risk of heart diseases, cancer and other diseases (Goffman and Galletti 2001).

Influence on Blood Pressure and Cardiovascular Reactivity

A study on test persons with high blood pressure of borderline values examined the effect of blackcurrant seed oil on blood pressure measured at rest and cardiovascular reactivity to psychological stress. Over a period of eight consecutive weeks, the test persons received a supplement of 6 g blackcurrant seed oil per day. At the beginning and end of the study, the reactivity of the blood pressure and heart rate was measured by a standardised mental arithmetic test of 5 minutes' duration. Every week, the blood pressure was measured at rest. The supplement of blackcurrant seed oil resulted in an inhibition of the reactivity of blood pressure by more than 40%.

Preparations rich in γ-linolenic acid therefore have an influence on the control of the cardiovascular system, but the mechanisms of how this works are not yet entirely clear (Deferne and Leeds 1996).

Protection against the Effects of Alcohol

Chronic abuse of alcohol induces a deficiency in linoleic acid, essential fatty acids and prostaglandins in the blood-stream. In rats that were administered alcohol, there was ineffective transport of zinc over the placenta from the mother to the foetus as well as foetal alcohol syndrome, which manifests itself in clinical symptoms like hypoplasia of the jaw bones and anomalies of the gums. A study examined the effects of dietary supplements containing blackcurrant seed oil and zinc sulphate on rats in a state of advanced pregnancy that had been inebriated with alcohol over longer periods of time, and on their offspring. The study showed that the content of PGI_2 produced by the gastric mucosa was not significantly different from that of the control group. The results suggest that a suitable diet protects the gastric mucosa of the offspring of mothers that were given alcohol against its effects. The efficiency of this protection can be visualised by a PGI_2-dependent thrombocyte aggregation test (Seri and D'Alessandro 1997).

Positive Influence on the Lipid Metabolism

Omega-3 polyunsaturated fatty acids (PUFAs) from vegetable sources have anti-atherogenic potential. Blackcurrant seed oil is one of the few vegetable oils that contain both ω-3 PUFAs and ω-6 PUFAs. A study on rats examined the effect of blackcurrant seed oil on antioxidative parameters, the lipoprotein profile and the lipids of the liver. The rats were put on a diet containing 10% of the oil; the control group was fed with lard. After 3 weeks, the diet with blackcurrant seed oil induced a significant decrease in glutathione in the blood-stream and an increase in the oxidisability of serum lipids induced by Cu^{2+}, but showed no effect on the glutathione level in the liver and the lipid peroxidation of liver microsomes induced by t-butylhydroperoxide. These results suggest that blackcurrant oil has a positive effect on the lipid metabolism of the liver and does not cause fatty liver. The heightened sensibility of LDL to oxidation induced by copper in vitro has a negative influence on the antioxidative blood status and should be taken into account (Vecera et al. 2002).

No Mutagenicity

Blackcurrant seed oil rich in γ-linolenic acid has a prophylactic effect against neoplasia. To examine the mutagenicity, the following tests were carried out: Ames test, SCE micronucleid test on the spinal marrow of mice and CHL (Chinese hamster lung cell) chromosome aberration test. The results of all the tests were negative, which proves that blackcurrant seed oil is not mutagenic (Wang et al. 1998).

In Cosmetics

γ-Linolenic acid and oil concentrate containing γ-linolenic acid from blackcurrant seed oil are used in cosmetics (Hager 1994, p. 472). The oil is applied to dry, flaky and ageing skin, and in cases of neurodermatitis and psoriasis. In combination with β-carotenes and tocopherols that are also part of the oil, γ-linolenic acid acts as a scavenger.

Structure of the Stratum Corneum

One of the main functions of the skin is to form a protective layer, the *stratum corneum*. The protection is affected if the skin is dry, flaky or chapped. Examinations suggest that the reason for dryness is that the amount of linoleic acid in dry skin is lower and the amount of palmitoleic acid is slightly higher than normal. A study examined the effect of blackcurrant seed oil, which is rich in linoleic acid. An emulsion of palm oil, shea butter and blackcurrant seed oil was applied to the dry skin on the legs of five women on 14 consecutive days, in the same way as commercially available cream. As a consequence, the amount of palmitoleic acid decreased, while the amount of linoleic acid increased, which led to an improvement in skin appearance. At the end of the test, 2% γ-linolenic acid was detected, which is not detectable in normal skin. Polyunsaturated fatty acids can be hydroxylated to form polyoxyacyl lipids. These peroxidised lipids are thought to influence the differentiation of epidermal cells into an effective *stratum corneum* (Brod et al. 1988).

As a Dietary Supplement

A study on healthy, elderly patients given blackberry seed oil (675 mg γ-linolenic acid/d) as a dietary supplement over a period of 2 months showed a slight increase in the immune reaction, an increased proliferation of lymphocytes and a decrease in PGE_2 (Wu et al. 1999).

Borage Seed Oil

Oleum Boraginis synonyms: Borretschsamenöl (D); huile de bourrache (F)

1 Source Plant

Borago officinalis L. (Boraginaceae), borage.

S. Krist, *Vegetable Fats and Oils*, https://doi.org/10.1007/978-3-030-30314-3_21

B

Habitat

Originally, borage was a native plant of the Mediterranean area from Spain to Turkey. Today, it is also found in the warmer regions of Western, Central and Eastern Europe, in West Asia and in the USA, often growing wild. Borage is fairly undemanding regarding the soil, and grows in moist, fertile, limey, sandy as well as loamy soil, in both sunny areas and in half shade. In light soil, a good supply of water is essential. Borage is an excellent garden plant.

Description

Borage is an annual plant. The stem is up to 60 cm tall and up to 1 cm thick, upright, unbranched or ramified, crenate, densely covered with rough hair, hollow and juicy. The leaves are alternate. The lower ones are densely arranged, up to 12 cm long, occasionally over 6 cm broad, obovate, oval or elongated and heart-shaped, and either blunt or pointed. All leaves are nearly entire or slightly recessed. They are dark green on the upper side, paler on the lower side, and usually densely covered with stiff or bristly hair on both sides. The flowers sit on stems 0.5–2 cm long, sticking out or nodding on partly leafy cincinni, which do not have many flowers, but are often arranged in extensive umbellate panicles. The calyx separates into linear, quite pointed, densely haired lappets. The corolla is a pure sky-blue, more rarely whitish, 20–26 mm broad, with protruding throat scales.

2 Cultivation and Extraction

Cultivation

The cost-effectiveness of cultivating herbs, medicinal plants and aromatic plants depends mainly on the quality achieved and on the sales potential. Borage seed oil is one of the few vegetable oils that contain a high proportion of the pharmacologically interesting fatty acid γ-linolenic acid, which is why an increasing demand for borage oil is expected. In 1995, the demand for borage oil in Germany amounted to 8000–15,000 l. Since there is a potentially increasing demand, studies examined the yield of *Borago officinalis* as well as data concerning cultivation techniques, processing of the harvested crop, and selection of suitable oil extraction methods, in order to prepare for a large-scale agricultural cultivation of borage.

Extraction of the Oil

Borage oil is cold pressed from borage seeds, and in some cases subsequently refined (Kerschbaum and Schweiger 2001, p. 17).

3 Character

Colour

Golden yellow.

Refined: colourless to yellowish.

Odour

Green, herbal, slightly woody, reminiscent of cucumbers.

Flavour

Fresh, slightly acetous, reminiscent of cucumbers, more distinct, slightly bitter aftertaste.

4 Ingredients

Composition of Fatty Acids

Fatty acids	Content in % acc. to Hänsel et al. (1999), p. 267	Content in % acc. to Nature certificate of analysis/28.08.03	Content in % acc. to Kerschbaum and Schweiger (2001), p. 16
Palmitic acid	11.8	9.94	11.1
Linoleic acid	38.1	38.75	37.5
Palmitoleic acid	–	0.16	0.1
Oleic acid	16.3	16.28	16.6
α-Linolenic acid	–	0.2	0.2
γ-Linolenic acid	22.8	21.63	21.1
Stearic acid	–	3.58	3.7
Arachidic acid	–	0.25	0.3
Eicosenoic acid	–	–	3.9
Eicosadinic acid	–	–	0.3
Gondoic acid	–	4.02	–
Behenic acid	–	0.19	0.2
Vaccenic acid	–	–	0.7
Erucic acid	–	2.66	2.5
Nervonic acid	–	1.64	1.6
Myristic acid	–	–	0.1
Others	–	0.7	–

B

Composition of Sterols

Sterols	Content in % acc. to Firestone (1999), p. 16 et seq
Cholesterol	–
Brassicasterol	0–1.6
Campesterol	25–30
Stigmasterol	–
β-Sitosterol	22–42
Δ5-Avenasterol	15–28
Δ7-Stigmasterol	–
Δ7-Avenasterol	1
24-Methylen-cholesterol	15–20

Composition of Tocopherols

(concentration and content in mg/100 g)

Tocopherols	Borage seed oil acc. to Kerschbaum and Schweiger (2001), p. 36	Borage seed oil acc. to Firestone (1999), p. 17
Total content	87.7	73.2–111.1
α-Tocopherol	–	0–4.6
β-Tocopherol	–	–
γ-Tocopherol	–	3.3–27.2
δ-Tocopherol	31.9	69.0–101.3

Other Ingredients

(Kerschbaum and Schweiger 2001, p. 43)

Vitamin E
Flavonoids
Mineral nutrients
Protein
Mucilage
Tanning agents
Saponins

Physical Key Figures of Borage Seed Oil

refractive index: n_D^{20} 1.4756
density: d_{20}^{20} 0.9197
saponification number: 191
acid number: 0.17
iodine number: 141
peroxide number: 6
unsaponifiable: 1.83%

Shelf Life

Borage seed oil has a shelf life of 10–12 weeks.

Characteristics of the Oil

Of all the vegetable oils examined, borage seed oil contains the largest portion of γ-linolenic acid (about 21%). In comparison, hempseed oil contains about 3% of this fatty acid. Unfortunately, the composition of fatty acids in borage seed oil also has negative aspects. The oil contains 15% saturated fatty acids as well as erucic acid, which may cause vascular damage if ingested in large amounts. The herb and the flowers contain pyrrolizidine alkaloids, which are toxic for the liver. The oil is almost free from these alkaloids, however, and thus non-hazardous (Kerschbaum and Schweiger 2001, p. 18).

5 Application

As a Dietary Supplement

Borage seed oil is advertised for diets and as a dietary supplement. Its usefulness has not, however, been proven for healthy persons not suffering from a deficit in Δ-desaturase. Δ-Desaturase is an enzyme produced naturally in the body; it catalytically transforms linoleic acid, which is consumed with food, into γ-linolenic acid. The activity of desaturase depends on several factors. Insulin and a diet rich in protein increase its activity, whereas glucagon, glucocorticoids and thyroxine inhibit it. In addition, desaturase may be entirely or partly blocked by diabetes, excessive carbohydrate intake, or because of age. Decreasing activity of the enzyme leads to an overbalance of prostaglandin, prostacyclin and thromboxane of the ω-3 series. The prostaglandin and leukotriene pattern may shift, which is of pathophysiological significance (Hänsel et al. 1999, p. 265 et seqq).

Borage seed oil does not have a long shelf life and must not be heated, which is why it is often sold in the form of capsules (borage seed oil capsules).

It is not authorised as a medicine (Austria Codex 2004).

To Treat Skin Diseases

Dry skin and itching may be caused in some cases by a deficiency in γ-linolenic acid. This fatty acid is an important component of the skin and a pre-stage for the synthesis of prostaglandin E_1, which has anti-inflammatory properties and reduces itching. Borage seed oil, which contains an extraordinarily high proportion of γ-linolenic acid, is applied topically in the following cases (Hunnius 1998, p. 228):

- for skin regeneration;
- to treat ageing, wrinkly, dry and chapped skin;

B

- to treat neurodermatitis;
- to treat atopic eczema;
- to treat psoriasis;
- as an antipruritic;
- for moisture regulation;
- in anti-wrinkle creams;
- for skincare and nail care; and
- for improving the activity of the sebaceous glands.

For Treatment of Premenstrual Syndrome (PMS)

As mentioned above, γ-linolenic acid is a pre-stage of prostaglandin E, which is responsible for the interaction of female hormones (oestrogen, progesterone, etc.). Some women suffer from physical and psychic complaints from the tenth day before menstruation starts; these complaints are summarised under the term PMS. It is assumed that women suffering from PMS have a deficiency in γ-linolenic acid, causing an oversupply of the female hormone prolactin. γ-Linolenic acid is known to counterbalance oscillations of blood glucose and can therefore create the prerequisites necessary for the organism to protect itself effectively against these monthly complaints. As a result of its high content of γ-linolenic acid, borage seed oil also helps to treat hormonally induced menopausal depression (Ulmer 1996, p. 55 et seq; Löw 2003, p. 118; Kerschbaum and Schweiger 2001, p. 17).

To Treat Rheumatic Arthritis

Double-blind studies have shown that γ-linolenic acid, which is contained in borage seed oil in considerable quantities, eases rheumatoid arthritis. The supposed reason is the effect of γ-linolenic acid of increasing the level of prostaglandin E_1, which also increases the production of intracellular cAMP. This suppresses the synthesis of tumour necrosis factor alpha, which is an important inflammatory mediator for rheumatoid arthritis.

As a Substitute for Breast Milk

Breast milk is known to be the perfect nourishment for newborn babies. For mothers who cannot or do not want to breastfeed their children, it is necessary to find the best substitute available, as nourishment not only affects the growth and functional development of babies in general, but also that of premature babies, in particular. In utero, the foetus does not depend on the synthesis of long-chain polyunsaturated fatty acids (LC-PUFAs) because it is nourished by the placenta. This results in a special problem for premature babies, although in principle, they are able to synthesise LC PUFAs endogenously. Breast milk contains considerable amounts of γ-linolenic acid. Babies fed on breast milk have a higher level of LC PUFAs than babies not fed on breast milk. It is assumed that premature babies cannot meet the demand for the respective fatty acids by endogenous synthesis alone, so that breast milk must supply them with the sufficient quantities. By way of trial, formula food has been developed that contains – in addition to milk fat, coconut oil, soya bean oil and sunflower oil – borage seed oil because of its high content of γ-linolenic acid. The study has not proven beyond doubt that formula food containing borage seed oil is more advantageous than formula food containing only other vegetable oils, but further research in this direction is encouraged and advocated (Feldl 2000).

Borneo Tallow Nut Oil

synonyms: Borneo Talg, Borneo Tallow Nussöl (D); beurre d'illipé de Bornéo (F)

1 Source Plant

Shorea spp. (Dipterocarpaceae), Tengkawang tree, engkabang illipe, Borneo illipe.

S. Krist, *Vegetable Fats and Oils*, https://doi.org/10.1007/978-3-030-30314-3_22

Habitat

Shorea is a wild growing tree in Malaysia, the Philippines, Java and Borneo. It preferably grows in bog forests on river banks, in alluvial soil or shallow peat (Axtell 1992, p. 29).

Description

Shorea is a tree of high-quality wood. The fruit is used for oil production. It is oval and winged. The size varies depending on the species, but a length of about 4 cm is common. The shell is ligneous and either brown or black. Within, there are seeds containing between 45% and 70% oil. The tree first bears fruit at an age of 18–25 years (Axtell 1992, p. 29).

Shorea consists of about 357 species. For the extraction of Borneo tallow, the species *pachycarpae*, in particular, is used. The species are selected according to seed size, fruit production, accessibility of the fruit and amount of crop. Ninety-five percent of the seeds used for oil extraction belong to *Shorea macrophylla* (Blicher-Mathiesen 1994).

Shorea stenoptera is another species frequently used for oil extraction. The seeds weigh up to 45 g when whole, and break lengthwise into 3–6 segments. This presumably happens either because of drying or because the seeds prepare themselves for germination (Kershaw 1987).

Tengkawang is the Indonesian name for trees of the *Dipterocarpaceae* family, usually of the genus *Shorea*. The common Malaysian name is engkabang or illipe. The latter term is unfortunately confusing, as it is also the Tamil name for several plants of the *Bassia* species, usually *Bassia longifolia* and other *Sapotaceae* (Seibert 1996, p. 616–617).

2 Cultivation and Extraction

Cultivation

Shorea is usually a wild-growing plant. In peak years, up to 41,000 tons are exported from Indonesia. The fruit are only gathered after they have fallen from the tree, only once in 5 years provided that the harvest has been good. After collecting the fruit, they are dried until their moisture content is reduced to 7%. The shell is then removed. There are several methods to do this. Either the nuts are collected in containers and dipped into water for 2–4 weeks until germination starts and the shell breaks open by itself. The seeds can then be taken out by hand. The same method can be applied without water: The nuts are put into a hollow in the soil and brought to germination in this way. The shell can also be broken manually with a sharp tool or by heating. After the breaking of the shell, the seeds are dried in the sun for 5–7 days (Axtell 1992, p. 30–31).

Extraction of the Oil

In rural areas, the oil is extracted using a simple method. The seeds are heated in a pan and afterwards put in a rattan bag, which is placed between two boards of hardwood and pressed. For industrial amounts, the oil is extracted with hexane as a solvent; it is subsequently processed and bleached (Axtell 1992, p. 31).

3 Character

Colour

Light green; if exposed to air for some time: yellow gradually turning white (Geitel 1887).

Odour

Similar to cocoa butter (Geitel 1887).

Flavour

Reminiscent of cocoa butter (Geitel 1887).

4 Ingredients

Composition of Fatty Acids

Fatty acids	Content in % acc. to Roth and Kormann (2005), p. 154	Content in % acc. to Axtell (1992), p. 31	Content in % acc. to Lipp and Anklam (1998)	Content in % acc. to Blicher-Mathiesen (1994)	Content in % acc. to Kershaw (1987)
Lauric acid	–	–	0.1	–	–
Myristic acid	1.5	–	0.1	–	–
Palmitic acid	21.5	18.0	5.7–23.1	4.1–25.5	13.3–13.8
Palmitoleic acid	–	–	0.2–1.1	–	0.2–0.3
Stearic acid	39.0	43.3	39.4–46.7	39.5–54.6	48.0–49.0
Oleic acid	38.0	37.4	33.2–49.0	32.4–41.7	33.5–34.2
Linoleic acid	–	0.2	0.4–4.3	0.2–2.4	0.9–1.3
Linolenic acid	–	–	0.2–1.8	–	0.4–0.6
Arachidic acid	–	1.1	0.1–1.3	0.6–3.6	1.9–2.1

B

Composition of Triglycerides

Triglycerides	Content in % acc. to Ali et al. (1998)	Content in % acc. to Blicher-Mathiesen (1994)
POP	9.6	5.2–15.4
POS	31.1	32.4–47.2
SOS	38.9	35.0–53.8
SOA	–	0.8–5.0

Composition of Sterols

Sterols	Content in % acc. to Lipp and Anklam (1998)
Cholesterol	1.5–1.8
Brassicasterol	0.3–9.6
Campesterol	12.7–19.5
Stigmasterol	4.2–7.5
β-Sitosterol	44.2–71.8
Δ5-Avenasterol	0.8–2.4
Isofucosterol	3.6
Δ7-Stigmasterol	1.0
Δ7-Avenasterol	0.1

Physical Key Figures

refractive index: n_D^{40}	1.4559–1.4573
density: d_{99}^{99}	0.851–0.857
saponification number:	190–197
iodine number:	27–34
point of solidification:	22–30 °C
melting point:	28–37 °C
(Roth and Kormann 2005, p.154)	
unsaponifiable:	0.4–2.0%
Reichert value:	0.1–0.5
Polenske value:	0.3–0.4
titre of fatty acids:	51–53 °C
(Seibert 1996, p. 620)	

5 Application

Technical Uses

The first trade recorded with Borneo tallow nut oil dates from 1856, when 651,586 kg of the fat was transported to Singapore. It was used for the production of candles and as a lubricant for steam engines and tramways (Blicher-Mathiesen 1994).

Today, Borneo tallow is a substitute for cocoa butter. It is one of the permitted "cocoa butter equivalents" (CBEs). It has a similar distribution of fatty acids and, to a great extent, the same kinds of monounsaturated triglycerides as cocoa butter. The two fats are thus completely compatible and can be mixed. CBEs are used because they are less expensive than cocoa butter, but have the same physical properties (Blicher-Mathiesen 1994).

Another reason why CBEs are important is that cocoa butter is not heat resistant. Chocolate made of cocoa butter melts at temperatures of over 30 °C. If we compare Borneo tallow nut oil with cocoa butter, it is conspicuous that both have a high proportion of the triglycerides POS. In addition to this, Borneo tallow nut oil has a high content of SOS, but not much POP. Chocolate made exclusively from cocoa butter remains solid at temperatures below 31.5 °C, but if the cocoa butter is mixed with Borneo tallow nut oil, the temperature can be increased to 32.5 °C before the chocolate melts. Another advantage of the mixture of both fats is that it delays chocolate bloom. This blotched, whitish film on chocolate forms because of bad storage (either too cold, too warm, or too much temperature fluctuation). If Borneo tallow nut oil is added, the chocolate's surface stays shiny for longer. Regarding hardness, there is no significant difference (Ali et al. 1998).

In Cosmetics

Borneo tallow from *Shorea stenoptera* is used in cosmetic products because of its convenient composition of fatty acids, triglycerides and phytosterols. It moisturises and protects the skin. The natural lipid layer is preserved (Athar and Nasir 2005).

Tengkawang butter not only substitutes cocoa butter in the chocolate industry, but also in the cosmetics industry. The fat remains fresh for longer without becoming rancid or starting to smell bad. This is important for its use in lipsticks and creams, which are usually stored in open containers and at room temperature. Another advantage is the melting point of Tengkawang butter, which is similar to body temperature. The cosmetic products therefore have a solid to creamy texture in the storage container, and are soft and easy to spread on the skin when they come into contact with the body (Seibert 1996, p. 619).

As Food

The native people of Borneo use Tengkawang as an edible fat and consider it a delicacy. Usually, they add it to other food, such as rice (Seibert 1996, p. 619).

Brazil Nut Oil

Oleum Bertholletiae synonym: Paranussöl (D)

1 Source Plant

Bertholletia excelsa Humb. et Bonpl. *(Lecythidaceae)*, Brazil nut

S. Krist, *Vegetable Fats and Oils*, https://doi.org/10.1007/978-3-030-30314-3_23

B

Habitat

The Brazil nut is a native tree of the tropical regions of northern South America. It is common in the rain-forests of the Amazon and the Orinoco, in Brazil, Bolivia, Peru, Ecuador, Colombia and Venezuela. The German name *Paranussbaum* derives from the Brazilian state Pará, where the tree has its origins. The native population used Brazil nuts as food long before the Portuguese explorers arrived. Alexander von Humboldt (1769–1859) is thought to have offered the native population one ounce of gold (about 31 g) for an offshoot of the tree after he had tasted the delicious seeds in 1800. His attempt at cultivation failed, however, because the offshoots did not develop any flowers (Bärtels 1996, 4th edition, p. 312).

Description

The Brazil nut is one of the largest jungle trees of tropical South America. It grows to a height of 50 m and has a very large trunk with a large crown. The leaves are up to 60 cm long, simple, elongate, oval and entire. The flowers are very large, yellow, with four petals and many stamina in dense panicles. The fruit are capsules that weigh up to 1.5 kg. They are 30 cm in diameter, round and flattened. Within each capsule, there are 15–30 triangular seeds ("nuts") about 4 cm long, with hard shells. An enlarged part between the cotyledons and primary roots serves as a eutrophic parenchyma (Bärtels 1996, 4th edition, p. 312).

2 Cultivation and Extraction

Cultivation

The Brazil nut does not grow in the so-called barrancas, i.e. flooded locations on higher elevations. As the tree only bears its first fruit after 30 years, and the fruit need 12 months to ripen, the Brazil nut is rarely cultivated. As there are large wild-growing populations, cultivation is not yet necessary. The blooming period begins after the rainy season, from September to February. For pollination, the Brazil nut needs a particular species of bee that only exists in the tropical regions of South America. For this reason, attempts at cultivating Brazil nuts in other countries have failed. The harvest takes place from November to early June. After the ripe fruit have fallen from the trees, workers collect them, wash and dry them; drying is necessary because the fruit are prone to be affected by mould that produces aflatoxins. The capsules are opened with a special kind of knife, and the nuts removed. The nuts are soaked in water for 24 hours, then boiled for 5 minutes and subsequently peeled by hand. The peeled nuts need to be stored in a cold, dark, well-ventilated place, to prevent them from becoming rancid (Axtell 1994, p. 35).

Extraction of the Oil

Brazil nut oil is obtained from the seeds of *Bertholletia excelsa* in two different ways:

Extraction with Solvents

In this method, the oil is extracted from the cleaned and peeled seeds with petroleum ether under pressure, and subsequently cleaned (Elias and Bressani 1961).

Pressing

The cleaned and peeled seeds are pressed in a screw press and subsequently cleaned. The quality of pressed oil is better than that of the extracted oil. In the latter, there may be remnants of the solvent, and all liposoluble vitamins have been removed (Elias and Bressani 1961).

3 Character

Colour

Clear, yellowish (Elias and Bressani 1961).

Pressed oil: colourless to light yellow (Schuette and Enz 1931).

Oil extracted with solvents: dark yellow (Schuette and Enz 1931).

Light yellow (Schuette et al. 1930).

Rancid oil: dark (Von Wiesner 1927, p. 747).

Light brown (Roth and Kormann 2000, p. 143).

Odour

Pleasantly sweet smell.

Nutty smell (Schuette et al. 1930).

Odourless (Roth and Kormann 2000, p. 143).

Flavour

Nutty taste (Schuette et al. 1930).

Pleasant taste (Roth and Kormann 2000, p. 143).

B

4 Ingredients

Composition of Fatty Acids

Fatty acids	Content in % acc. to Roth and Kormann (2000), p. 143	Content in % acc. to Firestone (1999), p. 18	Content in % acc. to Schuette et al. (1930)
Myristic acid	1.79	0.6	7.94
Palmitic acid	13.55	14–16	60.48
Palmitoleic acid	–	0.3	–
Stearic acid	2.58	6–10	11.57
Oleic acid	55.64	29–48	51.26
Linoleic acid	21.65	30–47	18.84
Arachidic acid	–	0.3	–
Phytosterol	0.04–0.08	–	–

Composition of Sterols

Sterols	Content in % acc. to Firestone (1999), p. 18
Cholesterol	1
Brassicasterol	–
Campesterol	2
Stigmasterol	9
β-Sitosterol	85
Δ5-Avenasterol	–
Δ7-Stigmasterol	2
Δ7-Avenasterol	–

Other Ingredients

(Gomes Da Silva et al. 1997)

α-Tocopherol
β-Tocopherol
γ-Tocopherol
Δ-Tocopherol
Sitostanol
α-Amyrin
β-Amyrin

Physical Key Figures of Brazil Nut Oil

refractive index: n_D^{20} 1.4678–1.4711
density: d_{15}^{15} 0.914–0.917
point of solidification: 0–4 °C

saponification number: 193–202
iodine number: 94–106
unsaponifiable: 0.5–1%
(Roth and Kormann 2000, p. 143)

Shelf Life

Brazil nut oil becomes rancid quickly and does not have a long shelf life (Roth and Kormann 2000, p. 143).

5 Application

In Pharmacy and Medicine

Brazil nut oil contains a high amount of selenium (50–250 μg*) a*nd can therefore be used in selenium production. Selenium is an essential trace element as well as an intracellular antioxidant, and can counterbalance a vitamin E deficiency syndrome to a certain extent. Administration of additional selenium is not, however, recommended, as the toxic level begins just above the optimal amount. Chronic poisoning manifests itself in symptoms like headaches, paleness, depression, fatigue, skin damage, indigestion and a typical garlic smell of the breath, which is due to the production of dimethyl selenide. Brazil nut oil also contains several amino acids like cysteine, methionine, glutamine, glutamic acid and arginine. From amino acids, proteins and peptides are built, and in their free form, they are among the most important substances of living cells (Burger and Wachter 1993, 7th edition, p. 1278).

In Cosmetics

Brazil nut oil moisturises the skin and prevents it from drying out after showering. For this reason, it is frequently added to lipid-replenishing soap (Schuette et al. 1930), shampoo, conditioners, hair sprays and fluids, to prevent the tips of the hair from drying out.

Thanks to its moisturising properties, Brazil nut oil is ideal for dry skin. It forms part of creams, balms and body lotions intenden to make the skin smooth and supple. Due to its selenium content, Brazil nut oil has antioxidant properties.

As Food

In Brazil, in particular, Brazil nut oil is used as an edible oil. Unfortunately, it becomes rancid easily and does not keep long, which is why it is not particularly well suited as an edible oil (Schuette et al. 1930).

The nutty taste of Brazil nut oil makes it suitable for refining cakes, biscuits and other bakery products.

6 Possible Unwanted Side Effects

In particular, persons suffering from allergies to other nuts, for example peanuts, hazelnuts or walnuts, may be affected by allergies in consequence of ingestion of or contact with Brazil nuts. The most frequent reaction is atopic dermatitis, followed by itching, nausea, diarrhoea, and swelling of the larynx; even anaphylactic shock is possible. Cross-reactions may occur with rapeseed, cocoa beans, sunflower seeds and sesame seeds (Markson 1942).

Since Brazil nuts contain 0.24–0.26% barium, there may be toxic effects. Barium is an alkaline earth metal and poisonous in the form of soluble barium salt. The barium ion causes excitation and contraction of the smooth, skeletal and heart muscles. Oral administration results in diarrhoea with vomiting, vertigo and an increase in blood pressure. If the poisoning makes rapid progress, symptoms like ascending paralysis and finally apnoea follow. If patients are given artificial respiration, they may die of cardiac complications. 2–4 g barium chloride or barium carbonate are regarded as a lethal dose (List and Hörhammer 1973, 4th edition, 3rd volume, p. 427; Burger and Wachter 1993, 7th edition, p. 186).

Cactus Pear Seed Oil

synonyms: Kaktusfeigenöl (D); prickly pear seed oil (E)

1 Source Plant

Opuntia ficus-indica L. *(Cactaceae)*, cactus pear, Indian fig opuntia.

S. Krist, *Vegetable Fats and Oils*, https://doi.org/10.1007/978-3-030-30314-3_24

Habitat

The cactus pear originates in Mexico. In the early seventeenth century, it was imported to the Mediterranean area and South Africa. In South Africa, the conditions were good and it spread rapidly, soon developing into a pest (Lieberei and Reisdorff 2007, p. 217).

Description

The cactus pear grows up to 5 m high. Its stem parts are oval and flat, with a maximum length of 50 cm and a width of up to 20 cm. On the flattened stem parts sit glochids, which consist of many fine, prickly bristles with barbs; they easily detach from the plant and lodge in the skin. *Opuntia* is bred either without thorns or with one or two small, whitish-yellow thorns on the glochids. The yellow, perfect flowers sit on the upper edges of the stem parts. The fruit is an orange- or red-coloured berry with a thin, smooth peel, which also bears glochids and thorns. The juicy fruit pulp lying beneath is about 1 cm thick. The pulp surrounds many black seeds, which are flat, oval, and about 5 × 3.5 mm in size. If the fruit is eaten raw, it is cut lengthwise, and the fruit pulp, pulp and seeds are spooned out (Nowak and Schulz 1998, p. 214).

The fruit consists of 15% seeds, which in turn contain 98.8 g/kg oil, and of 85% pulp containing 8.7 g/kg oil. For this reason, it is the seeds that are used as a source of oil, even though the pulp contains lipids as well (Ramadan and Mörsel 2003a, b [1]).

Opuntia ficus-indica tolerates dry conditions, does not make demands on the soil and uses its water supply efficiently. The oil content of the seeds is relatively low, but due to the ideal composition of fatty acids, it is nevertheless potentially useful for the health-care market (Labuschagne and Hugo 2010).

2 Cultivation and Extraction

Cultivation

The cactus pear grows wild in dry and semi-dry regions. Vegetative propagation is possible via cuttings. The fruit are harvested shortly before or when they are ripe (Nowak and Schulz 1998, p. 215). In South Africa, the cultivation of cactus pears is increasing, as the costs are minimal and the plant is undemanding (Labuschagne and Hugo 2010).

The fruit of the cactus pear are versatile, and several million kilograms of seeds accumulate as a by-product of processing. Due to the problem of waste disposal, there is considerable interest in using the seeds, which are a source of high-quality oil. If conditions are ideal, the annual production of cactus pears can reach 50 t/ha (Ramadan and Mörsel 2003a, b).

Extraction of the Oil

When the fruit are ripe, they are selected according to their colour, the pH value of the fruit pulp (ideally 6.05) and the total titratable acid (ideally 0.39%). The sseds are separated by pressing the fruit pulp. They are then lyophilised and subsequently ground. The oil is extracted with chloroform/methanol (Ramadan and Mörsel 2003a, b [1]).

The seeds can also be cold pressed after they have been separated, washed and dried in the sun.

3 Character

Colour

Light yellow (Coskuner and Tekin 2003).

Flavour

Characteristically mild (Coskuner and Tekin 2003).

4 Ingredients

Composition of Fatty Acids

Fatty acids	Content in g/kg acc. to Coskuner and Tekin (2003)	Content in g/100 g acc. to Ennouri et al. (2005)	Content in % acc. to Labuschagne and Hugo (2010)	Content in % acc. to Ramadan and Mörsel (2003a, b) [1]
Lauric acid	–	–	–	–
Myristic acid	1.3–1.9	–	–	–
Palmitic acid	132.1–156.0	9.32	11.44–15.89	20.1
Palmitoleic acid	14.4–18.5	1.42	0.36–1.01	1.80
Stearic acid	33.1–47.9	3.11	2.80–4.01	2.72
Oleic acid	210.5–256.0	16.8	12.38–16.51	18.3
Linoleic acid	522.5–577.6	70.3	61.40–68.92	53.5
Linolenic acid	2.9–9.7	–	0.49–0.73	2.58
Arachidic acid	4.2–6.6	–	–	–
Behenic acid	2.1–3.0	–	–	–

C

Composition of Sterols

Sterols	g/kg acc. to Ramadan and Mörsel (2003a, b) [1]
Total content	–
Cholesterol	–
Ergosterol	–
Campesterol	1.66
Stigmasterol	0.30
Lanosterol	0.28
β-Sitosterol	6.75
Δ5-Avenasterol	0.29
Δ7-Stigmasterol	–
Δ7-Avenasterol	0.05

Composition of Tocopherols

Tocopherols	g/kg acc. to Ramadan and Mörsel (2003a, b) [1]	% acc. to www.oleador.com, 20.04.2011
Total content	0.403	720 mg/l
α-Tocopherol	0.056	13.9
β-Tocopherol	0.012	3
γ-Tocopherol	0.330	81.9
δ-Tocopherol	0.005	1.2

Other Ingredients: (Ramadan and Mörsel 2003a, b [1])

β-Carotene	0.047 (g/kg)
Vitamin K_1	0.525 (/kg)

Physical Key Figures: (Ennouri et al. 2005)

solids content (%):	93.00
oil (%):	10.90
protein (%):	5.40
ash (%):	1.10
viscosity at 20 °C (Pa s):	0.0531
refractive index at 20 °C:	1.475
density at 20 °C:	0.903
saponification number:	169.0
iodine number:	101.5

5 Application

In Pharmacy and Medicine

Cactus pear seed oil has positive effects on the cholesterol level. It lowers the total cholesterol, LDL and VLDL, but has no effect on HDL. The effect is probably due to β-sitosterol. It is assumed that phytosterols cause a decrease in the solubility of cholesterol and in the absorption through the intestinal barrier. The consequence is lower plasma cholesterol values. Phytosterols also have a protective function against arteriosclerotic lesions or at least delay them. The β-carotenes and vitamin E in cactus pear seed oil protect against structural changes in the lipoproteins (Ennouri et al. 2006).

A current study by Ennouri et al. on rats asserted that a dietary supplement of 25 g/kg^{-1} cactus pear seed oil administered over a period of 9 weeks led to a significant decrease in the serum glucose concentration and glycogen of the liver. On the other hand, the glycogen value in the muscles increased. The authors of the study ascribe this to an increase in insulin secretion, as insulin stimulates the transformation to glycogen in the skeletal muscles and the liver (Ennouri et al. 2006).

In Cosmetics

The oil of the cactus pear is used frequently in cosmetics. Since it contains a high amount of tocopherols, it acts as an antioxidant, protecting the skin from premature ageing, preventing wrinkles and tightening the skin. The skin looks healthy, and dark circles around the eyes are reduced. The oil is also suitable for haircare and nail care. It moisturises dry, brittle hair and it is thought to help in the treatment of brittle, chapped nails.

As Food

The composition of fatty acids in cactus pear seed oil is similar to that of sunflower oil and grapeseed oil. Cactus pear seed oil is a high-quality edible oil with many essential fatty acids. It is especially of economic importance in Tunisia, where traditional oil crop is difficult to cultivate (Ennouri et al. 2005).

Capsicum Seed Oil

synonyms: Paprikakernöl (D)

1 Source Plant

Capsicum annuum L. *(Solanaceae)*, pepper.

S. Krist, *Vegetable Fats and Oils*, https://doi.org/10.1007/978-3-030-30314-3_25

Habitat

The wild varieties of the species *Capsicum* originate in South and Central America. Of the five varieties cultivated today, *Capsicum annuum* is the one used most often (Teubner 1993, p. 15 et seqq). *Capsicum* is cultivated in tropical and subtropical countries in Asia, Africa, South and Central America and Southern Europe (Krishna De 2003, p. 5), especially in Hungary, Spain, southern France, Italy, Turkey and Bulgaria (Hunnius 1998, p. 267).

Description

Peppers are mostly annual, upright plants that can grow up to 1.5 m tall. The leaves are broad lanceolate to oval-elliptical. The flowers are solitary, pendant and have a calyx with characteristically short lappets; they are bell- to star-shaped and white. The fruit is a usually pendant, longish to cylindrical, cavernous berry that tapers at the top and contains a great quantity of seeds. Unripe, it is green, and its colour changes to yellow or red when ripe. The light-coloured seeds are kidney- or disc-shaped and 2.5–5 mm in size (Teubner 1993, p. 15 et seqq).

2 Cultivation and Extraction

Cultivation

Peppers are mainly grown in India, Nigeria, China, Indonesia and Korea (Krishna De 2003, p. 5). The plant needs sunny locations, warmth, moisture and soil that is rich in humus.

Extraction of the Oil

The seeds are removed from the fruit, dried for 1 week, ground and then the oil is extracted (El-Adawy 2001). In Korea, the kernels are roasted first, and the oil is extracted by means of a mechanical press (Jung et al. 1999).

3 Character

Colour

Golden yellow oil (Marion and Dempsey 1964).

Odour

Weak pepper odour (Marion and Dempsey 1964).

When the seeds have been roasted before extraction: reminiscent of nuts or peanut butter (Jung et al. 1999).

Flavour

Spicy, pleasantly nutty (Jung et al. 1999).

4 Ingredients

Capsicum seeds contain about 24.43% protein, about 25.61% oil and about 34.91% fibre (El-Adawy 2001).

Additionally, they contain about 0.97% hydrocarbonates, 94.16% triglycerides, 1.48% free fatty acids, 0.89% sterols, 0.30% diglycerides, 1.35% monoglycerides and 0.85% phospeolipids (El-Adawy 2001).

Composition of Fatty Acids

Fatty acids	Content in % acc. to El-Adawy (2001)	Content in % acc. to Krstiae et al. (2001)	Content in % acc. to Firestone (1999), p. 86	Content in % acc. to Marion and Dempsey (1964)	Content in % acc. to Tarandzhiiska et al. (1991)
Myristic acid	–	–	0.3	0.2	0.3
Palmitic acid	13.84	15.42	11.4	11.3	11.4
Palmitoleic acid	0.12	–	0.3	0.5	0.3
Hexadecanoic acid	–	–	–	0.2	–
Margaric acid	–	–	0.1	0.1	0.1
Stearic acid	3.71	3.92	2.5	4.4	2.5
Oleic acid	14.56	12.80	8.9	14.8	8.9
Linoleic acid	67.77	61.00	75.8	67.4	75.8
α-Linolenic acid	–	–	0.3	0.2	0.3
Arachidic acid	–	–	0.1	0.5	0.1
Behenic acid	–	–	0.1	0.2	0.1
Lignoceric acid	–	–	0.2	–	0.2

Physical Key Figures of Capsicum Seed Oil

density: d^{25} 0.912 ± 0.007

refractive index: n_D^{25} 1.4715 ± 0.001

saponification number: 168 ± 1.90
iodine number: 131 ± 3.49
unsaponifiable: 2.3%
point of solidification: −14 °C
(El-Adawy 2001; Roth and Kormann 2000, p. 163)

5 Application

Capsicum seed oil is a very nutritious oil that lowers the cholesterol level in the serum because of its high amount of unsaturated fatty acids. It is a good edible oil and salad oil and is also used to produce margarine. Furthermore, it can be a substitute for other oils with a high amount of unsaturated fatty acids (El-Adawy 2001).

In Korea and other countries, the oil is used like sesame oil or perilla oil for seasoning, as the seeds are roasted before extraction (Jung et al. 1999).

Caraway Seed Oil

Oleum Carvi synonym: Kümmelöl (D)

1 Source Plant

Carum carvi L. (Apiaceae), caraway.

S. Krist, *Vegetable Fats and Oils*, https://doi.org/10.1007/978-3-030-30314-3_26

Habitat

Caraway originates from Eurasia. The plant is cultivated especially in Poland, the Netherlands, Egypt and Central Europe. *Carum carvi* is a relatively undemanding and robust plant that can be cultivated at high altitudes. It usually grows in meadows, neglected grassland and by the roadside, and prefers sunny locations (Hager 1978, volume 3, p. 727).

Description

Carum carvi is a biennial herb growing to a height of 75 cm. In the first year, the plant only develops a rosette of leaves. It bears flowers and fruit in the second year. Caraway has a spindle-shaped root and an upright, branched, angular, furrowed stem. The leaves are doubly pinnate, with feathery leaflets; the leaf tips are arranged in whorls. The flowers are small and white, often pink in mountainous regions, with inversely heart-shaped petals. They are arranged in terminal umbels with seven to ten rays, without a perianth. The fruit is a bipartite schizocarp that appears compressed when viewed from the side. It is 3–6.5 mm long and 1–1.5 mm thick, and brown or greyish brown. The five noticeably protruding ribs are typical. When ripe, the fruit disintegrates into two bow- or crescent-shaped fruitlets with a characteristic smell (Hager 1978, volume 3, p. 728).

2 Cultivation and Extraction

Cultivation

Although caraway is nearly ubiquitary, it is not often cultivated. In small quantities, it is cultivated in England, the Netherlands, Germany, Austria, Finland, Norway, Egypt and Morocco, but also in the USA (California, Idaho), Argentina and Chile (Hager 1978, volume 3, p. 727).

Extraction of the Oil

The fat oil of caraway is obtained either by pressing or extraction (Reiter et al. 1998).

3 Character

Colour

Slightly yellow tint, greenish yellow, dark.

Odour

Carvone, aromatic-pungent.

Flavour

Intense flavour of caraway.

4 Ingredients

Composition of Fatty Acids

Please note the different specifications of Hager (1978) and Reiter et al. (1998). There is a significant shift towards petroselinic acid, which is much in demand as vegetable raw material.

Fatty acids	Content in % acc. to Hager (1978)	Content in % acc. to Reiter et al. (1998) (origin Szilas)	Content in % acc. to harvestfield.netfirms.com/04.11.03
Palmitic acid	3	4.2	3.4
Linoleic acid	31	36.3	28
Petroselinic acid	26	42.4	63 (oleic acid + petroselinic acid not differentiated)
Oleic acid	40	15.2	
Stearic acid	–	1.2	0.8
Vaccenic acid	–	0.6	–

Other Ingredients

Phytosterols
Flavonoids (>0.1%)
Coumarin (in traces)

Physical Key Figures of Caraway Seed Oil

refractive index: n_D^{35} 1.471
saponification number: 178
iodine number: 129
unsaponifiable: 2.7%
point of solidification: −7 °C
(Roth and Kormann 2000, p. 161)

5 Application

In Pharmacy and Medicine

Caraway seed oil promotes digestion and is an effective remedy for wind. Scientific studies still need to determine to what extent carvone, which is the most important ingredient of the essential oil, is contained in fat caraway seed oil and to what extent it may be responsible for the effects mentioned above.

Technical Uses

Like coriander seed oil, caraway seed oil contains petroselinic acid. It therefore has the same uses as those described for coriander seed oil, but caraway seed oil does not appear to be used to the same extent.

In the Food Industry

Since caraway seed oil contains a relatively high amount of the essential fatty acid linoleic acid and has an intense, aromatic taste, it is used increasingly in the food industry.

Carnauba Wax

Cera Carnauba synonym: Karnaubawachs (D)

1 Source Plant

Copernicia prunifera (Mill.) H.E. Moore *(Arecaceae)*, carnauba palm.

Habitat

The carnauba palm is a native plant of northeastern Brazil. It grows especially in the state Ceará, in the valleys of Assu and in Apodi, Rio Grande do Norte (Roth and Kormann 2000, p. 56).

Description

The carnauba palm is a fan palm 10–15 m tall. Its cicatricose trunk is up to 60 cm thick. The leaves are fan-like and nearly round, with a diameter of about 1.5 m; they are blue-green, frosted, and up to one third cut into stiff segments sticking out. The stems of the leaves are as long or longer than the leaf blade, and covered with thorns. The inflorescences are located between the leaves; they are triple-branched and up to 2.5 m long. The flowers are perfect, small and white. The sepals and petals are half grown together, forming a tube; the petals are slightly longer than the sepals. The three separate compartments of the carpels end in a common style. Only one of the carpels develops into a fruit, which is oval, about 2.5 cm long and 2 cm broad. The albumen of the seed is heavily fissured (Hänsel and Sticher 1992, volume 4, p. 993).

S. Krist, *Vegetable Fats and Oils*, https://doi.org/10.1007/978-3-030-30314-3_27

2 Cultivation and Extraction

Cultivation

The carnauba palm prefers sunny, hot climates, for example savannahs or open forests that are occasionally flooded. The palm is very adaptable, surviving droughts as well as floods. The carnauba palm does not require a particular soil type, and even grows in slightly salty soil. As there are many wild-growing palms, the palm is only seldom cultivated, apart from a few plantations in Brazil. As protection against transpiration, the leaves of the carnauba palm segregate the wax in delicate flakes. The harvest takes place in the dry season, from September to March; 6–8 leaves are carefully cut three times, with a time span of 2 weeks in-between cuts. The leaves are spread on mats to dry. In the drying process, they shrink, thereby releasing flakes of wax, which can then be knocked off or brushed off. The yield is highest when the trees are about 50 years old (Hänsel and Sticher 1992, volume 4, p. 994).

Extraction

Carnauba wax is the wax gathered from the surface of the leaves of *Copernicia prunifera*. The wax is subsequently cleaned (Hänsel and Sticher 1992, volume 4, p. 994).

3 Character

Colour

Good quality: light yellow (Hänsel and Sticher 1992, volume 4, p. 994).

Inferior quality: yellow to greenish-yellow (Hänsel and Sticher 1992, volume 4, p. 994).

Hard, yellowish to greyish-green pieces (Roth and Kormann 2000, p. 129).

Bleached and cleaned wax: white (Roth and Kormann 2000, p. 129).

Odour

Pungent, not unpleasant smell (Roth and Kormann 2000, p. 129).

Odourless or weak, insipid, but not rancid smell (Hänsel and Sticher 1992, volume 4, p. 994).

Flavour

Tasteless (Hänsel and Sticher 1992, volume 4, p. 994).

4 Ingredients

Composition

Main components	Content in % acc. to Cosmetic, Toiletry and Fragrance Assoc. (1984)
Aliphatic esters	38–40
ω-Hydroxyesters	12–14
ρ-Hydroxy cinnamic acid esters	20–23
Free wax alcohols	10–12
ρ-Methoxy cinnamic acid esters	5–7
Hydrocarbons	0.3–1
Carnaubadiol	0.5
Free acids	5–7

Other Ingredients

(Roth and Kormann 2000, p. 129)

Cerotic acid myricyl ester
Free carnauba acid
Cerotic acid
Cerotic acid melissyl ester
Melissyl alcohol
Ceryl alcohol

Physical Key Figures of Carnauba Wax

refractive index: n_D^{90} 1.4500
density: d_4^{20} 0.990–0.999
melting point: 82–85.5 °C
saponification number: 78–89
iodine number: 10–15
unsaponifiable: 54–55%
(Roth and Kormann 2000, p. 129)

Shelf Life

Carnauba wax has a shelf life of 3–4 years.

5 Application

In Pharmacy and Medicine

The bleached wax is used in the production of enteric-coated medicines such as pills, dragées and capsules (Cosmetic, Toiletry and Fragrance Assoc. 1984).

In dental medicine, carnauba wax is an ingredient of impressive masses; it heightens plasticity (Hänsel and Sticher 1992, 4th volume, p. 995).

In Cosmetics

Carnauba wax adds stability to make-up products such as lipsticks, kohl and mascara, even at high temperatures. Waterproof make-up also contains carnauba wax. The wax forms a waterproof film, preventing the make-up from smearing when it comes into contact with water. This property is also useful for waterproof sun-block. Producers have also discovered that carnauba wax can be used for haircare and skincare products: The wax gives shine and springiness to damaged hair and, for this reason, it is added to conditioners, hair-tinting lotions and styling products. If applied to the skin, the wax has a smoothing effect; it increases skin elasticity and is therefore added to creams and lotions, especially anti-wrinkle creams. Since it has a hard texture, it is also used in soap production (Cosmetic, Toiletry and Fragrance Assoc. 1984).

Industrial Uses

Carnauba wax is used as food additive E 903. It is a separating agent that prevents the ingredients from sticking together. In addition, it is an authorised coating material for citrus fruit and potatoes (Roth and Kormann 2000, p. 129).

If carnauba wax is added to other waxes, it decreases their tendency towards crystallisation as well as their plasticity and viscidity. This results in a higher melting point, increased firmness and more gloss. In the detergents industry, carnauba wax is added to shoe polish, floor wax and furniture wax. It is also used as a polishing wax for car finish (Cosmetic, Toiletry and Fragrance Assoc. 1984).

Other Uses

Carnauba wax is added to candles because of its solid texture; it also increases burn time (Cosmetic, Toiletry and Fragrance Assoc. 1984).

In painting, it is used for the production of dry-bright emulsions and carbon paper, and to smooth special kinds of paper, for example art prints (Hänsel and Sticher 1992, volume 4, p. 995).

Carnauba wax also forms part of records, matches and isolating material (Cosmetic, Toiletry and Fragrance Assoc. 1984).

6 Possible Unwanted Side Effects

No unwanted side effects have been observed. The use of carnauba wax is regarded as non-hazardous. External application did not result in phototoxic effects, skin irritation or allergisation. Internal use did not cause any pathological changes, but only a decrease in free fatty acids in the serum. The suspicion that the long-chain fatty acids of carnauba wax might enter the blood circulation has not been substantiated so far. On the one hand, this effect proved independent of dosage; on the other hand, the phenomenon of the lowered fatty acid level did not occur in studies on dogs as experimental animals (Cosmetic, Toiletry and Fragrance Assoc. 1984).

Carrot Oil

Oleum Dauci synonym: Karottenöl (D)

S. Krist, *Vegetable Fats and Oils*, https://doi.org/10.1007/978-3-030-30314-3_28

1 Source Plant

Daucus carota L. *(Apiaceae)*, carrot.

Habitat

C

The carrot has two subspecies: *Daucus carota ssp. carota* (wild carrot) and *Daucus carota ssp. sativus* (cultivar). Oil can be extracted from both subspecies (Kilibardia et al. 1989). The carrot is a widespread cultivated plant; wild forms grow in Europe and Asia. The exact region of origin is unknown. The vegetable is cultivated practically everywhere; there are more than 60 cultivars and hundreds of varieties. The annual harvest amounts to about 13 million tons.

Description

The carrot is a biennial plant with pinnate leaves. It grows to a height of 0.3–1 m. The white flowers appear from June to September (Schoenfelder and Schoenfelder 2001, p. 90). The roots vary in form, depending on the varieties; they can be globular, conical or cylindrical with a ligneous core, and are usually orange. Carrots are taproots; the phloem of the root serves as storage tissue. Wild carrots contain 7.42–10.83% fat oil, cultivated carrots 7.3–12.44% (Kilibardia et al. 1989). Additionally, carrots contain considerable amounts of carotene, which is responsible for their colour (Hänsel et al. 1999, p. 607). The content of carotene amounts to 54 ppm, relating to the dry weight (Belitz and Grosch 1999).

2 Cultivation and Extraction

Cultivation

Different varieties are sown at different times. New carrots can be sown as early as late March, other varieties at the end of April. Carrots are sown in rows; there should be about 15 cm distance between the plants. Since many plants germinate, they must be thinned out so that one plant remains every 2–3 cm. Carrots need a regular water supply in summer and manuring in spring. If manuring is not desirable, it is sufficient to enrich the soil with compost in spring.

Extraction of the Oil

Carrot oil can be produced in several ways. Carrot extract can be mixed with soya bean oil, or carrot seeds can be extracted with hexane (Anon. 1967; Roth and Kormann 2000, p. 162; Kilibardia et al. 1989). After extraction, the hexane is removed via vacuum distillation (Anon. 1967).

3 Character

Colour

If carrot extract is mixed with soya bean oil, the resulting carrot oil is clear and a deep-red colour.

The extracted oil is a yellowish olive-green colour (Anon. 1967).

Odour

Ethyl acetate, smell of grated carrots.

Flavour

Neutral in the case of mixed oil; bitter in the case of extracted oil (Anon. 1967).

4 Ingredients

The following tables refer to extracted carrot oil:

Composition of Fatty Acids

Fatty acids	Content in % acc. to Parker et al. (2003)	Content in % in wild carrots acc. to Kilibardia et al. (1989)	content in % in cultivated carrots acc. to Kilibardia et al. (1989)	Content in % acc. to Roth and Kormann (2000), p. 162
Lauric acid	–	1.56	2.34	–
Capric acid	–	2.08	6.08	–
Palmitic acid	3.71 ± 0.05	3.75	3.81	4
Palmitoleic acid	–	0.31	0.57	–
Stearic acid	0.42 ± 0.38	0.41	0.33	–
Petroselinic acid	–	76.25	64.08	58
Linoleic acid	13.19 ± 1.30	14.37	10.75	24
г-Linolenic acid	–	0.62	–	–
α-Linolenic acid	0.28 ± 0.16	0.31	–	–
Arachidic acid	0.33 ± 0.02	0.31	–	–
Myristic acid	–	–	12.02	–
Oleic acid	82.08 ± 1.41	–	–	14

The considerable differences between carrot oil from *Daucus carota* and carrot oil from *Daucus carota sativus are conspicuous.*

Composition of Sterols

(content in %)

Sterols	Oil from Daucus carota ssp. carota acc. to Kilibardia et al. (1989)	Oil from Daucus carota ssp. sativus acc. to Kilibardia et al. (1989)
Unsaponifiable portion	8.11	19.17
Total share of sterols in the unsaponifiable portion	1.47	0.88
Stigmasterol	12.6	23.1
β-Sitosterol	35.4	67.0
Δ5-Avenasterol	19.6	9.8
Δ7-Avenasterol	32.2	–

Composition of Tocopherols

Extracted carrot oil has a tocopherol content of 0.5–1.0% (Heftmann 1947).

Other Ingredients

(Tung et al. 1981; Kilibardia et al. 1989)

β-Carotene 0.32%
β-Bisabolene
Squalene
Fibres
Mineral nutrients

Physical Key Figures of Carrot Oil

(Kilibardia et al. 1989; Roth and Kormann 2000, p. 162)

Carrot Extract Mixed with Soya Bean Oil

refractive index: 1.4760
relative density: 0.9195

Extracted Carrot Oil

refractive index: 1.472–1.477
relative density: 0.9114–0.9413
iodine number: 104.82–110.32

saponification number (wild carrot): 170.63–175.08
(cultivated carrot): 210.45–210.76
acid number (wild carrot): 4.51–4.59
(cultivated carrot): 5.83–5.52
point of solidification: −6 °C
unsaponifiable: 1.5%

5 Application

In Pharmacy and Medicine

Carrot seed oil can be used as a diet oil because of its high amount of monounsaturated oleic acid. It has been observed that a diet rich in monounsaturated fatty acids is a good alternative to a low-fat diet; it lowers cholesterol levels in the blood and supports the immune function. Monounsaturated fatty acids reduce the risk of arteriosclerosis more effectively than polyunsaturated fatty acids because they prevent local accumulation of oxidised LDL in the aorta, which constitutes a risk factor and a possible trigger for arteriosclerosis and narrowing of the blood vessels (Parker et al. 2003).

In Cosmetics

Carrot oil is part of skin oils and emulsions for dry, chapped and flaky skin and scalp. It is also suitable for ageing skin and can be added to sunblock and after-sun lotions. It is thought to improve cell division, reduce wrinkle depth, increase skin metabolism and regulate keratin formation.

Industrial Uses

Extracted carrot oil is used to colour food (Anon. 1967).

6 Possible Unwanted Side Effects

Carrot oil tints the skin and should only be applied in diluted form mixed with another carrier oil (10–15%), as different parts of the skin have different absorption capacities, and the skin may thus become blotchy. In experiments with mice, the administration of 0.05 ml/kg reduced fertility and led to miscarriages. This result is ascribed to β-bisabolene (Tung et al. 1981).

Cashew Oil

Oleum Anacardiae synonyms: Kaschuöl, Acajouöl (D); huile d'acajou (F)

1 Source Plant

Anacardium occidentale L. *(Anacardiaceae)*, cashew tree, *acajú*.

S. Krist, *Vegetable Fats and Oils*, https://doi.org/10.1007/978-3-030-30314-3_29

C

Habitat

The tropical cashew tree is a native plant of north-eastern Brazil; the Portuguese transferred it to Africa and India in the sixteenth century, where it is still common today and is cultivated in large plantations. Today, it grows in South and Central America as well as in Asia (Roth and Kormann 2000, p. 38).

Description

Cashew trees are bushes or trees similar to walnut trees, and grow up to 12 m tall. The leaves are elongated, oval, leathery, rough and 15–20 cm long. The small flowers are pale green with red stripes, arranged in terminal panicles. The kidney-shaped, one-seeded nuts grow on pear-shaped, fleshy fruit stalks, the so-called cashew apples, which are a red or yellow colour when they are ripe (Roth and Kormann 2000, p. 38). Between the skins of the receptacle, there is a pungent black oil, which must be removed during the process of roasting (brasilien.de/20.06.06). The seeds contain 45–60% fat oil, the proper cashew oil. In addition, they contain protein (10–29%), saccharose (approx. 6%), starch (9–20%), nitrogen compounds (8–23%), cellulose (0.5–4%) and water (3.5–5%). The oil fruit is kidney-shaped, compressed, and weighs 4–5 g (Roth and Kormann 2000, p. 112, 117).

2 Cultivation and Extraction

Cultivation

Today, the cashew tree is mainly cultivated in India, Mozambique and Brazil, generally in tropical rainforest regions with 500–3500 mm rain/year. Under ideal conditions, the tree bears up to 70 kg nuts each year. It is propagated from seed, vegetatively from slips, or – in the case of especially productive species – by grafting (Roth and Kormann 2000, p. 38 et seq). The plant grows in poor, sandy soil and is hardened against salty winds from the sea (Behrens 1996, p. 2). The global production of cashew nuts is thought to amount to 0.38 Megatons. World Bank Data[1] estimates that 97% of the harvest is from wild trees, and that only 3% originates from plantations (Axtell and Fairman 1992, p. 41 et seq).

Extraction of the Oil

The ripe cashew apples are either plucked from the tree or – to avoid picking unripe fruit – collected from the ground. As the nuts decay quickly, they are picked within a week when the weather is fine, but when the weather is rainy, they are picked every day (Axtell and Fairman 1992, p. 41 et seq).

[1] World Bank Group, supports developing countries and collects data.

When picked, the nut remains with the cashew apple, which is removed with a twisting movement of the hand, together with any remains of the apple. Directly after the harvest, the nuts are dried in the sun for several days, until they have a moisture content of 8%, or until the kernel can be heard inside the nut when it is shaken. After drying, the seeds are stored in sacks or as bulk goods for a few days until they are processed (Axtell and Fairman 1992, p. 41 et seq). The shell of the fruit becomes brittle by gentle roasting and can then be removed easily. The oil is extracted by squeezing the seeds (Roth and Kormann 2000, p. 112, 117).

3 Character

Colour

Light-yellow oil, liquid at room temperature, non-drying.
(Roth and Kormann 2000, p. 112, 117).

Odour

Odourless (Roth and Kormann 2000, p. 117).

Flavour

Mild taste (Roth and Kormann 2000, p. 117).

4 Ingredients

Composition of Fatty Acids

Fatty acids	Content in % acc. to Rossell and Pritchard (1991), p.293	Content in % acc. to Firestone (1999), p.22	Content in % acc. to Gallina Toschi et al. (1993)
Myristic acid	Traces	–	–
Palmitic acid	11	4–17	9.0–14.2
Palmitoleic acid	0.5	0.3–0.5	0.3–0.4
Heptadecanoic acid	–	0.1	0.1–0.2
Stearic acid	8	2–11	6.3–11.6
Oleic acid	61	57–80	57.3–65.1
Linoleic acid	19	16–22	15.6–18.6
α-Linolenic acid	0.3	0.1–0.3	Traces
Arachidic acid	0.3	0.3–0.8	0.3–0.8
Eicosenoic acid	Traces	–	–

Composition of Triglycerides

Triglycerides	Content in % acc. to Firestone (1999), p. 138	Content in % acc. to Gallina Toschi et al. (1993)
PSO	3–6	3.1–5.9
SOS	2–3	2.0–3.2
PPO	2–5	2.2–4.6
POO	15–19	15.2–18.6
SOO	11–12	10.6–12.1
PPL	1–2	1.0–2.2 (PLP + POPo)
OOO	19–29	19.0–29.4
POL	8–11	8.3–10.6
SOL	4–5	3.5–4.9
OOL	12–17	11.8–16.7
PLL	2–3	1.7–2.6
OLL	3–5	3.6–5.1
LLL	0.5	0.4–0.5

Composition of Sterols

Sterols	Content in % acc. to Gallina Toschi et al. (1993)
Cholesterol	0.3–1.3
Campesterol	6.0–6.8
Stigmasterol	0.1–0.3
β-Sitosterol	76.2–82.7
Fucosterol	0.6–0.9
Δ + 5-Avenasterol	6.4–10.6
Others	2.1–4.9

Composition of Tocopherols

Tocopherols	Content in mg/100 g acc. to Gallina Toschi et al. (1993)
α-Tocopherol	2.8–8.2
γ-Tocopherol	45.3–83.5
δ-Tocopherol	2.0–5.9

Physical Key Figures of Cashew Oil

density: d_4^{15} 0.911–0.918
refractive index: n_D^{40} 1.4623–1.4633
saponification number: 187–200
iodine number: 77–89
60–89 (According to Mensier)
unsaponifiable: 0.4–1.5%
(Roth and Kormann 2000, p. 117)

Characteristics of the Oil

Cashew nuts contain two different kinds of oil. In the fruit skin, between the pericarp and the nut, is the so-called cashew nut shell liquid (CNSL), a poisonous, vesicatory, viscous, brownish-black liquid, which contains anacardic acid and a phenol compound called cardol (Roth and Kormann 2000, p. 39).

Hot Oil Bath-Method

CNSL is a byproduct in cashew oil production and is usually extracted via the "hot oil bath"-method. The roasted nuts are immersed in a hot oil bath with a temperature of 192 °C for about 90 seconds. About 25% of the CNSL flows into the bath. The CNSL remaining in the outer skin is removed using sawdust (Behrens 1996, p. 5 et seqq).

Extraction with Supercritical Gas

Another method to remove the cashew nut shell liquid is to extract it with supercritical CO_2 (Smith Jr. et al. 2003). Natural CNSL contains 74.1–90% anacardic acid, 10–20.1% cardol, 1.2–9.2% cardanol and 1.7–2.6% 2-methyl cardol. It is used as a wood preservative against insects (raw CNSL), as germicide, fungicide, insecticide, as a developing agent in photography, in paint and varnish, in resin, as a waterproof ingredient of chemically resistant cement, in brake linings, in filter paper or as a raw material source for cardanol[2] (Behrens 1996, p. 6 et seqq). In Japan, CNSL is used as an oil or extract in topical preparations to treat acne (Shimomura and Nakamura 1991). In Egypt, it is also used to treat acne, as well as in pharmaceutical products to treat dandruff and psoriasis (Abd Elreheim 2002).

5 Application

As Food

The average content of unsaturated fatty acids, given in percent of the total content of fatty acids in cashew oil, is 78.9%, of which 99.6% is oleic acid and linoleic acid. Both are nutritionally important. Oleic acid is one of the fatty acids that metabolise most quickly, and linoleic acid is a precursor of prostaglandins and thus an essential fatty acid. Its oxidative stability makes cashew oil better suited for cooking and baking than, for example, peanut oil (Gallina Toschi et al. 1993). "Cajuna", a low-alcohol wine, is produced from cashew nuts.

[2] Cardanol is a phenolic compound formed out of anacardic acid by decarboxylation in the commercial production of CNSL, and after extraction via the "hot oil bath"-method has a percentage share of 60–65%.

In Cosmetics

In Japan, the oil from the seed and extracts containing cardol are used in cosmetics to bleach the skin, and as a dentifrice (Koizumi and Shimomura 1991).

In Industry

CNSL is used in brake linings, waterproof components, preservative agents, and to produce paint and synthetic materials (Axtell and Fairman 1992, p. 40).

6 Possible Unwanted Side-Effects

Precancerous Diseases

A study on workers of a cashew oil-producing factory in Kerala (India) showed an increase in the risk of oral precancerous diseases, such as fibrosis of the oral submucosa. For workers coming in contact with the oil, the incidence of fibrosis or oral cancer is 7.85%, whereas it is only 1.12% for workers who do not come into contact with the oil (Rajendran et al. 1988).

CNSL

Cashew nut shell liquid (CNSL) is found in the mesocarp of the cashew shell. It accounts for about 15–30% of the nut's weight. It is a viscous, oily or balmy substance with a weight density of 1.013 g/cm^3. Its colour varies from light yellow to dark brown; it tastes bitter and is caustic (Behrens 1996, p. 6 et seqq).

Castor Oil

Oleum Ricini virginale synonyms: Ricinusöl (D); castor seed oil (E); huile de ricin (F)

1 Source Plant

Ricinus communis L. *(Euphorbiaceae)*, ricinus, castor bean.

S. Krist, *Vegetable Fats and Oils*, https://doi.org/10.1007/978-3-030-30314-3_30

Habitat

Ricinus has its origins in the tropical regions of India and Africa, but is now cultivated in many countries as an ornamental or an economic plant. In the Mediterranean area, *Ricinus* often grows wild by the roadside and on waste disposal sites. In Central Europe, *Ricinus* is grown as an attractive garden plant. The seeds often form part of Indian and African necklaces.

Description

The size of *Ricinus communis* depends on the climate; in temperate zones, the plant can reach a height of about 1–2 m; in the tropics, it can grow to 13 m tall. The stem is branched. It is a greenish to reddish colour, often frosted blue. The leaves are up to 1 m broad, palmate, and of the same colour as the stem. They are long-stemmed, 5- to 11-lobed, dented and alternate. The male flowers are arranged in dense clusters below the terminal petiolate female flowers. The side shoots often surmount them. The fruit are a fluorescent-red colour; they are 2.5 cm thick and spiked. Wild forms of *Ricinus* eject the seeds when the achenes are ripe. The fruit of today's cultivated varieties do not split open. The seeds, of which there are three in each fruit, are the size of beans. They are oval, mottled, and up to 17 mm long (Hänsel et al. 1999, p. 248).

2 Cultivation and Extraction

Cultivation

Ricinus communis is cultivated as an ornamental plant as well as for economic purposes. It is a plant of the tropical regions of summer rain and needs relatively high temperatures (20–25 °C) to thrive. The duration of growth amounts to about half a year. The plant prefers a high amount of precipitation, but cultivation is also possible in relatively dry conditions. *Ricinus* is grown in all tropical countries for oil production, and also in India, the USA, the Mediterranean area, South Africa and Mexico. The main producers are India and Brazil (Roth and Kormann 2000, p. 147 et seq).

Extraction of the Oil

Castor oil is cold pressed from the peeled seeds of *Ricinus communis*. To ensure that the toxic ricin remains in the press cake, the oil is degummed, neutralised and steamed. This refinement guarantees oil free of ricin for therapeutical use (Kubelka and Länger 2001, p. 394; Hunnius 1998, p. 1190; Hänsel et al. 1999, p. 248).

3 Character

Colour

Colourless, sometimes slight yellow tinge.

Clear, nearly colourless to slightly yellow, viscous liquid (Europäisches Arzneibuch 2002).

Odour

Smell of wet paper, slightly herbaceous, slightly woody, reminiscent of the oil coating of paintings.

Weak but characteristic (Hänsel et al. 1999, p. 248).

Nearly odourless (Hunnius 1998, p. 1190).

Volatile Compounds

SPME-GC-MS analysis (Krist et al. 2008).

Acetous acid	Heptanal
Octane	Alpha-pinene
Hexanal	t-2-heptenal
Valeric acid	Limonene
Styrene	1,8-cineol
Propenoic acid butyl ester	2-hexylfurane

Flavour

Nearly flavourless (Hunnius 1998, p. 1190).

Characteristic taste (Roth and Kormann 2000, p. 147 et seq).

4 Ingredients

Composition of Fatty Acids

Fatty acids	Content in % acc. to Roth and Kormann (2000), p. 148	Ccontent in % acc. to Kerschbaum and Schweiger (2001), p. 25
Palmitic acid	–	1.6
Linoleic acid	3	5.0
Ricinoleic acid	77	82.7
Oleic acid	9	3.6
α-Linolenic acid	–	0.4
Vaccenic acid	–	0.7
Stearic acid	3	1.5
Arachidic acid	–	0.1
Eicosenoic acid	–	0.3

Composition of Sterols

Sterols	Content in % acc. to Firestone (1999), p. 25	Content in % acc. to Karleskind (1996), p. 216
Cholesterol	–	0.1
Brassicasterol	–	0.1
Campesterol	10	10.3
Stigmasterol	22	21.8
β-Sitosterol	44–56	52.9
Δ5-Avenasterol	11–21	12.6
Δ7-Stigmasterol	0–2	1.6
Δ7-Avenasterol	1	0.5

Composition of Tocopherols

Tocopherols	Content in mg/100 g acc. to Kerschbaum and Schweiger (2001), p. 36
Total content	70.0
γ-Tocopherol	37.7
Δ-Tocopherol	32.3

Physical Key Figures of Castor Oil

refractive index: n_D^{25} 1.4731–1.4754
density: d_{15}^{15} 0.922–0.938
saponification number: 186–203
iodine number: 140–150
unsaponifiable: 0.5–1.5%
melting point: −5 °C
point of solidification: −13 to −20 °C
(Roth and Kormann 2000, p. 148)

Shelf Life

Castor oil has a shelf life of 6–8 months.

5 Application

In Pharmacy and Medicine

As an Oil for Injections

Castor oil is most frequently used to administer lipophilic medicine that cannot be taken perorally. In particular, sexual hormones with a repository effect are offered as oily solutions (ampullae intramuscular application). Castor oil is thus an important auxiliary material for pharmaceutical preparations (Hänsel et al. 1999, p. 248).

For Eye Drops

Sterilised castor oil is often added to oleaginous eye drops, in which case it must meet the requirements of parenteral preparations. Due to the higher viscosity of oily solutions, the ingredient solution does not flow off as quickly as watery solutions after it has been put into the eye, so that there is a deposit effect. In addition, oily solutions are less suitable as nutrient media for microorganisms than watery solutions, so that it is not necessary to add preservatives. The disadvantage of oily solutions is that they temporarily impair vision (Hänsel et al. 1999, p. 248).

As a Solvent

The polarity of *Oleum Ricini* is higher than that of other oils, which makes it a good solvent for many medicines (Hager 1978, volume 7b, p. 203).

As a Laxative

Castor oil has been used as a laxative since the Middle Ages. Like all oils, it causes rapid emptying of the gall bladder and, as a consequence, an increase in peristaltic contractions of the small intestine. Due to the interaction with bile salts and pancreatic lipase in the process of fat splitting in the small intestine, unesterified ricinoleic acid is generated, which is responsible for the effect. Castor oil is thus not effective in cases of insufficiency of the physiological digestion of fats. Ricinoleic acid is a laxative with an anti-absorptive effect, and a secretagogue impeding the absorption of water and ions. The effect depends on the dosage and begins 2–8 hours after administration. There are molecular-pharmacological indications that these effects are due to an impeding of adenine nucleotide transferase and a stimulation of the biosynthesis of prostaglandins (Hänsel et al. 1999, p. 248 – Kubelka and Länger 2001, p. 394).

Topically

Castor oil easily penetrates the intercellular areas of the callus. External application is useful in the treatment of dandruff, age spots and haemorrhoids, and to promote scarring. In dermatology, it is used as an emollient. *Oleum Ricini also* forms a mechanical barrier to water and hydrophilic harmful substances, which is why cracks and fissures heal better underneath the protective layer provided by the oil (Hänsel et al. 1999, p. 248; Hager 1978, volume 7b, p. 201; Roth and Kormann 2000, p. 148).

As a result of its solubility in ethanol, castor oil is used as a fat additive to formulations against dandruff and other external alcoholic products (Hunnius 1998, p. 1190; Roth and Kormann 2000, p. 148; Kerschbaum and Schweiger 2001, p. 26; Hager 1978, volume 7b, p. 201).

In Cosmetics

Due to the high viscosity of castor oil combined with its solubility in ethanol, the oil is well suited to be added to cosmetic products. It is found in the following preparations (Roth and Kormann 2000, p. 148):

brilliantine;
hair-care products;
care products for eyelashes; and
bath oils with emulsifiers to treat dry skin.

Technical Uses

Due to its constant viscosity, which is largely independent of the temperature, castor oil is used as a lubricant for engines, for example jet engines, or as a brake fluid. Furthermore, it is used in the production of sebacic acid, which is used in the production of emollients. The oil is an important raw material for polyamides, polystyrenes, lubricants and emollients. Castor oil is also used in the production of Turkey red oil, which is a mordant in dye works (Hänsel et al. 1999, p. 248; Roth and Kormann 2000, p. 148).

In Folk Medicine

Castor oil is thought to have the following effects (Ulmer 1996, p. 87 et seqq):

- loosening stiffened arthritic joints;
- stimulation of lymph glands;
- helping to treat spasms and obstructions in the abdomen;
- a positive influence on colitis and ulceration if applied as a hot oil poultice;
- strengthening the immune system;
- helping to treat herpes; and
- a soothing effect in cases of mastitis.

6 Possible Unwanted Side Effects

Although castor oil is a well-known laxative, it is not an entirely unproblematic one. High doses may cause stomach irritation, nausea, vomiting, colic and violent diarrhoea. Individual cases of allergic skin reactions have been observed. The oil should not be applied during pregnancy or lactation, as a high dose may induce labour (Hänsel et al. 1999, p. 248; Kubelka and Länger 2001, p. 394).

Castor oil has been used in so-called “labour-inducing cocktails” to hasten birth since antiquity, and is still used for this purpose today (Gilad et al. 2018).

The partaking of castor oil activates histamines, which induce a prostaglandin synthesis E2 (PGE2), which in turn leads to contractions of the colon and uterus. It seems that the activation of the prostaglandin receptor EP3 in muscles of the uterus and colon caused by the ricinoleic acid in the oil is responsible for this (Tunaru et al. 2012). How much PGE2 a woman releases seems to differ individually and depends on the dose. Caution when using castor oil is thus recommended.

In addition, the following interactions should be noted:

The absorption of liposoluble vitamins is inhibited; increased loss of potassium may intensify the effects of cardiac glycosides. The effect of the oil may be minimised by concurrent administration of antihistamines. Care must be taken with *Ricinus* seeds: They contain the glycoprotein ricin, which is absorbed from the bowels into the body cells, where it inhibits ribosomal protein synthesis. The cell dies as a consequence. Ricin is one of the most potent poisons. Six chewed seeds are sufficient to cause the death of a child (Kubelka and Länger 2001, p. 394).

Catappa L. Oil

Oleum Terminaliae synonyms: Talisayaöl, Indisches Mandelöl (D); Indian almond oil (E); huile de bandanier (F); *Terminalia catappa* seed oil (E); *Terminalia catappa* L. oil (E)

1 Source Plant

Terminalia catappa L. (Combretaceae), Bengal almond, sea almond.

S. Krist, *Vegetable Fats and Oils*, https://doi.org/10.1007/978-3-030-30314-3_31

Habitat

The origins of the Bengal almond lie in South-East Asia (India, Indonesia) (flora-toskana.de/10.10.06). Today, it is also cultivated in northern Australia, Polynesia, the Indian subcontinent, in East and West Africa, in Madagascar and in the lowlands of Central and South America (Hager 1994, p. 918).

C

Description

The Bengal almond is a deciduous tree that is 10–25 m tall. In form, it resembles a pagoda. Its trunk reaches a diameter of up to 1.5 m and has a greyish-green, dark, cracked bark. The young branches stand in whorls. They are thickened and densely covered with hair, which they lose later. The alternate leaves sit on short stems. They are oblanceolate to elliptic, 8–38 cm long, 5–24 cm broad, and parchment-like to leathery. The leaf surface is shiny, almost hairless, and covered with small warts. The Bengal almond has mostly male flowers, and only a few bisexual flowers in the lower segments. The flowers are arranged in axial spikes about 8–16 cm long; they are small and greenish white. The fruit of the Bengal almond is an oval to elliptical stone fruit; it is slightly flattened and about 3.5–7 cm long. The unripe fruit are green or yellow and turn red when they ripen. The stone of the fruit is surrounded by juicy fruit pulp about 3–6 mm thick. In between, there is a layer of air-ducting cells serving as a floating tissue, which helps the tree to spread along seacoasts (Hänsel et al. 1994, volume 6, p. 918).

2 Cultivation and Extraction

Cultivation

The Bengal almond prefers tropical and subtropical regions with high air humidity. It grows in sandy as well as loamy soil, but also on largely infertile ground. The harvest is best if the tree grows in the half shade in sandy soil. The Bengal almond blooms from November to March, and the fruit ripen from January to April, depending on the location.

Extraction of the Oil

Catappa L. oil is cold pressed from the seeds of *Terminalia catappa* (Conrado and Goyco 1943).

3 Character

Colour

Non-drying, liquid, light-yellow oil (Roth and Kormann 2000, p. 121).
Very pale, yellowish (Conrado and Goyco 1943).

Odour

Pleasant almond smell (Conrado and Goyco 1943).

Flavour

Like almond oil (Roth and Kormann 2000, p. 121).
Pleasant taste of almonds (Conrado and Goyco 1943).

4 Ingredients

On average, the seeds of *Terminalia catappa* contain 55.05% lipids, 22.41% protein and a high proportion of mineral nutrients (ash 4.20%) (Oderinde and Ajayi 1998).

Composition of Fatty Acids

Fatty acids	Content in % acc. to Roth and Kormann (2000), p. 121	Content in % acc. to Kalayasiri et al. (1996)	Content in % acc. to Sotheeswaran et al. (1994)	Content in % acc. to Conrado and Goyco (1943)
Palmitic acid	28.5	35.0	32.4 ± 2.3	36.52
Palmitoleic acid	–	0.3	–	–
Stearic acid	4.0	5.0	5.3 ± 0.4	3.79
Oleic acid	40.9	30.3	37.3 ± 2.2	35.05
Vaccenic acid	–	–	0.5 ± 0.1	–
Linoleic acid	22.9	27.9	24.5 ± 0.6	17.87
α-Linolenic acid	–	–	–	–
Arachidic acid	0.8	0.4	–	0.65

Composition of Lipids

Terminalia catappa seed oil contains 0.32% phytosterols and 11.00% diacylglycerol; at a detection limit of 100 ppm, no tocopherols could be found (Sotheeswaran et al. 1994, p. 12).

Physical Key Figures of Catappa L. Oil

density: 0.920 (15 °C)
refractive index: 1.464 (30 °C)
saponification number: 185–194
iodine number: 75–82
unsaponifiable: 0.5–2%
melting point: 3.5 °C
cetane index: 57.7
(Kalayasiri et al. 1996; Roth and Kormann 2000, p. 121)

Shelf Life

Since catappa L. oil has a high content of glycerides of saturated fatty acids, it has a very long shelf life (Roth and Kormann 2000, p. 121).

5 Application

In Folk Medicine

In India, catappa L. oil is used to treat colic and headaches, and is used topically, boiled together with the leaves, to treat leprosy, scabies and other skin diseases. These applications are not clinically proven (Hager 1994, p. 920).

As Fuel

The oil and its methyl esters can be used as a substitute for diesel fuel (Kalayasiri et al. 1996, p. 471 et seqq) and also have technical uses.

As Food

Catappa L. oil is a popular edible oil because it has a long shelf life. It can be used as a substitute for almond oil (Roth and Kormann 2000, p. 121).

In Cosmetics

Since catappa L. oil has similar properties to almond oil, it is often added to cosmetic lotions, balms and facial creams as a substitute (Roth and Kormann 2000, p. 121).

6 Possible Unwanted Side Effects

In a study, rats fed with flour made from the seeds of *Terminalia catappa* showed haemagglutination and trypsin-inhibitory activity. The erythrocytes agglutinate, and complex formation inhibits the activity of trypsin. The rats maintained their body weight, but suffered from hypertrophy of the stomach, intestine and pancreas and from atrophy of the spleen. Hypertrophy means that tissue or organs are enlarged because the cells become larger; atrophy leads to loss of tissue in organs (Oliveira et al. 2000).

Cedar Oil

synonyms: Zedernussöl (D); Siberian pine seed oil (E)

1 Source Plant

Pinus sibirica (Pinaceae), Siberian pine.

Habitat

Pinus sibirica grows in Siberia, in the Ural Mountains and in Northern Europe (Shikov et al. 2008).

S. Krist, *Vegetable Fats and Oils*, https://doi.org/10.1007/978-3-030-30314-3_32

Description

The Siberian pine is an evergreen tree that can reach a height of 35–40 m and a diameter of about 1.8 m (Shikov et al. 2008).

The crown is narrow and cone-shaped. Young trees have an ellipsoid crown. The cones are 6–12 cm in length (Küssmann 1979, p. 186).

The short branches are densely covered with bunches of needles. There are five needles in one bunch; depending on the climate, they are 6–10 cm long. The inner surface of the needles is grey.

The main distribution area is Siberia. The Siberian pine does not make many demands on the soil.

The seeds, of which there are up to 150 in one cone, are called cedar nuts (Shikov et al. 2008).

2 Cultivation and Extraction

Cultivation

Pinus sibirica has been cultivated since 1803, but cultivation has never been widespread. On acidic soil, it is a small bush, but it is a tall tree if the climate and the soil conditions are favourable.

In Buryatia, more than 1000 hectares of *Pinus sibirica* are cultivated for forestry use. 100 kg/ha of pine seeds are gathered annually. The main product of processing is cedar oil (Khanturgaev et al. 2003).

Extraction of the Oil

The oil is mainly obtained by pressing, although the yield is not high and alternatives are greatly sought after. Extraction with solvents leads to better yields, but the solvents have to be carefully removed after the process, as they are toxic. An exception is ethanol, which is edible and has no contraindications with the use in food, pharmaceutics or cosmetic products. It also has the advantage of good solubility in oil at temperatures of more than 120 °C, and settles after cooling down to 16–24 °C. The disadvantage is that the yield is not high – only 45–48% of the oil can be extracted with ethanol, whereas with hexane, 63–65% of the oil can be extracted. A study by Khanturgaev et al. examined extraction with ethanol in a high-frequency electromagnetic field; the result was a very high yield of 62–64% (Khanturgaev et al. 2003).

3 Character

Colour

Clear, light yellow (Khanturgaev et al. 2003).

Odour

Weak odour of cedar nuts (Khanturgaev et al. 2003).

Flavour

Nutty.

4 Ingredients

Composition of Fatty Acids

Fatty acids	Content in % acc. to Averina et al. (2010)	Content in % acc. to Destaillats et al. (2010)	Content in % acc. to Khanturgaev et al. (2003)	Content in % acc. to Skakovskii et al. (2007)	Content in % acc. to Wolff et al. (2000)
Myristic acid	0.03	–	0.31		–
Palmitic acid	4.61	4.35	6.82	4–4.2	4.31–4.4
Palmitoleic acid	0.04	0.08	0.63	–	0.06–0.1
Stearic acid	2.98	2.49	3.23	2.6–3	2.38–2.6
Oleic acid	27.33	25.48	26.57	22–22.6	25.07–25.1
Vaccenic acid	0.37	–	–	–	0.3–0.5
Linoleic acid	41.33	43.45	40.41	45.8–47	43.2–43.71
γ-Linolenic acid	18.73	18.29	19.65	18–18.9	18.1–18.49
α-Linolenic acid	0.16	0.2	–	0.3	0.2
Arachidic acid	0.33	0.28	0.31	–	0.27–0.3
Gondoic acid	1.12	–	–	1–1.2	1.3
Gondo acid	0.09	1.27	–	–	–
Arachidonic acid	0.51	–		–	–

Composition of Triglycerides

Triglycerides	Content in % acc. to Deineka (2003)
LnLnLn	0.33
LnLnL	3.34
LnLL	66.4
LLL	30.0

Composition of Sterols

Sterols	Content in % acc. to Shikov et al. (2008)
Campesterol	2.98
Sitosterol	8.95
Fucosterol	0.92

Composition of Tocopherols

Tocopherols	Content in % acc. to Shikov et al. (2008)
α-Tocopherol	1.49
γ-Tocopherol	4.72

Other Ingredients

Squalene: 0.47%.
(Shikov et al. 2008)

Physical Key Figures of Cedar Oil (Khanturgaev et al. 2003)

	Extraction with ethanol	Extraction with hexane
Refractive index	1.477–1.479	1.475–1.4768
Acid number	0.55–0.60	3.22–3.43
Peroxide number	3.0–3.5	3.0
Iodine number	158.62–159.12	107.87–113.58
Saponification number	197.15–199.10	186.60–189.98
Colour index	7–8	–

5 Application

In Pharmacy and Medicine

Burns, furuncles, eczema, psoriasis and gastric ulcers can be treated with cedar oil. It has the capacity to normalise the blood lipid spectrum, lower cholesterol levels and counterindicate arterial hypertension (Skakovskii et al. 2007).

The oil is also used to stimulate cell proliferation. According to Averina et al., it lowers the lipid level in the blood as well as blood sugar levels (Averina et al. 2010).

A study by Shikov et al. examined the anti-inflammatory properties of cedar oil. Rheumatoid arthritis is a widespread disease is currently treated with non-steroidal antirheumatics, but these have serious side effects and are not recommended for long-term use. It is therefore necessary to find alternative treatment and medicine that is effective yet has a low toxicity. Against this background, cedar oil was tested for its anti-inflammatory effects, which were compared to those of the non-steroidal antirheumatic phenylbutazone. Rats were fed either 300 mg/kg cedar oil or 56 mg/kg phenylbutazone. The result showed a significant analgesic effect of cedar oil, which lasted longer than that of phenylbutazone. An antipyretic effect was also verified (Shikov et al. 2008).

In Cosmetics

The oil is added to various cosmetic products, for example to cleansers. It is also used as a massage oil (Averina et al. 2010).

As Food

Due to its low smoke point, cedar oil is not very suitable as a frying fat, but it can be used to flavour cold dishes.

Charlock Mustard Oil

Oleum Sinapis arvensis synonym: Ackersenföl (G)

1 Source Plant

Sinapis arvensis L. (Brassicaceae), charlock, charlock mustard, wild mustard.

S. Krist, *Vegetable Fats and Oils*, https://doi.org/10.1007/978-3-030-30314-3_33

Habitat

Charlock is a native plant of the temperate zones of Europe. The Greeks and Romans cultivated it as one of the first plants, and it soon spread to Asia and North Africa. In the Middle Ages, the Arabs brought charlock to Spain, but it was only introduced to Germany and England in the seventeenth century. In the nineteenth century, it was cultivated in the Russian steppes as well as in Kirghizia. During the Second World War, charlock reached America, which became its main growing area. Today, charlock is cultivated in Canada, California, Montana, Ontario, but also in Asia, Denmark, Hungary, France and North Africa (Hornok 1992, p. 232).

Description

Charlock is an annual plant, grass-green, tinged violett in some parts, and has a thin, spindle-shaped tap root. Its stem is 30 to 60 cm tall, grass-green, leafed and usually branched, square-edged, striated, with stiff hair near the base, but often hairless near the top. The leaves have coarse hair; the lower leaves are stalked, the upper leaves sessile or short petiolate, irregularly dented to pinnatifid, with a large end section. The flower stalks on the stem and branches are arranged in terminal racemes. The four sepals are hairless, more rarely covered with coarse hairs. The sepals are 5–6 mm long, 2.5–3 mm broad, and stick out horizontally. The four petals are nearly twice the length of the calyx and are a sulphuric yellow colour. The fruit is a silique sitting on a short, more or less protruding stem. It is 2.5–4 cm long, 2.5–3 mm thick, square-edged, rarely nearly cylindrical, either hairless or covered with stiff, backwards-pointing hairs. The seeds are a dark-red or dark-brown colour and have a nearly smooth, round seed coat with a diameter of 1–1.3 mm. The flowers appear from May to July (Hänsel et al. 1994, volume 6, p. 713 et seq).

2 Cultivation and Extraction

Cultivation

Charlock prefers dry habitats. As it is not susceptible to frost, it is usually sown in early spring. The plant prefers sunny, sandy, clayey soil; it grows in acidic, neutral as well as saline soil. The yield is, however, much higher in eutrophic soils. Too much heat at the beginning of the growth period has a negative effect on the plants, but long dry spells affect growth and yield, too. Charlock should not be sown on the same field in consecutive years or on a field on which black mustard has previously been grown, as the two plants are affected by the same pests and plant diseases (Hornok 1992, p. 233).

Extraction of the Oil

Charlock mustard oil is extracted from the ripe seeds at low temperatures (40–60 °C) and under low pressure (cold pressed oil).

3 Character

Colour

Colourless to light yellow.

Odour

Mild, only mildly irritating to the eye (Von Schoenebeck 1936).

Volatile Compounds

SPME-GC-MS analysis (Krist et al. 2008):
Allyl-isothiocyanate; 1-Butene-4-isothiocyanate;

Flavour

Pungent taste, irritant (Von Schoenebeck 1936), nutty.

4 Ingredients

Composition of Fatty Acids

Fatty acids	Content in % acc. to Hänsel et al. (1994), volume 6, p. 715
Oleic acid	8–39
Linoleic acid	12–25
Linolenic acid	8–18
Erucic acid	6–39
Eicosenoic acid	12–19

With regard to the content of linoleic acid, linolenic acid and oleic acid and their ratio, charlock is more similar to *Brassica nigra* than to *Sinapis alba.* In contrast to *Sinapis alba, the oil has a high content of eicosenoic acid, and the quotient of oleic acid and linoleic acid is* <1 (Hänsel et al. 1994, volume 6, p. 715).

Composition of Sterols

Sterols	Content in % acc. to Hänsel et al. (1994), volume 6, p. 714
Cholesterol	Traces
Brassicasterol non-esterified/esterified	7.8/5.9
Campesterol non-esterified/esterified	25.8/32.5
β-Citosterol non esterified/esterified	66.4/56.1
Δ + 5-Avenasterol esterified	4.4
Δ + γ-Stigmasterol esterified	1.1

Other Ingredients (Hänsel et al. 1994, volume 6, p. 715)

Glucotropaeolin 0.06%
Sinigrin 0.004%
Gluconapin 0.002%
Sinalbin 3.9%
Flavonoids
Vitamin C
Sinapine 0.6%

Physical Key Figures of Charlock Mustard Oil

refractive index: n_D^{40} 1.4646–1.4662
density: d_{30}^{30} 0.907–0.910
saponification number: 187–200
iodine number: 77–89
unsaponifiable: 0,4%
(Roth and Kormann 2000, p. 157)

5 Application

In Pharmacy and Medicine

As Mustard Plaster in Cases of Acute Cardiac Pain and Angina Pectoris

The sinalbin contained in charlock mustard oil, in particular, warms the skin, stimulates blood circulation, and thus eases breathing and alleviates pain.

To Treat Arthritis, Rheumatism, Neuralgia, Gout and Inflammation

Since charlock mustard oil causes light skin irritation, it warms the skin and stimulates blood circulation, which alleviates pain and improves mobility.

To Improve the Appetite and for Digestive Purposes

In very small doses, charlock mustard oil stimulates the secretion of gastric acid and pancreatic juice, and thus has a digestive effect.

In Cases of Constipation and Poisoning with Barbiturates

Charlock mustard oil stimulates the activity of the gastrointestinal tract and supports the effect of other emetics to induce vomiting.

For Use against Bacteria, for Example *Bacillus anthracis*

Allyl mustard oil has been proven to have bactericidal and germicidal properties, even in a dilution of 1:330,000. For this reason, it has always been used for preserving food and forage (Von Schoenebeck 1936).

In Cosmetics

As Massage Oil

As charlock mustard oil warms the skin and stimulates blood circulation, it should only be used as massage oil in a diluted form in a ratio of 1:3 with other vegetable oils. It is suited as an embrocation for athletes to prevent myogelosis.

As Food

The Indian cuisine, in particular, uses charlock mustard oil for seasoning.

6 Possible Unwanted Side-Effects

Skin Irritation

If charlock mustard oil is applied for longer periods of time, it can cause erythema and blistering. The oil should never be applied in concentrated form and especially not on sensitive areas of the skin, or on the skin of babies or small children.

Inflammatory Reactions

According to a study, a topical solution containing 5–20% charlock mustard oil caused acute inflammatory reactions, escape of plasma into the tissue and oedema formation on the ears of mice 30 minutes after it had been applied. These reactions were triggered by histamine and 5-hydroxytryptamine (Inove et al. 1997).

Gastrointestinal Disorders

If swallowed, high doses of charlock mustard oil can result in vomiting, diarrhoea and gastrointestinal disorders.

Chaulmoogra Oil

Oleum Chaulmoograe, Oleum Hydnocarpi synonym: Gynocardiaöl (D); Chaulmoograöl (D)

1 Source Plant

Hydnocarpus kurzii Warb. *(Achariaceae), syn. Taraktogenos kurzii* King *and other species of Hydnocarpus*, hydnocarpus.

Habitat

Hydnocarpus is a native tree of south-western India, but it is cultivated across the whole of India and Sri Lanka (Hager 1994, p. 865).

Description

The tree is up to 15 m tall, with a thick trunk and a smooth bark. The leaves are elongated, pointed and entire. The flowers are usually dioecious, light yellow to light orange in colour, and grow in axial cymes. The fruit is a large capsule reminiscent of an orange, with a brown, woody shell and many seeds resembling hazelnuts (Roth and Kormann 2000, p. 70).

2 Cultivation and Extraction

Cultivation

Apart from Sri Lanka, cultivation has also been tried in Uganda, Cameroon, Nigeria and Brazil. Hydnocarpus grows in flood areas, along rivers and in wet, evergreen rain forests (Hager 1994, p. 865).

S. Krist, *Vegetable Fats and Oils*, https://doi.org/10.1007/978-3-030-30314-3_34

Extraction of the Oil

The oil is either cold pressed or extracted from the fresh, ripe seeds (Roth and Kormann 2000, p. 121).

3 Character

Colour

White apricot-coloured, viscous at room temperature.

Yellowish-brown, soft fat with crystalline exudations (Roth and Kormann 2000, p. 121).

Odour

Like vinegar, sweetish, like tropical fruit, woody.

Volatile Compounds

(Krist et al. 2008) SPME-GC-MS analysis.

Acetic acid	Cyclopentanol acetate
Propanoic acid	Sabinene
Cyclopentanol	Myrcene
Cyclopentanone	p-cymene
Hexanal	Limonene
2-Cyclopenten-1-one	Linalool

Flavour

Bitter, pungent (Roth and Kormann 2000, p. 121).

4 Ingredients

Composition of Fatty Acids

The oil extracted from seeds largely consists of glycerides of cyclopentenyl fatty acids (Hager 1998, p. 863 et seqq):

Composition of Fatty Acids in Chaulmoogra Oil

Cyclopentenyl fatty acids/fatty acids	Content in % in Oleum Chaulmoograe acc. to Hager 1998, p. 863 et seqq	Contant in % in Chaulmoogra Oil (*Hydnocarpus wightiana)* acc. to nature in Bottle (2019)	Content in % in Oleum Hydnocarpii acc. to Hager 1998, p. 863 et seqq
Hydnocarpic acid	20–30	23.3	35–45
Chaulmoogric acid	20–30	29.1	25–35
Gorli acid	20–25	25.1	Approx. 10
Palmitic acid	Approx. 5	8.3	Approx. 5
Stearic acid	Approx. 10	–	Approx. 3
Oleic acid	Approx. 4	5.4	Approx. 1
Palmitoleic acid	–	6.2	–
Linoleic acid	–	1.7	–
Myristic acid	–	0.5	–

Physical Key Figures of Chaulmoogra Oil

density: approx. 0.957 (at 15 °C)
refractive index: 1.4751–1.4771 (at 40 °C)
saponification number: 197–215
iodine number: 95–105
unsaponifiable: 0.3%
point of solidification: 9–14 °C (pressed)
18–20 °C (extracted)
melting point: 22–26 °C (pressed)
33–39 °C
(Roth and Kormann 2000, p. 121)
solubility: completely in hot 90% ethanol, soluble in diethyl ether, chloroform, benzene and carbon disulphide, but practically insoluble in water.

5 Application

In Pharmacy and Medicine

In Cases of Hypopigmentation

Chaulmoogra oil and the esters of its fatty acids induce pigmentation of hypomelanotic skin by enlarging already pigmented areas, and can therefore be effective in treating hypopigmentation (Leclere 1994).

C

To Treat Mycobacterium Leprae

For years, chaulmoogra oil has been employed in the treatment of leprosy. Levy examined the effect of the oil against *mycobacterium leprae in a study. Mice infected with mycobacterium leprae* on the pads of their hind legs were treated with 5.44 mg/ml chaulmoogric acid intraperitoneally three times a day over a period of 10 weeks. The results showed that the reproduction of bacteria was inhibited during treatment, but the effect stopped as soon as treatment ceased. In a second experiment, the infected mice were treated with 0.28 mg/ml hydnocarpic acid once a week for a period of 13 weeks. Hydnocarpic acid did not, however, inhibit the reproduction of bacteria. In another experiment, 0.5 mg/ml chaulmoogra acid and dihydro-chaulmoogric acid was administered five times a week over a period of 52 days. The results prove that chaulmoogric acid and its salts as well as dihydro-chaulmoogric acid have an anti-microbic effect against *mycobacterium leprae* (Levy 1975).

As a Vulnerary Drug

Another study examined the effect of chaulmoogra oil on open wounds of rats. A round piece of skin with a diameter of 2.5 cm, at a distance of 5 cm from the animals' ears, was removed. The oil was applied topically. The parameters for wound healing were set as contraction of the skin around the wound and epithelialisation of the wound. Results showed no effect on the contraction of the skin, neither after oral nor after topical application, but a significant advancement of epithelialisation. The oil can thus be applied in cases of cancer of the sole of the foot and other open wounds of leprosy patients (Oommen 2000).

In Folk Medicine

In India, *oleum chaulmoogra* is applied internally to treat leprosy and externally to treat many chronic skin diseases, rheumatism, contusions, sprains and sciatica. The effects are not clinically proven (Hager 1994, p. 866 et seq).

6 Possible Unwanted Side Effects

When humans take large doses of chaulmoogra oil orally, they suffer from anorexia, vomiting, diarrhoea, as well as from initial excitation and subsequent paralysis of the central nervous system. After oral or intravenous administration, animals show symptoms of anorexia, nausea and vomiting. The LD value of chaulmoogra oil for cats is 4300 mg/kg body weight, if administered orally. The oil is legally classified as poisonous (Hager 1994, p. 866 et seq).

Cherry Kernel Oil

synonyms: Prunus avium seed oil (E); Kirschkernöl (D); huile de cerise (F)

1 Source Plant

Cerasus avium L., *Prunus avium* L.(Rosaceae), sweet cherry.

S. Krist, *Vegetable Fats and Oils*, https://doi.org/10.1007/978-3-030-30314-3_35

Habitat

From the sweet cherry (*Cerasus avium*), which originates in Persia and is also a wild plant, the cultivars heart cherry (*Cerasus avium* ssp. *juliana*) and Bigarreau cherry (*Cerasus avium* ssp. *duracina*) developed. They were probably already cultivated several centuries B.C. in the Black Sea region. Eventually, they were imported to Italy; the Romans brought them to Germany (Lieberei and Reisdorff 2007, p. 189).

C

Description

Prunus avium is a tree that can reach a height of 20 m. The leaves are oval and pointed, the edges disparately serrate. The upper side of the leaves is hairless. The lower side is paler than the upper side and is covered with soft hair along the leaf veins. The white flowers sit on long stems and form umbels on short shoots in the axils of the cataphylls. The ovary is half-inferior. After bees have pollinated it, it develops into a stone fruit. A ligneous endocarp surrounds the seed. Sweet cherries can be dark or light red, or even yellow. The fruit are small, between 8 and 10 mm in diameter. Cultivars usually have larger fruit (up to 20 mm in diameter) and contain more mesocarp (Guimpel et al. 1815, p. 84; Lieberei and Reisdorff 2007, p. 189).

2 Cultivation and Extraction

Cultivation

For the cultivation of sweet cherries, all kinds of soil are suitable, although the plant prefers good, thick, gravelly soil. The seeds are sown in autumn; the tree blossoms from the end of April to early May. The fruit harvest takes place in summer. Sweet cherries ripen earlier than sour cherries. They have only half the acid content of the latter and are cultivated more frequently (Guimpel et al. 1815, p. 84; Lieberei and Reisdorff 2007, p. 189–190).

Production of sweet cherries in 1000 t (Lieberei and Reisdorff 2007, p. 433)

	1990	2005
Global	1396	1864
Turkey	143	260
USA	142	250
Iran	85	224
Germany	141	120
Russian Federation	–	110
Italy	100	108

Extraction of the Oil

Sweet cherries are eaten as well as processed. During processing, large amounts of seeds accumulate as waste material. This is not only a waste of a potentially useful substance, but also a serious disposal problem. A possible use for the kernels is to produce oil from them. Cherry kernels are either pressed or extracted with solvents (hexane). Another method is supercritical fluid extraction, with carbon dioxide acting as a solvent. The advantage of this method is that the oil need not be refined subsequently (Bernardo-Gil et al. 2001).

3 Character

Colour

Light yellow.

Cherry kernel oil extracted via supercritical fluid extraction is clearer than the oil extracted with solvents (Bernardo-Gil et al. 2001).

Odour

Slight odour of bitter almond (Wimmer et al. 2003).

Flavour

Pleasant touch of almond (Wimmer et al. 2003).

4 Ingredients

Composition of Fatty Acids

Fatty acids	Content in % acc. to Comes et al. (1992)	Content in % after supercritical fluid extraction acc. to Bernardo-Gil et al. (2001)	Content in % after extraction with solvents acc. to Bernardo-Gil et al. (2001)	Content in % acc. to Kamel and Kakuda (1992)
Lauric acid	–	–	–	–
Myristic acid	–	–	–	–
Palmitic acid	6.8–7.6	5.26	6.02	7.6
Palmitoleic acid	0.4–0.5	0.27	0.22	0.3
Stearic acid	2.1–2.6	2.15	2.41	2.3
Oleic acid	32.8–37.5	32.64	31.97	52.9

Fatty acids	Content in % acc. to Comes et al. (1992)	Content in % after supercritical fluid extraction acc. to Bernardo-Gil et al. (2001)	Content in % after extraction with solvents acc. to Bernardo-Gil et al. (2001)	Content in % acc. to Kamel and Kakuda (1992)
Trans-linoleic acid	–	0.59	0.33	–
Linoleic acid	40.0–44.7	40.84	41.05	35.0
α-Linolenic acid	0.3	1.10	1.10	–
α-Elaeostearic acid	9.9–13.2	10.11	10.48	–
Arachidic acid	1.2	1.50	1.74	1.4
11-Eicosenoic acid	–	1.21	1.08	
Eicosanoic acid	0.4	1.14	1.09	–

Composition of Triglycerides

Triglycerides	Content in % acc. to Comes et al. (1992)
EEL	0.5
ELL	12.9
LLL	3.3
ELO	6.9
ELP	3.9
LLO	18.5
LLP/EOO	9.8
LOO	15.4
LOP	7.7
LPP	1.0
OOO	16.1
OOP	4.0

Composition of Sterols

Sterols	Content in % after supercritical fluid extraction acc. to Bernardo-Gil et al. (2001)	Content in % after extraction with solvents acc. to Bernardo-Gil et al. (2001)
Cholesterol	0.36	0.36
Campesterol	1.03	1.13
Campestanol	2.98	2.98
Stigmasterol	3.85	3.45
Chlesrosterol	0.80	0.77
β-Sitosterol	83.41	83.71
Δ5-Avenasterol	7.55	6.58

Physical Key Figures

acid number:	0.75
saponification number:	192
iodine number:	113
hydroxyl value:	5.5
unsaponifiable:	0.66%
(Kamel and Kakuda 1992)	

Shelf Life

Oxidative stability tests according to Bernardo-Gil et al. showed that the oil is not significantly protected against oxidation (Bernardo-Gil et al. 2001). It does not keep long.

5 Application

In Cosmetics

If suitable antioxidants are added, the oil can be used in the production of cosmetics (Bernardo-Gil et al. 2001).

It serves as an emollient and emulsion stabiliser (Bährle-Rapp 2007, p. 331).

Cherry kernel oil is also used as a massage oil and bath oil (Wimmer et al. 2003) as well as in soap production.

As Food

The oil can be used as an edible oil (Wimmer et al. 2003).

Chia Oil

synonyms: Chiaöl (D); huile d'olive (F)

1 Source Plant

Salvia hispanica (Lamiaceae), chia.

S. Krist, *Vegetable Fats and Oils*, https://doi.org/10.1007/978-3-030-30314-3_36

C

Habitat

Salvia hispanica originates in South America and in northern Guatemala. Chia is cultivated in tropical and subtropical regions and, as the plant is frost-susceptible, in greenhouses in Europe (Ixtaina et al. 2008).

Description

Chia is an annual, herbaceous plant that can reach a height of 2 m (Peiretti and Gai 2009; Palma et al. 1947).

The plant produces oval, narrow seeds 2 mm long and 1 mm broad, which are either white or black. Most populations only produce a low percentage of white seeds. The small share of white seeds on the market originates from plants that produce exclusively white seeds. As *Salvia hispanica* has similar uses to those of sesame, white seed coats might be preferred in modern commerce and would therefore gain in importance. The white seeds are slightly larger than the black ones (Ixtaina et al. 2008; Peiretti and Gai 2009; Palma et al. 1947).

Chia is well suited as forage and contains many high-quality nutrients, especially if the seeds are harvested shortly before they sprout (Peiretti and Gai 2009).

2 Cultivation and Extraction

Cultivation

Before Columbus, chia was the staple food of the Mesoamericans, who also used it as a medical plant. The plant was an important article of commerce, and the seeds were used as food, medicine and a source of oil. It is assumed that chia was as important as, and, in some regions, valued even more highly than maize. The sixteenth century codices of the Aztecs contain a wealth of ethnobotanical information and list large agricultural acreages reserved exclusively for chia cultivation. The codex Mendoza and La Matricula de los Tributos show that 21 of the 38 tributary provinces paid chia as a tribute. After the conquest by Spain, cultivation declined dramatically (Cahill 2003).

In the twentieth century, *Salvia hispanica* was only cultivated in the regions of Los Altos and Jalisco. The total cultivation area amounted to less than 450 hectares annually, and on average, the fields were smaller than 6.3 hectares. All the work was done by hand. The trend was to substitute chia for maize, which had a fixed price and was subsidised by the government (Ayerza 1995).

At the beginning of the third millennium, chia is grown in Mexico, Bolivia, Argentina, Ecuador and Guatemala. In Argentina, it is a summer and autumn plant cultivated instead of traditional corn in the north-western regions, where corn is not profitable (Ixtaina et al. 2008).

In South America, the cultivation of oilseed plants is limited due to lack of water, and chia therefore presents a good alternative. *Salvia hispanica* grows well in arid and semi-arid climates (Peiretti and Gai 2009).

Extraction of the Oil

The seeds of *Salvia hispanica* are ground and the oil is subsequently extracted with solvents. The seeds can also be cold pressed or pressed by adding heat. In this case, there are almost no differences in the quality of the oil (Palma et al. 1947).

Extraction with solvents increases the yield by about 30% in comparison to pressing. The composition of fatty acids is the same for both methods of extraction (Ixtaina et al. 2011a).

An environment-friendly method is extraction using supercritical carbon dioxide. If the pressure and temperature are optimal, the yield is nearly the same as when the oil is extracted with solvents. There is almost no difference in the composition of fatty acids, either. The only exception is linoleic acid, of which there is more when the oil is extracted with supercritical carbon dioxide. The physical key figures are similar for both extraction methods, but the content of tocopherols is lower when the oil has been extracted with supercritical carbon dioxide. The lower amount of tocopherols and phospholipids is presumably the main reason why the oil extracted with supercritical carbon dioxide is not as oxidatively stable as the oil extracted with hexane (Ixtaina et al. 2011b).

3 Character

Colour

Light yellow (Palma et al. 1947).

The colour of the oil depends on the total amount of pigments. Chia oil contains carotenoids, but no chlorophyll pigments (Ixtaina et al. 2011a).

Odour

Odourless to slightly nutty (http://de.wikipedia.org/wiki/Chiaöl).

Flavour

Neutral (http://de.wikipedia.org/wiki/Chiaöl).

4 Ingredients

Composition of Fatty Acids

Fatty acids	Content in g/kg acc. to Peiretti and Gai (2009)	Content in % acc. to Taga et al. (1984)	Content in % acc. to Ayerza (1995)	Content in % acc. to Palma et al. (1947)
Palmitic acid	71	5.2–9.9	6.2–7.1	–
Stearic acid	33	2.9–16.2	3.1–3.7	–
Oleic acid	60	7.6–21.3	7.3–8.2	4.0
Linoleic acid	188	15.3–46.3	19.8–20.8	26.0
α-Linolenic acid	641	6.3–69.0	52.0–63.4	54.1

The ratio of polyunsaturated fatty acids and saturated fatty acids is 7.9, and the ratio of omega-6 and omega-3 fatty acids is 0.29 (Peiretti and Gai 2009).

The plant contains a high amount of polyunsaturated fatty acids, which decreases, however, according to a study by Peiretti and Gai, from 752 g/kg to an average of 623 g/kg as the plant develops. The proportion of α-linolenic acid decreases from 649 to 499 g/kg, while the amounts of all other fatty acids increase with continuing growth of the plant. The quality of chia therefore depends on the age of the plant. The seed contains a high proportion of polyunsaturated fatty acids, its main components being α-linolenic acid and linoleic acid. The seed contains 250–390 g oil per kg fresh seeds (Peiretti and Gai 2009).

Composition of Triglycerides

Triglycerides	Content in % in Argentinean chia oil acc. to Ixtaina et al. (2011a)	Content in % in chia oil from Guatemala acc. to Ixtaina et al. (2011a)
LnLnLn	32.8	47.0
LnLnL	20.3	20.8
LnLL	13.8	11.7
LnLnP	7.7	6.7
LnLO	7.0	3.9
LnLP	5.3	4.7
LnOO+LnOP	8.3	4.1
LnPP	0.8	0.2
LLS	1.1	0.2
LnOS	2.1	0.6
LnSP	1.0	0.1

Composition of Tocopherols

Tocopherols	Content in mg/kg after extraction with solvents acc. to Ixtaina et al. (2011a)	Content in mg/kg after extraction with $SC\text{-}CO_2$ acc. to Ixtaina et al. (2011b)
Total content	238–427	36–95
α-Tocopherol	0.4–9.9	–

Tocopherols	Content in mg/kg after extraction with solvents acc. to Ixtaina et al. (2011a)	Content in mg/kg after extraction with $SC\text{-}CO_2$ acc. to Ixtaina et al. (2011b)
β-Tocopherol	–	–
γ-Tocopherol	>85%	–
δ-Tocopherol	–	4.17

Other Ingredients: (Ixtaina et al. 2011a)

Carotenoids: 0.53–1.21 mg/kg.

Physical Key Figures: (Palma et al. 1947)

refractive index:	1.4812
density:	0.9330
saponification number:	192
acid number:	2.0
iodine number (Hanus ½ hour):	186.7
iodine number (Wijs 1 hour):	191.4
iodine number (Wijs 24 hours):	198.7
unsaponifiable:	1.21%
cloud point:	−13.0 °C
yield point:	−16.5 °C
titre:	−14.7 °C
relative viscosity:	34.1

5 Application

In Pharmacy and Medicine

The use of chia oil in tumour treatment is discussed. In a study, Espada et al. examined the effects of a diet containing chia on mice with carcinoma of the thymus gland. The diet resulted in a reduction of tumour weight as well as in an increase in apoptosis and a decrease in mitosis in the tumour tissue. A diet containing polyunsaturated omega-3 fatty acids caused an increase in eicosapentaenoic acid and a decrease in arachidonic acid. The growth of the cancer was stopped, and metastasis could be prevented (Espada et al. 2007).

In Cosmetics

In cosmetics, chia oil is a softening agent (Cahill 2003).

Technical Uses

Chia oil is an excellent drying oil. Heat causes practically no darkening, which is why the oil can be added to light-coloured varnish (Palma et al. 1947).

Chia oil is a component in paint and varnish and a basic component of body paintings (Cahill 2003).

As Food

Chia oil is an edible oil, for cooking purposes as well as for salads (Cahill 2003).

As Forage

Chia oil can also be added to forage to make the forage richer in polyunsaturated fatty acids. Rabbits fed with chia supply high-quality meat with an advantageous composition of fatty acids for the consumers. This rabbit meat could also be an alternative to fish or oil seeds as a source of polyunsaturated omega-3 fatty acids (Peiretti and Meineri 2008).

Chia seeds are an important vegetable source of α-linolenic acid. In a recent study, Ayerza and Coates observed the effects of a chia diet on rats. The result was a considerable decrease in the triglyceride level and an increase in high-density lipoprotein and omega-3 fatty acids in the serum of the rats. The ratio of TG/HDL and omega-6/omega-3 was reduced without any disadvantageous effects on the rats' growth. The study indicates that chia oil is a good alternative source of omega-3 fatty acids for persons who do not like fish, have a fish allergy, or are vegetarians (Ayerza and Coates 2005).

The dark meat of chickens fed chia contains less fat than that of chickens raised on a conventional diet. The content of palmitic acid in the meat decreases, whereas the content of α-linolenic acid increases. In total, the meat contains fewer saturated fatty acids, without any consequences for the flavour (Ayerza et al. 2002).

Feeding hens chia has consequences for their eggs. The egg-yolks contain a higher proportion of polyunsaturated omega-3 fatty acids, and the ratio of saturated and unsaturated fatty acids is more favourable. Again, there is no significant difference in flavour. A study by Ayerza and Coates established, however, that the control group that did not receive the chia diet produced more eggs with larger egg yolks. The chia diet caused a decrease in saturated acids like palmitic acid and an increase in α-linolenic acid, however, which would help to make the eggs more acceptable to health-conscious consumers (Ayerza and Coates 1999).

Cocoa Butter

Oleum Cacao synonyms: Kakaobutter (D); beurre de cacao (F)

1 Source Plant

Theobroma cacao L. (Sterculiaceae), cocoa tree.

S. Krist, *Vegetable Fats and Oils*, https://doi.org/10.1007/978-3-030-30314-3_37

C

Habitat

The cocoa tree originates in the tropical areas of Central and South America in the upper Amazonas area, but also in Belize, Guatemala and southern Mexico. Over the course of the millennia, cultivation of the cocoa tree, especially in Central America, led to many varieties. In the seventeenth century, the Spanish and the Portuguese transported cocoa trees to the islands in the Gulf of Guinea. Jamaica and Haiti were important producers by the mid-seventeenth century. The Spanish and the Dutch brought the cocoa tree to South-East Asia in the seventeenth century; the Germans transported it to Samoa and New Guinea in the eighteenth century. Today, the main growing areas are Panama, Guatemala, Honduras, Nicaragua, Costa Rica, Haiti, Jamaica, Puerto Rico and Grenada. In South America, the cocoa tree is cultivated in Venezuela, Guyana and in the northern part of Brazil. In Asia, it is grown especially in Sri Lanka, Java and the Philippines. The main growing areas in Africa are Lagos, Cameroon, Ghana and the Congo Basin. In the South Pacific region, cocoa trees are cultivated in Samoa and New Guinea (Warburg and Van Someren Brand 1909, p. 257; Hänsel et al. 1994, volume 6, 5th edition, p. 945).

Description

The cocoa tree can reach a height of 10–15 m. It is an evergreen tree usually cultivated as a shrub. Its bark is very thin, and cinnamon brown with a silvery shine. The wood is very porous and slightly pink. The trunk is about 30 cm thick and has many slender branches. The leaves are alternate and first peach-coloured, then a smooth dark green on the upper side and a duller colour on the lower side, which is slightly hairy. The leaves are 20–30 cm long, sit on short stems, and have a lanceolate, oval, pointed form. The edge is smooth and slightly curved. The flowers of the cocoa tree are cauliflorous, which means that they appear in dense, forked groups from the stem and branches, outside the axils. The small, pink flowers are perfect, of regular form, and consist of five lanceolate sepals. The unripe fruit are green, and the ripe ones are either yellow, orange or red, depending on the variety. Their colour only changes to maroon when they are dried. The fruit sit on short stems and resemble short, thick, pointed cucumbers about 12–14 cm long and 6–8 cm in diameter. Each fruit has a pentagonal pericarp with ten longitudinal ribs. The pericarp is made of firm fruit pulp, which later dries to form a leathery mass. The fruit pulp is colourless, juicy, and has a pleasant, sweet and sour taste. It contains five to eight longitudinal rows with about 60 seeds. The fresh seeds are white, the dried ones brown with a brittle skin (Warburg and Van Someren Brand 1909, p. 262).

2 Cultivation and Extraction

Cultivation

Cultivating cocoa trees is laborious. The plants are cultivated in seedbeds for 8–10 months and then bedded out. Shade trees like bananas or cotton plants are planted between the cocoa plants. When the cocoa trees have reached a height of about 1 m, all the branches are cut except the three or four uppermost ones, so that the trees form a broad, pyramid-shaped crown. Provided that good care is taken of the plant, that the soil is suitable and the conditions are good, the tree will be fully grown after 12 years and will bear fruit until it is 50. The ripe fruit are either carefully picked from the stems by hand with a twisting motion, or cut with a knife. This must be done with care so as not to damage the buds and flowers that ensure the next harvest (Warburg and Van Someren Brand 1909, p. 259).

Extraction

Cocoa butter is the fat of cocoa kernels or cocoa mass that has been pressed, filtered or centrifugalised. The butter is produced from roasted seeds or peeled and roasted kernels that have not been pre-treated with alkalis, or roasted and peeled seeds that are pressed by adding heat (Hänsel et al. 1994, volume 6, 5th edition, p. 946).

3 Character

Colour

Pale yellow (Hänsel et al. 1994, volume 6, 5th edition, p. 947).
Yellowish-white (Roth and Kormann 2000, p. 128).
Inferior quality: deep yellow to brownish-yellow (Bauer 1928, p. 250).
Ivory (Von Wiesner 1927, volume 1, p. 744).

Odour

Weak, pleasant odour of cocoa (Hänsel et al. 1994, volume 6, 5th edition, p. 947).

Flavour

Pleasant, mild taste (Roth and Kormann 2000, p. 128).
Strong cocoa flavour (Von Wiesner 1927, volume 1, p. 744).

C

4 Ingredients

Composition of Fatty Acids

Fatty acids	Content in % acc. to Roth and Kormann (2000), p. 128	Content in % acc. to Firestone (1999), p. 28	Content in % acc. to augustus-oils.ltd.uk/01.09.05
Myristic acid	–	0.1	–
Palmitic acid	24–32	25–27	27.3
Palmitoleic acid	–	0.1–0.3	–
Margaric acid	–	0.1	–
Stearic acid	31–38	31–37	34.7
Oleic acid	30–38	31–35	32.5
Linoleic acid	1.4–4.2	2.8–4.0	2.5
α-Linolenic acid	–	0.1	–
Arachidic acid	–	0.2–1.0	–
Oleodistearin	10	–	–
Oleopalmitin	52	–	–
Oleodipalmitin	6	–	–
Stearodiolein	12	–	–
Palmitodiolein	9	–	–

Composition of Triglycerides

Triglycerides	Content in % acc. to Karleskind (1996), p. 209	Content in % acc. to Firestone (1999), p. 138
OOL	0.1–0.2	–
POL	0.3–2.5	–
PPL	1.5–2.5	1–2
OOO	0.2–2.0	–
SOL	0.3–1.0	–
POO	1.5–7.0	3–4
PSL	2.5–3.5	3
PPO + MSO	14–21	14–18 (without MSO)
PPP	0.1–0.2	–
PMgO	0.1–0.3	–
SOO	2.0–9.0	3–6
SSL	1.5–2.0	2
PSO	32–41	36–40
PPS	0.3–0.6	1
MgSO	0.1–0.3	–
AOO	0.2–0.5	–
SSO	20–30	23–26
PSS + APP	0.2–1.0	1 (without APP)
SAO	0.8–1.5	1–2
SSS + PSA	0.1–0.5	–
AOP	–	0–1

Composition of Sterols

(concentration in mg/100 g and content in %)

Sterols	Cocoa butter acc. to Karleskind (1996), p. 211	Cocoa butter acc. to Firestone (1999), p. 28
Total content	140–160	–
Cholesterol	1–2	1
Brassicasterol	–	–
Campesterol	7–9	8–11
Stigmasterol	25–28	24–31
β-Sitosterol	59–63	58–63
Δ5-Avenasterol	2–4	3–5
Δ7-Stigmasterol	1–2	1
Δ7-Avenasterol	–	–

Composition of Tocopherols

(concentration in mg/100 g and content in %)

Tocopherols	Cocoa butter acc. to Karleskind (1996), p. 211
Total content	16–37
α-Tocopherol	5–8
β-Tocopherol	–
Γ-Tocopherol	84–94
Δ-Tocopherol	2–7

Physical Key Figures of Cocoa Butter

refractive index: n_D^{40} 1.456–1.458
density: d_{15}^{15} 0.970–0.998
melting point: 30–35 °C
saponification number: 192–197
iodine number: 33–42
acid number: max. 2.0
peroxide number: max. 3.0
unsaponifiable: max. 0.4%
(Roth and Kormann 2000, p. 128)

Shelf Life

Cocoa butter has a shelf life of up to 2 years.

C

Characteristics of the Fat

Cocoa butter is a very expensive fat, which is why substitutes have become important. These substitutes for cocoa butter can be subdivided in two groups:

Cocoa Butter Equivalents

The composition of fatty acids in and the glyceride structure of cocoa butter equivalents are comparable to those of cocoa butter. These fats partly substitute cocoa butter in cocoa products without having a negative influence on the sensory and processing properties. They are used to various extents in several countries, depending on the national legislation. Cocoa butter equivalents are either natural fats, for example Borneo tallow (Indonesia) and sal fat (India), or products created by fractionation of vegetable fats (especially palm oil) that should have a content of symmetrical monounsaturated glycerides that is higher than 50% of the total glyceride content.

Cocoa Butter Substitutes

The composition of fatty acids in and the glycerine structure of the substitute fats is significantly different from those of cocoa butter. The substitute fats are only used for the complete substitution of cocoa butter in confectionery and products containing cocoa. A differentiation is made between products containing lauric acid (laurics) and products free from lauric acid (non-laurics). Laurics are produced from virgin and/or hydrogenated or fractionated coconut or palm kernel oil, whereas non-laurics are produced from fractionated, hydrogenated or transesterified vegetable oils (sich as soya bean oil or palm oil) or animal fats (such as tallow or lard) and their mixtures.

5 Application

In Pharmacy and Medicine

For the Production of Suppositories and as an Ointment Base

Cocoa butter is used to produce bacilli, globuli vaginales and suppositories, and more rarely as an ointment base. Due to its many modifications, cocoa butter has very ambiguous melting behaviour. The fact that there is no volume contraction at solidification and that preparations containing water have a short shelf life is another reason why cocoa butter is seldom used (Hänsel et al. 1994, volume 6, 5th edition, p. 948).

For the Production of Patches

Patches containing cocoa butter have improved sticking properties, and metallic ingredients, for example mercury, are bound more securely (Warburg and Van Someren Brand 1909, p. 279).

For the Production of Pastilles

Formerly, it was customary to use chocolate in the production of pastilles, among other reasons in order to neutralise the unpleasant taste of some medicines. For example, chocolate was mixed with quinine, rhubarb, iron preparations, magnesium, calomel, ipecacuanha or santonin. Santonin, in particular, was used as a worm cake for children to treat ascaris in China (Warburg and Van Someren Brand 1909, p. 281).

In Cosmetics

In Soaps

Together with alkali, cocoa butter forms a good, firm and very mild white soap. Cocoa butter makes soaps firm and is used for super-fatted soaps and as a base oil for luxury soaps (Warburg and Van Someren Brand 1909, p. 279).

In Lip Balms and Body Care Products

Cocoa butter melts at body temperature and leaves the skin soft. It is especially suited for dry and chapped skin and prevents skin irritations. For these reasons, cocoa butter is added to lip balms and numerous balms, lotions and body creams.

Other Cosmetic Uses

Cocoa butter reduces wrinkles around the eyes and mouth, which is why it is added to antiwrinkle creams.

The butter is also useful for pregnant women: Since it contains a high amount of fat, it increases skin elasticity, and is thus used to treat and prevent stretch marks.

In the Food Industry

Cocoa butter has a good texture and is used in the production of chocolate candy as a covering material for almonds and liqueurs dissolved in sugar. Chocolate is a mixture of cocoa, sugar and spices (Warburg and Van Someren Brand 1909, p. 279).

According to the "Kakaoverordnung" (Cocoa Regulation, 2738 German Federal Law Gazette volume 2003, part I, no. 63, Bonn 23.12.2003), chocolate must contain at least 4% cocoa dry mass in total, of which at least 26% must be cocoa butter. Apart from cocoa butter, other vegetable fats may also be added, but they must not exceed 5% of the final product, retaining the minimum content of cocoa butter and cocoa dry mass, and after subtracting the total weight of the ingredients. Apart from cocoa butter, illipe, Borneo tallow, kokum butter, mango seed oil, palm oil, sal butter and shea butter may be used.

In Perfumery

Enfreurage is a method in perfumery. Flowers are placed between two layers of fat (in this case, cocoa butter). The fat absorbs the fragrance, from which it is extracted by alcohol (Von Wiesner 1927, 1st volume, p. 744).

As a Repellent

A study in Côte d'Ivoire showed that cocoa butter is a good insect repellent. Repellents are products applied to the skin; their smell repels blood-sucking insects (Pitroipa et al. 2002).

6 Possible Unwanted Side Effects

Cocoa products may cause allergic reactions of the skin, respiratory tract and gastrointestinal tract. The allergens seem to be proteins; their effect decreases during processing. With progressing age, the allergy disposition decreases. Persons sensitive to cocoa products may get migraines and headaches after consuming chocolate (Hänsel et al. 1994, volume 6, 5th edition, p. 951).

As cocoa butter contains a high amount of fat, cosmetic products containing cocoa butter are not suitable for persons with greasy skin and increased activity of the sebaceous glands (Valentino et al. 1984).

Since cocoa butter contains a high amount of palmitic acid, it increases the concentration of LDL (low-density lipoprotein) cholesterol to a greater extent than many other vegetable oils. Low-density lipoproteins are built using lipoprotein lipase from VLDL (very low-density lipoprotein) and contain about 80% serum cholesterol. A high serum level of LDL cholesterol may lead to arteriosclerotic vascular diseases and, as a result, to cardiac infarction (Denke 1994).

Coconut Oil

Oleum Cocos synonyms: copra oil (E); Kokosnussöl (D); huile de coprah (F)

1 Source Plant

Cocos nucifera L. (Arecaceae), coconut palm.

S. Krist, *Vegetable Fats and Oils*, https://doi.org/10.1007/978-3-030-30314-3_38

Habitat

The coconut palm is thought to originate in the Malay Archipelago, where it has been cultivated for 3000 or 4000 years. Today, it is common throughout the tropics, especially along coasts, since the palm prefers salty sea air. *Cocos nucifera* also needs constant warmth, with average temperatures between 20 °C and 25 °C; even in the coldest months, the temperature should not drop below 20 °C. The coconut palm grows in all kinds of soil, as long as the soil ensures sufficient moisture penetration and a good air and water supply (Roth and Kormann 2000, p. 53 et seqq; Hager 1978, volume 4, p. 181).

Description

Under normal conditions, the coconut palm reaches a height of 30–35 m and develops a lean, columnar, unbranched stem that often bends slightly from the ground upwards. A full-grown coconut palm has between 25 and 40 pinnate palm fronds with broad leaf stems clasping the stem. The fronds are 4–6 m long, with a ligneous midrib and 150–180 leaflets that are leathery and 90 cm long. The inflorescences are located laterally in the axils, in groups of three on the base with one yellowish-green female and two yellowish male flowers each. As the flowers blossom at different times, insects and the wind pollinate them with pollen from other inflorescences of the same or a neighbouring tree. Only about a fourth to a third of the pollinated flowers develop into ripe fruit. The fruit is a stone fruit, which is fully ripe about 13 months after the spathe has opened. The state of ripeness is discernible from the dry shell, the exocarp. At this time, the nuts contain the maximum amount of copra, which is the fruit pulp. Depending on the variety, ripe fruit weigh 1.5–2.5 kg. Sixty to one-sixty nuts ripen on a full-grown palm (Roth and Kormann 2000, p. 53 et seq; Hager 1978, volume 4, p. 181).

2 Cultivation and Extraction

Cultivation

Coconut palms gained commercial importance as late as the nineteenth century. The Dutch in Ceylon established the first plantations in the mid nineteenth century. Today, coconut palms are cultivated on both sides of the equator in the tropics, in coastal areas as well as inland along riverbanks. The global production is difficult to estimate, as a great part is used locally and does not appear in commercial statistics. Additionally, the harvest takes place not only in plantations, but also on the side of roads and in gardens. It is therefore difficult to register production figures and growing areas. Nevertheless, the following table gives an overview of the cultivation and the main producers, and shows that since 1980, production has doubled. This is a result of the manifold and increasing uses of copra (Roth and Kormann 2000, p. 130; Hager 1978, volume 4, p. 181; Salunkhe et al. 1992, p. 281 et seq).

The most important producing countries and their coconut production given in 1000 kg (fao.org/29.04.04):

Country	1980	1990	2000
Indonesia	8,660,000	12,120,000	16,235,000
Philippines	9,141,000	11,023,000	12,562,879
India	4,250,000	7,230,000	9,151,500
Sri Lanka	1,540,000	1,924,000	1,950,000
Brazil	341,820	477,372	1,880,325
Thailand	671,100	1,426,300	1,373,162
Mexico	892,871	1,063,600	1,213,000
total global production	32,215,261	42,482,756	50,074,327

Copra is mainly imported by countries with large oil mills like the Netherlands, France and Germany. The USA, on the other hand, imports mainly coconut oil, since it is not interested in accumulating more press residues that can be used as forage in addition to its own oleiferous fruit such as soy, peanut and cotton. The coconut itself does not play a significant role in export trade, at least not if compared with its products copra, coconut oil and coconut fibres. For more than 400 million people living in the tropics, however, the coconut palm is the most important cultivated plant and an essential source of fat and protein.

Extraction of the Oil

Coconut oil is also called coconut fat, as it is solid at room temperature. The reason for this is its high content in saturated fatty acids, which amounts to about 90%. However, since it has a low melting point of about 23 °C, it liquefies easily. The cut-up and dried endosperm of the coconut, the copra, is pressed to produce the oil. Fresh copra is dried, pressed, and the oil is subsequently refined. It can only be used as an edible oil after it has been refined and deodorised (Hänsel et al. 1999, p. 248; Roth and Kormann 2000, p. 130; Salunkhe et al. 1992, p. 305 et seq).

3 Character

Colour

White to yellowish (Hunnius 1998, p. 350).

Odour

Weakly fatty, waxy, fresh, slight touch of coconut, mild.

Usually slightly rancid (Hunnius 1998, p. 350).

Δ-Lactones of aliphatic hydroxycarboxylic acids with chain lengths between 8 and 14 C-atoms are responsible for the typical coconut smell. The most important individual compound for this is 5-pentyloxan-2-one. However, the odour of coconut oil is often termed "rancid", although it contains a high percentage of saturated fatty acids and does not spoil easily. *Kellard* et al. have tried to isolate the flavouring substances responsible for the smell, which they have named "perfume or ketone rancidity" (Kerschbaum and Schweiger 2001, p. 26; Kellard et al. 1984).

C

Volatile Compounds

Aroma analysis by enrichment with a Soxhlet extractor and GC-MS analysis of off-flavour compounds in coconut oil (Kellard et al. 1984):

Pentan-2-one	Heptan-2-ol
Pentan-2-ol	Octan-2-one
Hexan-2-one	Nonan-2-one
Hexan-2-ol	Nonan-2-ol
Heptan-2-one	Undecan-2-one

The aroma compounds listed above are not typical "off-flavours" (undesired flavouring substances), so the off-flavour probably results from a higher concentration in the coconut oil examined (Kellard et al. 1984).

Flavour

Like coconut (Kerschbaum and Schweiger 2001, p. 26).

4 Ingredients

Composition of Fatty Acids

Fatty acids	Content in % acc. to Roth and Kormann (2000), p. 130	Content in % acc. to Nature certificate of analysis 28.8.03	Content in % acc. to Kerschbaum and Schweiger (2001), p. 25
Caprylic acid	9	4.8	10
Capric acid	10	5.1	6.7
Lauric acid	45	46.6	52.1
Myristic acid	20	19	17.5
Palmitic acid	7	10.3	7.3
Stearic acid	5	3.4	1.7
Oleic acid	5	8.2	3.9
Linoleic acid	–	2.1	0.8
Caproic acid	–	0.2	–
Arachidic acid	–	0.1	–

Composition of Sterols

(concentration in mg/100 g and content in %)

Sterols	Coconut oil acc. to Firestone (1999), p. 29	Coconut oil acc. to Karleskind (1996), p. 201	Refined coconut oil acc. to Carstensen (2001), p. 42	Coconut oil acc. to Pantzaris and Basiron (2002), p. 164
Total content	47–114	80–230	262.4	–
Cholesterol	0.6–3.0	1–3	1.5	1.8
Brassicasterol	0–0.9	–	–	0.4
Campesterol	3.1–11.2	7–8	8.0	8.8
Stigmasterol	5.4–15.6	12–15	14.0	12.7
β-Sitosterol	19.7–50.7	42–47	43.8	46.1
Δ5-Avenasterol	13–40.7	25–34	27.6	27.4
Δ7-Stigmasterol	0–3.0	<3	–	1.6
Δ7-Avenasterol	0–3.0	<0.1	1.0	0.9
Others	0–3.6	–	–	–

Composition of Tocopherols

(concentration in mg/100 g and content in %)

According to Kerschbaum and Schweiger (2001), coconut oil does not contain vitamin E as a result of the necessary refinement. According to Salunkhe et al. (1992), however, it contains 20 mg α-tocopherol/kg oil (Kerschbaum and Schweiger 2001, p. 26; Salunkhe et al. 1992, p. 297).

Tocopherols	Coconut oil acc. to Karleskind (1996), p. 201	Refined coconut oil acc. to Carstensen (2001), p. 42
Total content of tocopherols	2–8	2.9
α-Tocopherol	–	–
β-Tocopherol	–	–
γ-Tocopherol	–	11
δ-Tocopherol	11	–
Tocotrienols	89	13.5

Other Ingredients

(Kerschbaum and Schweiger 2001, p. 14, 43; Salunkhe et al. 1992, p. 297 et seq)

Calcium
Potassium
Sodium
Phosphorus
Copper
Iron
Amino acids

Lactones
Vitamin E
Riboflavin
Niacin
Phytosterols

C

Physical Key Figures of Coconut Oil

refractive index: n_D^{40} 1.4485–1.4495
density: d_4^{15} 0.92–0.93
saponification number: 242–263
acid number: not over 6
iodine number: 7–10
unsaponifiable: 0.5%
point of solidification: 21 to 25 °C
(Roth and Kormann 2000, p. 130)

Shelf Life

Coconut oil has a shelf life of 1–2 years if stored in the refrigerator.

5 Application

In Pharmacy and Medicine

In Pharmaceutical Technology

Medium-chain triglycerides from coconut oil are used for pharmaceutical technological purposes. Medicines that are administered orally and are unstable in water can be dispensed or suspended in them. *Oleum Cocos* also serves as a base for ointments and creams and as a substitute for cocoa butter (Roth and Kormann 2000, p. 130; Hänsel et al. 1999, p. 250; Hunnius 1998, p. 350; Hager 1978, volume 4, p. 184).

In Parenteral Nutrition

As coconut oil contains caprylic acid and capric acid, it can be enriched after saponification and processed to form medium-chain triglycerides, which have a chain length of 6–12 C-atoms. They are easy to digest, as they can be absorbed without pancreatic lipase and bile acids, and are transported via the portal vein directly to the liver. They serve as an energy source in parenteral nutrition of seriously ill patients, and as a dietary supplement for customary edible fats in conditions of insufficient fat absorption, for example in cases of steatorrhea, enteritis, and after small intestinal resection. After having been absorbed, they cannot, however, be stored. Unwanted side effects often include stomachache and diarrhoea (Hänsel et al. 1999 p. 250; Fleischhacker 2002, p. 60; Roth and Kormann 2000, p. 130).

As Food

Coconut fat can be used as an edible fat for baking, roasting and frying. It is solid at room temperature, but liquefies easily due to its low melting point. When it melts, it collects considerable amounts of fusion heat and therefore has a pleasant cooling effect in the mouth (Roth and Kormann 2000, p. 130; Hunnius 1998, p. 350; Kerschbaum and Schweiger 2001, p. 45).

Technical Uses

Since coconut oil has a cooling effect, the sweets industry uses it in wafer fillings, ice cream and as a coating material. It is also an important addition to vegetable margarine and chocolate. Fatty acids accumulating during refinement are used to produce synthetic resins, insecticides and medicines. Higher saturated and unsaturated fatty alcohols, fatty alcohol sulphonates and other alcohol derivates are extracted from coconut oil. From capric acid methyl esters and octyl alcohol contained in coconut oil, octylaldehyde is obtained, which is used in the production of synthetic rose oil and lemon oil. *Oleum cocos* is also used in the production of soap and candles (Roth and Kormann 2000, p. 130; Hager 1978, volume 4, p. 184).

In Cosmetics

Although *Oleum cocos* hardly penetrates the skin, it is used frequently in cosmetics due to its moisturising properties and its cooling effect on the skin. It is added to the following products (Roth and Kormann 2000, p. 130; Hager 1978, volume 4, p. 184):

- hair grease;
- haircare products;
- sunblock;
- after-sun products;
- bath oils;
- creams, and
- massage oils.

Coffee Seed Oil

Oleum Coffeae synonyms: Kaffeebohnenöl (D); huile de café (F)

1 Source Plant

Coffea arabica L. (Rubiaceae), coffee

S. Krist, *Vegetable Fats and Oils*, https://doi.org/10.1007/978-3-030-30314-3_39

Habitat

The origins of coffee lie in the mountain forests of southwestern Ethiopia. The Arabs and afterwards the Turks cultivated coffee plants, and in the mid-16th century, the first coffee houses were established in Constantinople. The Turks brought coffee to Europe when they besieged Vienna in 1683. In the seventeenth century, pilgrims to Mecca brought coffee to eastern India, and Dutch and French merchants to Asia and America. Today, the main growing areas of coffee are Central and South America (Brazil, Mexico, Costa Rica, Guatemala, Ecuador, Venezuela, Peru and Colombia). In Africa, coffee is cultivated in Ethiopia, Liberia, Cameroon, Angola and Congo. The main producers in Asia are Malaysia and Indonesia. A special kind of coffee called pearl coffee is grown in the Arabian peninsula (Roth and Kormann 2000, p. 55).

Description

Coffee is an evergreen shrub or small tree that grows to a height of 5–6 metres. The leaves are alternate, light green, lanceolate-elliptical and about 7–10 cm long. Their edges are wavy; their upper side has a leathery shine. The small white flowers are arranged in groups of four to sixteen in the axils; they smell like jasmine. The fruit are stone fruit, and the size of a cherry. They are first dark green, but become yellow and red in the ripening process. When they are fully ripe, they are violet-black in colour. The outer layer is composed of juicy, sweet fruit pulp, which contains the seeds; each seed is surrounded by a parchment-like, dry, smooth, straw-coloured shell (Warburg and Van Someren Brand 1909, p. 172).

2 Cultivation and Extraction

Cultivation

Although coffee is not demanding, it only thrives if the climate and soil meet certain requirements. The plant needs a constant temperature between 15 and 30 °C, shade and moisture. Coffee grows up to a height of 1000 metres. If the soil is too dry, artificial irrigation is necessary. The yield is greatest on slightly acidic, loose, permeable soil of volcanic origin. Coffee is damaged by temperatures that are too low or too high, and by dry winds. To protect the plant from the sun and wind, shade trees are planted. Growing coffee is a protracted business, as the first yield can only be expected after 4 or 5 years. Coffee then, however, goes on to bear fruit for up to 35 years. If there are no great differences in the temperatures of the seasons, coffee flowers throughout the year, which means that fruit and blossoms in all stages of development are found on the plant at all times. As soon as the fruit are a violet-black colour, the harvest begins. The fully ripe fruit must be picked and further processed under either wet or dry conditions (Warburg and Van Someren Brand 1909, p. 119).

Extraction of the Oil

Coffee seed oil is extracted from the seeds of *Coffea arabica* with either diethyl ether or light petroleum, sometimes by previous reduction of the wax with tetrachloroethane (Roth and Kormann 2000, p. 127).

3 Character

Colour

Greenish-brown (Roth and Kormann 2000, p. 127).
From raw coffee beans: light yellow (Von Wiesner 1927, 1st volume, 4th edition, p. 781).
From roasted beans: dark brown (Von Wiesner 1927, 1st volume, 4th edition, p. 781).

Odour

Smells of green coffee beans (Roth and Kormann 2000, p. 127).
From raw coffee beans: without any particular odour (Von Wiesner 1927, 1st volume, 4th edition, p. 781).
From roasted beans: characteristic smell of coffee (Von Wiesner 1927, 1st volume, 4th edition, p. 781).

Flavour

Since coffee seed oil has toxic side effects when ingested, it is not suitable as food. There is no information available on the taste.

C

4 Ingredients

Composition of Fatty Acids

Fatty acids	Content in % acc. to Roth and Kormann (2000), p. 127	Content in % acc. to Firestone (1999), p. 30	Content in % acc. to augustus-oils.ltd.uk/ 01.09.05
Palmitic acid	20.2–23.6	35–42	35–42
Stearic acid	1.1–9.1	7–11	7–11
Oleic acid	12.4–20.2	8–10	8–10
Linoleic acid	25.5–37.6	36–43	36–43
Arachidic acid	>2	4–7	4–7
Icosenoic acid	–	4–7	–
Behenic acid	–	4–7	4–7
Cetoleic acid	–	4–7	–
Lignoceric acid	–	4–7	–
Nervonic acid	–	4–7	–
Carnauba acid	10–14.3	–	–

Composition of Sterols

Sterols	Content in % acc. to augustus-oils.ltd.uk/01.09.05
Cholesterol	–
Brassicasterol	–
Campesterol	19
Stigmasterol	20
β-Sitosterol	54
Δ5-Avenasterol	6
Δ7-Stigmasterol	1
Δ7-Avenasterol	–

Physical Key Figures of Coffee Seed Oil

refractive index: n_D^{25} 1.4678–1.4691
density: d_{15}^{15} 0.928–0.952
melting point: 8–9 °C
point of solidification: 3–11 °C
saponification number: 149–195
iodine number: 76–101
unsaponifiable: 6–10%
(Roth and Kormann 2000, p. 127)

Shelf Life

Coffee seed oil has a shelf life of 1 year.

5 Application

In Cosmetics

As coffee seed oil has a high unsaponifiable portion, a density similar to water and sun-protective properties, it can be added to sunscreen. Additionally, its high amount of phytosterols increases skin moisture, and the oil therefore forms part of various moisture creams and body lotions.

In India, in particular, coffee seed oil is used in soap production because of its moisturising properties (Singh et al. 1970).

In the Chemical Industry

Since it contains high amounts of sterols and vitamin D, coffee seed oil is suitable for extracting sterols, sterol derivates and vitamin D (Singh et al. 1970).

6 Possible Unwanted Side Effects

A study on rats about the biological, physiological and histopathologic effects of raw coffee seed oil substantiated the oil's toxicity. The study indicates toxic substances in coffee seed oil that not only have damaging effects on the metabolism, but also on the germinal epithelium. They also lower the body temperature. A diet with 5% coffee seed oil was lethal for rats within 10 days. Chronic administration of a diet with 2.5% coffee seed oil led to a significant inhibition of growth, an increase in the absolute water absorption, as well as liver enlargement. Physiological examinations showed a heavy impairment of the excretory functions of the liver. From a histological point of view, a diet with 5% coffee seed oil resulted in a fatty degeneration of individual liver cells to various degrees and heavy atrophy of the testicles. Smaller doses of the oil did not result in detectable histopathological changes. In addition, an increase in the plasma cholesterol level was registered, which was due to cafestol contained in coffee seed oil (Terpstra et al. 2000).

Cohune Oil

synonyms: Cohuneöl (D); huile de cohune (F), Olio di cohune (I), aceite de cohune (ESP)

1 Source Plant

Attalea cohune Mart. *(Arecaceae)*, cohune palm

S. Krist, *Vegetable Fats and Oils*, https://doi.org/10.1007/978-3-030-30314-3_40

Habitat

The cohune palm is a native plant from southern Mexico to Honduras, growing especially in marshy areas (Bärtels 1996, p. 51).

Description

The cohune palm is a feather palm that can reach a height of 15–18 metres. The trunk is about 30–40 cm in diameter. The leaves at the top stand upright and are 9 metres long and 2.5 metres broad. The plant is monoecious, the male and female flowers developing separately. The inflorescences are pendant, 1–1.5 metres long, and branched. The palm bears many fruits, which are about 6 cm long and resemble coconuts. There are about 800–1000 fruits per inflorescence (Bärtels 1996, p. 51).

The fruit is oval, about 5–8 cm long and about 5 cm in diameter. It has a hard shell; the mesocarp beneath is sludgy and contains a nut. The seed in the nut is 300 mm long and 18 mm in diameter (Axtell 1992, p. 51). The seeds of the cohune palm contain 65–75% fat (Bärtels 1996, p. 51).

2 Cultivation and Extraction

Cultivation

Cohune palms are mainly cultivated in Mexico, Honduras, Belize, Paraguay, Guatemala, El Salvador and Nicaragua. The palm needs tropical climates and high amounts of precipitation. It is a characteristic plant of the South American rainforests; there are usually 15 wild palms per hectare. In 1987, the harvest amounted to 20,000 tons of nuts (Axtell 1992, p.51–52).

Cohune palms are robust and survive lumbering and forest fires. They prefer deep, well-drained soil. As coconut oil is more popular, only small amounts of cohune oil are produced. In Belize, cohune oil is preferred because it is believed to have a longer shelf life and to be more profitable (McKillop 1996).

Extraction of the Oil

Since the nuts have a hard shell, it is not easy to reach the seeds. As it is difficult to crack the nuts manually, a machine was developed especially for this purpose. The seeds are subsequently heated or flame scarfed, and the oil is extracted using an expeller (Axtell 1992, p. 52).

3 Character

Colour

Yellow (McSweeney 1995).

Odour

The smell is similar to that of coconut oil (Holmes and Deuel 1920).

Flavour

Similar to coconut oil, but cohune oil is slightly smokier (McSweeney 1995).

4 Ingredients

Composition of Fatty Acids

Fatty acids	Content in % acc. to Axtell (1992), p. 52	Content in % acc. to Idiem'Opute (1979)	Content in % acc. to Babayan (1987)	Content in % acc. to Collin (1933)
Caproic acid	–	–	0.3	–
Caprylic acid	7.5	7.0	8.7	7.5
Capric acid	6.5	5.9	7.2	6.6
Lauric acid	46.5	50.8	47.3	46.4
Myristic acid	16.0	18.4	16.2	16.1
Palmitic acid	9.5	9.0	7.7	9.3
Stearic acid	3.0	3.2	3.2	3.3
Oleic acid	10.0	5.6	8.3	9.9
Linoleic acid	1.0	–	1.0	0.9

Physical Key Figures

iodine number: 9.8
saponification number: 251
(Squibb et al. 1951)

5 Application

Industrial Uses

The oil is used as a lubricant (Bärtels 1996, p. 51).
It is also used to produce soap (Squibb et al. 1951).

As Food

Cohune oil is an edible oil (McSweeney 1995).
The refined oil can be used for margarine production and is well-suited for baking (Axtell 1992, p. 51).

Coriander Seed Oil

Oleum Coriandri synonym: Korianderöl (D)

1 Source Plant

Coriandrum sativum L. (Apiaceae), coriander

S. Krist, *Vegetable Fats and Oils*, https://doi.org/10.1007/978-3-030-30314-3_41

Habitat

Coriander is an old spice plant already mentioned in Sanskrit manuscripts, papyri and the Old Testament. The plant originates in the eastern Mediterranean region and the Orient. Today, coriander grows in Europe, India, China, Japan, the USA and Argentina. *Coriandrum sativum* is a frugal plant, growing in cool and moist as well as in dry and warm conditions, and tolerating all kinds of soil from sandy to loamy. It grows wild as a weed among the seeds, in gardens and in fields on which crops or potatoes are cultivated, in vineyards, on rubble and on railway station grounds (Hager 1978, volume 4, p. 300; Diederichsen 1996).

Description

Coriandrum sativum is hairless with an upright stem, and can reach a height of 20–50 cm. The leaves are light green, the basal ones long-stalked, entire or pinnately compound, the middle ones usually bipinnately compound, the upper leaves stalkless. The typical *Apiaceae* umbels have long stalks, 3–5 rays. They are compound umbels, which either have no covering or only an inconspicuous leaf. The flowers are small and white or slightly pink. The fruit consists of two fruitlets firmly connected. It is either round or slightly elliptical, 7 mm long and 4–5 mm thick, hairless, smooth, and straw-coloured to yellowish-brown. The 10 ribs (5 each) protrude slightly and are twisted. The 8 straight ancillary ribs protrude more prominently. The fresh herb smells of bugs (Hager 1978, volume 4, p. 300; Pahlow 1985, p. 405).

2 Cultivation and Extraction

Cultivation

Coriander is now cultivated in Belarus, Hungary, the Czech Republic, France, northern Italy, the Middle East, China, Japan and North America. As the fruit drop off easily, coriander also grows wild by the wayside and on fallow land. There are still some cultivation goals to be achieved in order to use coriander for commercial cultivation on farmland, such as improvement of stability and vitality, higher yields, reduction of the amount of pericarp and higher amounts of essential oil and fat oil in the fruit (Schuster 1992, p. 174–175).

Coriandrum sativum prefers light soil with a high amount of lime. Heavy, wet soil is not ideal. Seedlings need a minimum temperature of –9 °C, and the plant at rosette stage a minimum of –16 °C to thrive. At the beginning of the vegetative period, a sufficient water supply is essential; afterwards, warm and dry conditions are ideal. 16–20 kg seeds per hectare are sown in late March; the sowing depth is 20–60 cm. The vegetation period is 80–120 days. The harvest takes place from mid-July onwards; the seeds are then a yellowish-brown colour, but are not yet fully ripe. Combining must, however, be carried out carefully, to prevent the fruit from falling apart. The ideal time is in the morning, after the dew has dried. After the harvest, the fruit are dried. The crop amounts to 1300–1500 kg/ha; crops up to 2300 kg/ha are possible in peak years (Dachler and Pelzmann 1999, p. 206–208).

Extraction of the Oil

The fat oil, especially petroselinic acid, is obtained by extraction or by pressing the coriander seeds (Ramadan et al. 2003; Diederichsen 1996).

Coriander seed oil can be obtained directly from the seeds by finely grinding them and subsequent extraction in a Soxhlet extractor with hexane (Msaada et al. 2009).

It is also possible to extract the oil with carbon dioxide or propane in supercritical or subcritical conditions. Varying the conditions of extraction or the type of solvent only leads to minor changes in the composition of fatty acids (Illés et al. 2000).

Coriander seeds contain a valuable essential oil, which is obtained by steam distillation. The residues can then be processed further to extract the fat oil (Schuster 1992, p. 174).

3 Character

Colour

Colourless with a slight yellow tint

Odour

Fresh, sweet, smell of lemons/limes, aromatic, peppery.

Flavour

Pleasantly aromatic, warm, nutty, touch of orange.

C

4 Ingredients

Composition of Fatty Acids

Fatty acids	Content in % acc. to Schuster (1992), p. 11	Content in % acc. to Ramadan et al. (2008)	Content in % acc. to Moser and Vaughn (2010)	Content in % acc. to Illés et al. (2000)	Content in % acc. to Ramadan and Mörsel (2002a, b)
Myristic acid		0.85	–	–	–
Palmitic acid	6–10	4.47	5.3	2.1–5.7	3.96
Palmitoleic acid		0.44	0.3	–	0.41
Stearic acid		0.65	3.1	–	2.91
Petroselinic acid	50–70	70.2	68.5	65.3–80.1	65.7
Oleic acid	25–30	8.2	7.6	0.4–5.0	7.85
Linoleic acid	6–8	14.4	13.0	14.0–20.1	16.7
Linolenic acid	–	0.36	–	–	1.22
Arachidic acid	–	0.27	–	–	0.25
Eicosenoic acid	–	–	–	–	0.30
Eicosadienoic acid	–	0.26	–	–	–
Erucic acid	–	–	–	–	0.16
Nervonic acid	–	0.09	–	–	–

Petroselinic acid is a position isomer of oleic acid. The general formula of oleic acid is C18:1, and that of petroselinic acid is C18:1. They only differ in the position of the double bond.

This explains why the Indian Institute of Spices Research did not differentiate between the content of oleic acid and petroselinic acid.

Composition of Sterols

Sterols	Content in g/kg acc. to Ramadan et al. (2008)	Content in µg/g acc. to Ramadan and Mörsel (2002a, b)
Total content	–	5186
Ergosterol	–	94
Campesterol	–	508
Stigmasterol	0.803	1548
Lanosterol	1.456	93
β-Sitosterol	2.644	1464
Δ5-Avenasterol	1.511	1235
Sitostanol	0.112	–
Δ7-Stigmastenol	0.141	–
Δ7-Avenasterol	0.192	244

Composition of Tocopherols

Tocopherols	Content in g/kg acc. to Ramadan et al. (2008)	Content in ppm acc. to Moser and Vaughn (2010)	Content in µg/g acc. to Illés et al. (2000)
Total content		229	–
α-Tocopherol	0.028	196	2.1–8.0
β-Tocopherol	1.080	0	–
γ-Tocopherol	0.0016	18	–
β + γ-Tocopherol	–	–	3.9–12.6
δ-Tocopherol	0.013	15	0.8–3.6

Other Ingredients

(Adikari et al. 1991; Ramadan et al. 2003)

Phytosterols
β-Sitosterol (predominant)
Phosphatidylcholine
Fat soluble vitamins (especially vitamin E)

Physical Key Figures of Coriander Seed Oil

refractive index: n^{35}_{D}	1.471
density: d^{15}_{15}	0.9267
saponification number:	177
iodine number:	109
unsaponifiable:	1.1%
point of solidification:	–2 °C
(Roth and Kormann 2005, p.155)	
molecular weight:	880.43
acid number:	2.66
flow point:	–26 °C
absolute viscosity:	
25 °C:	$25.13 mm^2s^{-1}$
40 °C:	$15.84 mm^2s^{-1}$
100 °C:	$4.91 mm^2s^{-1}$
viscosity index:	273
oxidative stability:	
induction period:	21.0 h
decomposition temperature:	214.9 °C
relative density (25 °C):	0.886
relative density (40 °C):	0.877
(Moser and Vaughn 2010)	

5 Application

In Pharmacy and Medicine

A study by Ramadan, Amer and Awad examined the effects of coriander seed oil on the plasma lipid profile of rats fed a diet containing cholesterol. The result showed that the oil has a positive effect on the composition of plasma lipids. Despite its low vmic effect on animals. It is assumed that minor components like sterols and tocopherols are involved in this effect. The study indicates that coriander seed oil can probably inhibit the development of coronary arteriosclerosis, but further tests on humans are necessary to confirm this (Ramadan et al. 2008).

As a Source of Antioxidants
Since coriander seed oil contains a high amount of tocopherols, it is a source of antioxidants. The high-quality oil can be used like a natural antioxidant in food containing lipids (Ramadan and Mörsel 2002a, b).

Technical Uses

Coriander seed oil contains a considerable amount of petroselinic acid. Oxidative cleavage results in a mixture of lauric acid and adipic acid, which can be used for technical purposes. Adipic acid is used in the production of nylon and emollients, emulsifiers, detergents and soap (Ramadan and Mörsel 2002a, b).

Petroselinic acid can be cleaved into adipic acid (C6) and lauric acid (C12). Today, adipic acid – a 1,4-butanedicarboxylic acid – is produced from mineral oil under release of N_2O, which damages the ozone layer. The global production of adipic acid currently amounts to 2.5 million tons; it is used:

- as an intermediate product for nylon. This synthetic material is a condensate of adipic acid and hexamethylenediamine (which is also produced from adipic acid) under release of water;
- as acidifier E355 in food;
- as a substitute for tartaric acid in baking powder and lemonade;
- for monomer and polymer PVC-emollients, and
- as a component in synthetic lubricants.

Coriander seed oil is a natural, environment-friendly source for these products. Petroselinic acid is used:

- to produce medium-chain lauric acid, and
- as an addition to cosmetic products, soap, surfactants, detergents and emulsifiers.

Coriander seed oil can also be used to produce biodiesel. The oil is transesterified with methanol and sodium methoxide to coriander seed oil methyl ester. The yield amounts to 94%. Before transesterification, an acid catalysed pretreatment is necessary in order to reduce the oil's acid number from 2.66 to 0.47 mg KOH g^{-1}. Coriander seed oil methyl esters have excellent fuel properties. They possess good oxidative stability and a low iodine number. The derived cetane number is 53.3, kinematic viscosity amounts to 4.21 $mm^2\ s^{-1}$ (40 °C) and the induction period is 14.6 h (110 °C) (Moser and Vaughn 2010).

As Food

The populations of several African and Latin American countries use coriander seed oil for seasoning food such as vegetables, gravy and sweets.

Corn Oil

Oleum Maydis synonyms: Maisöl (D); maize oil (E); huile (de germe) de mais (F)

S. Krist, *Vegetable Fats and Oils*, https://doi.org/10.1007/978-3-030-30314-3_42

1 Source Plant

Zea mays L. *(Poaceae), maize*

Habitat

Corn maize has its origins in Central and South America, in the region between Mexico and Peru. In Europe, it was only cultivated from the seventeenth century onwards. Via Italy, the Balkans and Russia, it reached India and China. After the Second World War, the originally tropical and subtropical plant was cultivated worldwide. In the USA and other industrialised countries, in particular, it is mainly used as forage and as a source of protein and starch. Only a small amount of the harvest is used to obtain oil. The main growing area is North America (Roth and Kormann 2000, p. 109).

Description

Maize is an annual plant that grow to a height of up to 2.5 m, with a pithy stem. The leaves are broadly lanceolate, light green and can reach a breadth of 12 cm and a length of 120 cm. The male spikes are two-flowered and joined in spiral pseudo-spikes to form a panicle at the top. The female inflorescences are located on the same plant; they are cobs with protruding styli wrapped in leaf sheaths, from which the fruit develops, which is also formed like a cob. The corn is about the size of a pea and flat on the side; inside the corn is the germ from which the oil is extracted. The leathery pericarp of the corn varies in colour and can be white, yellow, brown or nearly black (Roth and Kormann 2000, p. 109). Each corn weighs 150–600 mg and consists of the pericarp, endosperm and germ. The latter takes up 10–14% of the corn's total weight and contains about 34% oil. In varieties especially rich in oil, the germ can amount to up to 24.7% of the corn's total weight and contain up to 50% oil (Salunkhe et al. 1992, p. 405 et seq).

2 Cultivation and Extraction

Cultivation

Maize prefers sunny locations and deep, rich soil. The seeds are sown in early May. The plant blossoms from June to August; the harvest takes place from August to September. The corn is harvested by machines (Roth and Kormann 2000, p. 109).

Extraction of the Oil

The oil is obtained from the maize germs after they have been separated from the corn (Ulmer 1996, p. 63). This usually occurs in the process of producing starch. The germs are then either pressed or extracted with solvents such as hexane (Salunkhe et al. 1992, p. 413). The raw oil (yield about 15–20%) decomposes quickly and must be refined immediately (Roth and Kormann, 2000, p. 135) to remove insoluble substances, free fatty acids, phosphatides, gum and mucilage (Salunkhe et al. 1992, p. 413).

3 Character

Colour

Yellow (Roth and Kormann 2000, p. 135) to reddish oil (Salunkhe et al. 1992, p. 415).

Odour

Green and nutty, smell of the leaf sheaths of the corn cobs, full.
Pleasant (Salunkhe et al. 1992, p. 415).

Volatile Compounds

The following volatile compounds were analysed in corn oil using gas chromatography after storage for 8 days at 60 °C (Snyder et al. 1985):

Ethane	Octane	Hexane
2-Butenal	Octene	Hexanal
Pentane	Octanal	t-2-Hexanol
Pentanol	Octenal	Heptanal
Pentenal	1-Octan-3-ol	c-2-Heptenal
Pentyl-furan	Propane	t-2-Heptenal
Nonanal	Propenol	c-2-Decenal

If corn oil is heated before storage, many more volatile compounds emerge after storage for 1 week at 55 °C (Wu and Chen 1992):

n-Propanal	n-Pentanol	(E)-2-Nonenal
(E)-2-Propenal	2-Octanone	3.5-Octadien-2-one
Ethyl acetate	(Z)-2-Heptenal	(E)-2-Decenal
n-Pentanal	(E)-2-Heptenal	2,3-Octadione
n-Hexanal	n-Hexanol	(E,E)-2,4,Nonadienal
(E)-2-Pentenal	n-Nonanal	(E)-2-Undecanal
1-Penten-3-ol	(Z)-2-Octenal	(E,Z)-2,4-Decadienal
2-Heptanone	1-Octen-3-ol	(Z,Z)-2,4-Decadienal
2-Pentylfuran	(E,Z)-2,4-Heptadienal	(E,E)-2,4-Decadienal
(E)-2-Hexenal	(E,E)-2,4-Heptadienal	n-Hexanoic acid

The longer the storage period, the more volatile compounds are created. It is significant that from a storage period of 9 weeks onwards, the acid content increae considerably, whereas the proportion of aldehydes decreases continually (Wu and Chen 1992).

Flavour

Weak, characteristic (Roth and Kormann 2000, p. 135).

4 Ingredients

Composition of Fatty Acids

Apart from conventional corn oil, there is a variety of corn oil with a high amount of oleic acid and a variety with few saturated fatty acids:

Fatty acids	Content in % acc. to Firestone (1999), p. 31 et seq	Content in % (high amount of oleic acid) acc. to Firestone (1999), p. 31	Content in % (few saturated FA) acc. to Firestone (1999), p. 32	Content in % acc. to the Deutsche Gesellschaft für Fettwissenschaft (2003)
Lauric acid	0–0.3	–	–	0–0.3
Myristic acid	0–0.3	–	–	0–0.3
Palmitic acid	9.2–16.5	10–16	6–8	8.6–16.5
Palmitoleic acid	0–0.4	–	–	0–0.5
Margaric acid	–	–	–	0–0.1
Stearic acid	0–3.3	2	1	0–3.3
Oleic acid	20–42.2	44–64	25–31	20–42.2
Linoleic acid	39.4–65.6	20–38	58–64	34–65.6
α-Linolenic acid	0.5–1.5	0.8–1.0	0.8–0.9	0–2
Arachidic acid	0.3–0.7	1	0.5	0.3–1
Eicosanoic acid	0–0.4	–	–	0.2–0.6
Behenic acid	0–0.5	–	–	0–0.5
Erucic acid	0–0.1	–	–	0–0.3
Lignoceric acid	0–0.4	–	–	0–0.5

Composition of Sterols

(Concentration in mg/100 g and content in %)

Sterols	Corn oil acc. to Firestone (1999), p. 32	Corn oil acc. to Carstensen (2001), p. 42	Corn oil acc. to Karleskind (1996), p. 142
Total content	795–2215	780.6	830–2550
Cholesterol	0.2–0.6	0.9	<1
Brassicasterol	0–0.2	0	–
Campesterol	18.6–24.1	21.9	16–21
Stigmasterol	4.3–7.7	6.2	4.3–7.7

Sterols	Corn oil acc. to Firestone (1999), p. 32	Corn oil acc. to Carstensen (2001), p. 42	Corn oil acc. to Karleskind (1996), p. 142
β-Sitosterol	54.8–66.6	69.6	63–70
Δ5-Avenasterol	4.2–8.2	2.3	1–9
Δ7-Stigmasterol	1.0–4.2	0.2	<1
Δ7-Avenasterol	0.7–2.7	0.2	<1
Ergosterol	–	–	1–2

Composition of Tocopherols and Tocotrienols

(Concentration and content in mg/100 g)

Tocopherols/ tocotrienols	Corn oil acc. to Firestone (1999), p. 32	Corn oil acc. to Carstensen (2001), p. 44	Corn oil acc. to Karleskind (1996), p. 142	Corn oil acc. to Belitz and Grosch (1987), p. 223
Total content of tocopherols	33.1–371.6	104.3	113–183	–
α-Tocopherol	2.3–57.3	26.3%	8–22%	27.2
β-Tocopherol	0–35.6	2.0%	<3%	0.2
γ-Tocopherol	26.8–246.8	68.7%	68–89%	56.6
Δ-Tocopherol	23–75	0.5%	2–7%	2.5
Total content of tocotrienols	0–70.9	2.7	–	–
α-Tocotrienol	0–23.9	2.6%	–	5.4
β-Tocotrienol	–	–	–	1.1
γ-Tocotrienol	0–45	–	–	6.2
Δ-Tocotrienol	0–2	–	–	–

Other Ingredients

(ISF 1995, p. 167; Karleskind 1996, p. 142)

Carotenoids: 1–2 ppm
Phosphatides: 1–2%
Hydrocarbons: 40 mg/100 g
Of these squalene: 13–24 mg/100 g

Physical Key Figures of Corn Oil

refractive index: 60 °C 1.4596.
40 °C 1.465–1.468
25 °C 1.470–1.474
density: 0.916–0.925
iodine number: 103–135
saponification number: 187–196

acid number: 0.6–4
melting point: −18 to −10 °C
ignition temperature: 393 °C
flash point: 321 °C
unsaponifiable: 0.8–3%
(Salunkhe et al. 1992, p. 415; Firestone 1999, p. 31; Roth and Kormann 2000, p. 135; Karleskind 1996, p. 142)

C

5 Application

In Pharmacy and Medicine

The pharmaceutical industry uses corn oil for coating pills and tablets (Salunkhe et al. 1992, p. 417) and as a carrier for immunosuppressives (Carstensen 2001, p. 3).

Industrial Uses

In the Cosmetics Industry

From corn oil, soap and hair-care products are produced (heess.de/07.04.06), as well as face and body powders (Salunkhe et al. 1992, p. 417).

As a Raw Material

Corn oil is used for lubricants and leather-care products. Smaller amounts of raw and refined oil are used for the production of ammunition, insecticides, substitutes for paint and varnish, anti-rusting agents and textiles (Salunkhe et al. 1992, p. 417).

In the Food Industry

Corn oil is added to food products (margarine, mayonnaise, dietetics, baby food).

As Food

Corn oil is frequently used as an edible oil due to its flavour stability in storage and during cooking. In the USA, 60–65% of corn oil produced is used as either salad oil or cooking oil. It is also suitable for frying, as it does not transfer flavour from one food item to another. Frequently (this applies to about 30% of the corn oil produced), margarine (Salunkhe et al. 1992, p. 417), mayonnaise and other kinds of food are made from it (Roth and Kormann 2000, p. 135).

As Dietary Oil

Corn oil is regarded as dietetically valuable: The amount of sitosterol and tocopherols and the favourable ratio of unsaturated fatty acids have a less negative influence on LDL cholesterol levels than other fats and oils. Sitosterol is a substance that is chemically similar to cholesterol. It is assumed that sitosterol interferes with the absorption and reabsorption of cholesterol and thus contributes to the lowering of the cholesterol level (Hänsel et al. 1999, p. 261).

Cottonseed Oil

Oleum Gossypii synonyms: Baumwollsamenöl (D); huile de coton (F)

1 Source Plant

Gossypium herbaceum L. and *Gossypium hirsutum* L. (Malvaceae), cotton plant

S. Krist, *Vegetable Fats and Oils*, https://doi.org/10.1007/978-3-030-30314-3_43

Habitat

The cotton plant originates in southern Africa. *Gossypium herbaceum* and *Gossypium hirsutum* were already cultivated 3000 years ago in India, as well as in America at the time of the Inca civilisation. Today, cotton is mainly grown in the so-called Cotton Belt, as the plant requires wet soil in a hot climate with at least 200 frost-free days.

Description

The species *Gossypium* consists mostly of shrubs that are nearly arborescent and up to 2 m tall, but they are usually cultivated as annual, herbaceous plants. The leaves are mostly 3–7 lobed and heart-shaped at the base. The flowers are a pale dark yellow, purple or white colour, and stand solitary in the axils. The fruit is a capsule about the size of a walnut, bursting open in 3–5 flaps and containing 5–10 seeds. The seed is kidney-shaped or oval, 3–5 mm thick and blackish. On the surface, it is covered with long, usually white but sometimes also yellowish or brownish monocellular hair, which together form the wads of wool that protrude from the ripe fruit and can reach the size of a fist. The length of the fibres varies from 20 to 46 mm. They consist almost exclusively of cellulose (Hager 1978, volume 4, p. 1182).

2 Cultivation and Extraction

Cultivation

The regions that meet the climatic prerequisites for the cultivation of cotton plants mostly lie between the 35th degree of southern latitude and the 45th degree of northern latitude. In this Cotton Belt are the most important countries for the cultivation of cotton plants and cotton export: China, the USA and India. Other cultivated areas are located in Mexico, Colombia, Argentina and Australia.

The most important countries cultivating cotton plants in million hectares:

Country	Cultivated area in million hectares 2003
USA	5.6
China	4.8
India	8.5
Argentina	0.25
Australia	0.6
Mexico	0.07
South Africa	0.05
Indonesia	0.012
Colombia	0.055

An average crop of 1000 kg cotton (West Africa) contains:

- 430 kg fibres (cotton wool)
- 200 kg cake of cotton seed or cotton seed meal
- 100 kg oil
- 200 kg shells
- 30 kg retained seeds
- 40 kg impurities

Extraction of the Oil

Cotton factories used to regard the seeds of the cotton plant as a disposal problem. Today, they are a valuable by-product. The seeds of *Gossypium hirsutum L.* and other *Gossypium* species contain 20–25% fat oil, which is pressed and then refined. The oil is a dark colour and is cleaned using sulphuric acid and subsequently with leach. Volatile compounds are removed by steam. At low temperatures, solid cottonseed-margarine settles, which is used as a foodstuff. The liquid portion is put on the market as cleaned cotton seed oil (Hager 1978, volume 7b, p. 164; Roth and Kormann 2000, p. 118).

Gossypii oleum hydrogenatum (hydrogenated cotton seed oil) Ph.Eur. is produced by cleaning and hydrogenation (Ph.Eur. 2002).

3 Character

Cotton seed oil is available in three grades:

Type 1: press oil – whole oil: light-brown colour, free fatty acids <1.5%
Type 2: dark-brown colour, free fatty acids <3.0%
Type 3: fuliginous colour, free fatty acids >3.0%

Colour

The pressed oil is always dark, and the colour intensifies due to oxidation if it is exposed to air. Oil from moist seeds is especially dark in colour. The poisonous gossypol colours the oil dark to blackish-red and must be removed before use. The refined oil is light yellow (Roth and Kormann 2000, p. 118).

Weakly yellow – the raw oil is a dark, often deep-red colour (Hager 1978, volume 7b, p. 164)

Pale yellow (Hunnius 1998, p. 616).

Odour

Refined: weakly sweetish, reminiscent of oblates, slightly muffled, weakly oily.
The odour is characteristic but not unpleasant, and after refining hardly perceptible (Hager 1978, volume 7b, p. 164).
Odourless (Hunnius 1998, p. 616).

C

Volatile Compounds

Analysis of the volatile compounds in cotton seed oil with capillary gas chromatography (Snyder et al. 1985):

Ethane	Octane
Propane	t-2-Hexenal
Propenal	Heptanal
Pentane	c-2-Heptenal
Hexane	t-2-Heptenal
2-Butenal	1-Octen-3-ol
Pentanal	Pentylfuran
Pentenal	Octanal
Pentanol	Octenal
Octene	Nonanal
Hexanal	t-2-Decenal

Flavour

Mild taste (Hager 1978, volume 7b, p. 164). Light to moderate aroma.

4 Ingredients

Composition of Fatty Acids

Fatty acids	Content in % acc. to Hager (1978), volume 7b, p. 164	Content in % acc. to dgfett. de/11.09.03
Palmitic acid	23	21.4–26.4
Linoleic acid	34–55	46.7–58.2
Palmitoleic acid	–	1.2
Oleic acid	15–36	14.7–21.7
α-Linolenic acid	–	0.4
Stearic acid	1.1	2.1–3.3
Arachidic acid	1.3	0.2–0.5
Behenic acid	–	0.6
Erucic acid	–	0.3
Myristic acid	1.4	0.6–1.0

Composition of Triglycerides

Triglycerides	Content in % acc. to Karleskind (1996), p. 193
PLL	24.5
POL	17.1
LLL	13.5
OLL	13.1
PPL	9.0
OOL	5.7
PPO	3.4
POO	2.5
SLL	2.1
PSL	1.7
SOL	1.5
OOO	0.8

Composition of Sterols

(Concentration in mg/100 g and content in %)

Sterols	Cotton seed oil acc. to Firestone (1999), p. 32 et seq	Cotton seed oil acc. to Karleskind (1996), p. 195
Total content	269–643	375–580
Cholesterol	0.7–2.3	–
Brassicasterol	0.1–0.9	–
Campesterol	6.4–14.5	4–8
Stigmasterol	2.1–6.8	<1
β-Sitosterol	76.0–87.1	86–93
Δ5-Avenasterol	1.8–7.3	–
Δ7-Stigmasterol	0–1.4	–
Δ7-Avenasterol	0.8–3.3	–
Others	0–1.5	–

Composition of Tocopherols

(Concentration in mg/100 g and content in %)

Tocopherols	Cottonseed oil acc. to Firestone (1999), p. 33	Cottonseed oil acc. to Karleskind (1996), p. 195
Total content	39–143	80–140
α-Tocopherol	13.6–67.4	32–49
β-Tocopherol	0–3	–
γ-Tocopherol	13.8–75.0	31–57
δ-Tocopherol	0–2	1–8

Other Ingredients

Vitamin E
Calcium
Potassium
Magnesium
Sodium
Phosphorus
Copper
Iron
Manganese
Zinc

Physical Key Figures of Cotton Seed Oil

Refractive index: n_D^{40} 1.4645–1.4655
density: d_{25}^{25} 0.915–0.921
saponification number: 190–198
iodine number: 101–117
unsaponifiable: 0.5–1.5%
melting point: −2 °C
point of solidification: −6 to +4 °C
(Roth and Kormann 2000, p. 118)

Characteristics of the Oil

The seeds of cotton seed oil contain up to 1.5% gossypol, a yellow triterpene that is particularly reactive due to its two aldehyde groups and may trigger unwanted reactions in the organism.

The Toxicology of Gossypol

Like the seeds, the raw oil contains toxic gossypol in its unsaponifiable portion. It must be removed in the process of refining. The toxic effect of gossypol is as follows: The acute toxicity results in circulatory failure, the subacute toxicity in pulmonary oedema and the chronic toxicity in malaise and malnutrition. Pregnant women may have miscarriage due to the antiprogesterone and anti corpus luteum effect of gossypol. In a dosage of 10 mg/day, it hems spermatogenesis and for this reason was tested as a "pill for men". It turned out to have considerable side effects, for example loss of libido, and thus proved unsuitable (Hänsel et al. 1999, p. 260; Roth and Kormann 2000, p. 118).

5 Application

Technical Uses

Cotton seed oil is used in the production of (Salunkhe et al. 1992, p. 274; Roth and Kormann 2000, p. 118):

- margarine;
- leather-care products;
- canned fish and canned smoked meat;
- soap, and
- cosmetic and pharmaceutical ointments and creams.

As a Raw Material for the Production of Hardened Cotton Seed Oil

Gossypii oleum hydrogenatum (hardened cottonseed oil) is used in pharmaceutical technology, for example as a retarding agent in the production of time-release drugs or as a lubricant and forming agent for direct compression (Hänsel et al. 1999, p. 259).

As a Source of Vitamin E

Cotton seed oil is rich in tocopherol and, for this reason, is used for vitamin E retrieval and to treat vitamin E deficiency symptoms (Roth and Kormann 2000, p. 118; Hunnius 1998, p. 616).

As Food

Oleum Gossypii must not be excessively heated because of its high proportion of unsaturated fatty acids. Instead, it should be used for cold meals such as starters and salads. Its mild aroma harmonises with other oils, with which cottonseed oil can be mixed in amounts varying from 25% to 50%. The press residue is rich in protein and after the poisonous gossypol has been removed, it can be used not only as forage, but also in human nutrition (Roth and Kormann 2000, p. 118; Hunnius 1998, p. 616; Salunkhe et al. 1992, p. 274).

Crambe Oil

Synonyms: Krambenöl (D); huile de Crambe (F), olio del Crambe (I), aceite de Crambe (ESP)

1 Source Plant

Crambe abyssinica L. *(Brassicaceae)*, crambe

Habitat

Crambe abyssinica is a native plant of Ethiopia and was originally used as a vegetable and for the extraction of edible oil. Crambe oil is no longer used as an edible oil because of its high amount of erucic acid, but rather for technical and pharmaceutical purposes (Ebermann and Elmadfa 2008, p. 543).

Description

Crambe abyssinica is an annual plant reaching a height of 50–150 cm. The taproot reaches deep into the ground. The stem is branched from the base. Its bottom part is covered with hairs, but it is hairless in the upper part. The plant has different forms of leaves. The bottom ones are long-stemmed, roundish-oval to heart-shaped, with an irregular edge. The upper leaves are smaller and lanceolate. The flowers are arranged in racemes. They have white petals as well as 4 long and 2 short stamina, as is usual in the cabbage family. Bees and bumblebees pollinate the plant, but autogamy is also possible. The fruit is a jointed pod with two parts and a diameter of 1–4.5 mm. The seed contained within is greyish-green to yellow, round, and 1.8–2.5 mm in diameter. The harvest consists of naked seeds as well as pods containing the seeds. The shells make up 30% of the total weight (Schuster 1992, p. 50–51).

S. Krist, *Vegetable Fats and Oils*, https://doi.org/10.1007/978-3-030-30314-3_44

C

2 Cultivation and Extraction

Cultivation

Crambe abyssinica was first cultivated in Russia in 1930. Cultivation was continued during the Second World War and spread to Ukraine, Poland and Germany. *Crambe abyssinica* is mainly important as a result of its high oil content, as well as its short vegetation period of 80–90 days, its modest requirements and its resistance to dry periods. The low yields of only 2000–3000 kg/ha are, however, disadvantageous (Schuster 1992, p. 51–52).

In Europe, the harvest amounts to about 2353 kg seeds per hectare, respectively 846 kg oil with an average amount of 57.8% erucic acid. Ideal conditions for *Crambe abyssinica* include a precipitation rate of 350–1200 mm, an annual mean temperature between 5.7 and 16.2 °C and soil with a pH value between 5.0 and 7.8 (Falasca et al. 2010).

As a result of its modest requirements and resistance to dry periods, *Crambe abyssinica* is ideal for locations with dry pre-summers. It prefers calcareous, sandy soil. It should be sown as early as possible, between the end of March and the beginning of April. The plant can be cultivated as an alternative to rape if the aim is a high amount of erucic acid. The harvest takes place from the end of July until the beginning of August with a combine harvester. After the harvest, the seeds are dried until their water content drops to 9% (Löw 2003, p. 21–22).

Extraction of the Oil

The seeds of *Crambe abyssinica* can be pressed or extracted with solvents. If amounts of 10–60 tons per day are processed, which is usual for *Crambe abyssinica*, pressing is the more economic method (Cuperus et al. 1996).

The advantages of using a screw press rather than extraction with solvents are that the seed capital is lower, the quality is better and the processing is more environment-friendly. The disadvantage is a lower yield. To increase the yield, the seeds can be cooked before pressing. In this way, less oil remains in the flour. Optimal yields are obtained by cooking the seeds for 10–15 minutes at 100 °C (Wiesnborn et al. 2001).

3 Character

Colour

The oil is yellow; if it is refined, it is light yellow. This is due to the decomposition of carotenoids and chlorophyll during bleaching (Vargas-Lopez et al. 1999).

Flavour

Crambe oil is similar to soya bean oil in taste (Mustakas et al. 1965).

4 Ingredients

Composition of Fatty Acids

Fatty acids	Content in % acc. to Schuster (1992), p. 10	Content in % acc. to Downey (1971)	Content in % acc. to Princen and Rothfus (1984)	Content in % of bleached oil acc. to Vargas-Lopez et al. (1999)	Content in % acc. to Mustakas et al. (1965)
Myristic acid	–	0.1	–	–	0.3
Palmitic acid	2–10	1.7	–	3.5	2.0
Palmitoleic acid	–	0.3	–	–	0.3
Stearic acid	2–4	1.0	–	1.7	0.7
Oleic acid	12–18	16.7	17	14.1	15.0
Linoleic acid	8–12	7.8	9	10.8	10.4
α-Linolenic acid	8–10	6.9	6	6.3	6.6
Arachidic acid	–	1.3	–	0.7	0.7
Eicosenoic acid	0–2	2.9	5	6.6	3.0
Eicosadienoic acid	–	–	–	0.4	0.1
Behenic acid	–	2.7	–	0.6	1.6
Erucic acid	55–62	55.7	55	52.8	55.0
Lignoceric acid	–	–	–	1.4	0.7
Nervonic acid	–	2.9	–	–	3.2

Triglycerides

886 g/kg (Gurr et al. 1974).

Phospholipids

Phospholipids	Content in g/kg acc. to Gurr et al. (1974)
Total content	110
Phosphatidylcholine	48
Phosphatidylethanolamine	19
Phosphatidylinositol	48

C

Other Ingredients

Mineral substance	Content in mg/kg in raw oil acc. to Vargas-Lopez et al. (1999)	Content in mg/kg in degummed[a] oil acc. to Vargas-Lopez et al. (1999)	Content in mg/kg in bleached oil acc. to Vargas-Lopez et al. (1999)
Calcium	129	95	74
Magnesium	70	46	29
Potassium	64	4	9
Sodium	3	31	18
Phosphorus	201	129	57
Selenium	99	89	55
Copper	3	2	2
Iron	6	4	4

[a]by adding citric acid and by subsequent neutralisation with NaOH, the greater part of the phospholipids is removed

Physical Key Figures

Key figures acc. to Mustakas et al. (1965)	Pressed oil	Oil extracted with solvents from pre-pressed seeds
Density at 25 °C	0.9394	0.9408
Refractive index at 20 °C	1.4713	1.4719
Melting point in °C	6.5	6.0
Viscosity at 25 °C in centipoise	85	85
Iodine number	93	96.1
Saponification number	173	173
Unsaponifiable in %	0.47	0.59
Peroxide number	1.9	1.0
Hydroxyl number	1.87	3.08

iodine number: 111–113
(Löw 2003, p. 22)

5 Application

Industrial Uses

Crambe oil is used as a lubricant. Additionally, biodiesel can be produced from the methyl esters of its fatty acids. Most notably, the oil is a source of erucic acid. Erucic acid amide is an important lubricant in the production of cling film. Erucic acid and its trans isomer brassidic acid are also used in the production of nylon (Ebermann and Elmadfa 2008, p. 543).

Brassidic acid is obtained by oxidative cleavage of erucic acid at a double bond. Pelargonic acid is a by-product of this process (Princen and Rothfus 1984).

Crambe oil is also used in the production of emollients, glue and electric insulation (Falasca et al. 2010).

In Cosmetics

Erucic acid amide obtained from crambe oil is added to cosmetic products (Falasca et al. 2010).

Croton Oil

Oleum Crotonis synonyms: Krotonöl, Crotonöl, Granatillöl (D); *Oleum (Crotonis) Tiglii, Oleum Tilii* (L); croton seed oil (E); huile de croton (F)

1 Source Plant

Croton tiglium L., *syn. Croton acutus Thunb. (Euphorbiaceae)*, purging croton

Habitat

The genus *Croton* consists of about 50 species growing in the tropics of Asia, America and Africa. The name is derived from the Greek ‚/kroton‘/ = dog louse, since the seeds look similar to a louse or a tick. The species name *tiglium* has its origins in the Portuguese language; one of the Molucca Islands on which the small tree or bush was found was named ‚/tilho‘/. The plant also grows in India and Bengal as well as – apparently in a cultivated form – in West Africa, Mauritius, Java, the Philippines and New Guinea. It is likely that croton seeds spread to the Arabian peninsula in the thirteenth century via the Silk Road.

Description

Croton tiglium is either a tree reaching a height of 4–6 m, or a bush. It has obovate or ovate elliptical leaves with a serrate edge and long, thin stems. The flowers are small and white; they are arranged in terminal, upright racemes. On the upper end are the male flowers, in the bottom part the female flowers. The obovate fruit is bluntly triangular, yellowish on the outside and brown on the inside. It contains oval, elongated, shiny black seeds flattened on one side and arched on the other (Hager 1998, p. 469).

S. Krist, *Vegetable Fats and Oils*, https://doi.org/10.1007/978-3-030-30314-3_45

2 Cultivation and Extraction

Cultivation

Apart from the regions named above, Croton tiglium is cultivated in Sri Lanka, on the Malabar Coast, in China, East Africa and on Ambon Island. In tropical and subtropical Asia, the tree is planted to shade coffee and cocoa plantations, and as a hedge.

Extraction of the Oil

Croton oil is the fat oil pressed from the peeled seeds, according to the German Pharmacopoeia 6 by cold pressing, according to Pharmacopoeia Helvetica V by pressing the ripe, lightly roasted seeds at mildly warm temperatures (Hager 1998, p. 469).

Another species of croton from which fat oil can be extracted is *Croton penduliflorus Hutch.* From its seeds, *Croton penduliflorus seed oil is produced* (Hager 1998, p. 467).

3 Character

Colour

Clear, slightly viscous, brownish-yellow, semi-drying oil (Hager 1998, p. 471).

Odour

Weak, peculiar, unpleasant (Hager 1998, p. 471).

Flavour

Mild at first, very pungent finish, persistently scraping and painfully burning (Hager 1998, p. 471).

4 Ingredients

The seeds of *Croton tiglium* contain 53–57% of the so-called croton oil and 18% protein (Huang 1999). The main ingredients of the oil are glycerides of oleic acid, linoleic acid, palmitic acid, myristic acid, lauric acid, tiglic acid and other esters of tiglic acid and crotonic acid with the diterpene alcohol phorbol (Roth and Kormann 2000, p. 131).

Composition of Fatty Acids

Fatty acids	Content in % acc. to Firestone (1999), p. 34	Content in % acc. to Hager (1998), p. 471
Myristic acid	–	7.5
Palmitic acid	1	Traces
Stearic acid	0.5	Traces
Oleic acid	56	37–56
Linoleic acid	29	19–29
Arachidic acid	–	1.5

Other Ingredients

(Hager 1998, p. 471)

Formic acid
Crotonic acid
Acetic acid
Isobutyric acid
Isovaleric acid
Lauric acid

Key Figures of Croton Oil

density: 0.936–0.956
0.940–0.950
refractive index: 1.4710–1.4724
saponification number: 212
iodine number: approx. 102
melting point: −16 °C
(German Pharmacopoeia 6 – Pharmacopoeia Helvetica V – Austrian Pharmacopoeia 8 – Hager 1998, p. 471)

Characteristics of the Oil

Croton oil contains several phorbol esters, among them the tumour-promoting 12-tetradecanoylphorbol-13-acetate (TPA). Tumour promoters are substances that are not carcinogenic themselves, but increase cancer incidence after exposure to carcinogenic (initiating) substances. TPA is not mutagenic itself, but it influences cell growth and proliferation by activating protein kinase C. The tumour-promoting effect is, however, very specific, and even minor changes in the ring system result in ineffective derivates.

5 Application

In Pharmacy and Medicine

C

Croton oil is heavily irritating; contact with the skin causes inflammation and oedema. It is used in the croton oil test for triggering skin oedemas, which tests the efficacy of anti-inflammatory substances (Huang 1999, p. 237).

As a Laxative

Croton oil is one of the strongest laxatives (single dose 0.05 g, daily dose 0.15 g; lethal: 4–20 ggt oil), which has become obsolete because of its co-carcinogenic properties. Even in the past, it was only prescribed in cases of especially persistent constipation (Hunnius 1998, p. 370).

In Homoeopathy

In homoeopathy, croton (D4-D12) is prescribed in cases of acute gastroenteritis with watery stool as well as in cases of acute blistery eczema, especially of the genitalia.

In Folk Medicine

In folk medicine, the diluted oil is used together with other oils as a skin irritant, for example to treat rheumatic complaints, debility, conjunctivitis, chronic otitis media, chronic tonsillitis, bronchial asthma and amenorrhoea. Effectiveness has not, however, been sufficiently proven, (Hager 1998, p. 472 et seq).

Baunscheidtism

Heavily diluted croton oil as an ingredient of *Oleum Baunscheidtii* was formerly used in Baunscheidt therapy, during which the oil was applied to the skin, which had been minimally perforated to improve blood circulation.

In Cosmetics

The oil is a banned substance according to the Cosmetics Regulation (Roth and Kormann 2000, p. 131); it is classified as highly poisonous and has to be kept apart from other medicines. In the USA, an emulsion of croton oil with phenol and a detergent is used in plastic surgery for a method of skin peeling (Hager 1998, p. 472 et seq).

6 Possible Unwanted Side Effects

Tumour-Promoting Potential (See Characteristics of the Oil)

In animal tests, croton oil was found to have relatively high tumour promoting activity. A diterpene ester (phorbol ester) that is mainly part of croton resin is responsible for this (Hager 1998, p. 473).

Damage to Epithelial Cells

A study tested the direct effect of croton oil on human intestinal epithelial cells (HIEC). The effect on HIEC proved to be dependent on dosage. High doses (80 mg/L) of croton oil resulted in delayed cell growth or cell death; low doses (4 mg/L) had no effect on the cells. Long-term application with increasing doses may stimulate cell proliferation, increase the concentration of heteroploid DNA and induce conversion into malignant cells. The release of COX-2 was lowered significantly and the expression of the COX-2 gene was increased significantly. Simultaneously, the expression of many other genes changed (Wang et al. 2002).

Pro-Inflammatory Effect

A study on mice examined how croton oil triggers inflammatory processes. A single dose of 0.25% applied to the ears of mice induces oedema. The maximum oedematous reaction was triggered by applying 4% croton oil once. Independently of dosage, the oedema reached a maximum 6–7 hours after application, and fell to the level of the control group after 30 hours. The inflammatory effect caused by croton oil is characterised by epidermal hyperplasia and infiltration of polymorphonuclear leukocytes into the dermis (Shwaireb 1995).

Cupuacu Butter

Synonyms: Cupuassu, Copoasubutter (D)

1 Source Plant

Theobroma grandiflorum Schum. (Sterculiaceae), cupuaçu

Habitat

Cupuaçu grows in the Amazon rainforest and is a native plant of Colombia (Boulanger and Crouzet 2000, p. 251).

Description

Wild trees can reach a height of 20 m, cultivars only grow to a height of 6–8 m. A full-grown tree produces 20–30 fruit. They have a length of 15–20 cm and weigh 800 g to 2 kg. Each fruit contains 30–50 seeds surrounded by creamy yellow fruit pulp (Boulanger and Crouzet 2000, p. 251). The fruit are large, dark brown, and similar to pumpkins in form.

2 Cultivation and Extraction

Cultivation

Cupuaçu is found throughout the Amazon rainforest. It is cultivated by native tribes and local communities.

S. Krist, *Vegetable Fats and Oils*, https://doi.org/10.1007/978-3-030-30314-3_46

Extraction

The fruit ripen from January to April; cupuaçu butter is produced by pressing the bean-shaped seeds.

3 Character

Colour

Light beige, solid at room temperature.

Odour

Herbaceous and fresh, fruity, aromatic, acetous.

4 Ingredients

Composition of Fatty Acids

Fatty acids	Content in % acc. to Firestone (1999), p. 37	Content in % acc. to Kelly et al. (2003)
Myristic acid	0.8	–
Palmitic acid	6–12	6.84
Palmitoleic acid	–	0.06
Stearic acid	22–35	33.4
Oleic acid	39–47	41.5
Linoleic acid	3–9	3.68
α-Linolenic acid	0–1	0.14
Arachidic acid	10–12	11
Eicosenoic acid	–	0.03

Composition of Sterols

(Concentration in mg/kg and content in %)

Sterols	Cupuaçu butter acc. to Firestone (1999), p. 37
Total content	245
Campesterol	4
Stigmasterol	9
β-Sitosterol	80

Composition of Tocopherols

Tocopherols	Content in mg/kg acc. to Firestone (1999), p. 37
Total content	128
Γ-Tocopherol	122
Δ-Tocopherol	6

Physical Key Figures of Cupuaçu Butter

refractive index: 1.4563 (at 40 °C)
saponification number: 189
iodine number: 44–45
(Firestone 1999, p. 37)

5 Application

In Cosmetics

If applied externally, the high amount of phytosterols in cupuaçu butter regulates the lipid production of the skin and acts as an UV filter. Additionally, cupuaçu butter serves as a natural emulsifier in creams and lotions because of its capacity to absorb water and because it is antibacterial to some extent.

Due to the balanced ratio of saturated and unsaturated fatty acids, which are responsible for the low melting point and the fact that cupuaçu butter melts quickly when applied to the skin, as well as the high amount of phytosterols (2%) and vitamin E, cupuaçu butter is an ideal skin moisturiser and is added to sunscreens. Both vitamin E and phytosterols have anti-inflammatory properties and support the regeneration of the skin's natural protective barrier. A study examined the effects of cupuaçu butter on skin irritations induced by SLS (sodium lauryl sulphate) solution. The skin was treated with 7.5% cupuaçu cream and, after 2 and 22 hours respectively, the moisture and reddening of the skin and the transepidermal water loss were measured using a corneometer, chromameter and tewameter. The results showed that a cream containing 7.5% cupuaçu butter significantly increases skin moisture and significantly decreases skin reddening and transepidermal water loss (Kelly et al. 2003).

In the Food Industry

A fat similar to cocoa butter is produced from cupuaçu butter. Their compositions of fatty acids are very similar, and the fats only differ externally (Lannes et al. 2003).

Sweets, stewed fruit and ice cream can be made from the fruit pulp of cupuaçu fruit.

Esparto Wax

Cera Esparto synonym: Espartowachs (D)

1 Source Plant

Stipa tenacissima L. *(Poaceae)*, esparto grass

S. Krist, *Vegetable Fats and Oils*, https://doi.org/10.1007/978-3-030-30314-3_47

Habitat

Esparto grass is a native plant of the Mediterranean area. It is especially common on the sandy coasts of North Africa from Morocco to Tunisia, but it also grows on the southern coast of Spain and Portugal (John 1937).

Description

Esparto grass grows 60–80 cm tall and consists of a bundle of close, fascicular, cylindrical stems, from which long, feathery leaves protrude. The leaves are 20–30 cm long and very hard, flexible and robust, which is why they were used for centuries to produce ropes, shoes, baskets and mats (John 1937).

2 Cultivation and Extraction

Cultivation

Esparto grass prefers light, sandy, loamy soil with a high moisture content. The soil can be acidic, neutral or alkaline. Esparto grass does not tolerate shade, but strong winds are not a problem, although the plant should not be grown near the coast.

Extraction

Esparto wax is a by-product of producing fine paper from esparto grass. It is extracted (Roth and Kormann 2000, p. 123).

3 Character

Colour

Hard, brown wax (John 1937).

Odour

Odourless (John 1937).

Flavour

As esparto wax is unfit for consumption, no information on its taste is available.

4 Ingredients

Ingredients of Esparto Wax

(Roth and Kormann 2000, p. 123; Schneidermaier 2020)
Wax acids 15–17% (e.g. Cerotinacid and Tricotanacid)
esters 63–65%
alcohols 20–22%
hydrocarbons (mostly Hentricontan)
(Saad et al. 2016)
alcohols 7.64–9.35%
alkanes 27.7–47.15%
Alkenes 2.02–7.54%
ester of fatty acides 17.6–28.13%
aldehydes 1.03–1.74%
fatty acids 0.47–0.96%
phenols 11.83–24.78%

Physical Key Figures of Esparto Wax

density: d_{20}^{20} 0.994
melting point: 75 °C
acid number: 23.9
ester value: 45.9
saponification number: 69.8.
iodine number: 15
unsaponifiable: 16.8%
(John 1937)

5 Application

Industrial Uses

In the shoe industry, esparto wax is used as a polish because of its attractive price and characteristic shine. It can be mixed with other waxes and makes the texture of the polish supple. Since it has a strong, steady shine, it is also used as a polish for heavily used floors and furniture in public buildings, offices and shops (John 1937).

6 Possible Unwanted Side Effects

Since esparto wax is not ingested, no unwanted side effects are to be expected.

The plant fibres themselves may cause asthma attacks and allergies when inhaled by workers coming into contact with them. This so-called extrinsic allergic alveolitis is a type III immune reaction in the lungs triggered by precipitated IgG antibodies circulating in the serum. If exposure does not cease, this disease frequently takes a chronic exacerbative course. More often, it results in fibrosis. Fibrosis leads to an abnormal increase in connective tissue in organs, for example in the lungs, referred to as pulmonary fibrosis (Hinojosa 2001).

Evening Primrose Oil

(Oenotherae biennis oleum) synonyms: Nachtkerzenöl (D); huile d'onagre (F)

1 Source Plant

Oenothera biennis L. *(Onagraceae)*, evening primrose

S. Krist, *Vegetable Fats and Oils*, https://doi.org/10.1007/978-3-030-30314-3_48

Habitat

The origins of the evening primrose lie in southern North America and Mexico, where the plant occurs in 200 species. At the beginning of the seventeenth century, it was naturalised in Europe, and today also grows in Asia Minor, New Zealand (Roth and Kormann 2000, p. 65), South America and southern Africa (Hager 1993, p. 930). In Central Europe, the evening primrose typically grows in little noticed locations such as railway embankments, earth banks, piles of debris, by the roadside or on riverbanks.

Description

E

The byname *biennis* shows that the evening primrose is a biennial plant. In the first year, only a rosette of 20–30 leaves is visible on the ground. In the second year, a stem densely covered with leaves emerges, which can reach a height of up to 150 cm. The shoot is angular, covered with glandular hairs, and branched in the upper part. The inflorescence is a raceme with many large, approx. 3 cm long, light-yellow to golden-yellow flowers. The plant blooms from July to September. The flowers only open in the evening. The fruit is a cylindrical, upright capsule up to 3 cm in length, which contains more than 200 seeds (Roth and Kormann 2000, p. 82). The seeds, which contain the oil, have a length of 1.2–2.2 mm and a breadth of 0.5–1.0 mm, depending on the species (Karleskind 1996, p. 170). They are dark grey to brownish-black in colour and contain 15–20% oil (Roth and Kormann 2000, p. 82).

2 Cultivation and Extraction

Cultivation

The evening primrose is now cultivated in 23 or more countries (Hager 1993, p. 930). It does not make any special demands on the soil, but it prefers sandy soil and sunny locations. In North America, it is cultivated on plantations. It is sown in July, and the harvest takes place in the September of the following year (Roth and Kormann 2000, p. 82). The quality and size of the seeds can be increased by low nitrogen dosage, suitable tending of plants, and breeding. One hectare of cultivated area yields up to 1600 kg seeds (Hager 1993, p. 930).

Extraction of the Oil

The oil is obtained by cold extraction with hexane in steel tanks or tanks lined with glass. The extract is cleaned with water, and the solvent is removed under reduced pressure. Extraction of the oil with supercritical carbon dioxide may also be tried in future (Hager 1993, p. 930 et seq). About 10,000 seeds result in one gram of evening primrose oil.

The following table shows the differences in the extracted quantitative amounts of three triterpene esters of caffeic acid, namely betulinic acid, moronic acid and oleanolic acid, in evening primrose oil that has been extracted with carbon dioxide, cold pressed or refined (nature.de/24.04.06):

Method of production	Betulinic acid (mg/100 g)	Moronic acid (mg/100 g)	Oleanolic acid (mg/100 g)
CO_2 extraction	24.5	22.6	10.6
Cold pressed	11	10.3	4.5
Normal/refined	0.7	0.5	0.05

This table shows that customary production methods either remove or destroy important ingredients, due to the high temperatures for pressing and the subsequent cleaning of the oil. The ideal method is extracting the oil with carbon dioxide (nature.de/24.04.06).

3 Character

Colour

Clear, light yellow to yellow.

Odour

Nutty and flowery, heavy.

Flavour

Slightly nutty.

4 Ingredients

Composition of Fatty Acids

Fatty acids	Content in % acc. to Karleskind (1996), p. 172	Content in % acc. to Firestone (1999), p. 40 et seq	Content in % acc. to Rossell and Pritchard (1991), p. 296
Lauric acid	–	0.03	–
Myristic acid	<0.1	0.07	–
Palmitic acid	5–9	6–10	7.0
Palmitoleic acid	<0.1	0.04	–
Stearic acid	1–2	1.5–3.5	2.0

Fatty acids	Content in % acc. to Karleskind (1996), p. 172	Content in % acc. to Firestone (1999), p. 40 et seq	Content in % acc. to Rossell and Pritchard (1991), p. 296
Oleic acid	8–12	5–12	9.0
Linoleic acid	70–79	65–80	72
α-Linolenic acid	<0.1	0.2	–
γ-Linolenic acid	8–13	8–14	10
Arachidic acid	<0.3	0.3	–
Eicosanoic acid	<0.1	0.2	–
Behenic acid	<0.1	0.1	–
Lignoceric acid	-	0.1	–

E

Composition of Sterols

The unsaponifiable portion of the oil is between 1.5% and 2.5%. Of this, 44% are sterols (Hager 1993, p. 931). The following table lists the composition of sterols in percent (Firestone 1999, p. 41):

Sterols	Amount (%) in the total content of sterols
Campesterol	8–9
β-Sitosterol	87–90
Δ5-Avenasterol	4
Δ7-Stigmasterol	2

Composition of Tocopherols

(Concentration and content in mg/kg)

Tocopherols	Evening primrose oil acc. to Firestone (1999), p. 41
Total content of tocopherols	263–661
α-Tocopherol	76–356
γ-Tocopherol	187–358
δ-Tocopherol	0–19

Other Ingredients

(Hager 1993, p. 931)

Amino acids
Minerals
Vitamins
4-Methyl sterols (8%)
Hydrocarbons
Triterpene alcohols (13%)
Alcohols

Physical Key Figures of Evening Primrose Oil

refractive index: 1.4757 to 1.4791
density: 0.9254 to 0.932
iodine number: 147 to 155
saponification number: 187 to 198
acid number: 0.58
point of solidification: −10 °C
unsaponifiable: 1.5–2.5%
peroxide number: 1.2
(Firestone 1999, p. 40; Roth and Kormann 2000, p. 139)

Shelf Life

Evening primrose oil does not keep long. It must be stored in the refrigerator in closed containers and should be used within 6 months.

Characteristics of the Oil

The linolenic acids mainly responsible for the effect of evening primrose oil (especially γ-linolenic acid) are involved in nearly all metabolic processes of the body as raw material of the tissue hormones (prostaglandins). As the body cannot produce these essential fatty acids, they have to be ingested in sufficient amounts to prevent deficiencies. The composition of evening primrose oil is unique with regard to its high amount of γ-linolenic acid.

5 Application

In Pharmacy and Medicine

As Dietary Food

Linoleic acids play a role in oxygen transport via the alveoli and erythrocytes, which contain haemoglobin, to the mitochondria of each cell. Together with sulphur-containing enzymes, they activate oxygen and the production of haemoglobin, support the decomposition of lactic acid and speed up the regeneration process. They are also important components of the lipid double membrane of cells and their organelles. Mammals convert linoleic acid into γ-linolenic acid using the enzyme δ-6-desaturase (Hager 1993, p. 931). Alcohol, cholesterol, diabetes mellitus, saturated fatty acids, trans fatty acids, viral infections and a zinc deficiency all inhibit this process of conversion. Evening primrose oil and other fat oils containing ω-6-fatty acids can bridge the shortage triggered by a lack of activity of this enzyme (Hager 1993, p. 931 et seq). In this way, another conversion of γ-linolenic acid into dihomo-γ-linolenic acid (DGLA) can take place, using vitamin B_6, zinc and magnesium. DGLA is in turn converted to prostaglandin E1, an important eicosanoid, or arachidonic acid.

Prostaglandin E1 has anti-inflammatory effects and inhibits thrombocyte aggregation. Ingestion of vegetable oils rich in gamma-linolenic acid, such as evening primrose oil, supports the production of prostaglandin E1 at the expense of the prostaglandin E2-series and thus counteracts inflammation and strain on the heart and blood circulation.

External Application

Clinical studies have shown that a treatment with evening primrose oil improves cases of atopic eczema, dry skin, itching and flaky skin. The oil is thought to support the treatment of neurodermatitis, premenstrual syndrome, hyperactivity in children, high cholesterol levels, diabetes mellitus and multiple sclerosis. Its effect on acne, psoriasis and rheumatoid arthritis is not yet proven, but there are indications of its effectiveness in various studies on the properties of ω-6-fatty acids (Hager 1993, p. 931 et seqq).

E

Internal Application

The composition of breast milk during lactation could be changed by administering evening primrose oil. For mothers in the 2nd to 6th month, the amounts of total fat, linoleic acid, γ-linolenic acid, dihomo-γ-linolenic acid and arachidic acid increased in comparison to those of mothers who were given a placebo. Neurological symptoms of diabetic polyneuropathy (i.e. perception of cold, heat, pain) improved. In diabetics, the viscosity of the membrane of erythrocytes is characterised by decreased deformability. This could be improved to standard values by administering evening primrose oil over a period of 4 months. The liver function of alcoholics in detoxification improved with daily administration of 4 g evening primrose oil; this was measured using γ-glutamyl transferase. Psychomotor values improved more quickly than in the group given placebos (Hager 1993, p. 931 et seqq).

Adjuvant in Cancer Therapy

The size of the tumour of patients with a primary liver cell carcinoma could be reduced by administering evening primrose oil and vitamin C. Intracerebral injections of γ-linolenic acid (1 mg/day over a period of 10 days) reduced the size of malignant cerebral glioma (Mills and Bone 2001, p. 370).

In Cosmetics

The nurturing properties of evening primrose oil have a soothing effect, especially on irritable skin. Evening primrose oil has a soothing effect, especially on irritable skin. Ointments and oils containing evening primrose oil with its many mineral nutrients, vitamins and unsaturated fatty acids are beneficial for dry, flaky and itching skin.

As Food

The oil can be used as salad oil, for example in combination with sunflower oil or other edible oils. Evening primrose oil should only be used for cold or lukewarm dishes, to preserce the valuable γ-linolenic acid.

6 Possible Unwanted Side Effects

Ingestion of evening primrose oil may occasionally lead to nausea, digestive disorders and headaches, in rare cases also to hypersensitivity reactions (Frohne 2002, p. 383) with symptoms such as exanthema and abdominal pain. The oil may cause temporal lobe seizures that have not yet been diagnosed to manifest themselves, especially in the case of schizophrenic patients who are taking epileptogenic drugs such as phenothiazine. In these cases as well as in cases of epilepsy, the effects of evening primrose oil must be monitored closely (Hager 1993, p. 934).

Until now, overdoses of evening primrose oil have only resulted in abdominal pain and soft stools. Countermeasures have not been necessary (Hager 1993, p. 934).

False Flax Oil

Oleum Camelinae, Oleum Sesami vulgaris synonyms: Leindotteröl, Saatdotteröl, Dotteröl, Rüllöl, Deutsches Sesamöl (D); cameline oil (E); huile de cameline (F)

1 Source Plant

Camelina sativa (L.) Crantz *(Brassicaceae)*, false flax, gold of pleasure

S. Krist, *Vegetable Fats and Oils*, https://doi.org/10.1007/978-3-030-30314-3_49

Habitat

False flax is an old cultivated plant, with its origins in the Neolithic Age. There are many findings of false flax from the Bronze Age and the pre-Christian Iron Age, especially in Eastern and Southern Europe. The heyday of false flax was the Iron Age between 400 B.C. and 500 AD, when cultivation and use of the seeds was common. Today, the plant grows in Europe, Central Asia and North Africa (Roth and Kormann 2000, p. 47).

Description

False flax is an annual plant with a stem 30–100 cm tall, with side shoots and lanceolate leaves. The small flowers are pale yellow to yellow, more rarely white; they are arranged in lengthened racemes. The fruit are small, upright, pear-shaped siliques about 7–9 mm long, with a hard shell. The seeds that they contain are elongate, yellowish-brown and 1.5–2.5 mm long (Roth and Kormann 2000, p. 47). Like rape, false flax has a winter and a summer variety.

F

2 Cultivation and Extraction

Cultivation

False flax is cultivated in Russia and several Central European countries. It does not make many demands on the soil and is inured to dry conditions and heat. The plant is also resistant to pests that attack other cruciferous plants. Its vegetation cycle is only 90–100 days (Roth and Kormann 2000, p. 47), which makes it interesting as a catch crop. The summer variety is sown in March or April; sudden frosts do not influence its growth. Cultivation is also possible on sandy or calciferous soil, as long as the nutrient supply is sufficient. As false flax is a robust and competitive plant, it does not need much in the way of chemical plant protection.

The harvest takes place in July when the plants develop the typical seed heads and 80% of the infructescences have ripened to a yellow or light-brown colour. Harvesting is carried out by pulling out or cutting the plants. The dried plants are threshed, and the seeds are cleaned, dried and stored. The yield amounts to 400–800 kg/ha (Rode 2002).

Extraction of the Oil

False flax oil is taken from the seeds, which contain 30–35% fat oil (Roth and Kormann 2000, p. 47). The seeds are either cold pressed or extracted. From 100 kg seeds, 27 kg raw oil can be obtained by pressing, 30 kg by extraction (Roth and Kormann 2000, p. 132). The seeds are first ground to a rough mass, then an equal amount of water is added, and the mixture is roasted at a temperature of 60–90 °C. The sandy mass is subsequently pressed. The oil is left standing until it is clear and is then filtered through gauze. Since it contains a high amount of unsaturated fatty acids, it does not keep long. For this reason, the seeds are only processed shortly before sale (Rode 2002).

3 Character

Colour

Deep-yellow oil, liquid at room temperature.
Thin, golden-yellow, semi-drying oil (Roth and Kormann 2000, p. 132).
Light-yellow, quick-drying oil. Dark-yellow, slightly orange colour (Krist et al. 2006b)

Odour

Green odour of pyrazine, freshly cut meadows, slight odour of lemons (Krist et al. 2006b).
Herbaceous (Roth and Kormann 2000, p. 132).

Volatile Compounds

The following components were detected using SPME-GC-MS (solidphase-microextraction – gas chromatography – mass spectrometry) (Krist et al. 2006b):

Odorous substance	Odour quality	Content in %
Acetic acid	Strong, pungent, poignant	9.3
Ethyl acetate	Pineapple, ethereal	2.2
Trans-2-butenal	–	9.8
Trans-3-penten-2-one	Fruity smell	0.2
Trans-2-pentenal	Pungent, green, apple, orange, tomato	1.3
Butyric acid	Pungent, cheeselike, rancid, sweaty, putrid, sour	0.1
Isovaleric acid	Rancid, cheeselike, sweaty, putrid	0.3
Trans-2-hexenal	Sweet, fragrant, almond, fruity, green, apple, plum, vegetables	0.2
Hexanol	Herbaceous, woody, fragrant, mild, sweet, green	0.6
2-Heptanone	Fruity, aromatic, cinnamon	0.3
Styrol	Synthetic	0.8
Heptanal	Oily, fatty, heavy, woody, penetrating, sweet	0.6
Trans,trans-2,4-hexadienal	Fresh, green, flowery, citrus	0.4
γ-Butyrolactone	Weak, sweet, caramel	0.5
α-Pinene	Pungent, pine	1.6
Trans-2-heptenal	Pungent, green	2.3
Benzaldehyde	Bitter almond, fragrant, aromatic, sweet	0.4

Odorous substance	Odour quality	Content in %
Sabinene	Warm, oily-peppery, woody-herbaceous, aromatic	0.6
1-Octen-3-ol	Herbaceous, earthy	1.9
6-Methyl-5-hepten-2-one	Herbaceous, green, oily, pungent	0.9
β-Myrcene	Sweet, balmy	0.9
Trans,trans-2,4-heptadienal	Fatty, cinnamon, hazelnut	3.6
Octanal	Fatty, citrus	0.7
3-carene	Sweet, penetrating, spreading	0.5
ρ-Cymene	Weak, citrus smell	0.2
Limonene	Mild, citrus, sweet, oranges	1.5
Trans-3-octen-2-one	Fruity, citrus	0.8
Trans,trans-3,5-octadien-2-one	–	3.8
Nonanal	Flowery, citrus, orange, rose, fatty, waxy	2.2
Decanal	Penetrating, sweet, waxy, flowery, citrus	0.2

Flavour

F

Slightly poignant taste.
Reminiscent of mustard oil (Roth and Kormann 2000, p. 132).

4 Ingredients

Composition of Fatty Acids

The seeds of false flax contain on average 45% false flax oil, in which the dominating fatty acid is α-linolenic acid with a share of about 37.8%, as well as oleic acid (13.4%), eicosenoic acid (15.4%) and erucic acid[1] (2.76%).

Fatty acids	Content in % acc. to Shukla et al. (2002)	Content in % acc. to Budin et al. (1995)	Content in % acc. to Firestone (1999), p. 19	Content in % acc. to Eidhin et al. (2003a)	Content in % acc. to Zubr (2002)
Lauric acid	–	–	0–0.1	–	–
Myristic acid	–	–	0–0.1	–	0.20
Palmitic acid	5.3 ± 0.1	7.0	5–6	5.5	5.33 ± 0.05
Palmitoleic acid	–	–	0–2	–	0.10
Stearic acid	3.0 ± 0.0	2.5	2–3	2.3	2.65 ± 0.02
Oleic acid	18.7 ± 0.1	15.4	12–24	14.9	13.4 ± 0.13
Linoleic acid	16.0 ± 0.1	21.5	15–16	15.8	14.8 ± 0.12
α-Linolenic acid	38.1 ± 0.1	33.2	33–38	38.9	37.8 ± 0.17
Arachidic acid	1.4 ± 0.1	–	1	0.4	1.25 ± 0.02

[1] Decree on erucic acid: BGBl.No.468/1994: Food with a total fat content of more than 5% must not contain more than 5% erucic acid in relation to the total content of fatty acids in the fat phase.

Fatty acids	Content in % acc. to Shukla et al. (2002)	Content in % acc. to Budin et al. (1995)	Content in % acc. to Firestone (1999), p. 19	Content in % acc. to Eidhin et al. (2003a)	Content in % acc. to Zubr (2002)
Eicosenoic acid	11.6 ± 0.1	12.6	14–16	16.2	15.4 ± 0.16
Eicosadienoic acid	–	–	2	2.1	2.02 ± 0.02
Eicosatrienoic acid	–	–	1–2	1.3	1.84 ± 0.02
Behenic acid	–	–	0.2–0.3	–	0.30
Docosadienoic acid	–	–	–	–	0.17 ± 0.01
Erucic acid	2.5 ± 0.1	2.2	3	2.4	2.76 ± 0.07
Lignoceric acid	–	–	0.2–0.3	–	0.20
Nervonic acid	–	–	–	–	0.64 ± 0.01
Other fatty acids	3.4 ± 0.1	5.6	–	–	0.26

Composition of Triacylglycerides in False Flax Oil from Austria, Content in % (Krist et al. 2006b)

Triacylglycerides	
LnPP	0.7 ± 0.5
LPP	0.9 ± 0.1
OPP	0.1 ± 0.2
LnLnPo	0.2 ± 0.4
LnLnP	4.0 ± 0.3
LnLP	3.8 ± 0.2
LnOP	5.0 ± 0.5
LLP	+
LOP	2.8 ± 0.4
LnSP	+
OOP	1.5 ± 0.1
LSP	+
LnLnLn	3.4 ± 0.5
LnLnL	4.1 ± 0.3
LnLnO	7.1 ± 0.5
LnLL	+
LnLnS	6.7 ± 0.3
LnLO	+
LLL	+
LnLS	6.2 ± 0.5
LnOO	+
LLO	+

Triacylglycerides	
LOO	6.0 ± 0.2
LnOS	+
GLnP	+
OOO	3.9 ± 0.3
LOS	+
GLP	+
OOS	1.2 ± 0.3
LSS	+
GOP	+
LnLnG	9.3 ± 0.9
LnLG	9.3 ± 0.5
LLG	6.2 ± 0.5
LnOG	+
LOG	4.2 ± 0.4
LnSG	+
LLA	+
LnOA	+
OOG	3.0 ± 0.7
LSG	+
LOA	+
OSG	0.2 ± 0.4
OOA	+
Ln,20:2,20:2	2.4 ± 0.1
Ln,20:2,G	2.7 ± 0.2
LnGG	2.4 ± 0.2
LGG	1.8 ± 0.2
LGA	1.0 ± 0.3

+ Component present but could not be quantified

Composition of Sterols

For a vegetable oil, false flax oil contains a relatively high amount of cholesterol (118 ± 8 *µg/g)* (Shukla et al. 2002).

The unsaponifiable portion of false flax oil amounts to 0.54%, its main components being desmethylsterols (Shukla et al. 2002).

Sterols	Content in µg/g oil acc. to Shukla et al. (2002)
Cholesterol	188 ± 8
Brassicasterol	133 ± 9
Campesterol	893 ± 74
Stigmasterol	103 ± 18
Sitosterol	1884 ± 144
Δ5-Avenasterol	393 ± 41

Composition of Tocopherols

Tocopherols	Content in mg/100 g acc. to Budin et al. (1995)
α-Tocopherol	1.75
α-Tocopherol-3-isomer	0.04
Β-Tocopherol	0.99
Δ-Tocopherol	0.34

Other Ingredients (Zubr 2002)

Arginine 8.15%
Glycine 5.44%
Isoleucine 3.96%
Leucine 6.63%
Lysine 4.25%
Phenylalanine 4.19%
Proline 5.09%
Threonine 4.25%
Valine 5.42%

Physical Key Figures of False Flax Oil

density: 0.919–0.926 (at 15 °C)
refractive index: 1.4761–1.4774 (at 20 °C)
saponification number: 185–194
iodine number: 124–153
unsaponifiable: approx. 1%
point of solidification: −11 to −18 °C
(Roth and Kormann 2000, p. 132)

5 Application

In Pharmacy and Medicine

Possible Cholesterol-Lowering Properties

A study on pigs given a diet containing 5–10% false flax oil resulted in a significant decrease in triglycerides in the serum, an increase in ω-3 long-chain fatty acids and an improvement of the ratio of ω-6 to ω-3 fatty acids. The cholesterol level of the pigs receiving a diet with 10% false flax oil was high at the beginning, and considerably reduced by the 33rd day of the study. The lipid metabolism of humans is very similar to that of pigs, which would mean that the daily consumption of false flax oil would have a positive effect on patients with high cholesterol levels (Eidhin et al. 2003a).

A study of 68 patients with high cholesterol levels showed that when 30 g false flax oil/d were administered over 6 weeks, the amount of α-linolenic acid in the serum lipid was significantly higher at the end of the study than when olive oil or rapeseed oil was used; the concentration of serum LDL cholesterol decreased significantly by 12.2%. The cholesterol-lowering effect of false flax oil is thus comparable to that of olive oil or rapeseed oil (Karvonen et al. 2002).

Production of Wax Esters

A study examining the production of wax esters from false flax oil showed that when a stoichiometric mixture of long-chain/very long-chain fatty acids and corresponding alcohols from false flax oil is esterified with lipase as a catalyst, the yield of long-chain/very long-chain wax esters amounts to about 90%. Novozyme 435 and papaya lipase acted as biocatalysts. The esterification took place in a vacuum. The wax esters obtained in this process are potential raw material for applications in pharmacy and cosmetics, and for technical lubricants (Steinke et al. 2001).

F

Source of ω-3 Fatty Acids

A study by Eidhin et al. examined the oxidative stability of a spread made from false flax oil in comparison with spreads produced from other vegetable oils and fish oil. Comparative parameters were the peroxide number, ρ-anisidine value, oxidation number, conjugated dienes, conjugated trienes and substances reacting with thiobarbituric acid. Results showed that false flax oil is susceptible to lipid oxidation, but less so than other sources of ω-3 fatty acids like fish oil or linseed oil. During 16 weeks of storage, conditions were met with regard to stability and sufficient quality; this corresponds to the normal period for which spreads are used. In comparison to a spread with a sunflower oil basis with 0.2 g α-linolenic acid, a spread with false flax oil basis contains 5.2 g α-linolenic acid if 25 g are consumed on average. False flax oil is therefore a good alternative to fish oil as a source of ω-3 fatty acids (Eidhin et al. 2003b).

In Folk Medicine

False flax oil has long been used as a household remedy and is thought to have pain-relieving, antiseptic and healing effects. It is administered internally in cases of gastric and intestinal ulcers, gastritis, colic and indigestion, and externally to treat contusion, skin abrasions, bruises, effusions, sprains, skin diseases, acne, frostbite and inflammation of all kinds. On animals, it is used to treat trichophytia and to support wound healing (Piroutz 2003).

Industrial Uses

False flax oil is used to produce soap, paint and as a lamp oil that burns brightly and soot-free (Roth and Kormann 2000, p. 132). Since it contains a high amount of polyunsaturated fatty acids, it is a quick-drying oil, and is used in oleochemistry for the production of environment-friendly polymers, paint and varnish (Löw 2003, p. 35). In addition, it is being tested as a possibly suitable substitute fuel for diesel engines (Bernardo et al. 2003).

Fennel Oil

Oleum Foeniculi synonym: Fenchelöl (D)

1 Source Plant

Foeniculum vulgare Mill. (Apiaceae), fennel

S. Krist, *Vegetable Fats and Oils*, https://doi.org/10.1007/978-3-030-30314-3_50

Habitat

Fennel originates in the Mediterranean area. Today, it is cultivated throughout Southern Europe as well as in climatically suitable regions of Central Europe. In addition, it is grown in Egypt, parts of South America and in Asia, as far as China and India.

Description

Fennel is a biennial or perennial shrub up to two metres tall and belongs to the *Apiaceae* family. The umbels reach a maximum breadth of 15 cm and form up to 25 umbel sections with many small, individual flowers of a deep yellow colour. The plant blossoms from July to August. The stem is grooved, marrowy inside, and frosted with a bluish tint in the upper third. Its leaves are bipinnate to triple pinnate, with delicately cleaved points three to seven centimetres long and hardly a millimetre broad. Fennel develops oval, elongated schizocarps of a yellowish-green to yellowish-brown colour, which split into fruitlets of about 1 cm length. The plant always bears fruit of different degrees of ripeness, as the umbels that are first in bloom also ripen first. Each fruitlet has five distinctly protruding ridges of a lighter colour, and four darker areas in-between (Wenigmann 1999).

Garden fennel *(Foeniculum vulgare ssp. vulgare)* exists in three different varieties: *var. vulgare* = bitter fennel, *var. dulce* = sweet fennel and *var. azoricum* = bulb fennel. For medical and healing purposes, the fried fruit of bitter fennel (*foeniculi amari fructus*) are used, more rarely also that of sweet fennel (*foeniculi dulcis fructus*).

2 Cultivation and Extraction

Cultivation

In order to thrive, fennel needs rich, calciferous soil in warm, sunny locations (Wenigmann 1999).

Extraction of the Oil

To extract the oil, the fennel seeds are first treated with steam to remove the essential oil. They are then ground, and the oil is extracted with petroleum ether in a Soxhlet extractor. The result is a dark green oil, which corresponds to about 16.5% of the weight of the seeds (Popov and Stefanov 1969).

3 Character

Colour

Dark-green oil (Popov and Stefanov 1969).

Odour

Typical touch of fennel, touch of anise, green, aromatic, sweetish.
Aromatic, spicy (Fischer and Krug 1984, p. 77).

Flavour

Aromatic, sweetish (Fischer and Krug 1984, p. 77).

4 Ingredients

Composition of Fatty Acids

Fatty acids	Content in % acc. to Firestone (1999), p. 41	Content in % acc. to Roth and Kormann (2000), p. 159
Palmitic acid	4	4
Petroselinic acid	60	60
Oleic acid	22	14
Linoleic acid	14	14

The high amount of petroselinic acid (60%) is conspicuous.

Other Ingredients

Phenylpropane derivates (trans-anethole, estragole)
Phenolcarboxylic acids (mainly quinic acid derivates)
Monoterpenes (fenchone, limonene), coumarins, furanocoumarins, sterols
Flavonoids (kaempferol and quercetin derivates)

Physical Key Figures of Fennel Oil

refractive index: 1.4795–1.480
density: 0.9304

iodine number: 99
saponification number: 181
point of solidification: −2 °C
unsaponifiable: 3–4%
(Roth and Kormann 2000, p. 159; Firestone 1999, p. 41)

5 Application

In Pharmacy and Medicine

Fennel oil is mucolytic and anticonvulsant, promotes expectoration, has antiflatulent and diuretic effects and is anti-inflammatory. The sweetish taste of bitter fennel is reminiscent of anise and is mainly due to anethole, which is part of the essential oil.

F

As an Expectorant

Bitter fennel oil contains 50–70% anethole, which is thought to have secretolytic and spasmolytic effects, and up to 15% fenchone in the essential oil. Fenchone inhibits the growth of bacteria and fungi. The effect of fennel oil on the respiratory ducts is threefold: It relaxes the bronchial muscles, stimulates the movements of the cilia in the respiratory ducts and secretolytically dilutes the liquid film that removes thick mucus. It is therefore a remedy for coughs and catarrh.

As a Carminative

In the case of dyspeptic complaints, fennel oil stimulates the motility of the gastrointestinal tract. Higher doses have an antispasmodic effect on the smooth musculature of these organs. The oil can be applied to alleviate flatulence (especially of babies) and in cases of menstrual complaints and digestive disorders.

External Application

Fennel oil is used externally to treat skin irritations, conjunctivitis and inflammation of the eyelids.

Industrial Uses

Fennel is used in the industrial production of anethole, which is a flavouring substance for the cosmetics industry and spirits industry (for example for absinthe, ouzo and sambuca). Since fennel oil conitans a high amount of petroselinic acid, it is also an important source of this acid (Popov and Stefanov 1969).

6 Possible Unwanted Side Effects

Fenchone

Small children, babies and pregnant women should only use fennel oil in a highly diluted form, since a higher dosage of fenchone may lead to acute dyspnoea and excitation. Epileptics and persons suffering from a celery allergy should not use fennel oil at all. During pregnancy, sweet fennel is contraindicated, as it is one of the essential oils that may induce labour and, for this reason, must be avoided.

Allergic Reactions

The essential oil sectioned during extraction may lead to allergic reactions of the skin and respiratory ducts (Hänsel et al. 1999, p. 698). In case the sectioning has not been completed, such unwanted reactions are possible.

Gevuina Oil

Synonyms: Wildes Haselnussöl, Avellanaöl (D); Chilenisches Haselnussöl (D); Chilean hazelnut oil (E)

1 Source Plant

Gevuina avellana Molina *(Proteaceae)*, Chilean hazelnut

S. Krist, *Vegetable Fats and Oils*, https://doi.org/10.1007/978-3-030-30314-3_51

Habitat

Gevuina grows mainly in the southern parts of Chile and in Argentina. It produces nuts with edible seeds, similar to macadamia nuts (Bertoli et al. 1998).

Description

The tree is of medium height and evergreen; it grows to a height of 10–12 m and has sturdy branches with downy hair. The leaves are dark green, bipinnate, shiny and leathery, with a serrated edge. The flowers are a snowy-white to ivory colour and are arranged in pairs. The fruit is a round, wooden stone fruit, about 1.2–2.3 cm in size. At first it is coral red, but when it is ripe, its colour changes to brown or black. It contains one round seed in a smooth, hard shell.

2 Cultivation and Extraction

Cultivation

Commercial cultivation is tried in Chile and New Zealand (Bertoli et al. 1998).

Extraction of the Oil

Traditionally, the nuts are ground after the shell has been removed. They are then heated and subsequently pressed. It is also possible to extract the oil enzymatically, which results in a higher yield. The ground seeds are mixed with water and enzymes in a ratio of 1:1. After the enzymatic reaction, the enzymes are deactivated in boiling water, and the mixture is then centrifugalised (Santamaria et al. 2003).

3 Character

Colour

Light yellow oil, liquid at room temperature.

Odour

Like chlorinated water, citrus scent, slightly woody, nutty, intense, fatty, slightly green, waxy.

4 Ingredients

Composition of Fatty Acids

Fatty acids	Content in % acc. to Firestone (1999), p. 46	Content in % acc. to Aitzetmüller (2004)
Myristic acid	–	0.1
C 14:1n-5	–	Traces
Palmitic acid	1.9	1.6
Palmitoleic acid	22.7	24.2
Stearic acid	0.5	0.5
Oleic acid	39.4	29.3 (n-9)
Linoleic acid	5.6	8.8
α-Linolenic acid	0.1	0.2
Arachidic acid	1.4	1.4
Eicosanoic acid	9.7	8.5
Behenic acid	2.2	1.9
Cetoleic acid	9.5	9.8
Lignoceric acid	0.5	0.6
Nervonic acid	–	Traces (n-9), 0.3 (n-5)
Cerotic acid	–	0.1

Composition of Tocopherols and Tocotrienols

(Concentration and content in mg/kg)

Tocopherols/tocotrienols	Gevuina oil acc. to Firestone (1999), p. 47
α-Tocopherol	0.4
γ-Tocopherol	0.6
Total content of tocotrienols	132
α-Tocotrienol	130
β-Tocotrienol	1.3
γ-Tocotrienol	0.9
δ-Tocotrienol	0.1

5 Application

In Cosmetics

As a result of its physical and chemical properties, gevuina oil is well suited to cosmetic applications. It absorbs short-wave UV rays, letting radiation tanning the skin pass, but blocking skin-damaging radiation. As a result of gevuina oil's high proportion of unsaturated fatty acids, it is absorbed quickly by the skin. It is a main component of many products like sunscreens and other protective creams. In addition, it is used as a vehicle for skin nourishing, protecting and regenerating substances in creams, soap and shampoo (Santamaria et al. 2003). It stabilises elastic fibres and collagen fibres in the skin, and strengthens the connective tissue. It contains mink oil, which is otherwise only found in animal oils, and for this reason, it is added to antiwrinkle creams. Mink oil is usually extracted from the thick layer of fat directly under the skin of mink. Of all oils, mink oil is the most similar to the fat of the human skin. It also provides natural light protection, which is thought to correspond to a sun protection factor of 10.

As a Diet Oil

The oil contains a high proportion of α-tocotrienol (130 mg/kg). It has a high oxidative stability and for this reason is used in dietetics as a vitamin E source (Bertoli et al. 1998).

G

Grapeseed Oil

Oleum Vitis viniferae synonyms: Weintraubenkernöl (D); huile de pépins de raisin (F)

1 Source Plant

Vitis vinifera L. (Vitaceae), grape

S. Krist, *Vegetable Fats and Oils*, https://doi.org/10.1007/978-3-030-30314-3_52

Habitat

Vitis vinifera has two subspecies: *silvestris* (wild grape) and *vinifera* (cultivar). The wild grape is rare today. It only grows in alluvial forests on hilly land or on plains, on moderately dry to moist soil; in Central Europe, it is found near rivers, but threatened with extinction. Wild grapes are native plants in the Mediterranean area, central France, south-western Switzerland, the Upper Rhine lowlands, the river basin of the Danube, southern Russia and Asia Minor. The cultivar is an old cultivated plant of vineyards and also grows near walls and houses, preferably on dry soil and in sunny locations, at all altitudes. In addition, the grape can be found as a remnant of previous cultures in alluvial forests, on riverbanks, borders of woodland, defiles and bushes.

Description

Vitis vinifera is a climbing bush up to 30 m tall, with deep, branched roots and a ligneous stem up to 40 cm in diameter. The stem has a striped bark. The branches are a yellowish-brown to dark reddish-brown colour, either hairless or slightly hairy, and striped. The leaves' outline is circular, heart-shaped at the base. They are disparately serrated, their upper side hairless, their lower side covered with woolly or felty hair. The flowers are arranged in dense compound spikes consisting of 10–1000 flowers. They are fragrant, and either male, female or perfect. The fruit are elongated to round, 6–22 mm in size, juicy and sweet; their colour ranges from a dark bluish violet to red, green or yellow. The seed is pear-shaped, hard-shelled and has two elongated depressions on the side.

2 Cultivation and Extraction

Cultivation

Vitis vinifera is one of the most important fruit plants in the world. The sugar content of the grapes varies from 20% (natural ripening) to 40% (drying by noble rot). According to global annual production, the grape is eighth in the list of all useful plants. Wine is produced in Europe, in Asia beyond the eastern edge of the Caspian Sea to Beijing, in Japan, in western North America to northern San Francisco, and in Canada. It is not common in tropical and subtropical climates; in the southern hemisphere, it is cultivated on St. Helena, in South Africa, Australia (New South Wales, Victoria), New Zealand and South America (Argentina, Chile).

Extraction of the Oil

The kernels are cold pressed to obtain the oil. They are separated from the pomace by machines and dried by means of indirect heating systems until their residual water content drops to 10%. Subsequently, they are pressed in twin-screw extruders. About 800 kg of pomace yields 1 litre of oil (Roth and Kormann 2000, p. 154).

If the oil is obtained by hot pressing, it is olive green and has an unpleasant, gob flavour (Löw 2003, p. 65). Grapeseed oil can also be extracted with fuel (Hunnius 1998, p. 1394).

3 Character

Colour

Colourless, sometimes slight yellow tint.
Yellow to brownish yellow (Salunkhe et al. 1992, p. 458). Golden yellow to greenish (Basuki 2003, p. 7).

Odour

Sweet, woody, aldehyde. Odourless (Basuki 2003, p. 7).

Volatile Compounds

SPME-GC-MS analysis (Bail et al. 2008)

Acetous acid	Hexanol
Ethyl acetate	Isoamyl acetate
3-Hydroxy-2-butanone	Trans-2-heptenal
3-Methylbutanol	Ethyl hexanoate
2,3-Butandiol	Phenylethanol
Hexanal	Ethyl octanoate

Flavour

Slightly sweetish (Basuki 2003, p. 7).

4 Ingredients

Composition of Fatty Acids

Fatty acids	Content in % acc. to Basuki (2003), p. 11	Content in % acc. to Salunkhe et al. (1992), p. 458	Content in % acc. to Kerschbaum and Schweiger (2001), p. 16
Palmitic acid	7.4	7–9.5	7.3
Linoleic acid	72.2	46–74	63.0
Palmitoleic acid	0.6	–	0.1
Oleic acid	15.6	14–44	24.8
α-Linolenic acid	0.24	–	0.2
Stearic acid	3.9	3.5–5.5	3.3
Arachidic acid	–	–	0.2
Eicosenoic acid	–	–	0.3
Vaccenic acid	–	–	1.3
Myristic acid	0.08	–	0.1

Composition of Triglycerides

Triglycerides	Content in % acc. to Karleskind (1996), p. 144
LLL	35.7
OLL	21.0
PLL	17.0
SLL + LOP	15.9
LOO	10.3

Composition of Sterols

(Concentration in mg/100 g and content in %)

Sterols	Grapeseed oil acc. to Firestone (1999), p. 44	Grapeseed oil acc. to Karleskind (1996), p. 146	Grapeseed oil acc. to Basuki (2003), p. 9
Total content	580	259–418	–
Cholesterol	0–0.5	<0.5	0.1
Brassicasterol	0–0.2	–	–
Campesterol	9–14	11–15	–
Stigmasterol	9–17	8–12	10.1
β-Sitosterol	–	66–73	74.5
Δ5-Avenasterol	1–3	2–4	2.4
Δ7-Stigmasterol	1–3	<3	1.7
Δ7-Avenasterol	0–1	1–3	0.5
Ergosterol	–	<2	–
Fucosterol	–	<2	–
Sitostanol	64–70	–	–

Composition of Tocopherols

(Concentration in mg/100 g and content in %)

Tocopherols/tocotrienols	Grapeseed oil acc. to Karleskind (1996), p. 146
Total content	32–52
α-Tocopherol	26–39
β-Tocopherol	2–14
γ-Tocopherol	10–34
δ-Tocopherol	<2
α-Tocotrienol	10–26
γ-Tocotrienol	14–34

Other Ingredients

(Basuki 2003, p. 7 et seqq; Kerschbaum and Schweiger 2001, p. 43; Hänsel et al. 1999, p. 248)

Procyanidin
Lecithin
Minerals
Resveratrol
Polyphenols
Phytosterines
Vitamin E
Vitamin K
Glycerides of ricinoleic acid

Physical Key Figures of Grapeseed Oil

refractive index: n_D^{25} 1.471–1.476
density: d_{15}^{15} 0.910–0.956
saponification number: 171–206
iodine number: 94–157
unsaponifiable: 0.3–2%
melting point: −10 °C
point of solidification: −10 to −24 °C
(Roth and Kormann 2000, p. 154).

Shelf Life

If stored in a cool, dark place, grapeseed oil has a shelf life of about 12 months.

5 Application

In Pharmacy and Medicine

OPC Ingredients

The antioxidative effect of grapeseed oil is considerable and can be traced back to its high vitamin E content and the flavonoid procyanidin. As a scavenger, procyanidin is thought to be 20 times more effective than vitamin C and 50 times more effective than vitamin E. For this reason, it seems possible that grapeseed oil could be used in cancer prophylaxis and therapy (Basuki 2003, p. 37 et seq; Kang et al. 1998; Ulmer 1996, p. 85).

Favourable Composition of Fatty Acids

In the composition of fatty acids, the high amount of linoleic acid is conspicuous. Linoleic acid is essential for the organism of mammals and also, therefore, for humans. Several medical effects of grapeseed oil can be traced back to its high content of linoleic acid:

G

- cytoprotective effect (Fleischhacker 2002, p. 24);
- positive influence on the cardiovascular system (Nakamura and Tonogai 2002; Basuki 2003, p. 22 et seq);
- lowering of cholesterol and triglyceride levels: This effect is supported by the high amount of phytosterines in the oil. In aprticular, β-sitosterin competitively inhibits the synthesis of cholesterol in lipid metabolism (Nakamura and Tonogai 2002; Fleischhacker 2002, p. 185; Basuki 2003, p. 53 et seq – pflanzenoel.ch/10.09.03);
- supportive effect in cases of metabolic dysfunctions (Basuki 2003, p. 53 et seq; Nakamura and Tonogai 2002);
- for the treatment of skin diseases such as atopic eczema, acne and psoriasis, as these complaints are presumably accompanied by a deficiency in linoleic acid (Basuki 2003, p. 8, p. 23), and
- to substitute linoleic acid in premature babies, who have a high demand for this essential fatty acid, which is important for neurophysiological development and growth (Basuki 2003, p. 24).

Other Ingredients and their Characteristics

Like castor oil, Oleum vitis viniferae contains glycerides of ricinoleic acid, a D-12-hydroxy oleic acid. Oils hydrolytically releasing hydroxy acids are antiabsorptive and hydragogue (Hänsel et al. 1999, p. 248).

The polyphenols in grapeseed oil have a positive influence on lipid metabolism, especially by lowering the serum lipid level. Cholesterol levels are not influenced to the same degree (Nakamura and Tonogai 2002).

Grapeseed oil is also suited for improving mental performance, as it contains lecithin, which is part of neurons and supports stimulus transmission. Additionally, lecithin has a positive effect on haematosis (Fleischhacker 2002, p. 111).

Grapeseed oil contains resveratrol. A study by the universities of Illinois and Madrid found that depending on dosage, resveratrol counteracts the development of skin tumours in mice. It is also antioxidant, antimutagenic and inhibits cyclooxygenase, the enzyme catalysing the transformation of arachidonic acid into pro-inflammatory substances.

Additional Applications

Other medical effects and applications of grapeseed oil (Basuki 2003, p. 7 et seq, p. 39):

- antimicrobial;
- astringent;
- positive influence on the libido;
- palliative for symptoms of old age;
- can be used to treat hair loss;
- muscle stimulation, and
- strengthening the immune system.

In Cosmetics

Grapeseed oil penetrates the skin without leaving a film. It has regenerative and moisturising properties, and restores the elasticity of the tissue because it contains procyanidin, which resembles the collagen of the connective tissue in structure. Phytosterins, which are contained in grapeseed oil, are currently added to cosmetic products as a skin-lipid barrier and to protect sensitive skin from UV rays. *Oleum Vitis viniferae* is used especially for oily skin, combination skin with oily tendencies, and blemished skin. As a result of its high spreadability and good compatibility with other ingredients, it is used extensively in cosmetics. It is used in or can be added to the following preparations (Roth and Kormann 2000 p. 155; Löw 2003 p. 65; Basuki 2003, p. 7 et seq, p. 56):

- ointment base;
- sun protection products;
- base oil for essential oils;
- massage oil;
- body oils;
- shampoo;
- bath oils;
- hairstyling products;
- conditioners, and
- creams to treat cellulitis.

Technical Uses

For technical purposes, grapeseed oil is added to soaps and varnishes as well as to coating material, paint, spray varnish (mixed with linseed oil) and linoleum (on a small scale) (Salunkhe et al. 1992, p. 458; Löw 2003, p. 65)

As Food

Grapeseed oil was an important edible oil in medieval times. Today, it could regain importance because it is beneficial to health and contains many essential fatty acids. Grapeseed oil is an aromatic, slightly spicy oil that is suited as a salad oil, as an addition to mayonnaise and dressings, but also, because of its high boiling point of 220 °C, for frying and roasting (Basuki 2003, p. 7; Roth and Kormann 2000, p. 155; Hunnius 1998, p. 1394; Löw 2003, p. 65).

Hazelnut Oil

Oleum Coryli avellanae synonyms: Haselnussöl (D); huile de noisette (F)

1 Source Plant

Corylus avellana L. (Betulaceae), common hazel

S. Krist, *Vegetable Fats and Oils*, https://doi.org/10.1007/978-3-030-30314-3_53

Habitat

The hazel is common in Europe and Asia Minor and has been cultivated since antiquity for its oleaginous seeds and as a source of food. The main growing areas of the bush, which is cultivated in northern climates, are Turkey, Italy and Spain. The hazelnut is the only nut of European origin (Ulmer 1996, p. 83). Today, there are about 15 known varieties in North America, Europe, North Africa and Asia, as well as thousands of hybrid forms. A type of oblong hazelnut preferred for commercial uses is the filbert. Among the best-known roundish hazelnuts is the Turkish hazel (Dassler and Heitmann 1991, p. 193). In Central Europe, the common hazel grows on plains and at altitudes of up to 1700 m on mountains, on the edges of forests and on river banks. Since it can withstand exhaust fumes, it is often planted by the roadside (Roth and Kormann 2000, p. 57). Turkey (60%), Italy (20%), Spain, the USA and France have the largest share of the annual global harvest of 400,000 to 650,000 t (Dassler and Heitmann 1991, p. 194).

Description

H

The common hazel can reach a height of 5–7 m; it blooms in winter and bears fruit the following autumn. The flowers are a yellow, green or red, and the nuts are brown (Dassler and Heitmann 1991, p. 194). The bush has a reddish-grey, shiny bark. The leaves are ovate, roundish or heart-shaped. From February to April, the common hazel bears male, pendant catkins up to 10 cm long, and female flowers encapsulated in buds from which red stigmas protrude.

The nuts contain 62% fat, 14% protein, 10.6% carbohydrates, 7.4% fibres, 5.2% water, 2.43% mineral nutrients (636 mg potassium, 226 mg calcium, 333 mg phosphorus, 156 mg magnesium and 3.8 mg iron) and vitamins such as vitamin C (3 mg), carotene (0.3 mg), vitamin **B1** (0.39 mg) and vitamin B_2 (0.21 mg) (Dassler and Heitmann 1991, p. 367).

2 Cultivation and Extraction

Cultivation

The common hazel is hardy and does not make great demands on its environment. It needs eutrophic soil and, if possible, half-shade. As the plant is dioecious, it is necessary to plant several bushes for a safe yield (Roth and Kormann 2000, p. 57).

Extraction of the Oil

Depending on the cultivar and growing area, the harvest takes place from August to October. The nuts are usually shaken from the bushes after cloths have been spread underneath. The nuts are subsequently dried in an airy, dry location. If they are stored in a cool place, they can keep for about 12 months (Dassler and Heitmann 1991, p. 194).

To extract the oil, the hazelnuts are first roasted, then the shells are removed, and the nuts are pressed (Ulmer 1996, p. 83). They can also be extracted with supercritical CO_2 or n-hexane (Bernardo-Gil et al. 2002).

3 Character

Colour

Yellow to yellowish-brown, clear oil.
Light yellow, clear, non-drying oil (Roth and Kormann 2000, p. 57).

Odour

Nutty, sweet to fruity, noticeable touch of hazelnuts.

Volatile Compounds

Examination using aroma extract dilution analysis (AEDA) and gas chromatography-olfactometry of headspace samples (GCOH) detected the following potent aromatic substances (Matsui et al. 1998):

Aromatic substances	Odour quality
Ethyl-2-methylbutanoate	Fruity
2-Ethyl-3,5-dimethylpyrazine	Roasted odour
2,3-Diethyl-5-methylpyrazine	Roasted odour
(E)-ß-damascenone	Like cooked apple
(E,E)-2,4-Decadienal	Fatty, fried
Ethyl-isobutyrate	Fruity
3-Ethyl-2,5-dimethylpyrazine	Roasted odour
(E,Z)-2,4-Nonadienal	Green
(E,Z)-2,4-Decadienal	Fatty, green
3-Methoxy-4-hydroxybenzaldehyde (vanillin)	Like vanilla
d-Limonene	Like limes
2,5-(or 2,6-)Diethyl pyrazine	Sweet
(E)-2-Nonenal	Fatty, green
Phenyl acetaldehyde	Sweet, like honey
(E,E)-2,4-Nonadienal	Fatty, fried
(E)-2-Undecenal	Fatty
Hexanoic acid	Sweetish
2-Methoxy-4-vinylphenol	Pungent, phenolic
Hexanal	Green
2-Ethyl-3-ethylpyrazine	Roasted odour

Aromatic substances	Odour quality
3-Methylbuyric acid	Sweetish
Γ-Octanolide	Fruity, like coconut
(E)-5-Methyl-2-hepten-4-on	Sweet, like hazelnuts
Unknown	Sweet
Trans-4,5-epoxy-(E)-2-decenal	Metallic, green
Unknown	Sweet
(Z)-2-Octenal	Fatty
(E)-2-Decenal	Fatty, green
2-Methylpropanal	Malty
3-Methylbutanal	Malty
Propanal	Sweet
2-Methylbutanal	Green
2,3-Butanedione (diacetyl)	Like butter
Methyl-2-methylbutanoate	Sweet
Pentanal	Mouldy

As a result of storage and the effects of light and oxygen, other volatile compounds may form, such as alkane aldehydes, 2-alkenaldehyde and alkanoic acid. The most important oxidation products are hexanal and octanal, which may reach ten times their original concentration (Fallico et al. 2003).

As mentioned above, hazelnuts are usually roasted before being pressed. The Maillard reaction thereby triggered leads to a change in the chemical composition, which in turn results in the typical taste, odour and colour. It has been discovered that the content of hydroxymethylfurfural (HMF), which is often used as an index for browning intensity, rises in proportion to the duration of roasting (Fallico et al. 2003).

Another influence on HMF formation is the amount of hexanal, which increases with oxidation, as well as the size of the nut particles that are roasted. Nuts reduced to small pieces show the highest HMF values, whereas defatting leads to much lower values. This explains why the HMF value increases with a higher oil content (Fallico et al. 2003).

Flavour

Aromatic, like hazelnuts.

4 Ingredients

Composition of Fatty Acids

The composition of fatty acids varies depending on the growing area and the method of extraction:

H

Fatty acids	Content in % acc. to Firestone (1999), p. 46	Content in % acc. to Karleskind (1996), p. 164	Content in % at extraction with n-hexane acc. to Bernardo-Gil et al. (2002)
Palmitic acid	4.1–7.2	5–9	5.27
Palmitoleic acid	0.1–0.3	Up to 0.3	0.17
Margaric acid	0–0.2	Up to 0.1	–
Stearic acid	1.5–2.4	1–4	2.45
Oleic acid	71.9–84.0	66–83	85.18
Linoleic acid	5.7–22.2	8–25	6.27
α-Linolenic acid	0–0.2	Up to 0.6	0.08
Arachidic acid	0.1	Up to 0.3	0.13
Eicosanoic acid	0.1–0.3	–	0.17
Behenic acid	0.1	–	0.03
MUFA	–	–	85.52
PUFA	–	–	6.35

Fatty acids	Content in % at extraction with SFE acc. to Bernardo-Gil et al. (2002)	Content in %, growing area Italy, Mortarella acc. to Parcerisa et al. (1998)	Content in %, growing area USA, Ennis acc. to Parcerisa et al. (1998)
Palmitic acid	6.01	4.98	5.41
Palmitoleic acid	0.20	0.15	0.22
Margaric acid	–	–	–
Stearic acid	2.19	2.72	1.38
Oleic acid	82.65	78.9	77.08
Linoleic acid	8.42	12.83	15.55
α-Linolenic acid	0.09	0.11	0.02
Arachidic acid	0.08	0.01	0.01
Eicosanoic acid	0.14	0.19	0.16
Behenic acid	0.01	–	–
MUFA	82.99	79.24	77.47
PUFA	8.15	12.94	15.66

This table shows that the percentages of fatty acid components may fluctuate considerably, while the proportions of the individual fatty acids remain balanced. The fatty acids with the highest proportions are oleic acid (18:1) and linoleic acid (18:2), followed by palmitic acid and stearic acid. All other fatty acids form a subordinate part.

In addition to the growing areas, the roasting process plays an important role in the composition of fatty acids:

Composition of Fatty Acids in Unroasted Hazelnuts, Classified According to Countries of Production

(Crews et al. 2005, content in %)

Fatty acids	Croatia	Italy	Spain	Turkey
Palmitic acid	5.1	5.8–6.6	5.3–6.1	5.1–6.3
Palmitoleic acid	0.2	0.2–0.3	0.2–0.3	0.1–0.3
Stearic acid	2.7	2.4–2.8	1.9–2.5	2.0–3.1
Oleic acid	82.6	80.3–83.4	75.3–81.8	75.7–82.9
Linoleic acid	8.7	6.2–10.6	10.1–15.9	8.1–15.4
α-Linolenic acid	0.2	0.1–1.0	0.1–0.2	0.1–0.3
Arachidic acid	0.1	0.1–0.2	0.1	0.1
Eicosanoic acid	0.2	0.1–0.2	0.1–0.2	0.1–0.2

Composition of Fatty Acids in Roasted Hazelnuts, Classified According to Countries of Production

(Crews et al. 2005, content in %)

Fatty acids	Croatia	Italy	Spain
Palmitic acid	5.2	6.2–6.6	5.2–6.2
Palmitoleic acid	0.2	0.2–0.3	0.2
Stearic acid	2.7	2.4–2.6	2.1–2.5
Oleic acid	81.1	83.3–83.8	78.5–82.3
Linoleic acid	9.7	6.5–7.2	9.8–12.0
α-Linolenic acid	0.6	0.1	0.1–0.5
Arachidic acid	0.2	0.1	0.2
Eicosanoic acid	0.2	0.1	0.1–0.2

Composition of Triglycerides

Similarly to the composition of fatty acids, the composition of triglycerides many also vary, especially with regard to various methods of extraction:

Triglycerides	Content in % acc. to Firestone (1999), p. 139	Content in % after extraction with n-hexane acc. to Bernardo-Gil et al. (2002)	Content in % after extraction with SFE acc. to Bernardo-Gil et al. (2002)
PPP	0.1–2.7	0.1	0.1
PPO	1–2	0.3	0.3
POO	10–18	9.9	10.4
SOO	2–7	3.5	3.4
PPL	0–1	0.1	0.1
OOO	36–57	69.6	63.3
POL	3–6	1.7	2.6
OOL	10–24	11.6	15.5
PLL	0.5–2	0.2	0.2
OLL	2–11	2.0	2.7
LLL	0.5–4	0.7	0.8

Composition of Sterols

(Concentration in mg/kg and content in %)

Sterols	Hazelnut oil acc. to Firestone (1999), p. 47
Total content	1200–2000
Cholesterol	0–0.7
Campesterol	5–6
Stigmasterol	1
β-Sitosterol	82–93
Δ5-Avenasterol	2–8
Δ7-Stigmasterol	1–3
Δ7-Avenasterol	2–3

Composition of Tocopherols

Tocopherols	Content in mg/kg acc. Firestone (1999), p. 47
α-Tocopherol	200–409
β-Tocopherol	6–17
γ-Tocopherol	18–150
δ-Tocopherol	1–7

It is interesting to observe that the content of α-tocopherol may vary considerably depending on the growing area. Hazelnut oil from Segorbe (Spain), for example, contains twice the amount of α-tocopherol as hazelnut oil from Halls Giant (USA) (Parcerisa et al. 1998).

Other Ingredients

Protein
B-vitamins
Manganese
Sulphur
Potassium
Calcium
Carotenes

Physical Key Figures of Hazelnut Oil

refractive index: 25 °C 1.469–1.472
40 °C 1.456–1.463
density: 15.5 °C 0.914–0.920
25.0 °C 0.908–0.915
iodine number: 83–90
saponification number: 188–197
point of solidification: −20 to −10 °C

unsaponifiable: 0.2–0.3%
(Firestone 1999, p. 46; Roth and Kormann 2000, p. 57)

Shelf Life

Hazelnut oil has a shelf life of about 6 months if kept in a cool, dark place.

5 Application

As Food

Cold-pressed hazelnut oil can be heated and is a delicious edible oil. Since it has a slightly nutty taste, it is used to enhance the flavour of vegetable dishes and health food.

In Pharmacy and Medicine

Applications for the Nose

H

Hazelnut oil forms part of several nose drops, either in pure form to treat dryness of the nasal mucosa, or in combination with essential oils to treat sinusitis (Kircher 2002, p. 59).

Vasoconstrictive Properties

As hazelnut oil has vasoconstrictive properties, its use is recommended to support the treatment of respiratory problems. These properties also make hazelnut oil useful for the treatment of varicose veins, in which case it is employed as a base oil.

Lowering the Cholesterol Level

Since it contains a high amount of oleic acid, hazelnut oil lowers high cholesterol levels.

In Cosmetics

With its delicate scent, hazelnut oil is ideal for skin care, is well suited as a massage oil and is a popular base oil in aromatherapy. It is a good treatment for many skin problems, especially for sensitive and dry skin. As a skin-nourishing oil, it forms part of many skincare products (cold creams, massage oils, lipsticks) (Ulmer 1996, p. 83) and well suited for dry facial skin. Since it only penetrates the skin slowly, it leaves a fatty film on the skin, which protects it from bad weather (Carstensen 2001, p. 20).

In the Food Industry

Hazelnut oil is used for baking and to adulterate extra virgin olive oil to lower the production costs (Carstensen, 2001, p. 20).

6 Possible Unwanted Side Effects

Hazelnut allergy is one of the most common food allergies, acquired by patients with pollinosis caused by early-flowering trees such as birch, filbert, hazel and alder because of the structural similarities of their allergens in the pollen. The most common clinical manifestation is an oral allergy syndrome, which causes itching of the oral mucosa upon intake of hazelnuts or hazelnut oil (Senti et al. 2000).

Hempseed Oil

Oleum Cannabis synonym: Hanföl (D)

1 Source Plant

Cannabis sativa L. (Cannabaceae), cannabis

S. Krist, *Vegetable Fats and Oils*, https://doi.org/10.1007/978-3-030-30314-3_54

Habitat

The origins of this long-cultivated plant lie in Central Asia. Today, cannabis is common throughout the world because of its robustness and its remarkable adaptability. The EU subsidises the cultivation of cannabis species with a THC (tetrahydrocannabinol) content of less than 0.2%.

Description

There are two subspecies of cannabis: *Cannabis sativa ssp. indica* and *Cannabis sativa ssp. sativa* (Roth and Kormann 2000, p. 48). The annual plant can reach a height of up to seven metres. It is therefore a profitable plant, and has gained ground in many business areas thanks to its strengths. Cannabis is dioecious. The leaves are digitate with three to eleven lanceolate, long-acuminate, roughly serrated segments. The male plant develops an inflorescence with loosely arranged components on the upper part of the stem. The female plant has one or two flowers per axil, surrounded by a glandular, hood-shaped bracteole, in a dense, leafed pseudo-spike. The fruit are shiny nutlets about 3–5 mm in size (Schoenfelder and Schoenfelder 2001, p. 322). The fruit surface is greenish-grey, with light-coloured veins and darker spots. Underneath the brittle shell lies the greenish seed. Apart from fat oil (30–35%), the seed contains trigonelline, protein (approx. 25%), resin and sugar (Roth and Kormann 2000, p. 49).

2 Cultivation and Extraction

Cultivation

Cannabis is a robust plant that does not leach the soil. It its growth phase, it has a high carbon fixation, and does not need chemical fertilisers or pesticides to improve the yield. In fact, it does not make many demands on the soil apart from its relatively high water demand. Cultivation is also possible in rather unfavourable climates and soils, as long as the plant is not grown in poor sandy soil or cold, wet locations. Deep, humid, calciferous, neutral to slightly alkaline soil with a good water supply is best. The yield is highest in temperate climes, and the young plants survive night frosts. The minimum temperature for germination is 4–5 °C, so the seeds should not be sown too early in the year.

If the plants are cultivated for their seeds and flowers rather than for the production of fibres, the seeds should not be sown too close to each other. Monoecious species developing female and male flowers on one plant or hybrid varieties are usually preferred, as they produce more seeds than conventional varieties. The yield may amount to 1200 kg of seeds or more per hectare. After China, France is the largest producer of cannabis seeds, with a harvest of 5000 tons per year.

Extraction of the Oil

To produce a high-quality oil from cannabis seeds, it is necessary to harvest and press the seeds gently. The seed is embedded in a solid pericarp protecting it from environmental influences like sunlight and oxygen. To extract the oil, the seed must be pressed until the oil escapes from it. It is preferable to cold press hempseed oil, so that its valuable ingredients are not damaged by heat. The temperature for pressing is 40–60 °C. If the oil content amounts to 30–35%, the yield is about 180–350 l/ha.

3 Character

Colour

Cold pressed: greenish-yellow (Roth and Kormann 2000, p. 48).
Pressed by adding heat: dark green (Roth and Kormann 2000, p. 48).
Extracted with petrol and carbon disulphide: brownish-yellow (Roth and Kormann 2000, p. 48).
Extracted with diethyl ether: deep yellow (Roth and Kormann 2000, p. 48).

Odour

Herbaceous, aromatic, green and nutty.

Flavour

Nutty, herbaceous.

4 Ingredients

Composition of Fatty Acids

Fatty acids	Content in % acc. to Parker et al. (2003)	Content in % acc. to Rossell and Pritchard (1991), p. 293	Content in % acc. to ipp.boku.ac.at/08.03.06	Content in % acc. to Firestone (1999), p. 47
Palmitic acid	6.26 ± 0.34	5.6	6–7	6–12
Palmitoleic acid	–	0.3	0.1	–
Stearic acid	2.72 ± 0.03	2.6	2	1–2
Oleic acid	11.72 ± 0.04	10.6	11–12	11–16

Fatty acids	Content in % acc. to Parker et al. (2003)	Content in % acc. to Rossell and Pritchard (1991), p. 293	Content in % acc. to ipp.boku. ac.at/08.03.06	Content in % acc. to Firestone (1999), p. 47
Linoleic acid	59.96 ± 0.23	59.4	50–70	45–65
α-Linolenic acid	19.33 ± 0.08	19.4	20–25	15–30
Arachidic acid	–	1.9	–	2

Composition of Sterols

(Concentration in mg/kg and content in %)

Sterols	Hempseed oil acc. to Roth and Kormann (2000), p. 48
Total content	3720
Campesterol	17
Stigmasterol	15
Δ5-Avenasterol	2
Δ7-Stigmasterol	2
Δ7-Avenasterol	1

Other Ingredients

Phytosterols
Protein
Chlorophyll
Carotenoids
Vitamins

Physical Key Figures of Hempseed Oil

refractive index: 1.470–1.4767
relative density: 0.9230.931
iodine number: 143–166
saponification number: 190–195
acid number: 3.98
point of solidification: −25 °C to −15 °C
unsaponifiable: 0.5–1.5%
(Roth and Kormann 2000, p. 48; Firestone 1999, p. 47)

Shelf Life

Hempseed oil has a shelf life of 6–9 months if stored in a dark bottle in the refrigerator.

Characteristics of the Oil

δ-9-Tetrahydrocannabinol (THC)

The resin of female inflorescences contains cannabinoids, one of which is tetrahydrocannabinol. The amount depends on the variety as well as the climate (Schoenfelder and Schoenfelder 2001, p. 322). Cannabinoids are nitrogen-free phenolic derivates of benzopyran, which biogenetically derive from a monoterpene and a phenol. The basic structure consists of 21 carbon atoms. The hallucinogenic properties are attributed to δ-9-tetrahydrocannabinol (THC) alone (Hunnius 1998). THC is not part of the seed itself. Commercially available hempseed oil has a THC content of under 0.3%. Through improper production, THC, which is soluble in the oil, may, however, end up in the oil in relevant quantities, and may then be detected in the blood and urine of the consumer.

γ-Linolenic Acid (GLA)

The most important ingredients of hempseed oil are the polyunsaturated fatty acids linoleic acid, α-linolenic acid and the rare γ-linolenic acid, which makes up 2–4% of the oil. γ-Linolenic acid is an ω-6-fatty acid. Sources of GLA are rare; the substance containing the highest amount of GLA is breast milk. Evening primrose oil, blackcurrant seed oil and borage seed oil also contain GLA; but unlike hempseed oil, they cannot form part of an everyday diet as a result of their taste.

5 Application

In Pharmacy and Medicine

To Treat Vascular Diseases

As is the case with other vegetable oils, the positive health effects of hempseed oil, for example on high cholesterol levels and arteriosclerosis, can be traced back to its high amount of monosaturated and polyunsaturated fatty acids. As a result of its valuable polyunsaturated fatty acids α-linolenic acid and ω-linolenic acid, hempseed oil can be used directly in therapies. Polyunsaturated fatty acids have a positive influence on cell membranes, the elasticity of blood vessels and the blood flow.

Positive Effects of GLA

GLA is applied internally and externally in the treatment of skin diseases like neurodermatitis and psoriasis. In addition, it is used as an anti-inflammatory substance to treat rheumatic complaints and is effective in treating premenstrual syndrome because it has effects similar to hormones. Hempseed oil contains γ-linolenic acid, an essential ω-3 fatty acid found predominately in fish oil. This fatty acid is thuoght to lower the risk of heart attacks, cardiovascular diseases, diabetes and chronic inflammation. In the body, γ-linolenic acid is formed from linoleic acid by the enzyme desaturase and is a biogenetic precursor of prostaglandins with a double bond (Hunnius 1998).

Spasmolytic Properties of Hempseed Oil

Hempseed oil has an antispasmodic effect in cases of epileptic seizures, multiple sclerosis and chronic pain. It also has a positive influence on the well-being of cancer and AIDS patients (Ulmer 1996, p. 59).

As Food

As a result of its composition of fatty acids, hempseed oil is one of the best vegetable oils. Since it has a smoke point of 165 °C, it should not be used for roasting or frying, as the fatty acids are destroyed and the taste of the oil is impaired at the temperatures involved. Hempseed oil is, however, suitable for stewing and steaming.

The nutty taste of hempseed oil makes it ideal for use on salads and in dressings, gravies, marinades and spreads.

Industrial Uses

As hempseed oil contains a high amount of triacylglycerides, it lends itself to use as a raw material for the production of cleaning supplies. Various raw materials for the cleaning industry are produced from hempseed oil via chemical processes such as saponification, ethoxylation and sulphation. These raw materials include soap, emulsifiers, solvents, conditioners and surfactants, which are noteworthy for their biodegradability. As a result of its high slippage, hempseed oil is also used in the production of paint, varnish (Ulmer 1996, p. 59) and printing ink (Salunkhe et al. 1992, p. 457 et seq).

In Cosmetics

Hempseed oil has special properties that make it useful in cosmetics. The polyunsaturated fatty acids in its molecular structure give it excellent sliding properties, which make cosmetic products beneficial for the hair and the skin. Hempseed oil is especially good for rough, inflamed and flaky skin. It is a drying oil, and only a small amount (about 10%) of hempseed oil is therefore usually added to mixtures of other vegetable oils. It softens skin and hair and is added to massage oils, ointments, creams, soap and shampoo (Roth and Kormann 2000, p. 48).

6 Possible Unwanted Side Effects

As mentioned above, faults in production may lead to THC entering the oil. Depending on the amount of THC, this may have psychic, but also pharmacologic consequences, as THC may interact with other drugs. It is therefore absolutely necessary to comply with official threshold values.

Illipe Butter

Cera Illipe synonyms: Illipefett (D); Beurre d'illipe de Borneo (F)

Synonyms: Mowrahbutter (D); Mowrah butter, Indian illipe butter (E); Beurre d'illipe (F)

1 Source Plant

Madhuca longifolia (J. König) J.F. Macbr. *(Sapotaceae)*, mahuwa, iluppai

Habitat

The mahuwa is a native tree of India, Sri Lanka, Madagascar and Malacca. It is especially common in the woods of the southern slopes of the Himalayas (Bauer 1928, p. 240; von Wieser 1927, p. 758).

Description

The mahuwa is an evergreen, broad-leaved tree, up to 20 m tall, with a dense, overhanging crown. The cracked bark is grey or dark brown. The leaves are 7.5–23 cm long, leathery, linear-lanceolate, pointed and tapering towards the base. The flowers are arranged in dense panicles at the ends of the branches, first upright, later more or less pendant. The flower stalks are 2.5–3.8 cm long, pendant and covered with stiff, felty hair; the calyx is also densely covered with hair. The corolla is cream-coloured, the calyx-tubes are fleshy. The 20–30 stamina are arranged in three rows, the anthers are pointed and nearly sessile. The fruit is a small, fleshy, elongate, greenish berry about 2.5–5 cm long. At first, it is covered with stiff hair, but when

S. Krist, *Vegetable Fats and Oils*, https://doi.org/10.1007/978-3-030-30314-3_55

ripe, it is nearly glabrous and yellowish. It contains one to four seeds (List and Hörhammer 1976, 4th edition, volume 5, p. 626).

2 Cultivation and Extraction

Cultivation

The mahuwa is a hardy plant and survives dry spells and droughts. It grows in inaccessible regions that are not suitable for other cultivated plants. It bears its first fruit after 8–10 years and continues producing fruit for about 60 years. The harvest takes place from July to August, either by shaking the tree, by waiting until the fruit fall to the ground by themselves, or by plucking them by hand. To get to the seeds, the seed coat has to be removed by hand or in mills. As there are many wild-growing trees, the mahuwa is not cultivated on a grand scale. It is sometimes planted in towns as an alley tree (Salunkhe et al. 1992, p. 517; Axtell 1994, p. 79).

Extraction of the Oil

Illipe butter is obtained from the seeds of *Madhuca longifolia* by pressing or extraction with subsequent refinement (Axtell 1994, p. 80; Salunkhe et al. 1992, p. 518).

3 Character

Colour

Light-yellow (Roth and Kormann 2000, p. 125).
Greenish-yellow to yellow, bleaching when in contact with air (Von Wiesner 1927, volume 1, p. 753).
Pale yellow (List and Hörhammer 1976, 4th edition, volume 5, p. 626).
Raw fat: golden yellow (Roth and Kormann 2000, p. 138).
Refined fat: yellowish-white, like lard (Roth and Kormann 2000, p. 138).

Odour

Raw fat: acetous odour (Roth and Kormann 2000, p. 138).
Refined fat: neutral odour (Roth and Kormann 2000, p. 138).
Mild, nutty odour (Von Wiesner 1927, volume 1, p. 753).
Pleasant, nutty odour (Bauer 1928, p. 240).

Flavour

Bitter fat, similar to cocoa in taste (Roth and Kormann 2000, p. 125).
Mild, slightly nutty taste (Von Wiesner 1927, volume 1, p. 753).

4 Ingredients

Composition of Fatty Acids

Fatty acids	Content in % acc. to Roth and Kormann (2000), p. 125	Content in % acc. to Firestone (1999), p. 49	Content in % acc. to ieindia.org/11.08.05	Content in % acc. to Sen Gupta et al. (1982)
Caprylic acid	–	0.2	–	–
Capric acid	–	0.1	–	–
Lauric acid	–	0.2	–	–
Myristic acid	–	0.3	–	–
Palmitic acid	28	23	24.5	23.7–24.7
Palmitoleic acid	–	0.2	–	–
Stearic acid	14	23	22.7	19.3–29.9
Oleic acid	50	34	37	36.3–43.3
Linoleic acid	8	14	14.3	11.6–15.8
α-Linolenic acid	–	0.2	–	–
Arachidic acid	–	0.2	–	–

Composition of Triglycerides

Triglycerides	Content in % acc. to Karleskind (1996), p. 209 (2 measurements)	
SLL	–	–
OOL	–	–
POL	–	0.4
PPL	–	0.1
OOO	0.4	0.4
SOL	0.3	1.0
POO	1.5	2.0
PSL	1.0	0.3
PPO + MSO	6.5	6.5
PPP	–	0.1
PMgO	–	–
SOO	4.0	6.0
SSL	–	–
PSO	35	35
PPS	–	0.5
MgSO	–	–
AOO	1.5	–
SSO	47	45
PSS + APP	–	1.5
SAO	4.0	–
SSS + PSA	–	1.0

Composition of Sterols

(Concentration in mg/100 g and content in %)

Sterols	Illipe butter acc. to Firestone (1999), p. 49	Illipe butter acc. to Karleskind (1996), p. 211
Total content	550	120–460
Cholesterol	–	>1.5
Brassicasterol	–	–
Campesterol	16	12–20
Stigmasterol	7	5–8
β-Sitosterol	70	68–81
Δ5-avenasterol	6	–
Δ7-stigmasterol	1	–
Δ7-avenasterol	–	–
Isofucosterol	–	5–6

Composition of Tocopherols

(Concentration in mg/100 g and content in %)

Tocopherols	Illipe butter acc. to Karleskind (1996), p. 211
Tocopherol content	328
α-Tocopherol	68
β-Tocopherol	2
γ-Tocopherol	17
δ-Tocopherol	8
Tocotrienols	4

Other Ingredients

(Soulier et al. 1989; Karleskind 1996, p. 212)

Butyrospermol
Cycloartenol
24-methylen-cycloarthanol
Taraxasterol
α-Amyrin
β-Amyrin

(Roth and Kormann 2000, p. 138)

Oleodipalmitin 0.9%
Oleopalmitostearin 26.9%
Palmitodiolein 41.3%
Stearodiolein 29.7%

Physical Key Figures

Physical Key Figures of Illipe Butter (Roth and Kormann 2000, p. 125)

refractive index: n_D^{40} 1.459–1.462
density: d_4^{15} 0.9166
melting point: raw fat: 25–29 ° C
refined fat: 26–29 ° C
iodine number: 84–90
unsaponifiable: 0.5–0.7%

Shelf Life

Illipe butter has a limited shelf life and easily becomes rancid (Awasthi et al. 1975; Bauer 1928, p. 240;).

5 Application

In Pharmacy and Medicine

Against Skin Complaints

Illipe butter is applied to treat various skin complaints such as eczema, but also in cases of itching, dry or inflamed skin. It supports skin cell regeneration and has an anastatic effect (Awasthi et al. 1975).

Component of the Indian Pharmaceutical Codex

Illipe butter as well as the leaves, bark and blossoms of the mahuwa are of considerable pharmaceutical significance; they are frequently used in India especially. In the Indian Pharmaceutical Codex, they are listed under the name "Madhuca". An extract of the bark of mahuwa is used against rheumatism, ulcers, itching, gum bleeding, tonsillitis, leprosy and diabetes. As the bark contains 17% tannin, it is also used for dyeing and tanning. The flowers have a cooling, aphrodisiacal, expectorating, carminative effect and promote lactation. They are also used in the treatment of heart disease, burns, complaints of the gall bladder and earaches. The dried flowers are applied as a hot fomentation in cases of testitis. Persons suffering from haemorrhoids eat the flowers fried in butter. The fruit is regarded as a cure against bronchitis, consumption and diseases of the blood (Awasthi et al. 1975).

Ayurveda

In Ayurvedic medicine, illipe butter, if applied externally, is said to have an astringent, stimulating, softening, nourishing and pain-relieving effect (Awasthi et al. 1975).

In Folk Medicine (India)

Internal Application

Biliary fever is an intermittent or remittent fever appearing in cases of cholangitis, leading to catharral or ulcerous inflammation of the extrahepatical and/or intrahepatical bile ducts. In these cases, refined and cleaned illipe butter is applied internally. As it is a purgative, the population of India also uses it in cases of chronic constipation (Awasthi et al. 1975).

External Application

Illipe butter is applied externally in cases of rheumatism; it is said to improve mobility and relieve pain (Awasthi et al. 1975).

In Cosmetics

Illipe butter effectively moisturizes the skin, which is why it is added to body care products such as creams, lotions and ointments. It makes the skin soft and improves elasticity. Besides, illipe butter increases the stability of emulsions. It also supports skin cell regeneration. This is the reason why illipe butter is an ingredient of anti-wrinkle creams. Because of its relatively low melting point – illipe butter already melts at body temperature – it is used in various cosmetic products, very often in lipsticks, lip balms and other pencil-shaped body care products, for which fat with a higher melting point would not be suitable (Awasthi et al. 1975).

As Food

The poor population of India use illipe butter as an edible fat. It is not well suited for this purpose, however, as it becomes rancid very quickly (Von Wiesner 1927, volume 1, p. 753). Besides, it contains a relatively high amount of saponins that insufficient cleaning cannot remove and that may cause toxic reactions (Awasthi et al. 1975).

In the Food Industry

After refinement and complete removal of the toxic saponins, illipe butter can be used in chocolate production. According to the Cocoa Regulation (2738 German Federal Law Gazette volume 2003, part I, no. 63, Bonn 23.12.2003), illipe butter may be added only up to 5% of the total weight of the end product, retaining the minimum content of cocoa butter and cocoa dry mass.

In the mid-19th century especially, illipe butter was used in margarine production. Because of its short shelf life, it is not used for this purpose any more.

As Fuel

An Indian study explored the results of mixing illipe butter with diesel fuel and came to the conclusion that up to 20% can be added to diesel fuel without any significant differences being noted in comparison with pure diesel fuel. Illipe butter neither caused an increase in fuel consumption nor reduced engine power.

Industrial Uses

As a low-cost raw material, illipe butter is used in candle production (Axtell 1994, p. 79).

It is also used as a base in soap production. Normal soap often damages the skin's protective barrier and thus leads to dehydration, whereas illipe butter is moisturizing (Awasthi et al. 1975).

Other Uses

In India especially, illipe butter is used as lamp oil and in candle production by the poor population (Awasthi et al. 1975).

Possible Unwanted Side Effects

Toxic side effects can occur if illipe butter is insufficiently refined, as the fat contains a high amount of saponins (up to 9%). Oral application in small doses does not cause any side effects, but very high doses of saponins can be lethal. Especially when parentally fed, saponins have a strong toxic effect and may be dangerous. One of the characteristics of saponins is that they may cause haemolysis, which means that they have the capacity to destroy red blood cells (erythrocytes) (Awasthi et al. 1975).

A multi-generational reproduction study fed rats of both sexes on a diet containing 10% illipe butter. In the second generation, the female rats were completely infertile. Histological studies showed that the infertility was caused by atrophy of the oviduct. When illipe butter was removed from the food, fertility of the female rats could be restored. The intake of illipe butter thus leads to temporary infertility in female rats. In this study, no other side effects were found (Rukmini 1990).

Indian Butter

Oleum Diploknema synonyms: Phulwara butter, Fulwabutter, Fulwafett, Fulwabutter (D); beurre de fulwara (F)

1 Source Plant

Diploknema butyracea H.J. Lam., *syn. Bassia butyracea* Roxb., INCI[1]*: Madhuca butyracea (Sapotaceae),* Indian butter tree

Habitat

The tree is a native plant of northern India (Roth and Kormann 2000, p. 124).

Description

Diploknema butyracea, the Indian butter tree, grows up to 15–20 m tall. The leaves are 20–35 cm long, obovate or roundedly elongated, at the base blunt and rhomboid. The fragile white flowers are 2–2.5 cm in size and grow in dense bunches. The fruit are green berries, 2.4–5 cm long, ellipsoid and shiny. They usually contain one or two seeds each (Roth and Kormann 2000, p. 124).

[1] *INCI (International Nomenclature of Cosmetic Ingredients), founded by the American Cosmetic, Toiletry and Fragrance Association (CTFA) (ubka.uni-karlsruhe.de/29.08.05).*

S. Krist, *Vegetable Fats and Oils*, https://doi.org/10.1007/978-3-030-30314-3_56

2 Cultivation and Extraction

Cultivation

Apart from India, the tree is cultivated in tropical Asia and in Africa (Roth and Kormann 2000, p. 124).

Extraction

The fat is obtained by pressing or extracting the seeds, which contain 40–60% fat (Roth and Kormann 2000, p. 124).

3 Character

Colour

Almost white butter.

Non-drying, semi-solid, greasy, greenish-yellow fat (Roth and Kormann 2000, p. 124).

Odour

Pleasant odour.

Aromatic (Roth and Kormann 2000, p. 124).

Flavour

Pleasant taste.

Mild, bitter (Roth and Kormann 2000, p. 124).

4 Ingredients

Composition of Fatty Acids

Fatty acids	Content in % acc. to Roth and Kormann (2000), p. 124	Content in % acc. to Reddy and Prabhakar (1994)
Palmitic acid	56.6	61.0
Stearic acid	3.6	3.8
Oleic acid	36.0	30.0
Linoleic acid	3.8	0.2

Physical Key Figures of Indian Butter

density: 0.9245 (at 15 °C)
refractive index: 1.4552–1.4659 (at 40 °C)
saponification number: 170–200
iodine number: 38–43
unsaponifiable: 0.5–5.3%
point of solidification: 27 °C
melting point: 39–51 °C
(Roth and Kormann 2000, p. 124)

5 Application

In Cosmetics

Since Indian butter has softening properties, high amounts of palmitic acid and oleic acid and a natural sun protection factor, it is added to skincare products such as cleansers and moisturisers, make-up foundations, lipsticks, sunscreens and hair care products.

In the Food Industry

In India, Indian butter is used as an edible fat and sometimes for the production of chocolate and soap (Roth and Kormann 2000, p. 124).

A study examined whether Indian butter is suitable as a confectionery fat. It behaves like a semi-solid fat without showing any supercooling property. If cocoa butter is mixed with a middle stearin fraction of Indian butter produced by two-stage fractionation, the cooling curve is the same as that of pure cocoa butter. The cooling effect in the mouth is, however, higher than that of pure cocoa butter. This fraction can therefore be added to chocolate and other sweets (Reddy and Prabhakar 1994).

Japan Tallow

Cera Japonica synonyms: Japantalg (D); Japan wax (E); cire du Japon (F)

1 Source Plant

Rhus succedanea L. (Anacardiaceae), *Toxicodendron succedaneum* var. *succedaneum*, sumac

Habitat

The origins of *Rhus succedanea L.* lie in Japan. From there, the tree spread to the Indian subcontinent, Nepal, Punjab and the Ryukyu Islands. It is not, however, native to the Himalayas. Of all the species of *Rhus*, it is the most sensitive, only growing up to altitudes of 2000 m, whereas related species are found at much higher altitudes. In Japan, the tree is cultivated on plantations and parks. Since its leaves are a beautiful colour, it is also planted in parks in Australia and England, but has meanwhile been banned from these countries because of its allergenic potential.

Description

Rhus succedanea L. reaches a height of 3–6 m. It has a greyish-brown bark and hard, yellow wood containing a light-coloured juice that turns black when it comes into contact with the air. The crown is highly branched; the branches are only partly covered with leaves. The leaves sit predominantly at the end of the branches; they turn from green to red in autumn. They are bi-pinnate and have a hairless leaf stalk. The leaflets are oval to lanceolate, entire and hairless, 5–7 cm long and 2–2.5 cm broad. The upper side is shiny, the lower side dull. The young leaves are covered in

S. Krist, *Vegetable Fats and Oils*, https://doi.org/10.1007/978-3-030-30314-3_57

brown tomentum. The numerous flowers form panicles, which sit in the axils of the leaves heaped at the end of the branches. The panicles are about 6–8 cm long; the greenish-yellow flowers are either perfect or unisexual. The fruit is a roundish berry of a brownish-yellow colour; its form is compressed at the top and on the sides.

2 Cultivation and Extraction

Cultivation

The tree prefers light, loamy, wet soil and grows on acidic, neutral as well as alkaline soil. It does not, however, grow in the shade and in dry soil. The flowers appear in June and July and are pollinated by bees. The fruit harvest takes place in autumn.

Extraction

Japan tallow is obtained from the fruit and fruit pulp by pressing. The pressed material is filtrated, bleached in the sun and then cast in small slices or rectangular plates (Roth and Kormann 2000, p. 126).

3 Character

Colour

Light yellow (Roth and Kormann 2000, p. 126).
Whitish-yellow (Burger and Wachter 1993, 7th edition, p. 1498).

Odour

Rancid smell (Roth and Kormann 2000, p. 126).

Flavour

Rancid taste (Roth and Kormann 2000, p. 126).

4 Ingredients

Composition of Fatty Acids

Fatty acids	Content in % acc. to Roth and Kormann (2000), p. 126	Content in % acc. to ibiblio.org/02.09.04
Palmitic acid	77	77
Japanic acid	Traces	–
Stearic acid	Traces	5
Oleic acid	12	12
Arachidic acid	Traces	5
Palmitic acid-glycerol ester	Traces	25
Essential oil	Traces	–

Physical Key Figures of Japan Tallow

density: d_4^{20} 0.97–0.98
melting point: 53.5–55 °C
saponification number: 217–237
iodine number: 5–17
acid number: 22–23
unsaponifiable: 1–2.5%.
(Roth and Kormann 2000, p. 126)

5 Application

In Pharmacy and Medicine

The use of Japan tallow in patches and ointments plays a subordinate role. As a result of its unpleasant rancid smell, it is seldom used as an ingredient of patches, ointments, grease and analgesic unctions.

Industrial Uses

Since Japan tallow has a low price, it is used instead of bees-wax in wax polish and floor wax. Japan tallow is suitable for the treatment of leather and leather products, as it keeps the material supple and shiny. It is used for textile finishing.

In the candle industry, Japan tallow is a low-cost raw material.

6 Possible Unwanted Side Effects

Japan tallow itself does not cause any unwanted side effects, but the fresh leaves of sumac contain urushiol, which is one of the most potent contact allergens. Urushiol is a mixture of n-pentadecyl-2,3-dihydroxybenzol and other catechol derivates, which have mono to triple unsaturated C15 chains instead of a saturated C15 chain. The chyle that escapes when the leaves are rubbed or squeezed causes severe skin inflammation and a rash, accompanied by unbearable itching. This dermatitis is very persistent and may recidivate after several years have passed, even if the plant has not been touched again. The poison causes conjunctivitis and keratitis if it comes into contact with the eyes. Oral ingestion causes severe symptoms of poisoning, such as vomiting, colic, gastroenteritis, dizziness, drowsiness and haematuria (Burger and Wachter 1993, 7th edition, p. 1415).

Jatropha Seed Oil

Oleum Jatrophae synonyms: *Oleum infernale, Oleum Ricini majoris* (L); Jatropha-Curcas-Samenöl (D); huile de jatropha, huile de pignon d'Inde (F)

1 Source Plant

Jatropha curcas L. (Euphorbiaceae), purging nut

S. Krist, *Vegetable Fats and Oils*, https://doi.org/10.1007/978-3-030-30314-3_58

Habitat

Jatropha curcas originates in the regions between Mexico and Chile; Portuguese and Dutch seafarers brought it to Asia and Africa, and the plant is now common throughout the tropics. It is of commercial value on the islands of Cape Verde, especially on the Sotavento Islands, Fogo, Brava and Santiago, where large populations were planted for their seeds. In these locations, the cultivation of other plants is not possible due to extreme climatic and soil conditions.

Description

The purging nut is a sturdy bush or small tree that can reach a height of 6 m. It has oval leaves on long stalks that grow mainly in terminal, dense clusters. Female and male flowers sit together in small corymbs. They are yellow to green, campanulate and about 6–7 mm long. The ball-shaped, fleshy fruit are 2–4 cm long. In the process of ripening, they change from green and yellow to black. Each fruit contains three compartments with one black seed each. The seed is 2 cm long. All parts of the plant contain a transparent, yellowish latex, which becomes red when it comes into contact with the air (Hager 1994, p. 888).

2 Cultivation and Extraction

Cultivation

Purging nuts grow well in regions with a minimum annual precipitation of 400 mm. The tree is able to adapt to various kinds of soil, but prefers sandy soil. This is an important prerequisite for its cultivation in marginal locations in the dry tropics. The plant sheds the majority of its leaves during dry spells to reduce water loss. Propagation is possible via seedlings or cuttings, and is best done at the beginning of or shortly prior to the rainy season. Traditionally, the trees are planted in linear formation as a hedge, but sometimes also at a distance of 1–5 m from each other, depending on the growing area and country. Good precipitation conditions make a harvest in the first growing season possible; the maximum harvest is reached in approximately the fifth year. The main harvest takes place in November and December. Smaller amounts can be harvested throughout the year. The fruit are plucked by hand, either when ripe or at the end of the ripening period (Höggerl 1992, p. 15 et seq).

Jatropha curcas plantations are a pilot scheme for landscape protection in emerging economies like China, Brazil and several African countries. The test sites are stony ground and depleted soil on fallow land, which are considered unfit for agriculture. The dense, widespread root system of the purging nut protects the soil from erosion and is a valuable source of nutrients. In this way, *Jatropha* plants make barren lands fertile again.

Extraction of the Oil

The yield of seeds and oil vary depending on origin and conditions (climate, soil, planting distance, water supply, manuring). Dry climes, in particular, increase the oil content of the seeds.

For oil production, the seeds are either pressed or extracted (Hager 1994, p. 891).

3 Character

Colour

Light-yellow to colourless, semi-drying oil (Hager 1994, p. 891).
Very light yellow, semi-drying oil (Ajiwe et al. 1996).

Odour

Almost odourless (Hager 1994, p. 891).

4 Ingredients

Composition of Fatty Acids

Fatty acids	Content in % acc. to Hager (1994), p. 891	Content in % acc. to Kalayasiri et al. (1996)	Content in % acc. to Höggerl (1992), p. 19
Myristic acid	–	–	0.1
Palmitic acid	15–17	15.2	14.3
Palmitoleic acid	–	–	0.8
Stearic acid	Approx. 8	6.1	6.5
Oleic acid	30–44	41.8	29.0
Linoleic acid	30–52	35.8	34.6
α-Linolenic acid	Approx. 0.3	–	0.2
Arachidic acid	Approx. 0.3	–	0.2
Eicosanoic acid	–	–	0.1

Composition of Sterols

In the oil's unsaponifiable part, which amounts to about 1.6%, it contains phytosterols, the main component of which is β-sitosterol with a share of about 80% (Hager 1994, p. 891).

Physical Key Figures of Jatropha Seed Oil

density: 0.9141
refractive index: 1.4698
saponification number: 196
iodine number: 102
acid number: 10
hydroxyl number: 2.15
(Hager 1994, p. 891)

Characteristics of the Oil

Jatropha seed oil contains numerous diterpenes: 0.01% esters of the tiglian type, which are two intramolecular 12-desoxy-16-hydroxyphorbol-13,16-diesters, and two other diterpene esters (Hager 1994, p. 891).

The oil's toxicity (see below) is due, to a small degree, to curcuson A, but mainly to a lectin called curcin. Its structure is comparable to the ricin of *Ricinus communis.* Curcuson A causes skin irritations.

5 Application

In Pharmacy and Medicine

The oil shows molluscicide activity, for example against *Pomacea sp.* and the transmitter of bilharziasis (schistosomiasis). This is of interest because schistosomes are endemic in over 70 tropical and subtropical countries, and more than 200 million people contract the disease each year (Gübitz et al. 1998). Bilharziasis is a chronic infectious disease. The invasive phase, which starts after 24 hours, is characterised by cercarial dermatitis with itching, caused by schistosomes. Depending on the species and in cases of massive infestation, the acute phase leads to fever, urticaria, angioneurotic syndrome, bronchitis and pneumonia after 3–12 weeks. In the chronic phase, which can start 30–90 days after the infection, various symptoms and sequelae may develop for the duration of up to 10 years (Pschyrembel 1994).

The methanolic extract of raw jatropha seed oil has a higher activity than the oil itself, with an LD_{50} of 0.004% for *Biomphalaria glabrata* (transmitter of *Schistosoma mansoni)* and 0.00025% for *Oncomelania hupensis* (transmitter of *S. japonicum).* The molluscicide effect can probably be traced back to phorbol esters or saponins (Gübitz et al. 1998).

In Folk Medicine

In its growing areas, jatropha seed oil is used as a laxative and to treat diseases of the skin, such as scabies, eczema, herpes, as well as a liniment against rheumatism. The effect is not, however, clinically proven (Hager 1994, p. 891).

Technical Uses of Jatropha Seed Oil

After esterification with methanol, the oil can be used as a substitute for diesel fuel and as a renewable resource. The oil's fuel value is comparable to that of diesel (jatropha seed oil 9470, diesel 10,170), but the oil has a lower sulphur content (oil 0.13%, diesel 1.2%), which means that no sulphur dioxide is detectable in the exhaust fumes of an engine operated with jatropha seed oil. As fuel, the oil therefore has a comparable potential to diesel, but it is better for the environment (Kandpal and Madan 1995).

Other Applications

In Nigeria, the oil is used in agriculture to produce an arrow poison and as a bait for guinea fowl. In industry and technology, it is used as a lamp oil and to produce soap, candles, paint and lubricants (Hager 1994, p. 893). Since it has an insecticidal effect, it is a biological pesticide.

6 Possible Unwanted Side Effects

The oil made from the seeds of *Jatropha curcas* has an inflammatory activity with an ID_{50} of 25 µg/ear after 24 hours; a mixture of the isolated diterpene esters even has an ID_{50} of 0.02 µg/ear, its poisonousness being comparable to croton oil (Neuwinger 1998, p. 516). Oil made of the seeds of *Jatropha curcas* irritates the skin and mucosa, and is toxic. Its toxicity is ascribed to its content of diterpene esters. If the oil is applied to the skin, it causes erythema and pustular rash. An intake of only eight drops results in severe vomiting and diarrhoea. Pregnant women have a miscarriage if their abdomen is rubbed heavily with the oil (Hager 1994, p. 891 et seqq). The phorbol ester DHPB contained in the oil (intramolecular 13,16-diester of 12-deoxy-16-hydroxyphorbol) promotes tumours. It induces ornithine decarboxylase in the skin of mice, and activates protein kinase C in vitro ($ED_{50} = -36.0$ nM) (Hirota et al. 1988).

Jojoba Oil

Cera liquida Simmondsiae synonyms: Jojobaöl, Jojobawachs (D); simmondsia chinensis seed oil (E); huile de jojoba (F)

Jojoba seed oil is not an oil in the true sense of the word, but rather a liquid wax.

1 Source Plant

Simmondsia chinensis L. C.K. Schneid. (Simmondsiaceae), jojoba

S. Krist, *Vegetable Fats and Oils*, https://doi.org/10.1007/978-3-030-30314-3_59

Habitat

The exact origin of the jojoba bush is unknown; its name probably derives from the Native American name *jojowi*. It grows in the Sonoran Desert and other south-western regions of the USA, in Baja California and north-western Mexico (Salunkhe and Desai 1986, p. 189 et seq). Its habitat is slopes with grainy soil and the banks of rivers that have run dry (Roth and Kormann 2000, p. 103). Outside its countries of origin, the bush is also successfully cultivated in Israel, Argentina, Peru and Australia.

Description

The jojoba is evergreen and does not tolerate temperatures below 12–14 °C. It normally grows to between 2 and 2.5 m tall, but can grow to anything from 0.5 to 6 m tall (Salunkhe and Desai 1986, p. 190). The jojoba has a taproot of up to 3 m in length (Roth and Kormann, 2000, p. 103). It is a dioecious plant that produces seeds similar to nuts, which contain a liquid wax. Wild-growing jojoba bushes differ considerably in appearance. The leaves are hairy, oval, alternate and of various sizes (14–40 mm long, 6–20 mm broad). They are usually dark green, but (albino) species with woolly, yellow leaves also exist. The female flowers are usually solitary and a greenish colour, whereas the male flowers appear in small groups and are yellow (Salunkhe and Desai 1986, p. 190). The fruit of the jojoba is a green, elongated to oval capsule containing one to three seeds. The size and number of the seeds depends on the location of the plant. When the seed is ripe, the capsule opens and releases the seed. The seed is about 12–20 mm long and 6–12 mm broad. It contains an embryo and cotyledons in a hard, thick testa, which is a dark colour when the seeds are ripe, but has no or almost no endosperm. The cotyledons are large, and their parenchyma cells are filled with liquid wax (Salunkhe and Desai 1986, p. 190).

The wax content of the seeds increases rapidly in the first 4 weeks of ripening and can reach 60% of the seeds' weight (Salunkhe and Desai 1986, p. 192).

2 Cultivation and Extraction

Cultivation

Jojoba can be cultivated on plantations in the dry regions of many countries. The long taproot ensures a supply of water and nutrients from deeper ground layers. Fertilisation is not necessary, but additional irrigation increases the yield (Roth and Kormann 2000, p. 104). The bush is planted to form hedges, with the plants at a distance of two metres from each other. Fertile soil with a pH value higher than 5 is ideal. To ensure pollination, at least 5% of the plants should be male. The first harvest is possible after 3–5 years; the maximum yield is reached after about 12 years.

Drip irrigation via underground pipes minimises water loss by surface evaporation. An additional supply of water is only necessary from autumn to spring, as dry and hot summers are important for fruit formation. Jojoba plantations can be established together with citrus plantations. Their tolerance towards salinity makes it possible to convert salty soil into arable land by artificial irrigation (Roth and Kormann 2000, p. 104).

Extraction

The ripe seeds are ground and cold pressed. This is the easiest method to produce oil. Another method is extraction with solvents. Carbon tetrachloride, benzene, hexane, heptane, isopropyl alcohol and tetrachloroethylene can be used, as they have the ability to dissolve waxes (Maghsood 1991).

3 Character

Colour

Clear, golden yellow.

Odour

Cosmetic, soapy.

Flavour

Mild.

4 Ingredients

Composition of Wax Esters

Unlike most other vegetable oils, jojoba seed oil is not a mixture of triacylglycerides, but rather a mixture of wax esters with a chain length of 38–44 C atoms (Hänsel et al. 1999, p. 292).

Chain length of the esters	Ester content in %
C36	1–2
C38	6–7
C40	26–30
C42	49–52
C44	10–11
C46	1–2
C48	0–1

The wax esters contain 47–49% fatty acids and 50–52% fatty alcohols (Karleskind 1996, p. 315). The following table shows the amount of fatty acids and fatty alcohols of jojoba seed oil in per cent (relative to the respective component):

Chain length of fatty acids	Share of fatty acids in % of the total content of fatty acids acc. to Karleskind (1996), p. 315	Share of fatty alcohols in % of the total content of fatty alcohols acc. to Karleskind (1996), p. 315
Palmitic acid	0–2	–
Oleic acid	10–13	1–2
Eicosanoic acid	66–71	39–46
20:2	–	1–2
Behenic acid	0–1	1–2
Cetoleic acid	14–20	45–52
Lignoceric acid	–	0–1
Nervonic acid	1–3	6–11

Composition of Sterols

(Concentration in mg/100 g and content in %)

Sterols	Refined jojoba seed oil acc. to Carstensen (2001), p. 42	Cold pressed jojoba seed oil acc. to (Carstensen 2001), p. 42
Total content	339.8	339.8
Brassicasterol	0.5	–
Δ5-Campesterol	15.6	19.3
Δ5-Stigmasterol	6.0	7.6
Δ7-Campesterol	0.1	–
β-Sitosterol	74.5	67.4
Δ5-Avenasterol	2.8	4.8
Δ7-Stigmasterol	<0.1	0.8
Δ7-Avenasterol	0.3	–

Composition of Tocopherols

(Concentration in mg/100 g and content in %)

Tocopherols	Refined jojoba seed oil acc. to Carstensen (2001), p. 44	Cold pressed jojoba seed oil acc. to Carstensen (2001), p. 44
Total content	25.4	66.3
α-Tocopherol	23.0	85.0
β-Tocopherol	69.8	5.4
Γ-Tocopherol	5.6	2.9
Δ-Tocopherol	1.7	6.8

Other Ingredients

Provitamin A
Amino acids
Minerals
Squalene

Physical Key Figures of Jojoba Seed Oil

refractive index: 1.465
density: 0.8642
iodine number: 81.7–89
saponification number: 93
acid number: 0.25–0.57
ignition point: 338 °C
flash point: 295 °C
point of solidification: 10 °C
melting point: 6.8–7 °C
boiling point (under nitrogen): 398 °C
unsaponifiable: 37–49%
(Roth and Kormann 2000; Salunkhe and Desai 1986, p. 194)

Shelf Life

Jojoba seed oil does not become rancid and is thought to have a shelf life of up to 25 years.

5 Application

In Pharmacy and Medicine

In Pharmaceutical Technology

As a liquid wax, jojoba seed oil is used in pharmacy as a carrier and coating for drugs, in the production of penicillin, in appetite suppressants (Maghsood 1991, p. 35) and in reducing diets, since jojoba seed oil is not affected by digestive enzymes. Jojoba seed oil is relatively stable against rancidity and is thus a suitable vehicle for substances susceptible to oxidation (Frohne 2002, p. 508). It is used as a natural emulsifier instead of cetaceum.

In Dermatotherapy

In cases of sunburn and various skin diseases, jojoba seed oil supports the healing process and alleviates inflammation. Jojoba seed oil penetrates the skin quickly and does not leave a fatty film.

In Cosmetics

All-Purpose Jojoba Seed Oil

As a liquid wax, jojoba seed oil is used in the cosmetics industry in setting lotions, shampoo, hairsprays, hair oils, in products for the regeneration of the hair and scalp, hair colouring products, hair grease, soap, shaving cream, aftershave, tanning lotions, massage and body oils, bath oils, colognes, hand lotions, skin food, moisturisers, cleansers, make-up, make-up removers, face masks and lipsticks (Maghsood 1991, p. 35).

As a Moisturiser

Jojoba seed oil is anti-allergenic and suitable for all skin types. Its excellent penetration ensures moisture balance and makes the skin smooth and supple. It strengthens the connective tissue and prevents wrinkles. The oil has a natural sun protection factor of 4.

Industrial Uses

As a Lubricant

Jojoba seed oil is the raw material of many lubricants for precision instruments. By sulphonation, a high-quality lubricant is produced, which retains its lubricity even under high pressure and at high temperatures (Roth and Kormann 2000).

As a Hard Wax

Jojoba seed oil forms part of polishing wax for floors, shoes, furniture and cars, of protective covers for fruit and other foodstuffs, of liquid containers made from paper, candles, insulating material, matches, grease paint sticks and carbon paper (Maghsood 1991). The individual fatty alcohols and fatty acids are added to disinfectants, detergents, solvents, plastic (plasticiser), protective covers, fibres and anti-corrosive coatings (Maghsood 1991, p. 35).

As a Fuel

The Arabian peninsula and other regions undertook successful experiments to test jojoba seed oil as an environment-friendly fuel for diesel engines. Jojoba seed oil is low in carbon and rich in sulphur. Only methanol has to be added.

As Food

Jojoba seed oil can be used not only as a salad oil, but is also suitable for frying and can be heated to high temperatures (Roth and Kormann 2000). So far, it is hardly used as an edible oil in reduction diets; due to the lack of suitable enzymes, it cannot be digested. Since unwanted side effects are possible (see below), its use as an edible oil is not currently recommended.

6 Possible Unwanted Side Effects

In studies on animals, it was found that jojoba seed oil causes pathological changes to the blood count, the small intestine and especially the liver if it is ingested over long periods of time. The liver showed pale discolorations and had a similar appearance to a nutmeg, which indicates fatty infiltrations. As a result of these harmful side effects, it is not recommended to use liquid jojoba seed wax as food for particular nutritional uses (Hänsel et al. 1999, p. 292). An LD_{50} test could not be determined; it showed a value of >21.5 ml/kg body weight and was discontinued because of the high doses involved. Additionally, a clinical examination showed no indications that jojoba seed oil has locally irritating or sensitising properties if applied topically in undiluted form (Hager 1993, p. 703).

Kapok Seed Oil

Oleum Ceibae seminis synonyms: Kapoksaatöl, Kapoköl (D); silk-cotton tree oil (E); huile de kapok (F)

1 Source Plants

Ceiba pentandra L. *Gaertn., syn. Bombax ceiba* L. *(Bombacaceae)*, kapok, white silk-cotton tree

S. Krist, *Vegetable Fats and Oils*, https://doi.org/10.1007/978-3-030-30314-3_60

Habitat

Ceiba pentandra grows in rain forests, but also in the dry forests and gallery forests of South and Central America. *Bombax ceiba,* the Indian silk-cotton tree, is a native plant of India, Sri Lanka and Myanmar. It is also cultivated in other tropical regions (Roth and Kormann 2000, p. 52).

Commercially, *Ceiba pentandra is not* usually distinguished from *Bombax ceiba*, as they are both used for oil production (Roth and Kormann 2000, p. 52).

Description

The tree grows up to 60 m tall. The leaves are unevenly pinnate, with leaflets that are 10–18 cm long and red when they sprout. The flowers are yellowish white and have up to 3.5 cm long, woolly petals. The fruit are leathery, elliptical capsules 10–30 cm long, with many black seeds and seed fibres 2–4 cm in length (Roth and Kormann 2000, p. 52).

2 Cultivation and Extraction

Cultivation

The kapok is cultivated in its countries of origin, but also in Indonesia, Cambodia and Madagascar. It prefers rich, moist but well-drained soil and a minimum temperatures of 15 °C. Propagation is possible via large cuttings and branches of about 1 m in length planted at a distance of about 3 m from each other. The tree bears about 300–400 capsules annually from its 5th to its 60th year. The fruit are dried with the capsules (Roth and Kormann 2000, p. 52, 128).

K

Extraction of the Oil

After the outer capsule has been removed, the seeds are separated from the fibres, ground with roller frames and pressed by adding heat without removing the shells (Roth and Kormann 2000, p. 52, 128).

3 Character

Colour

Light-yellow, rather viscous, non-drying oil (Roth and Kormann 2000, p. 129).

Odour

Pleasant smell.

Volatile Compounds

SPME-GC-MS analysis (Krist et al. 2008):

Propanoic acid	Hexanal
3-Hydroxy-2-butanon	r-Butyrolaktone
3-Methylbutanol	Alpha-pinene
2-Methylbutanol	Decane
2,3-Butanediol	1,8-Cineol
Octane	β-Ocimene

Flavour

Pleasant, mild taste.

4 Ingredients

Composition of Fatty Acids

Fatty acids	Content in % acc. to Roth and Kormann (2000), p. 128	Content in % acc. to Rossell and Pritchard (1991)	Content in % acc. to Firestone (1999), p. 52
Myristic acid	0.5	–	–
Palmitic acid	16.1	10	10–28
Stearic acid	2.3	9	2–9
Oleic acid	50.6	46	45–65
Linoleic acid	29.7	34	7–35
Arachidic acid	0.8	1	1

Kapok seed oil also contains up to 15% cyclopropenoid fatty acids (Rossell and Pritchard 1991, p. 294).

Composition of Sterols

Sterols	Content in % acc. to Firestone (1999), p. 53
Campesterol	9
Stigmasterol	2
β-Sitosterol	86
Δ5-Avenasterol	2
Δ7-Stigmasterol	1

Physical Key Figures of Kapok Seed Oil

density: d_4^{20} 0.920–0.928 at 15 °C
refractive index: n_D^{20} 1.4685–1.4710
n_D^{30} 1.4878
saponification number: 189–195
iodine number: 92–98
unsaponifiable: approx. 1%
melting point: 26.2–31.6 °C
(Roth and Kormann 2000, p. 128)

5 Application

Kapok seed oil is an edible oil and is used in soap production. The kapok is more important for the production of fibres, which are smooth, water-repellent and very elastic. In water, they have a bearing strength of 50 times their weight (Roth and Kormann 2000, p. 52, 128).

A study examined the mutagenic activity of the oil by using the Ames mutagenicity assay[1] with *Salmonella Typhimurium* strains. Results show that the oil has no mutagenic activity (Polasa and Rukmini 1986).

[1]Ames mutagenicity assay: reverse mutation of *Salmonella Typhimurium;* mutagenic activity of substances is tested on *Salmonella typh.* Strains requiring histidine, in the presence and absence of a rat liver metabolising system.

Kiwi Seed Oil

Oleum Actinidiae chinensis synonym: Kiwiöl (D)

1 Source Plant

Actinidia deliciosa (A. Chev.) C.F. Liang et A.R. Ferguson, syn. *Actinidia chinensis*, (Actinidiaceae), kiwi fruit

S. Krist, *Vegetable Fats and Oils*, https://doi.org/10.1007/978-3-030-30314-3_61

Habitat

The name kiwi derives from the name of the flightless bird kiwi native to New Zealand. The region of origin of the kiwifruit is China, where it is still found on the edges of forests in the Yangtze Valley today. The kiwi is a plant growing between the 10th and 40th degree of northern latitude (Dassler and Heitmann 1991). The main growing area is now located between the 35th and 40th degree of northern latitude (Kratochvil 1995, p. 80). New Zealand is still the largest producer of kiwifruit, followed by France, the USA, Italy, Spain and Japan.

Description

The kiwi plant is a deciduous climbing plant with shoots that are several metres long and produce white or pink flowers. The arrangement and form of the styli gave the plant family its German name, *Strahlengriffelgewächse (Actinidiaceae,* Chinese gooseberry family*)*. The plant is dioecious (Dassler and Heitmann 1991), the only certain distinguishing feature between the male and female plants being the form of the flowers. Male flowers are, on average, 4 cm in size; first, they are a pure white colour, then yellowish. Their stamina are arranged like a powder puff; sometimes, they have a crooked style. Female flowers measure 5–6 cm in diameter. Their white colour changes to creamy yellow after 3 or 4 days. They have many styli, which are arranged like a corona (Kratochvil 1995, p. 81). One kiwi plant bears several hundred fruit. The fruit is an elongated, oval berry about the size of an egg, with a diameter of up to 5 cm and a weight of 60–100 g. The hairy, firm, inedible peel is first green, then becomes brown when the fruit is ripe. The juicy fruit pulp is emerald green (Dassler and Heitmann 1991). Depending on the degree of ripeness, it tastes either sourish or sweet.

2 Cultivation and Extraction

Cultivation

As male and female flowers grow on different trees, each kiwi garden is planted with one male plant for the pollination of eight female ones (Dassler and Heitmann 1991). Kiwi plants prefer sandy soil rich in humus, which should neither be too hard nor be subject to stagnant moisture. They need acid soil with a pH-value of 4.5–5.5. Young plants should be allowed to grow in sunny, protected locations – south-facing slopes are best. The distance between the plants should be a minimum of four metres (Kratochvil 1995, p. 82).

Extraction of the Oil

Kiwi seed oil is extracted with CO_2 from the seeds of the kiwifruit (de.wikipedia.org/28.03.06).

3 Character

Colour

Clear, yellow.

Odour

Varnish, fatty and rancid, slightly fruity, aromatic.

Flavour

Pleasant.

4 Ingredients

Composition of Fatty Acids

Fatty acids	Content in % acc. to Yao et al. (2001)	Content in % acc. to nature.de/28.03.06
Oleic acid	12.89	11.6
Linoleic acid	12.59	14.9
α-Linolenic acid	63.99	65.3

The unusually high amounts of the two unsaturated essential fatty acids linoleic acid and linolenic acid is conspicuous (Yao et al. 2001).

Composition of Tocopherols

The tocopherol content amounts to 0.81 mg/g kiwi seed oil.

Physical Key Figures of Kiwi Seed Oil

refractive index: 1.4830
density: 0.9308
iodine number: 123
saponification number: 196
acid number: 3.8
unsaponifiable: 1.3%

5 Application

In Cosmetics

Kiwi seed oil is added to antiageing products (Courtin 1994).

Kokum Butter

Oleum Kokum synonym: Kokumbutter (D)

1 Source Plant

Garcinia indica (Thouars) Choisy (Clusiaceae), Indian mangosteen

Habitat

Garcinia indica is a native plant of the western coastal area of the Indian subcontinent. It is also cultivated on a larger scale in some regions of eastern India (Bauer 1928, p. 255).

Description

The Indian mangosteen is an evergreen tree about 15–20 m tall. The stem is slender, with drooping branches. The bark is very thin and wrinkly; the lanceolate leaves are dark green. The fruit are round and about 4 cm in diameter; the ripe pericarp is a shiny black colour. The fruit contains 5–8 brown seeds. They are kidney-shaped, flattened, 1–2 cm long and up to 1 cm broad. The pericarp contains a secretion container, and the embryo contains the distinctly crystalline fat and several aleurone grains, of which one is conspicuously large (List and Hörhammer 1973, 4th edition, 4th volume, p. 1099).

S. Krist, *Vegetable Fats and Oils*, https://doi.org/10.1007/978-3-030-30314-3_62

2 Cultivation and Extraction

Cultivation

Growing Indian mangosteens is not easy. The tree only grows in very warm, tropical conditions, and needs a great deal of space. The trees must be planted individually and at a considerable distance from each other. The ripe fruit are plucked from the trees.

Extraction

In India, kokum butter is obtained in a very laborious process. The seeds are dried in the sun, chopped, and boiled in water. When the fat cools down, it solidifies, and can be lifted off and subsequently cleaned by remelting and filtering. The yield is, however, small (Bauer 1928, p. 255).

Today, kokum butter is extracted and subsequently refined.

3 Character

Colour

Greyish-white to white (List and Hörhammer 1973, 4th volume, p. 1099).
A yellowish-white to yellowish light-grey colour (Bauer 1928, p. 256).
Brownish if it has been stored for a long time or if it is rancid (Bauer 1928, p. 256).

Odour

Light smell (List and Hörhammer 1973, 4th volume, p. 1099).
Weak, not unpleasant smell (Bauer 1928, p. 256). Fresh, similar to cocoa butter.

Flavour

Mild taste (Bauer 1928, p. 256).

4 Ingredients

Composition of Fatty Acids

Fatty acids	Content in % acc. to Firestone (1999), p. 53	Content in % acc. to List and Hörhammer (1973), 4th volume, p. 1099	Content in % acc. to oilsbynature. com/10.08.05
Myristic acid	0–1	–	–
Palmitic acid	2–5	2.5–5	2.5–6
Stearic acid	52–56	52–56	50–62
Oleic acid	39–42	39–42	30–42
Linoleic acid	1–2	1.7	0–4
Linolenic acid	–	–	<0.2
Arachidic acid	–	–	<1.2
Eicosanoic acid	–	–	<0.5
Behenic acid	–	–	0.2

Composition of Triglycerides

Triglycerides	Content in % acc. to Firestone (1999), p. 139
PSO	8–14
SOS	59–68
PPO	2
POO	0.5–2
SOO	20–21
OOO	2

Composition of Sterols

Sterols	Content in % acc. to Hilditch and Murti (1941)
Oleodistearin	68
Stearodiolein	20
Oleopalmitostearin	8
Triolein	2
Tristearin	1.5
Palmitodiolein	0.5

Physical Key Figures of Kokum Butter

refractive index: n_D^{40} 1.4565–1.4571
density: d_{40}^{40} 0.895–0.899
saponification number: 186–190
iodine number: 35–37
acid number: max. 3
unsaponifiable: 1–2%

melting range: 39–42 °C
(List and Hörhammer 1973, 4th volume, p. 1099)

Shelf Life

Kokum butter has a shelf life of 1 year.

5 Application

In Pharmacy and Medicine

In the pharmaceutical industry, kokum butter as a transesterified fat is a substitute for cocoa butter. By adding 5–10% triethanolaminstearate, suppositories with a higher melting point are produced (List and Hörhammer 1973, 4th volume, p. 1099).

In Cosmetics

Kokum butter moisturises the skin and therefore forms part of many creams, balms and lotions for dry skin. Adding kokum butter to antiwrinkle creams results in improved skin elasticity and regeneration of skin cells (Thippeswamy and Raina 1989). As kokum butter is a cheaper raw material than shea butter, it is sometimes blended with the latter in cosmetic products (Bauer 1928, p. 256).

As a result of its firmness and moisturising properties, kokum butter can be used to produce soap (Thippeswamy and Raina 1989).

As Food

The poor population in India frequently uses kokum butter as a butter substitute (Thippeswamy and Raina 1989).

In Chocolate Production

Kokum butter is suitable for chocolate production. Cocoa butter has a relatively low melting point that causes chocolate to melt at higher temperatures, but the melting point of kokum butter is significantly lower. Adding 5% of kokum butter to the total mass significantly improves heat resistance, without any negative consequences for the chocolate's appearance and taste. According to the *Kakaoverordnung* (Cocoa Regulation, 2738 German Federal Law Gazette of 2003, part I, no. 63, Bonn, 23.12.2003), only other vegetable fats may be used apart from cocoa butter, while retaining the minimum content of both the cocoa butter and the total cocoa dry weight, so that the content does not exceed 5% of the end product after deducting the total weight of the ingredients. The vegetable fats that are permitted to be used are kokum butter, illipe fat, Borneo tallow, mango seed oil, palm oil, sal butter and shea butter (Maheshwari and Reddy 2005).

Technical Uses

During experiments with the fatty berries of the Indian tree *Garcinia indica*, Indian researchers developed a method to transform oil into hardened fat. If one part kokum butter is mixed with five parts sunflower oil, the mixture gelatinises in a reversible process at room temperature. This method could be useful, for example in the event of oil spills from tankers to prevent the oil slick from spreading and causing an environmental disaster. This method works for every saturated fat with 10–31 carbon atoms.

Since kokum butter contains a high concentration of oleic acid and stearic acid, it is also used to extract these fatty acids (Thippeswamy and Raina 1989).

Kukui Oil

Oleum Kukui synonyms: Kukuinussöl (D); kukui nut oil (E); lumbang oil (E); huile de bancoul (F)

1 Source Plant

Aleurites moluccana (L.) Willd. *(Euphorbiaceae)*, candlenut, candleberry, Indian walnut, kemiri, varnish tree

S. Krist, *Vegetable Fats and Oils*, https://doi.org/10.1007/978-3-030-30314-3_63

Habitat

The origin of Aleurits moluccana is not accuratly known, but it is distributed from India to China, throughout South-East Asia, to Polynesia and New zealand. today, it is cultivated in many tropical countries all over the world (uses.plantnet-project.org, 2016).

Description

Aleurites moluccana is a medium-sized tree with a large spreading crown that can reach 20 m in high and 0.9 m trunk diameter (Krisnawati et al. 2011). The candlenut is a deciduous tree with entire leaves that are 3 to 5-lobed and long-stemmed. The flowers are white, but may be delicate shades of red or pink. The male and female flowers are located on separate terminal inflorescences. In both kinds of flowers, the 2–3 segmented sepals and the 5 petals look the same (Roth and Kormann 2000, p. 37). The fruit are dark green to brownish in colour and have a shell about 6 mm thick. Inside, there are one or two hard nuts resembling stones (nature.de/29.03.06). These nuts, from which the oil is extracted, are as large as walnuts; they have a rough, wrinkly, often greyish-white surface and a shell that is 2–5 mm thick (Roth and Kormann 2000, p. 37). The seeds of Aleurites moluccana ("kemiri") are an indispensable spice in Indonesian and Malaysian cuisine. Per 11g dry seeds contain 60-62g fat (uses.plantnet-project.org, 2016; Krisnawati et al. 2011).

2 Cultivation and Extraction

Cultivation

K

The species *Aleurites* prefers slightly acidic soil; highly acidic soil is not suitable. In small amounts, nuts can be harvested from the third year of cultivation onwards; after 10 years, a full crop can be expected. The tree has a life span of about 30 years (Roth and Kormann 2000, p. 37).

Extraction of the Oil

Kukui oil is either extracted from the seeds with solvents, cold pressed or pressed by adding heat. Hot-pressed oil is less stable, as the oil's natural antioxidants are destroyed (Kehren 1950). After pressing, the oil is carefully filtrated and cleaned; to stabilise the product, natural antioxidants are added, for example vitamin C (ascorbyl palmitate), vitamin A (palmitate) or vitamin E (d-α-tocopherol). The remainable seed cake can be used as a fertiliser (Krisnawati et al. 2011).

3 Character

Colour

Light yellow to orange (Kehren 1950).

Odour

Sourish, grassy, muffled chemical smell.

4 Ingredients

Composition of Fatty Acids

Fatty acids	Content in % acc. to np-d.com/10.02.03	Content in % acc. to Firestone (1999), p. 21
Saturated fatty acids	>10	–
Unsaturated fatty acids	<90	–
Palmitic acid	6.4	5–9
Palmitoleic acid	0.1	–
Stearic acid	<0.3	3–7
Oleic acid	13.9	11–35
Linoleic acid	43.6	37–49
α-Linolenic acid	33.2	24–35
Eicosanoic acid	0.6	–

The high amount of unsaturated fatty acids such as linoleic acid and linolenic acid is conspicuous.

Physical Key Figures of Kukui Oil

refractive index: 1.437–1.479
relative density: 0.924–0.929
iodine number: 136–175
peroxide number: <5 meq/kg
saponification number: 185–202
acid number: <1.0
unsaponifiable: 0.3–1%
(Firestone 1999, p. 21 – np-d.com/10.02.03)

Shelf Life

If stored in a cool, dry place protected from air and light, kukui oil has a shelf life of at least 12 months.

5 Application

In Pharmacy and Medicine

Kukui oil is used successfully in the treatment of atopic dermatitis, neurodermatitis, psoriasis, acne, eczema and transepidermal water loss. It is sometimes used like castor oil (Krisnawati et al. 2011).

In Cosmetics

Skin-Tightening Oil

Since the oil has a high proportion of vitamins A and E, it supports the development of healthy skin. It also strengthens the connective tissue and prevents stretch marks.

Moisture Control

The skin quickly absorbs the oil, which is especially suitable for dry and stressed skin. The unsaturated fatty acids that it contains support the development of epidermal ceramides and the regeneration of the skin barrier. In this way, the body's own moisture balance is preserved. Kukui oil is absorbed easily and does not leave a fatty film on the skin. It has a natural sun protection factor of 10 and helps skin that has been irritated by solar radiation. It also serves as a base oil in the production of cosmetic products, for example face oils (preventing wrinkles) and conditioners (Mueller et al. 1995).

Industrial Uses

The oil serves as a basis for paint and fuel and soaps (Hata 1993). (uses.plantnet-project.org, 2016).

As Food

The residual oil cake is sometimes processed into a snack-food calles "dage kemiri" in Indonesia (uses.plantnet-project.org, 2016).

Kusum Oil

Synonyms: Khussambinussöl, Macassaröl (D); kon oil, paka oil (E)

1 Source Plant

Cussambium glabrum Buch.-Ham., syn. Schleichera trijuga Willd., Schleichera oleosa Merill., syn. Pistacia oleosa (Sapindaceae), kusum tree, Ceylon oak

S. Krist, *Vegetable Fats and Oils*, https://doi.org/10.1007/978-3-030-30314-3_64

Habitat

The tree grows in the dry woods of India, Myanmar and Sri Lanka (Salunkhe et al. 2003).

Description

Schleichera trijuga is a broad-leafed tree that grows to a height of 10–14 m (Axtell and Fairman 1992, p. 83), which blooms in early spring. The seeds contain up to 54% kernels, which in turn contain about 57% oil (Salunkhe et al. 2003).

2 Cultivation and Extraction

First, the seeds are cleaned. After the seed coat has been removed, the seeds are dried, decorticated, and either pressed or extracted with solvents (Salunkhe et al. 2003). Kusum oil contains poisonous glycosides of hydrocyanic acid, which must be removed by careful cleaning (Roth and Kormann 2000, p. 134).

3 Character

Colour

Non-drying, semi-solid, yellowish-brown fat (Salunkhe et al. 2003).
Yellowish-white oil (Roth and Kormann 2000, p. 134).

Odour

K

Pleasant smell (Roth and Kormann 2000, p. 134).

4 Ingredients

Composition of Fatty Acids

Fatty acids	Content in % acc. to Salunkhe et al. (2003)	Content in % acc. to Firestone (1999), p. 54
Myristic acid	1	1
Palmitic acid	5–8	5–8
Stearic acid	2–6	2–6
Oleic acid	60	57–62
Linoleic acid	3–4	2–5
α-Linolenic acid	–	–

Fatty acids	Content in % acc. to Salunkhe et al. (2003)	Content in % acc. to Firestone (1999), p. 54
Arachidic acid	20–25	20–25
Lignoceric acid	–	2–4
Total oil content	57	–

Physical Key Figures of Kusum Oil

density: 0.9099 (at 30 °C)
refractive index: 1.4607 (at 40 °C)
saponification number: 234
iodine number: 60
unsaponifiable: 2.1
acid number: 31.3

Characteristics of the Oil

Kusum oil contains 1.15% nitrogen. After saponification, 0.8–1.2% cyanide is generated. The main cyanogen compound consists of two fatty acids (behenic acid 40.6%, oleic acid 28.1%, gondoic acid 16%, palmitic acid 11.6%, stearic acid and palmitoleic acid 1.6% each) that are esterified with 1-cyano-2-hydroxymethylprop-en-1-ol (Kasbekar et al. 1971).

5 Application

In its countries of origin, kusum oil is used as lamp oil, as an addition to medicines, as a hair oil, in soap and to obtain fatty acids. Soap made from kusum oil is hard and has excellent emulsifying and cleaning properties (Salunkhe et al. 2003, p. 520). If refined, kusum oil can also be used as an edible oil (Roth and Kormann 2000, p. 134).

Lallemantia Oil

Synonyms: Öl des Ölziest (D); Lallemantiaöl (D); Drachenkopföl (D).

S. Krist, *Vegetable Fats and Oils*, https://doi.org/10.1007/978-3-030-30314-3_65

1 Source Plant

Lallemantia iberica *(Lamiaceae)*, dragon's Head

Habitat

Dragon's head originates in the Middle East, Transcaucasia, Iran and Turkmenistan. In the 1920s and 1930s as well as after the second World War, it was brought to southern Russia in an attempt to cultivate it there (Schuster 1992, p. 119).

Southern Europe is also among the countries of origin, from where the dragon's head spread to Central Europe after the second World War, when its cultivation as an oil plant began (Garber 1949).

Description

Dragon's head belongs to the *Lamiaceae* family, but is closely related to the mint family. It is an annual, highly branched plant 30–100 cm tall. The roots reach deep into the ground, which is why the plant is resistant to droughts. The leaves are alternate, narrow-lanceolate in form, and have a serrated edge. The flowers are arranged in the axils and are white, light pink or blue. Mainly bees and bumblebees pollinate the plant. The seeds are located in the calyx, which remains until they are ripe. Each calyx contains four seeds, which have a triangular, wedge-shaped, oval form. They are 4–6 mm long, 1.2–2.2 mm broad and 1.0–1.5 mm thick. The thousand-seed weight is 4.8–5.0 g (Schuster 1992, p. 118–119).

The seeds are dark brown. They have a dull shine, and the surface is corrugated. On the basal side, the white, anchor-shaped hilum is clearly visible (Garber 1949).

2 Cultivation and Extraction

Cultivation

Dragon's head is cultivated in all regions of the Commonwealth of Independent States because of its oil, which is used for technical, but also nutritional purposes. *Lallemantia iberica* makes hardly any demands on the climate. Warm, sunny locations are ideal, whereas too much precipitation may result in stem rot. Dragon's head needs calciferous soil to grow. Stagnant moisture, slightly acidic soil and soil with a high degree of clay are unsuitable. Dragon's head should be sown in mid-April; the minimum temperature for germination is 2–3 °C (www.tll.de (1), 08.09.2011).

The earlier the seeds are sown, the higher the yields are. Dragon's head does not require a high nutrient supply. Usually, no pest control is necessary, because of the young plants' rapid growth. They are, however, liable to fungal attacks, which may lead to considerable economic damage. Cold, wet weather, in particular, may be a problem in this respect. The plant is harvested using combine harvesters. The seeds do not ripen at the same time and fall to the ground easily, which is why the harvest should begin as early as possible, ideally when the seeds of the main shoot start turning brown. Depending on the weather conditions, yields amount to 10–25 dt/ha, with an oil content between 26% and 40% (www.tll.de (2), 08.09.2011).

To counteract irregular ripening, it is advisable to sow the seeds in a short distance from each other. If the plants do not stand close together, they produce too many side shoots, which in turn promotes seed ripening at different times (Hackbarth 1944, p. 344).

A study by Samadi et al. compared several varieties of *Lallemantia iberica* with regard to their physical and chemical characteristics. The cultivar *Shahin dej* has the highest oil content (35.50%), whereas the cultivar *Myandoab* has the highest content of linolenic acid (56.37%) (Samadi et al. 2007).

Extraction of the Oil

The seeds can either be cold pressed or extracted with solvents (Garber 1949).

Prior to extraction, the seeds are dried and ground (Samadi et al. 2007).

3 Character

Colour

Cold-pressed oil is light to dark yellow, whereas extracted oil has a greenish tint (Garber 1949).

Flavour

The taste is scratchy at first, then resembles linseed oil (Hackbarth 1944, p. 344).

4 Ingredients

Composition of Fatty Acids

Fatty acids	Content in % acc. to Schuster (1992), p. 10	Content in % acc. to Overeem et al. (1999)	Content in % acc. to Hagemann et al. (1967)	Content in % acc. to Samadi et al. (2007)	Content in % acc. to Garber (1949)
Palmitic acid	8–10	6.5	6.9	8.07	–
Stearic acid	1–3	1.8	2.2	2.48	–
Oleic acid	1–10	10.3	14	17.24	1.4
Linoleic acid	22–36	10.8	9.4	17.33	36.4
α-Linolenic acid	50–60	68.0	66	51.22	55.3

Physical Key Figures

refractive index: n^{40}_D	1.4758
iodine number:	200–201

(Hagemann et al. 1967)

saponification number:	181
iodine number:	162
point of solidification:	−35 °C

(Hackbarth 1944, p. 344)

5 Application

Technical Uses

Lallemantia oil is used for varnish, lighting, oil paintings and lubricant production. It is a substitute for linseed oil in the production of linoleum, as it contains a high amount of linolenic acid. The cultivar *Myandoab*, in particular, is used for this purpose (Samadi et al. 2007).

Since the oil has a high content of linolenic acid, it can also be used to produce aliphatic epoxides, which are cross-linkers in powder coating. Powder painting is 100% solvent-free; no emission of volatile organic compounds therefore takes place. Powder coating material usually consists of a resin, cross-linkers, colour pigments and auxiliary substances. Triglycidylisocyanurate (TGIC) is a frequently used cross-linker, but as it is harmful to health as well as to the environment, tests are being carried out to substitute it with oils that have a high content of linolenic acid (Overeem et al. 1999).

As Food

Lallemantia oil is an edible oil, for which the cultivar *Shahin dej*, in particular, is used, since it has a particularly high oil content (Samadi et al. 2007).

Laurel Oil

Oleum Lauri expressum synonyms: Lorbeeröl (D); oil of laurel (E)

1 Source Plant

Laurus nobilis L. *(Lauraceae)*, bay laurel

S. Krist, *Vegetable Fats and Oils*, https://doi.org/10.1007/978-3-030-30314-3_66

Habitat

Bay laurel has its origins in the Middle East, and from there, it spread to the Mediterranean area. As a tree, it can grow up to 10 m high. As it is not very hardy, it only survives in Central Europe in milder regions with winter protection. In harsher climates, it can only be grown as a container plant.

Description

Bay laurel is an evergreen shrub or tree, with alternate, lanceolate, pointed, leathery, dark-green leaves. The leaves have wavy edges and are shiny on the upper side; they have short stems and an aromatic smell. Bay laurel is dioecious. It has white flowers arranged in short-stemmed, nearly round inflorescences, which appear from April to May. The fruit from which the oil is gathered are deep black in colour and oval in form (Roth and Kormann 2000, p. 72 et seq). The fruit and leaves are used for seasoning because of their content of essential oil (Frohne and Pfänder 1997, p. 237). Apart from the essential oil (1%), the fruit contain 30–40% fat oil (Fischer and Krug 1984, p. 153).

2 Cultivation and Extraction

Cultivation

Bay laurel is frequently cultivated in the Mediterranean area, especially in Italy (Lake Garda), in former Yugoslavia, Greece and Turkey, as well as in the subtropical regions of Russia and in Central and South America (Fa. Bressmer 2003).

Extraction of the Oil

The fruit are reduced to small pieces and subsequently pressed in heated presses, or boiled with water (Diener 1987, p. 229).

3 Character

Colour and Texture

Dark-green or brownish-green, ointment-like mass of a granular nature, more seldom a thick liquid consisting of solid and liquid portions (Diener 1987, p. 229).

Odour

Like eucalyptus, ethereal, herbaceous, pungent.

Heavily aromatic (Diener 1987, p. 229).

Flavour

Spicy, bitter (Diener 1987, p. 229), aromatic.

4 Ingredients

Composition of Fatty Acids of Laurel Oil

Fatty acids	Content in % acc. To Marzouki et al. (2008)	Content in % acc. to Firestone (1999, p. 55)
Lauric acid	27.7	11–35
Myristic acid	1.0	0.1
Palmitic acid	17.1	10–18
Palmitoleic acid	0.3	1–2
Stearic acid	1.5	2
Oleic acid	27.2	33–41
Linoleic acid	21.5	18–32
α-Linolenic acid	1.2	2

Other Ingredients

(Bressmer 2003; Diener 1987, p. 229; Fischer and Krug 1984, p. 153)

Chlorophyll
Sterols
Hydrocarbons
Sugar
Bitterns
Lauric alcohol
Myricyl alcohol
Sesquiterpene lactones

Physical Key Figures of Laurel Oil

refractive index: 1.460–1.465
density: 0.921–0.941
iodine number: 75–99
 49–81
saponification number: 197–210
acid number: <12
melting point: 30 °C
unsaponifiable: 1–6%
(Firestone 1999, p. 55; Diener 1987, p. 230)

Characteristics of the Oil

Laurel oil is a mixture of essential (2–3%) and fat oil. The main component of the essential oil is 1,8-cineol, also called eucalyptol (Diener 1987, p. 229).

5 Application

In Pharmacy and Medicine

As a result of the hyperemising effect of products containing laurel oil, they are applied externally in the treatment of rheumatic complaints, as an antipsoriatic, to treat swelling and in cases of colic (Bressmer 2003). *Oleum lauri expressum* is occasionally still used as a local therapeutic agent in the treatment of furuncles and abscesses as a drawing salve.

In Veterinary Medicine

The oil is added to udder ointments and horse-care products, for example hoof grease.

In Cosmetics

Laurel oil is used as scent and scented oil, and also as an antiseptic addition to foot-care products, soap and facial oils.

Industrial Uses

For a long time, hatbands were impregnated with laurel oil (Frohne and Pfänder 1997, p. 237), and the oil is still used as a finishing agent in felt processing (Bressmer 2003).

L

6 Possible Unwanted Side Effects

Laurel oil is a heterogeneous product, and its allergenic potential differs. The most important allergenic components that are responsible for allergisation are the sesquiterpene lactones costunolide and dehydrocostus lactone (Hager 1993, p. 351). Other possible side effects are contact allergies and cross sensitivities to the *Asteraceae* family. For these reasons, the Cosmetics Regulation does not permit application of laurel oil in its pure form on humans.

Linseed Oil

Oleum Lini synonyms: Leinsamenöl (D); solin oil (E); huile de lin (F)

1 Source Plant

Linum usitatissimum L. *(Linaceae)*, flax

S. Krist, *Vegetable Fats and Oils*, https://doi.org/10.1007/978-3-030-30314-3_67

Habitat

Flax is one of the oldest cultivated plants. The Sumerians and Egyptians as well as the lake-dwelling populations of the later Stone Age, for example around Lake Constance, grew flax 6000–8000 years ago. The origins of the plant are unknown. It is cultivated all over the world and grows wild on all continents (Salunkhe et al. 1992, p. 506). Today's important cultivation areas are the countries of the Rio de la Plata basin (Argentina, Uruguay, Paraguay), the USA, Canada, India, Egypt, the former Soviet Union, and, to a smaller degree, Belgium, the Netherlands, France and Germany. In 2004, more than 2,030,607 Mt. of flax seeds were produced worldwide. The most important production area in Europe lies in Russia, where 58,000 Mt. were produced in 2004 (Krist et al. 2006b).

Description

Flax is usually an annual plant 60–120 cm tall (Salunkhe et al. 1992, p. 506), with hairless, upright stems densely covered with alternate, narrow, lanceolate leaves. The flowers are blue, sometimes white; the stamina and the stylus are the same colour. The round, locular capsule contains seeds (Roth and Kormann 2000, p. 73) consisting of the seed coat, endosperm, embryo and cotyledons. The seeds are 4–6 mm long and deep yellow to brown. The shell is hard and only contains small amounts of oil and protein. It is 4-layered, and the outmost layer is especially rich in carbohydrates. The endosperm consists of 2 to 6-layered parenchyma cells and contains oil drops and alleuron grains (Salunkhe et al. 1992, p. 506). The plant blossoms from July to August, and the seeds ripen from August to September.

2 Cultivation and Extraction

Cultivation

As a cultivated plant, flax grows everywhere except in equatorial countries, even in mountainous regions (Wenigmann 1999, p. 161). There are three varieties, which are all cultivated plants: *Linum usitatissimum convar. Crepitans (*Kulpa et Danert), *Linum usitatissimum convar. Mediterraneum (fat linseed,* Kulpa et Danert) and *Linum usitatissimum convar. Usitatissimum* (Salunkhe and Desai 1986, p. 172). Until the First World War, flax was cultivated in Central Europe in great quantities. Today, fat linseed is cultivated on 130 ha in Krappfeld (Carinthia, Austria) for the purpose of oil production. Weather conditions have a great influence on the oil content of the seeds: The oil content decreases in dry and hot weather (Salunkhe and Desai 1986, p. 172). The plant makes hardly any demands on the soil, but it is susceptible to wet conditions.

Extraction of the Oil

Linseed oil for pharmaceutical uses is cold pressed from the ripe, crushed seeds (Hager 1993, p. 673). This is carried out using a screw compressor; the seeds are pushed through a pressing cylinder at low pressure and a maximum temperature of 40 °C. If the oil is to be used for technical purposes, the seeds are pressed by adding heat, and the press cake is extracted with solvents. The oil is then refined to remove mucilage (Hager 1993, p. 673).

3 Character

Colour

Cold-pressed oil is golden yellow; unpressed oil is yellowish brown (Salunkhe et al. 1992, p. 508).
Refined oil is light yellow to golden yellow.

Odour

Odour of pyrazine, aromatic, hay, herbaceous to muffled, slightly roasted, slightly fishy.

Volatile Compounds

The following components were detected in SPME-GC-MS analysis (Krist et al. 2006b):

Odorous substance	Odour quality	Content in % of sample A	Content in % of sample B	Content in % of sample C
Acetous acid	Intense, sour, pungent smell	3.7	3.8	3.6
2-Butanol	–	–	2.2	–
2-Methyl-1-propanol	Penetrating, like wine	1.6	–	1.9
Trans-2-butenal	–	5.0	1.3	2.1
2-Buten-1-ol	–	4.3	–	–
1-Penten-3-ol	Butter, mild, green	–	0.4	–
Propionic acid	Pungent, rancid	1.9	–	–
2-Ethylfuran	Strong, sweet, ethereal, burnt smell		1.0	1.5
3-Hydroxy-2-butanone	Like butter	0.2	–	–
3-Methyl-1-butanol	Fusel oil, whisky	0.9	–	0.6
Trans-2-pentenal	Pungent, green, apple, orange, tomato	0.8	1.0	0.8

Odorous substance	Odour quality	Content in % of sample A	Content in % of sample B	Content in % of sample C
Pentanol	Strong, slightly sweet, balmy	2.8	–	3.0
Isovaleric acid	Rancid, cheese-like, sweat, putrid	0.3	–	–
Trans-2-hexenal	Sweet, fragrant, almond, fruity, green, leafy, apple, plum, vegetables	0.9	0.8	0.7
Ethylbenzene	–	–	0.3	–
Trans-2-hexen-1-ol	Strong, leafy, green, like wine, fruity	0.3	0.2	0.3
Hexanol	Herbaceous, woody, fragrant, mild, sweet, green	9.3	6.5	20.3
2-Heptanone	Fruity, aromatic, cinnamon	0.2	0.3	0.4
2-Heptanol	Earthy, oily	0.3	0.3	0.7
Heptanal	Oily, fatty, heavy, woody, penetrating, sweet		0.7	
Trans,trans-2,4-hexadienal	Fresh, green, floral, citrus	0.3	0.7	0.2
γ-Butyrolactone	Weak, sweet, caramel	0.8	0.6	0.5
α-Pinene	Poignant, pine	0.2	–	0.3
Trans-2-heptenal	Pungent, green	1.6	0.6	–
Benzaldehyde	Bitter almond, fragrant, aromatic, sweet	–	1.0	–
Trans-2-hepten-1-ol	–	0.5	–	–
Heptanol	Fragrant, woody, heavy, oily	0.8	0.5	1.0
Hexanoic acid	Offensive, like sweat, rancid, sour, poignant, pungent, cheese-like, fatty		0.2	0.7
1-Octen-3-ol	Herbaceous, earthy	1.2	2.0	1.5
2-Pentylfuran	Green beans, metallic, vegetables	0.9	1.0	2.7
Trans,trans-2,4-heptadienal	Fatty, cinnamon, hazelnut	4.3	0.9	–
2-Octanol	Fatty, oily, earthy	–	–	0.6
Octanal	Fatty, citrus, honey after diluting	–	1.0	–
3-Carene	Sweet, penetrating	–	–	0.3
Limonene	Mild, citrus, sweet, orange, lemon	1.3	0.7	2.8
Benzyl alcohol	Weakly aromatic, poignant, burning	0.5	–	–
Trans-3-octen-2-one	Fruity, lemon	–	–	0.3
γ-Hexalactone	Sweet, herbaceous, warm	1.6	1.7	1.2
Octanol	Poignant, fatty, waxy, citrus	0.6	1.0	1.6
Trans,trans-3,5-octadien-2-one	–	3.9	3.3	0.7
Nonanal	Flowery, citrus, orange, rose, fatty, waxy	0.6	2.8	0.6
2-Phenylethanol	Rose, honey, fragrant, flowery	0.8	0.4	0.4
Trans-2-nonenal	Penetrating, fatty, waxy	0.2	0.7	0.2

Odorous substance	Odour quality	Content in % of sample A	Content in % of sample B	Content in % of sample C
Octanoic acid	Unpleasant, oily, fatty, rancid	–	1.4	–
Nonanol	Rose, citrus	–	–	0.7
Decanal	Penetrating, sweet, waxy, flowery, citrus		0.5	
Trans-2-decenal	Orange, slightly fatty, flowery, green	–	0.2	–
Nonanoic acid	Cheese-like, waxy	–	0.7	–

Sample A: from Lower Austria, Austria; pressed at room temperature (20 °C).
Sample B: from Saxony, Germany; pressed at room temperature (20 °C).
Sample C: from Styria, Austria; pressed at room temperature from linseeds previously roasted (30 min at 60 °C).

Flavour

Reminiscent of freshly cut grass (Ulmer 1996, p. 61).

Fresh linseed oil tastes slightly nutty and hay-like; if stored for a long time it tastes bitter and scratchy.

4 Ingredients

Composition of Fatty Acids

The composition of fatty acids varies depending on the growing area of the flax. Flax produced in Canada and Europe is richer in linolenic acid than flax produced in Argentina. The following table illustrates the differences in the composition of fatty acids according to the region of origin:

Fatty acids	Content in % acc. to Krist et al. (2006b)	Content in %, Europe acc. to Karleskind (1996), p. 159	Content in %, Argentina acc. to Karleskind (1996), p. 159	Content in %, Canada acc. to Karleskind (1996), p. 159	Content in % acc. to Firestone (1999), p. 56
Palmitic acid	5.1–6.7	4–6	4–5	5–6	5.7–7
Palmitoleic acid	–	<0.5	<0.3	<0.3	–
Stearic acid	0.25–4.6	2–3	5–6	3–4	3–4
Oleic acid	17.8–24.3	10–22	19–21	19–20	20–20.3
Linoleic acid	16.3–20.0	12–18	15–24	14–16	17–17.3
α-linolenic acid	45.1–55.0	56–71	45–53	54–61	52–54
Arachidic acid	–	<0.5	<0.5	<0.5	0–0.1
Eicosanoic acid	–	<0.6	<0.6	<0.6	–

The table shows that linseed oil is especially rich in the essential fatty acids linoleic acid and linolenic acid (Ulmer 1996, p. 61).

Composition of Triglycerides

The greater part of fatty acids exists in the form of triglycerides (92%). Only 3% of the fatty acids are free fatty acids, and only 1% are phospholipids. The proportion of linolenic acid in the fatty acids of the triglyceride group amounts to 47%. The group of phospholipids, on the other hand, contains more saturated fatty acids than the triglyceride group (Hager 1993, p. 674).

Composition of Triacylglycerides in Linssed Oil, Content in % (Krist et al. 2006b)

Triacylglycerides	Linseed oil from Austria, cold pressed	Linseed oil from Germany	Linseed oil from Austria, cold pressed from previously heated seeds
LnPP	0.3 ± 0.3	0.4 ± 0.5	0.4 ± 0.3
LPP	0.3 ± 0.2	0.1 ± 0.2	0.1 ± 0.2
OPP	–	–	0.1 ± 0.1
SPP	–	0.1 ± 0.1	–
LnMoPo	–	0.2 ± 0.2	0.3 ± 0.2
LnMoP	–	0.1 ± 0.1	0.2 ± 0.2
LnMaP	–	0.10 ± 0.1	–
LMoP	–	*?*	–
LnLnPo	–	0.1 ± 0.2	–
LnLnP	5.7 ± 3.9	9.6 ± 0.6	7.4 ± 0.3
LnLP	5.2 ± 2.3	3.3 ± 0.4	2.7 ± 0.1
LnOP	4.9 ± 2.4	4.6 ± 0.3	5.2 ± 0.1
LLP	+	+	+
LOP	1.1 ± 0.6	1.3 ± 0.2	1.1 ± 0.4
LnSP	+	+	+
OOP	2.2 ± 1.0	1.0 ± 0.1	1.6 ± 0.3
LSP	+	+	+
SSP	–	0.1 ± 0.0	–
LnLnLn	22.3 ± 2.9	23.3 ± 0.1	20.0 ± 0.5
LnLnL	14.2 ± 2.0	16.2 ± 0.9	13.9 ± 0.1
LnLnO	15.2 ± 2.0	14.1 ± 0.6	16.1 ± 0.2
LnLL	+	+	+
LnLnS	10.1 ± 1.5	10.3 ± 0.6	10.5 ± 0.1
LnLO	+	+	+
LLL	+	+	+
LnLS	9.9 ± 1.4	8.3 ± 0.9	11.1 ± 0.5
LnOO	+	+	+
LLO	+	+	+
LOO	3.3 ± 0.5	3.1 ± 0.3	3.9 ± 0.7
LnOS	+	+	+
OOO	2.4 ± 0.4	1.7 ± 0.3	2.4 ± 0.4
LOS	+	+	+
OOS	0.7 ± 0.4	0.8 ± 0.3	0.6 ± 0.3
LSS	+	+	+
LnLnG	–	–	0.2 ± 0.2
LnLG	–	0.5 ± 0.2	0.5 ± 0.2
LLG	–	0.2 ± 0.0	0.4 ± 0.1
LnOG	–	+	+

L

Triacylglycerides	Linseed oil from Austria, cold pressed	Linseed oil from Germany	Linseed oil from Austria, cold pressed from previously heated seeds
LOG	–	0.2 ± 0.1	0.3 ± 0.2
OOG	–	0.3 ± 0.2	0.3 ± 0.1
OSG	–	0.1 ± 0.3	0.2 ± 0.1

+ Component present but could not be quantified

Composition of Sterols

(Concentration in mg/100 g and content in %)

Sterols	Refined linseed oil acc. to Carstensen (2001), p. 42	Cold-pressed linseed oil acc. to Carstensen (2001), p. 42	Linseed oil acc. to Firestone (1999), p. 56	Linseed oil acc. to Karleskind (1996), p. 160
Total content	338	377.3	–	200–410
Cholesterol	1.7	–	0–0.9	1–2
Brassicasterol	0.7	–	0.1–0.7	0–1
Δ5-Campesterol	0.3	2.9	25–31 (+ A7)	28–29 (+ A7)
β-Sitosterol	49.3	49.3	45–53	44–53
Δ5-Avenasterol	12.3	9.0	8–12	10–13
Δ7-Stigmasterol	2.3	1.4	0–3	1–4
Δ7-Avenasterol	0.9	1.9	0–0.6	<1

Composition of Tocopherols

(Concentration in mg/kg and content in %)

Tocopherols	Linseed oil acc. to Firestone (1999), p. 56	Refined linseed oil acc. to Carstensen (2001), p. 44	Cold-pressed linseed oil acc. to Carstensen (2001), p. 44
Total content	440–588	440–590	440–590
α-Tocopherol	5–10	1.1	2.3
β-Tocopherol	–	2.3	–
γ-Tocopherol	430–575	95.9	81.6
Δ-Tocopherol	4–8	0.7	13.4

Other Ingredients

(Karleskind 1996, p. 160):

Hydrocarbons: 80–100 (mg/100 g)
Of these squalene: 20–30 (mg/100 g)
Triterpene alcohols: 150 (mg/100 g)
Mucilage
Protein
Vitamin B

Physical Key Figures of Linseed Oil

refractive index: 1.472 to 1.485
density: 0.930 to 0.936
iodine number: 170 to 204
peroxide number: <20
saponification number: 188 to 196
acid number: <4.0
unsaponifiable: <2%
(Firestone 1999, p. 56; Hager 1993, p. 674; Karleskind 1996, p. 159)

Shelf Life

Linseed oil becomes rancid quickly and only keeps for about 2 months if stored in the refrigerator. All linseed products should be stored in dark, airtight containers.

Characteristics of the Oil

Atomised linseed oil has a tendency to be pyrophoric at room temperature. To prevent this, rags soaked in linseed oil should be stored in an airtight screw-top jar or incinerated under controlled conditions, and brushes with linseed oil soap should be scoured.

5 Application

In Pharmacy and Medicine

Lipid-Lowering Properties

The pharmaceutical effects of linseed oil are mainly based on its high content of linolenic acid. Linseed oil has lipid-lowering properties: Rats fed with a diet of linseed oil for 14 days showed a significant decrease in the plasma level of cholesterol and triglycerides. The cholesterol level of the liver did not, however, change (Hager 1993, p. 675 et seq).

Regulating Blood Pressure

30 days of a diet of linseed oil significantly reduced the blood pressure of hypertonic rats from the 15th day onwards. Patients suffering from a mild form of essential hypertonia who partook of a diet to which linseed oil had been added (60 ml/day) for 14 days experienced a significant reduction of systolic blood pressure from 175 mm Hg to 161 mm Hg during times of psychophysiological stress. Diastolic blood pressure remained the same. At the same time, linseed oil reduced the excretion of sodium in the urine by 27% (Hager 1993, p. 675 et seq).

Antibacterial and Anticarcinogenic Properties

Linseed oil has antibacterial properties: If it is hydrolysed at a concentration of 0.025% using alkalis, the free fatty acids inhibit the growth of methicillin-resistant *Staphylococcus aureus* in vitro at 37 °C for a duration of 18 hours by 100%. In addition, linseed oil has anticarcinogenic properties: It reduced the tumour weight of mastocarcinoma of mice fed with the oil for 8–10 weeks by 46%, compared to the mice fed with corn oil for 8–10 weeks (Hager 1993, p. 675 et seq).

In Formulations

In pharmacy, linseed oil is a base for the manufacture of ointments, powder dressings and viscous liniments.

As Food

Although the taste of linseed oil limits its uses as an edible oil (Ulmer 1996, p. 61), it is nevertheless used as such. In India, 35–40% of the linseed oil produced is used for cooking and frying. It is also used in combination with mustard, peanut oil and coconut oil. As linseed oil is not produced commercially in large amounts and is not very stable against oxidation, it is used relatively seldom as an edible oil (Salunkhe et al. 1992, p. 509). It can accompany vegetable dishes, potatoes, and raw fruit and vegetables.

In Folk Medicine

For external use, compresses with linseed oil have proved of value because they have emollient and pain-relieving properties. The cell-regenerating effect is beneficial in cases of painful split skin, burns and skin damage.

In Veterinary Medicine

When linseed oil is added to horse fodder, the horses' coats become shiny. For horses, sheep and dogs, linseed oil also has a digestive effect. It is therefore used as a laxative, especially for horses and sheep.

Industrial Uses

Linseed Oil in Chemical Engineering

Linseed oil is a raw material in the production of oil paint, varnish and linoleum, in smaller amounts also in the production of soft soap, sealants, printing ink and liniments. Due to its polymerising properties, it can be used as an ink-binding agent.

For Impregnation

Its water-repellent properties and diffusion openness make the oil suitable for outdoor application in all weather conditions. It forms a waxy film without cracks, and is therefore ideal for impregnation of wooden furniture and exterior facades.

As an Additive

Less frequently, linseed oil is used for the production of linings, packaging, polish, lubricants, oilcloth and pyrotechnic products. Furthermore, it is added to latex paint to improve adhesion. Epoxidised linseed oil is used as a stabiliser and to model plastics on vinyl base (Salunkhe et al. 1992, p. 509).

6 Possible Unwanted Side Effects

Allergic reactions to linseed oil were first described about 70 years ago. With increasing use of linseed in food (oil, cereal, etc.), this allergy gains in importance. The symptoms of an allergy range from rhinitis, urticaria, asthma and nausea to gastrospasms (Senti et al. 2000).

Macadamia Oil

Oleum Macadamiae synonyms: Macadamiaöl (D); huile de macadamia (F)

1 Source Plant

Macadamia ternifolia F. Muell. (Proteaceae), macadamia

S. Krist, *Vegetable Fats and Oils*, https://doi.org/10.1007/978-3-030-30314-3_68

Habitat

The origins of the macadamia lie in eastern Australia, in the rain forests of Queensland. In 1857, the scientist *Dr. John Macadam* discovered the macadamia nut, which was *subsequently named after him*. In 1880, the first seeds and nuts were brought to Hawaii, where a new industry emerged. At the beginning of the 1960s, the Australians also built an industry sector around the macadamia nut, which, after all, has its origins in Australia. Since then, the cultivation of macadamias has been increasing in the southern hemisphere. *Macadamia ternifolia* needs a moist, subtropical climate to grow, and is very frost-susceptible.

Description

Macadamia ternifolia is an evergreen, sclerophyllous tree that can grow up to 12–15 m tall and can reach an age of over 50 years. Like all *Proteaceae, its flowers have a relatively simple form*. The perianth is simple and divided into four segments. The filaments and petals have merged. Each carpel contains one or more ovules. Between the petals and the ovary, there is a gland consisting of four scales, which produces nectar. The relatively small flowers are usually arranged in extensive, showy, ornamental inflorescences. They are pollinated by insects, birds, but also marsupials. Pollination by mammals that do not fly was first proven for *Proteaceae*. The macadamia nut is a stone fruit with a diameter of 2–3 cm. It has a hard, brown shell coated with a thick, green protective layer. The fruit are arranged in long panicles reminiscent of grapes. The pericarp is heavily lignified. Macadamia nuts have the hardest shell of all nuts, and must be cracked open using machines.

2 Cultivation and Extraction

Cultivation

The main growing areas of *Macadamia ternifolia* apart from Hawaii and Australia are South Africa, New Zealand, Kenya, Costa Rica and Guatemala. The trees only bear fruit after 5 years, and reach their maximum capacity after about 10 years. The yield then amounts to about 40 kg per tree. As soon as the fruit are ripe, they fall from the trees of their own accord. Usually, the so-called picker crews then collect the nuts from the ground by hand. Machines, however, can also be used. The nuts are stored in well-ventilated rooms for several months, after which they are peeled and then stored in drying rooms for another few months. They are subsequently cooked or roasted and processed further. Fresh macadamia nuts are not commercially available. At the moment, it is mostly macadamia nuts that are on the market; the oil is still of comparably little interest, but its importance is increasing.

Extraction of the Oil

Macadamia oil is either cold pressed or extracted from the ripe seeds of *Macadamia ternifolia; if it is extracted, it is subsequently refined* (Hunnius 1998, p. 849; Alvarez and Rodriguez 2000).

3 Character

Colour

Brown (Roth and Kormann 2000, p. 134).
Light yellow (Kerschbaum and Schweiger 2001, p. 22).
Pale yellow (Alvarez and Rodriguez 2000).

Odour

Slightly nutty, like rancid nuts.
Mild (Alvarez and Rodriguez 2000).

Flavour

Nutty (Roth and Kormann 2000, p. 134).
Slightly nutty (Kerschbaum and Schweiger 2001, p. 22).
Mild (Alvarez and Rodriguez 2000).

4 Ingredients

Composition of Fatty Acids

Fatty acids	Content in % acc. to Maguire et al. (2004)	Content in % acc. to Cavaletto (fao.org/29.04.04)	Content in % acc. to Nature certificate of analysis 28.08.03	Content in % acc. to Kerschbaum and Schweiger (2001) (LAP-Forchheim)
Palmitic acid	8.37	9.1	8.1	8.5
Linoleic acid	2.31	1.9	1.9	1.7
Palmitoleic acid	17.28	21.9	19.6	20.2
Oleic acid	65.15	59.9	59.2	55.3
α-Linolenic acid	0.06	–	–	0.1
Stearic acid	3.17	2.2	3.6	3.1
Arachidic acid	2.28	1.8	2.7	2.2
Eicosenoic acid	–	2.0	–	1.9
Vaccenic acid	–	–	–	4.5
Myristic acid	0.95	0.7	0.8	1
Erucic acid	–	–	0.3	0.2

Composition of Sterols

Sterols	Content in µg/g acc. to Maguire et al. (2004)
Campesterol	73.3 ± 8.9
Stigmasterol	38.3 ± 2.7
β-Sitosterol	1506.7 ± 140.5

Composition of Tocopherols

Macadamia oil contains 19.0 mg γ-tocopherol in 100 g oil (Kerschbaum and Schweiger 2001, p. 36).

Physical Key Figures of Macadamia Oil

refractive index: n_D^{20} 1.4675–1.4698
density: d_4^{15} 0.912–0.916
saponification number: 193–196
iodine number: 74–76
unsaponifiable: 0.5%
melting point: −12 °C
(Roth and Kormann 2000, p. 134)

Shelf Life

If stored in a cool and dark place, macadamia oil has a shelf life of 6–8 months.

Characteristics of the Oil

A distinctive feature of macadamia oil is its chemical composition. It contains about 20% palmitoleic acid, a monounsaturated fatty acid that is mostly only found in noteworthy quantities in animal fats. Apart from a few exceptions, vegetables only contain less than 0.5% palmitoleic acid. Since macadamia oil contains a high amount of palmitoleic acid, it is especially skin-friendly (Alvarez and Rodriguez 2000; Kerschbaum and Schweiger 2001, p. 22).

5 Application

In Pharmacy and Medicine

A single case study tested the effect of macadamia oil on plasma lipids. 34 men suffering from hypercholesterolaemia received either a diet rich in oleic acid, or food rich in palmitoleic acid. The study showed that palmitoleic acid has a positive influence on both LDL and HDL cholesterol levels. It can therefore be assumed that macadamia oil will gain in importance as a dietary food, due to its composition (Ako et al. 1995).

In Cosmetics

As mentioned above, macadamia oil is skin-friendly due to its high amount of palmitoleic acid. The cosmetics industry values palmitoleic acid highly, since it penetrates the skin easily and ensures good dissemination of the triglycerides. In addition, macadamia oil protects the lipid layer of the skin, and has a regenerating and skin-tightening effect. It also has a natural sun protection factor. It is used as a substitute for mink oil in hair care products for split ends. Macadamia oil is part of the following beauty products (Hunnius 1998, p. 849; Kerschbaum and Schweiger 2001, p. 45; Alvarez and Rodriguez 2000):

- moisturisers;
- anti-wrinkle creams;
- sunscreens;
- body oils;
- ointments to regenerate the skin;
- massage oils;
- shampoo, and
- deep conditioners.

As Food

Macadamia oil can be used for stewing, roasting and as a salad oil. In Europe, it is, however, seldom used, as it is relatively expensive. In the USA, it is more common as an edible oil (Kerschbaum and Schweiger 2001, p. 43).

6 Possible Unwanted Side Effects

So far, only allergic reactions to cold-pressed macadamia oil have been described. The symptom is a reddening of the entire buccal cavity (Senti et al. 2000).

Madia Seed Oil

Oleum Madiae synonym: Madiaöl (D)

1 Source Plant

Madia sativa Molina. (Asteraceae), coast tarweed

S. Krist, *Vegetable Fats and Oils*, https://doi.org/10.1007/978-3-030-30314-3_69

Habitat

Coast tarweed originates in South America, especially in Chile, as well as in California, and was imported to South Africa. The Native Americans already used this plant long before the Spanish colonisation. Today, coast tarweed is still cultivated on the western coast of Chile. In the eighteenth century, the plant was imported to Europe, where it was mainly cultivated in the south-east. From there, it spread to southern Germany, but attempts at cultivation were soon abandoned because of small yields and the plant's unpleasant glandular hairs. It is now only cultivated in botanical gardens and plant production. Currently, interest in coast tarweed as a sustainable vegetable product is on the increase (Schuster 1992, p. 68).

Description

Coast tarweed is an annual, upright, herbaceous plant of the composite family. It grows to a height of 30–100 cm. The stem is densely foliated, slightly branched in the upper third, and densely covered with glandular hairs, which segregate a sticky essential oil with an unpleasant smell. The leaves are linear-lanceolate, entire and pointed, and amplexicaul at the base. The bottom leaves are oppositely arranged, the upper ones alternate; they are densely covered with glands and smell unpleasant. The flower heads are rather small and short-stemmed. A crown of bracts makes them appear nearly globular, with a length of 15–20 mm. The receptacle of coast tarweed is flat and 2–3 cm in diameter. Between the ray florets and the disc florets, there is a circle of broad, lanceolate paleas. The disc florets are yellow, dioecious and tubular. The ray florets are yellow, female, with a three-toothed, tongue-shaped corolla that hardly overhangs the perianth. The pappus is absent. The fruit are slightly bent, 6.5–7.5 mm long and broadened at the apex. The unripe fruit is black; due to the epidermis cells drying out, it later becomes grey and hairless, corrugated and weakly dotted. The pericarp is leathery (List and Hörhammer 1976, 4th edition, 5th volume, p. 628; Schuster 1992, p. 68).

2 Cultivation and Extraction

Cultivation

Coast tarweed has a relatively short vegetation period of only 3 months. It is thus a suitable plant for regions with short vegetation periods. Its only demand with regard to climate is that young plants are frost-susceptible. If the seeds are not sown too early in the year, the Central European climate should be suitable for cultivation. Neither are the demands of coast tarweed high with regard to the soil; rich soil is not ideal because the plant may then take longer to ripen. It prefers loamy, sandy soil in dry, sunny locations. The plant is sown from mid-May to early June, when there are no more night frosts. The harvest takes place when the majority of the seeds are black. Harvesting should not be delayed too long, as the plant tends to disseminate when it is fully ripe. In family farming, the dry plants can be laid out on sheets and shaken to separate the seeds (Hackbarth 1944, p. 179).

Coast tarweed has been cultivated in Europe since the beginning of the twentieth century, with yields of 250–300 kg seeds/ha, from which 70 to 110 kg oil can be obtained (Schmeda-Hirschmann 1995).

Due to its high amount of protein, the residue remaining after the extraction of the oil is high-quality forage. This ensures economical use of the seeds (Zardini 1992).

Extraction of the Oil

Madia seed oil is obtained from *Madia sativa Mol.* by cold pressing or hot pressing the seeds. The cold-pressed oil is better suited as an edible oil, as it is of higher quality and does not become rancid easily (Hackbarth 1944, p. 180).

The oil can also be extracted using a Soxhlet extractor (Schmeda-Hirschmann 1995).

Originally, the seeds were cooked and the oil was decanted from the surface. The oil obtained by this method has the disadvantage, however, of becoming rancid quickly. In addition, the yield is unsatisfactory. Another method was to crush the seeds in a stone mortar. The resulting flour was wrapped in a cloth, which was twisted to wring the oil from it. As the seeds were not heated, the oil did not become rancid so easily (Zardini 1992).

3 Character

Colour

Golden yellow, transparent (Zardini 1992).
Dark yellow (Hackbarth 1944, p. 180; Von Wiesner 1927, p. 744).
Brownish (Roth and Kormann 2000, p. 134).

Odour

Peculiar, pleasant smell (Hackbarth 1944, p. 180).
Mild smell (Roth and Kormann 2000, p. 134).

Flavour

Mild, reminiscent of nut oil
Nutty (Hackbarth 1944, p. 180).
Reminiscent of almonds (Roth and Kormann 2000, p. 134).

4 Ingredients

Composition of Fatty Acids

Fatty acids	Content in % acc. to Schuster (1992), p. 10	Content in % acc. to Angelini et al. (1997)	Content in % acc. to Schmeda-Hirschmann (1995)
Palmitic acid	12–14	–	12.9–14.0
Stearic acid	2–4	–	3.8–3.9
Oleic acid	10–12	23.3–40.8	7.9–10.2
Linoleic acid	50–60	43.3–61.2	71.4–72.4
α-Linolenic acid	12–14	4.0	–
Arachidic acid	–	–	0.6–1.0

Physical Key Figures of Madia Seed Oil

density: d_4^{15} 0.923–0.929
point of solidification: −10 °C
saponification number: 192–195
iodine number: 117–129
unsaponifiable: 0.5–1%
(Roth and Kormann 2000, p. 134)

refractive index: n^{20}_D	1.4750–1.4770
density (20 °C):	0.9120
molecular weight:	876.1–894.6
saponification number:	190–194
acid number:	0.15

(Schmeda-Hirschmann 1995)

The oil burns with a reddish-white flame and does not leave a residue (Zardini 1992).

Shelf Life

Since madia seed oil contains a high amount of saturated fatty acids, it has a very long shelf life (Hackbarth 1944, p. 180).

M

5 Application

In Pharmacy and Medicine

In South America, madia seed oil was originally used as a laxative. The purging effect can probably be traced back to the resin. Additionally, it formed part of ointments and salves (Zardini 1992).

In Cosmetics

From cheap, low-grade, hot-pressed madia seed oil, soap can be produced (Anonym 1915).

As Food

Since madia seed oil contains a high amount of unsaturated fatty acids, it is a high-quality edible oil. A study by Schmeda-Hirschmann could not detect any erucic acid, which is toxic in high doses and may only be contained in edible oils in small amounts (Schmeda-Hirschmann 1995).

Cold-pressed madia seed oil is of superior quality; the population of Chile uses it as an edible oil because of its long shelf life (Anonym 1915).

Industrial Uses

Madia seed oil is mainly used for technical purposes, for example to produce lubricating oil (Lieberei and Reisdorff 2007, p. 139).

Mango Seed Oil

Oleum Mangiferae synonym: Mangokernöl (D)

1 Source Plant

Mangifera indica J. König ex L. *(Anacardiaceae)*, mango tree

S. Krist, *Vegetable Fats and Oils*, https://doi.org/10.1007/978-3-030-30314-3_70

Habitat

The natural distribution area of mango trees is from India to Mayanmar, but they are cultivated in all tropical and subtropical countries. Mango trees are among the oldest cultivated plants and have been grown for over 4000 years. The main growing areas are India, Pakistan and Bangladesh (List and Hörhammer 1976, 4th edition, 4th volume, p. 695; Salunkhe et al. 1992, p. 453).

Description

The mango tree is an evergreen tree 10–15 m tall; in exceptional cases, it can grow to a height of 40 m. The stem measures up to 1 m in diameter. The tree may reach an age of several hundred years. It has a wide crest coronet of alternate, leathery, elongated leaves that have a minimum length of 10 cm if the tree is fully grown. They are coated with a thin layer of wax, which gives them a high-gloss effect. The flowers are small, white and fragrant. They have up to ten stamina, of which only a few are fertile. The flower stems are only slightly thickened. The fruit is a fleshy, kidney-shaped stone fruit with a compressed, bipartite shell consisting of thick fibres. The seeds have a delicate, paper-thin seed coat with a kidney-shaped embryo. There are two species of mango trees: the Indian mango tree, which produces yellow-reddish fruit, and the Philippine mango tree, the fruit of which remain green when they are ripe (List and Hörhammer 1976, 4th edition, 4th volume, p. 695).

2 Cultivation and Extraction

Cultivation

The mango tree requires a tropical to subtropical climate to thrive. As it needs much space, it must be planted at a great distance to other trees (Axtell 1994, p. 96). The soil should be well drained and slightly acidic, for larger plants also slightly loamy.

In India, the harvest takes place from April to September. Workers pick the fruit from the trees and dry them in the sun. Afterwards, the seeds are removed by hand and subsequently extracted with solvents (Axtell 1994, p. 96).

Extraction of the Oil

Mango seed oil is obtained from the seeds of *Mangifera indica* by extraction with hexane; it is subsequently refined (Salunkhe et al. 1992, p. 454).

3 Character

Colour

Cream-coloured (Sunday 1997).
Light yellow to ivory-coloured (Salunkhe et al. 1992, p. 454).
Light yellow (Axtell 1994, p. 96). White to yellowish (Hussain et al. 1983).

Odour

Mild smell.

Flavour

Mild taste.

4 Ingredients

Composition of Fatty Acids

Fatty acids	Content in % acc. to Gafur and Toregard (1984)	Content in % acc. to Badami and Alagadawi (1983)
Palmitic acid	9.8	12.7
Stearic acid	41.6	37.4
Oleic acid	40.6	38.8
Linoleic acid	5.4	6.8
α-Linolenic acid	0.3	–
Arachidic acid	1.9	2.2
Behenic acid	0.3	1.3

Composition of Sterols

Sterols	Content in % acc. to Gafur and Toregard (1984)
Campesterol	10.5
Stigmasterol	16.5
β-Sitosterol	73.9

Physical Key Figures of Mango Seed Oil

refractive index: n_D^{28} 1.4651.
density: d_{28}^{28} 0.90.

melting point: 34.5–35 °C.
iodine number: 33–48.9.
acid number: 0.28.
saponification number: 190.4.
unsaponifiable: 7.4%.
(Sunday 1997; Rukmini and Vijayaraghavan 1984)

Shelf Life

Mango seed oil has a shelf life of up to 1 year.

5 Application

In Pharmacy and Medicine

Due to its low melting point and its similar properties to cocoa butter, mango seed oil can be a substitute for cocoa butter in ointments and suppositories (Gafur and Toregard 1984).

A study established the oil's antibacterial effect on *Staphylococcus epidermidis*, but it promoted the growth of *Saccharomyces cerevisiae*. Due to this result, the use of mango seed oil cannot be recommended against bacteria (Raman et al. 1997).

In Cosmetics

Mango seed butter regenerates, heals, moisturises, and is often added to creams for babies, body creams, lotions and lip balms. Mango seed butter moisturises the skin, which feels softer and more supple as a result, and is especially useful for dry skin. It is also frequently added to hair-care products, for example shampoo, for dry hair, because of its nourishing and moisturising effects. As a result of its moisturising properties, mango seed butter is frequently used in the production of high-quality soap (Hussain et al. 1983). It is also added to sunscreens.

As a Substitute for Paraffin Wax

M

As paraffin can cause skin nodules (knotty granuloma), mango seed butter is a safe, compatible substitute for paraffin wax (Roche Lexikon Medizin 1993, 3rd edition, p. 1260).

As Food

As a result of its unobtrusive taste and similar properties to cocoa butter, mango seed oil is often employed in chocolate production. According to the *Kakaoverordnung*

(Cocoa Regulation, 2738 German Federal Law Gazette volume 2003, part I, no. 63, Bonn, 23.12. 2003), the content of certain vegetable fats may not exceed 5% of the final product, while retaining the minimum content of cocoa butter and cocoa dry mass, and after subtracting the total weight of the ingredients. The permitted vegetable fats are mango seed oil, illipe, Borneo tallow, kokum butter, palm oil, sal butter and shea butter (Gafur and Toregard 1984).

Since mango seed butter is produced in great quantities and is therefore a low-priced raw material, it is used in the production of margarine and shortening, especially in India (Sunday 1997).

6 Possible Unwanted Side Effects

In a study on rats, 10% mango seed oil was mixed into their food. No toxic or other unwanted side effects were observed; neither a change in organ weight or cholesterol, triglyceride or overall lipid levels in the serum or liver, nor a negative effect on mating habits, birth weight or mortality rate was observed. From this, it can be deduced that mango seed oil is safe to use (Rukmini and Vijayaraghavan 1984).

6 Possible Unwanted Side Effects

Marigold Seed Oil

Synonyms: Ringelblumenöl (D); calendula seed oil (E)

S. Krist, *Vegetable Fats and Oils*, https://doi.org/10.1007/978-3-030-30314-3_71

1 Source Plant

Calendula officinalis L. *(Asteraceae)*, Marigold, Ruddles

Habitat

The marigold's origins probably lie in the Mediterranean area. Today, it is a common ornamental plant, not least because it is undemanding. I sometimes also grows as a weed, as the seeds spread easily (Schuster 1992, p. 74).

Description

Calendula officinalis is an annual plant, rarely also wintering or biannual. It can grow up to 60 cm tall and has an upright, somewhat branched stem. The shoot is covered with felty hair; it branches in the upper third. The main root is spindle-shaped and protrudes deeply into the earth; it has many thin lateral roots. The leaves are alternate, and medium to light green in colour. They are sparsely covered with hair, sticky, entire and have an elongate, lanceolate form. The flowers are solitary on the stems. The husk leaves are patelliform. The ray florets are 15–20 mm long and dark yellow or orange. The flowerheads open in the morning between 8 and 9 am and close in the late afternoon around 5–6 pm. The flower lasts for about 4–5 days and, due to its essential oil, emits a slightly unpleasant smell. The pollen is large, with casually scattered spikes. Autogamy is common. The fruit is an achene, and may have different forms. It is usually bent like a ring and covered with short spikes on the back. The thousand-seed weight is 3.9–15.2 g (Dachler and Pelzmann 1999, p. 262; Schuster 1992, p. 73).

2 Cultivation and Extraction

Cultivation

The marigold does not make many demands on soil and climate, but extreme conditions such as droughts and extreme wetness should be avoided. It is not very frost-susceptible. The seeds are sown directly on location in early April at a distance of 40–60 cm from each other. The harvest takes place from mid-June to mid-August (Dachler and Pelzmann 1999, p. 263).

Marigolds are mainly cultivated for their flowerheads and the essential oil contained therein. In order for them to be cultivated for their fat oil on a large scale, several breeding goals still have to be achieved, for example a uniform time of ripening, improved failure strength and a higher yield of seeds (Schuster 1992, p. 75).

In Europe, *Calendula officinalis* produces between 700 and 2400 kg of seeds per hectare (Wilen et al. 2004).

Extraction of the Oil

The oil is extracted with a solvent in a Soxhlet extractor. Petroleum ether is a suitable solvent. Before extraction, the seeds are finely ground to obtain a higher oil yield (Chisholm and Hopkins 1960).

3 Character

Colour

Yellow (Saleem et al. 1986).

4 Ingredients

Composition of Fatty Acids

Fatty acids	Content in % acc. to Schuster (1992), p. 10	Content in % acc. to Nagao and Yanagita (2005)	Content in % acc. to Badami and Morris (1965)	Content in % acc. to Wilen et al. (2004)	Content in % acc. to Özgül-Yücel (2005)	Content in % acc. to Feder et al. (2009)
Lauric acid	–	–	–		1.4	–
Myristic acid	–	–	–		0.6	0.5
Palmitic acid	3–4	4.1	5	3.0–6.5	10.2	4.2
Stearic acid	2–3	2.1	2	1.5–4.0	3.8	2.0
Oleic acid	6–10	7.4	5.5	3.1–5.5	8.0	3.8
Linoleic acid	–	42.5	34	28–43	43.5	28.5
α-Linolenic acid	28–30	1.0	–	0.5–3.0	1.1	1.1
γ-Linolenic acid	2–4	–	–	–	–	–
Calendic acid	50–55	33.4	51–53	53–55	18.3	59.1
Arachidic acid	–	–	–	–	0.7	0.4
Behenic acid	–	–	–	–	0.9	–

Calendic acid is a fatty acid with three conjugated double bonds with the name trans-8:trans-10:cis-12-octadecatrienoic acid (Meier Zu Beerentrup and Röbbelen 1987).

Calendic acid possesses 18 carbon atoms and 3 double bonds, which is why it is also called conjugated linolenic acid. A study by Özgül-Yücel revealed another conjugated linolenic acid in addition to calendic acid, all the double bonds of which are in trans position. In total, the examined oil contains 29.5% conjugated linolenic acids, of which 18.3% are calendic acid, and 11.2% trans-8:trans-10:trans-12-octadecatrienoic acid (Özgül-Yücel 2005).

According to Badami and Morris, marigold seed oil contains about 5% D-(+)-9-hydroxy-10(*E*),12(*Z*)-octadecadienoic acid (Badami and Morris 1965). This acid is an inhibitor of aromatase, which is also found in the roots of the stinging nettle (*Urtica dioica*) (Kraus et al. 1991).

Composition of Tocopherols

Marigold seed oil contains 1.0–1.4% tocopherols (Feder et al. 2009).

Physical Key Figures

refractive index: n^{40}_{D}	1.4964 (Earle et al. 1964)
	1.4998 (Earle et al. 1962)
	1.4990 (Saleem et al. 1986)
iodine number:	154 (Earle et al. 1964)
	143 (Saleem et al. 1986)
saponification number: 189 (Earle et al. 1962)	saponification number: 189 (Earle et al. 1962)
	274.37 (Saleem et al. 1986)
cid number:	5.32
peroxide number:	31.25

(Saleem et al. 1986)

5 Application

Industrial Uses

Marigold seed oil is a drying oil and thus suitable for various industrial uses (Saleem et al. 1986).

The main ingredient of the oil is calendic acid, which resembles eleostearic acid, which is contained in tung oil. It can thus be used as an alternative to tung oil, which is a common coating agent. Calendic acid and eleostearic acid have three conjugated double bonds, which is the reason why they dry quickly. Coatings made from marigold seed oil or tung oil are resistant to water and penetration (Derksen et al. 1995).

By using calendic acid as a drying agent in varnish, paint and plastic, several hazardous volatile compounds can be omitted (Wilen et al. 2004).

Marigold seed oil proved to be a suitable substitute for wood oil in binding agents used in the paint industry. The suitability of *Calendula* compounds as emulsifiers for coating binders is currently being tested (Feder et al. 2009).

In Cosmetics

Marigold seed oil is not yet used in cosmetics, but may well be in the future, as it has skin-nurturing and occlusive properties (Andersen et al. 2010).

Marula Oil

Synonym: Marulaöl (D)

1 Source Plant

Sclerocarya birrea Hochst. *(Anacardiaceae)*, marula tree

S. Krist, *Vegetable Fats and Oils*, https://doi.org/10.1007/978-3-030-30314-3_72

Habitat

The marula tree grows mainly in warm, frost-free regions of subequatorial Africa, from Nigeria to Ethiopia and the whole of southern Africa (Ogbobe 1992).

Description

The stately tree can grow up to 20 m high and has alternate, imparipinnate leaves. The flowers are unisexual and arranged in spike-like racemes. The sepals are brownish red to red, the petals pink on the outside and white with a green base on the inside. The tree blossoms from January to May and bears fruit from April to June. The ovary has three segments and develops into a stone fruit. The pericarp is fleshy and, when the fruit is ripe, yellow and plum-like. A tree can bear up to 2 tons of fruit. The fruit pulp has a sweet-sour taste and is well liked by cattle and elephants. It contains high amounts of citric acid, malic acid, sugar and vitamin C. The kernels, which are edible, contain 50–60% oil rich in protein (Ogbobe 1992; Von Koenen 2001, p. 491).

2 Cultivation and Extraction

Cultivation

Even with minimum precipitation, the female marula tree bears many golden-yellow fruit for a good harvest. The tree is one of the most important food plants of the Ovambo people of Namibia (Von Koenen 2001, p. 491).

Extraction of the Oil

To obtain marula oil, the Namibian population pounds the seeds, places them in boiling water and decants the oil from the surface. Prior to this, the seeds must be removed from the fruit by hand (Von Koenen 2001, p. 491). This is also necessary in industrial production. Afterwards, the seeds are cold pressed and the oil subsequently filtered.

3 Character

Colour

Dark yellow to orange.

Odour

Fruity, fatty, floral, woody, sweet.

4 Ingredients

Composition of Fatty Acids

Fatty acids	Content in % acc. to marula.org.za/15.01.07	Content in % acc. to oilsbynature.com/16.01.07
Palmitic acid	11.2	9–12
Palmitoleic acid	0.1	<0.2
Stearic acid	7.3	5-8
Oleic acid	75.1	70–78
Linoleic acid	4.4	4–7
α-Linolenic acid	0.2	<0.6
Arachidic acid	0.6	<1.0
Eicosanoic acid	0.4	<0.5
Cetoleic acid	0.1	<0.5

Composition of Sterols

Sterols	Content in mg/100 g acc. to Burger et al. (1987)
Campesterol	181
Stigmasterol	1.1
β-Sitosterol	1494
Δ 3-Avenasterol	246

Composition of Tocopherols

Tocopherols/tocotrienols	Content in mg/100 g acc. to Burger et al. (1987)
α-Tocopherol	0.28
γ-Tocopherol	22.2
B-Tocotrienol	0.06
γ-Tocotrienol	0.04
Δ-Tocotrienol	0.59

Other Ingredients

(Burger et al. 1987)

Cycloartenol
24-Methylencycloartenal
Citrostadienol

Physical Key Figures of Marula Oil

density: 20 °C 0.905–0.918
refractive index: 1.460–1.474
acid number: <1.0
iodine number: 70–80
saponification number: 180–195
peroxide number: <5.0
melting point: 26–28 °C
(Ogbobe 1992)

5 Application

As Food

Marula oil is well suited for frying and baking.

In Cosmetics

Marula oil is rich in oleic acid, which is why it is ideal for stressed skin. Because it is heat-resistant and stable, it is used as a base for cosmetic formulations.

In Folk Medicine

In Namibia and South Africa, marula oil is used for skin care, as eardrops and for massage.

Mustard Seed Oil

Oleum Sinapis Pingue

Synonyms: Senfsamenöl (D); white mustard seed oil, black mustard seed oil (E)

1 Source Plant

Sinapis alba L. *(Brassicaceae)*, White Mustard

Brassica juncea L. (Brassicaceae), mustard greens, leaf mustard

Brassica nigra L. *(Brassicaceae)*, black mustard

S. Krist, *Vegetable Fats and Oils*, https://doi.org/10.1007/978-3-030-30314-3_73

Habitat

The origins of white mustard lie in the Mediterranean area and neighbouring regions. Leaf mustard was created in Asia by spontaneous crossing of *Brassica campestris* with *Brassica nigra*, and then spread to Egypt and Europe. The origins of black mustard are not as clear. There are two subspecies: *Brassica nigra occidentalis* presumably originates in North Africa and Southern Europe, *Brassica nigra orientalis* in the East. By the twentieth century, mustard was found growing wild throughout Europe, but also in Japan, New Zealand, North America and South America (Schuster 1992, p. 33, 37, 39).

Description

Sinapis alba reaches a height of up to 60–75 cm, but only 20–25 cm in dry conditions. The roots are weak and consist of a thin, spindle-shaped main root and many delicate lateral roots (Hackbarth 1944, p. 137, 138).

The shoot is angular, grooved and covered with stiff hair. It is highly branched especially in the upper third, so that the main shoot is difficult to distinguish. The leaves are pinnate, with serrated edges. The inflorescence is an umbel; the flowers consist of four yellow petals, four smaller sepals and six stamina. Propagation is via allogamy, but autogamy is also possible. The fruit is a husk. The largest husks are to be found in the middle of the plant. They are elongated, pointed and covered with prickly hair. Within are round, light-yellow seeds measuring 2–2.5 mm in diameter; they have a delicately crenate surface. The thousand-seed weight is 4–8 g (Schuster 1992, p. 39–40).

Brassica juncea can grow to a height of 1–1.50 m and is thus higher than *Sinapis alba*. The taproot is of medium strength and has many lateral roots. The lower leaves are lobate with a partly serrated edge. The upper leaves, on the other hand, are narrow and pointed. The seeds in the husks have a thousand-seed weight of 1.7–2.8 g and are thus smaller than the seeds of white mustard. The colour varies from brownish black and various shades of brown to golden yellow. Overall, leaf mustard shows great variability in its characteristics (Schuster 1992, p. 32).

Like leaf mustard, *Brassica nigra* can reach a height of 1–1.50 m. The roots are as delicate as those of white mustard (Hackbarth 1944, p. 148).

Unlike white mustard, black mustard has a pronounced main shoot with few bastard branches. Both leaves and flowers are very similar to those of white mustard. The husks are small and angular; they are arranged alternately on the shoots. The thousand-seed weight is 1.8–2.5 g and thus corresponds roughly to that of leaf mustard. The seeds of black mustard differ from those of leaf mustard inasmuch as they are a darker, reddish-brown colour and their surface has a coarse reticulated pattern (Schuster 1992, p. 35).

2 Cultivation and Extraction

Cultivation

Sinapis alba is an old cultivated plant that was originally regarded as a weed. In Ancient Greece and Rome, mustard was already used as a spice; the Romans brought it to Central Europe. Since the twentieth century, the plant has been cultivated in Southern, Western and Central Europe, North Africa and India. Mustard is undemanding regarding the climate and soil. Continental climates are ideal. Sandy, clayey or acidic soil as well as waterlogged areas should be avoided. The plant is also quite resistant to plant diseases, the flea beetle being the only threat. During the harvest, care should be taken that the husks do not burst prematurely. Overripeness should therefore be avoided. The harvest should take place when the husks turn yellow to brown, but not during the middle of a hot day (Hackbarth 1944, p. 140–144).

Brassica juncea is important economically throughout the world. In Asia, it is one of the most frequently cultivated vegetables, and in India, in particular, it is one of the main oil plants (Schuster 1992, p. 33).

Like white mustard, *Brassica nigra* is undemanding. Sowing is generally carried out in late May, when an infestation with flea beetles is less likely. In general, black mustard is more likely to be infested with flea beetles than white mustard (Hackbarth 1944, p. 149).

Extraction of the Oil

Mustard oil is cold pressed. One ton of mustard seeds produces about 250 litres of oil (Roth and Kormann 2005, p. 134).

3 Character

Colour (Vaughan and Hemingway 1959)

White mustard seed oil: golden yellow.
Black mustard seed oil: brownish.

Flavour (Roth and Kormann 2005, p. 135)

White mustard seed oil: acid, pungent.

Black mustard seed oil: mild, reminiscent of mustard.

4 Ingredients

Composition of Fatty Acids

Fatty acids of white mustard	Content in % acc. to Appleqvist (1970)	Content in % acc. to Roth and Kormann (2005), p. 135	Content in % acc. to Hackbarth (1944), p. 145	Content in % acc. to Appleqvist (1971)	Content in % acc. to Grynberg and Szczepanska (1966)	Content in g/100 g acc. to Cserhalmi et al. (2001)
Palmitic acid	2.8	2.0	2.0	2–3	3.0	2.43
Stearic acid	0.9	–		–	0.9	1.01
Oleic acid	24.8	28.0	28.0	16–28	21.9	26.03
Linoleic acid	9.0	19.5	14.5	7–10	10.2	9.69
α-Linolenic acid	9.7	–		9–12	7.6	9.10
Arachidic acid	–	1	1	–	0.9	0.62
20:1	10.6	–		–	28.5	10.30
Erucic acid	37.8	52	52.5	33–51	43.3	35.13
Lignoceric acid	–	1.0	1.0		–	
Nervonic acid	2.3	–		–	1.8	

Fatty acids of leaf mustard	Content in % acc. to Appleqvist (1970)	Content in % acc. to Appleqvist (1971)	Content in % acc. to Grynberg and Szczepanska (1966)
Palmitic acid	2.3–3.4	2–4	2.3
Stearic acid	0.8–1.4	–	1.0
Oleic acid	8.0–19.0	7–22	23.8
Linoleic acid	14.2–23.1	12–24	21.7
α-Linolenic acid	11.4–14.4	10–15	14.7
Arachidic acid	–	–	0.8
20:1	6.1–11.6	–	10.0
Erucic acid	22.2–47.6	18–49	22.7
Nervonic acid	1.4–1.9	–	0.5

Fatty acids of black mustard	Content in % acc. to Appleqvist (1970)	Content in % acc. to Roth and Kormann (2005), p. 135	Content in % acc. to Hackbarth (1944), p. 150	Content in % acc. to Ryan et al. (2007)
Palmitic acid	3.5	–	2.0	4.15
Stearic acid	1.3	–		1.4
Oleic acid	7.5	25	24.5	26.28
Linoleic acid	14.5	20	19.5	10.68
α-Linolenic acid	16.9	–	2.0	8.16
Arachidic acid	–	–		0.53
20:1	6.3	–		9.68
Erucic acid	40.5	50	50.0	38.76
Lignoceric acid	–	–	2.0	–
Nervonic acid	2.2	–		–

Composition of Sterols

Sterols	Content in % of white mustard seed oil acc. to Appleqvist (1971)	Content in mg/100 g of black mustard seed oil acc. to Ryan et al. (2007)
Brassicasterol	5.2	–
Campesterol	34.6	26.5
Cholesterol	3.2	–
β-Sitosterol	43.8	74.4
24-Methylidenecholesterol	13.2	–
Stigmasterol	–	2.5

Composition of Tocopherols

Tocopherols	Content in mg/100 g of black mustard seed oil acc. to Ryan et al. (2007)
Total content	–
α-Tocopherol	0.6
β – + γ-tocopherol	6.3
δ-Tocopherol	–

Other Ingredients: (Ryan et al. 2007)

Squalene: 0.5 mg/100 g

Physical Key Figures: (Roth and Kormann 2005, p. 135)

White mustard seed oil:

refractive index: n^{20}_D	1.4704
refractive index: n^{15}_D	1.4750
density: d^{15}_{15}	0.911–0.915
saponification number:	170–178
point of solidification:	−11 to −18 °C

Black mustard seed oil:

refractive index: n^{20}_D	1.4739
density: d^{15}_{15}	0.912–0.922
saponification number:	174–175
iodine number:	96–107
unsaponifiable:	1–1.5%
point of solidification:	−11 to −18 °C

Shelf Life

If stored in a cool place, the oil has a shelf life of up to 10 months (Roth and Kormann 2005, p. 135).

5 Application

Technical Uses

Mustard varieties with a high amount of erucic acid can be used in the production of grease and motor oil instead of winter oilseed rape. The residue of oil extraction is used as forage (Schuster 1992, p. 41).

In Cosmetics

Mustard seed oil is used in the production of soap and cosmetic products (Hackbarth 1944, p. 146).

As Food

According to the decree on erucic acid (*Erucasäureverordnung*, BGBl. No. 468/1994), it is prohibited to put into circulation edible oils, fats and foodstuffs intended for human consumption that have been produced by adding edible oils and fats and contain an overall fat content of more than 5%, if they contain more than 5% erucic acid – based on the total content of fatty acids in the fat phase (http://www.ris.bka.gv.at/GeltendeFassung/Bundesnormen/10010781/Erucas%C3%A4ureverordnung,%20Fassung%20vom%2005.05.2011.pdf, 04.07.2011).

The reason for this legal provision is the adverse health effects of erucic acid, which include pathological changes of the myocardium and a fatty degeneration of the heart. The cantonal laboratory of Basel examined the composition of fatty acids in mustard seed oils to verify the indications given on the labels. The measured content was 26.5% and therefore significantly exceeded the limit of tolerance. The label read "for external use only", but the laboratory suspected that this information was only added to keep the product marketable, and that the actual purpose was a different one. In conclusion, mustard seed oil is used in Asian cuisine, but is not marketable in Europe due to its high content of erucic acid (http://www.kantonslabor-bs.ch/files/berichte/Speiseoel.pdf, 04.07.2011).

There have been attempts to breed varieties with a low content of erucic acid, with the aim that they can be a source of mustard seed oil that may be used as an edible oil. A current study by Schuster-Gajzágó tested the effects of high-frequency heating on mustard seeds. The result showed that the treatment greatly inhibited myrosinase activity, which is responsible for the sharp, pungent taste, but the amount of erucic acid remained the same (Schuster-Gajzágó et al. 2006).

Myrica Wax

Cera Myrica

The term "wax" is misleading, as myrica wax is actually a fat (Roth and Kormann 2000, p. 139).

Synonym: Myricawachs (D)

1 Source Plant

Myrica pensylvanica Loisel (Myricaceae) wax myrtle, bayberry

Habitat

The wax myrtle is a native plant of Atlantic North America, occurring from Newfoundland to North Carolina and Ohio. It was imported to southern England and the Netherlands (Roth and Kormann 2000, p. 77).

Description

The wax myrtle is an evergreen, dioecious shrub or tree that can reach a height of up to 10 m. The leaves are up to 9 cm long, lanceolate, leathery and short-stemmed. The male flowers are arranged in scaled catkins, the female flowers in egg-shaped racemes. The fruit are round, grey and covered with wax. They measure about 1 cm in diameter and contain the seeds (Roth and Kormann 2000, p. 77).

S. Krist, *Vegetable Fats and Oils*, https://doi.org/10.1007/978-3-030-30314-3_74

2 Cultivation and Extraction

Cultivation

The wax myrtle prefers drained to wet, acidic soil in the sun or half-shade. The adaptable shrub also tolerates nutrient-poor soil in swampy areas. Propagation is possible by sowing in autumn or spring or by taking cuttings in spring. Similarly to the pea family, the Myrica species form a symbiotic relationship with nitrogen-fixing bacteria. This enables them to survive in extreme, nutrient-poor locations (Roth and Kormann 2000, p. 77).

Extraction

Myrica wax can be obtained from the fruit of *Myrica pensylvanica,* which are covered with a waxy fat, by two methods.

By Cooking in Hot Water

The fruit are placed in boiling water. The fat settles on the surface of the water and can be skimmed off after the water has cooled down (Harlow et al. 1965).

Extraction with Benzene

This is a quick and efficient method to obtain myrica wax (Harlow et al. 1965).

3 Character

Colour

Greyish green (Harlow et al. 1965).
Green (List and Hörhammer 1976, 4th edition, 5th volume, p. 917).

Odour

Aromatic smell (Harlow et al. 1965), weakly balmy.

Flavour

Pleasantly aromatic taste (List and Hörhammer 1976, 4th edition, 5th volume, p. 917).

4 Ingredients

Composition of Fatty Acids

Fatty acids	Content in % acc. to Harlow et al. (1965)
Myristic acid	33–61
Palmitic acid	37–52
Stearic acid	1

Physical Key Figures of Myrica Wax

melting point: 45 °C
point of solidification: 46.4 °C
acid number: 21.2
saponification number: 2.167
iodine number: 1.03
unsaponifiable: 0.4%
(Anonym 1918)

5 Application

In Pharmacy and Medicine

The characteristic texture of myrica wax was the reason why it was used in the last century as a base for medical poultices, into which various active components were incorporated. Poultices are called emplastrum saponatum or emplastrum saponatum salicylatum. Emplastrum is a formulation for external use; its base usually consists of lead salts of fatty acids with a high number of carbon atoms. The poultices are solid and kneadable at room temperature, and liquefy when heated. After heating, they can be spread on a suitable pad. This formulation used to be dermatologically important, but is no longer used today (Burger and Wachter 1993, 7th edition, p. 480).

In Folk Medicine

Myrica wax has an astringent and slightly narcotic effect and is used in cases of bacillary dysentery and amoebiasis. The water used in hot water extraction is also thought to be a cure for dysentery.

Externally, myrica wax is applied in cases of dermatophytosis, skin fungus and ulcers.

In Cosmetics

Since myrica wax has a pleasant, aromatic smell and a soothing effect on the skin, it is added to shaving foam. Its characteristic texture and moisturising effect are the reasons for its use in soap production.

Additional Uses

Candle Production

Myrica wax is used to produce candles with a pleasant, aromatic smell that do not fume when extinguished. The disadvantage is that they are brittle and break easily. In the USA, in particular, myrica wax is used as a substitute for bees-wax because of its similar properties (List and Hörhammer 1976, 4th edition, 5th volume, p. 917).

Industrial Uses

Since myrica wax has a special composition of fatty acids, it is suitable for use as a sealing wax to seal parquet floors.

6 Possible Unwanted Side Effects

Ingesting large amounts of myrica wax can lead to vomiting and stomach cramps.

M

Neem Oil

Oleum Melia synonyms: Nimöl, Neemöl, Niemöl (D); huile de neem (F)

1 Source Plant

Azadirachta indica A. Juss., (Meliaceae), syn.: Melia azadirachta L Sp., *Melia indica* Hooker *Melia indica* Brandis, neem, nimtree, Indian lilac

S. Krist, *Vegetable Fats and Oils*, https://doi.org/10.1007/978-3-030-30314-3_75

Habitat

Neem is a native tree in India and Myanmar, and is grown successfully in Africa, the Middle East, South America and Australia (Puri 1999, p. 10 et seqq).

Description

Neem is an evergreen tree that can reach a height of 30 m. It has pinnate leaves. The individual leaflets are lanceolate and 2–7 cm long. During droughts, young trees may shed their leaves, but they grow back again. Plants growing in regions with warm winters blossom earlier in the year, usually from January to May; some blossom from September to October and February to March. In the Murshidabad district in western India, the trees blossom throughout the year. They usually begin to blossom when they are 3–5 years old, and reach their maximum capacity at an age of about 10 years. The flowers are small, perfect, 4–5 mm long and of a whitish-pink colour. The fruit is an oval stone fruit 1–2 cm in length. When unripe, it is smooth and green; its colour changes to yellow, then brown when it is ripe. The seeds are 11–18 mm long and weigh 100–530 mg (Puri 1999, p. 10 et seqq).

2 Cultivation and Extraction

Cultivation

In dry and hot conditions, the tree grows extremely fast, even in the Sahara Desert. It is planted as a shade tree in countries with extremely hot temperatures. Neem tolerates temperatures of over 50 °C in the shade and near the equator grow to altitudes of 1000 m above sea level. An average annual precipitation of 400–1200 mm is ideal. Neem tolerates barren ground and sandy soil (Norten 1996, p. 13 et seqq).

Extraction of the Oil

A 10-year-old tree produces 50–150 kg fruit per year; on average, a tree bears 20–30 kg fruit. To obtain the oil, the kernel (of which there are sometimes 2 or 3) is separated from its seed coat and either cold pressed or extracted with hexane. Pressing is the gentler method and results in oil of higher quality than that obtained by extraction with solvents (Norten 1996, p. 13 et seqq). The residue (neem press cake) is used as a mild fertiliser with insecticidal properties and as forage.

3 Character

Colour

Greenish brown, viscous at room temperature.

At room temperature solid and spreadable, at about 23 °C a yellowish, sometimes brown oil with a green tint (Norten 1996, p. 19, 20).

Odour

Mouldy, pungent, nutty, touch of green.

Between peanut and garlic (Norten 1996, p. 20).

Volatile Compounds

SPME-GC-MS analysis (Krist et al. 2008):

Components	Peak area in %
Acetous acid	20.84
2-Methylbutanal	1.18
Heptane	7.79
Octane	6.82
Hexanal	22.54
Dihydro-2-methyl-3(2H)-furanone	1.36
Furfural	2.97
3,4-Dimethyl-thiophene	2.19
2-Heptanone	2.68
Gamma-butyrolactones	0.5
Heptanal	2.65
Alpha-pinene	0.6
t-2-Heptenal	1.63
5-Formyl-2-methyl-furane	0.5
1-Octen-3-ol	0.83
2-Pentylfurane	3.97
Octanal	2.08
Nonan-2-one	0.68

Flavour

Unpleasant due to bitterns and sulphur components (Norten 1996, p. 42).

4 Ingredients

Composition of Fatty Acids

Fatty acids	Content in % acc. to Salunkhe et al. (2003)	Content in % acc. to Kalayasiri et al. (1996)	Content in % acc. to Kaushik and Vir (2000)
Palmitic acid	13–15	21.4	17.3–34.3
Stearic acid	14–20	20.6	6.6–24.0
Oleic acid	49–62	35.1	25.4–57.9
Linoleic acid	10–16	17.7	6.2–16.7
α-Linolenic acid	–	0.6	–
Arachidic acid	1–2.4	1.1	–
Eicosenoic acid	–	0.6	–
Fatty acids in total	50	43.6	–

Other Ingredients

Azadirachtin: 820 mg/kg
Gedunin
Nimbin
Nimbidin
Nimbidol
Quercetin
Salannun
Sodium nimbinat

Physical Key Figures of Neem Oil

density: 0.9189 (at 30 °C)
saponification number: 193
iodine number: 75
unsaponifiable: 2.0%
acid number: 4–16
(Salunkhe et al. 2003, p. 504)

Characteristics of the Oil

An important ingredient of neem oil is the triterpenoid azadirachtin. Azadirachtin inhibits the hormone responsible for growth of insects called ecdysone. It thus prevents the development of larvae and inhibits the fertility of insects. Neem oil is effective as a natural anti-feedant and insecticide in pest management.

5 Application

In Pharmacy and Medicine

Stimulating the Immune System

A study on rats examined the effect of NIM-76, a volatile compound of neem oil isolated by means of steam distillation. It resulted in an increase in PMNs (polymorphonuclear leukocytes) in the blood of rats administered 120 mg NIM-76/kg body weight. An increased dose of 300 mg NIM-76/kg body weight caused an increase in the number of neutrophil granulocytes. The immuno-modulatory properties of NIM-76 can be explained with an increased activity of macrophages induced by NIM-76 and a boost of T-lymphocyte proliferation (SaiRam et al. 1997).

Spermicidal Effect

Another study examined the spermicidal effect of NIM-76. At a concentration of 25 mg/ml NIM-76, the mobility of human sperms decreased to zero within 20 s. The time span until the sperms are immobile depends on the dosage. The sperms immobilised by NIM-76 did not regain their mobility even after incubation with buffered glucose salt, which means that NIM-76 is spermicidal, not spermiostatic. Intravaginal application of NIM-76 on rabbits did not result in irritation of the vaginal mucosa, which is why NIM-76 is regarded as a promising vaginal contraceptive (Riar et al. 1990).

Antibacterial Effect

The antibacterial activity of neem oil was verified in a test with various microorganisms such as *E. coli*, *Pseudomonas aeruginosa*, species of *Salmonella*, species of *Shigella*, *Staphylococcus aureus* and other staphylococci as well as *Vibrio cholerae*. In a preliminary examination on agar, all strains of bacteria were susceptible to neem oil. A series dilution test showed that 21.42% of the strains were inhibited at a concentration of 500 µl/ml neem oil, 71.42% at a concentration of 125 µl/ml and 7.14% at a concentration of 250 µl/ml. A time-dependent study proved that the oil devitalised all bacteria within 4 h, staphylococci even within 15 min (Baswa et al. 2001).

NIM-76 also proved to be antimicrobially active. It inhibits the growth of *Escherichia coli* and *Klebsiella pneumoniae,* which are not greatly influenced by neem oil. Additionally, fungicidal activity against *Candida albicans* as well as antiviral properties were detected (SaiRam et al. 2000).

To summarise, the antimicrobial (antibacterial, fungicidal and antiviral) effects of NIM-76, together with its spermicidal properties, substantially contribute to hygiene and a healthy vaginal flora (SaiRam et al. 2000).

In Folk Medicine

In India, the oil is added to mixtures for contraception, as it has a spermicidal effect (Norten 1996, p. 43 et seqq). In many tropical countries, neem oil is sold as a household remedy for all kinds of indisposition and complaints, for example muscle pain, malaria, tuberculosis and even diabetes (Boeke et al. 2004).

As an Insecticide

In the form of a 2% solution, neem oil can be used in pesticides against vermin and plant diseases. It is even effective against severe plant diseases such as rust and blight (Norton 1996, p. 43 et seqq). Neem oil wards off insect pests such as leaf miners, fleas, beetle larvae, lice, caterpillars, mosquitoes, bugs, ticks and cicadas. House dust mites are repelled if bed sheets are treated with the oil. Neem oil products show no or almost no effect on useful animals like parasitic hymenopterans and honey bees.

Other Applications

In cosmetics, neem oil can be used in the production of soap, lotions, nail oil and creams (Norten 1996, p. 43 et seqq).

In India, it is a substitute for cooking salt in the leather industry. It has also been tried as bio fuel. Furthermore, it is used in the production of olein and, after it has been hydrogenised, as a high-melting stearin (Puri 1999, p. 73 et seq).

Neem oil is an edible oil if the raw oil is treated with an alkaline solution or hydrogen peroxide and subsequently distilled (Puri 1999, p. 73 et seq).

A study examined whether neem oil is a suitable agent to remove ammonium from brackish water. The experimental results showed that 90 mg/l neem oil removed ammonium effectively. At this concentration, the total amount of nitrogen decreased to 0.40–0.45 mg/l in 96 hours (Krishnani et al. 2002).

6 Possible Unwanted Side Effects

Effects on Humans

N

Until now, two cases have been described in which oral administration of neem oil led to acute poisoning of children. Even in small amounts, the oil caused toxidromes like nausea, vomiting, dizziness and hyperventilation. Laboratory tests show that neem oil causes leucocytosis and metabolic acidosis. As aflatoxins were found in these oil samples, the toxic effect could possibly be traced back to a synergistic interaction of aflatoxins and other toxic ingredients (Boeke et al. 2004).

Effects on Animals

Neem oil reveals its acute toxicity depending on dosage and time when given to rabbits and rats. It has an influence on movement, breathing and the sense of direction. The animals suffer from diarrhoea, cramps and tremor. Neem oil has a higher abortive effect on rats when it is given in the first days of pregnancy than when it is administered towards the end of the pregnancy. The study also showed reduced fertility of the rats as a subacute effect induced by a disruption of the oestrogen balance (Boeke et al. 2004).

Mutagenic Potential

A study examined the mutagenic activity of neem oil by varying out the Ames test[1] on *Salmonella Typhimurium* strains. The results showed that neem oil does not have mutagenic activity (Polasa and Rukmini 1986).

[1]Ames test: reverse mutation of *Salmonella typh.;* mutagenic activity of substances is tested on strains of *Salmonella typh.* Requiring histidine, in the presence and absence of a rat liver metabolising system.

Niger Seed Oil

Oleum Guizotiae synonyms: Nigersamenöl, Ramtillöl (D); huile de niger (F)

1 Source Plant

Guizotia abyssinica (L.f.)Cass., *(Asteraceae)*, niger, noog, ramtilla, inga seed

S. Krist, *Vegetable Fats and Oils*, https://doi.org/10.1007/978-3-030-30314-3_76

Habitat

Niger is a common plant in Ethiopia and East Africa, and has long been used for its oil. It spread to eastern India and the whole of Africa. It is also found sporadically in Central Europe, where it probably grew wild from refuse left behind from bird feeding. Since the nineteenth century, cultivation has been attempted in Russia, but as the fruit of niger ripen late in the year, the harvest is small in cold climates. According to Schuster, breeders still have much work to do in order for niger to become an oil plant that can be cultivated economically in Europe (Schuster 1992, p. 66).

Description

Niger is an annual, herbaceous plant that is highly branched and can reach a height of 1–2 m. The lower part of the stem is nearly hairless; the upper part is covered with sticky, rough hair. The leaves are lanceolate and serrate, opposite in the bottom part, alternate above. The inflorescences are medium-sized heads that are arranged at the axils of leaves and at the ends of the shoots on medium-length stems. They consist of yellow ray florets that are all female. The small composite flowers are heterogamous, with bracts resembling chaffy bracts that are enlarged like leaves on the outside. The corolla is hairy, but no pappus is formed. The fruit have a thin pericarp; they are black or shiny brown, slightly angular, pointed, 3.5–5 mm long and 1–1.5 mm thick (Schuster 1992, p. 66).

2 Cultivation and Extraction

Cultivation

Niger is a very adaptable plant. It prefers loamy soil, but also grows on moist and very moist soil. The plant does not make any demands with regard to pH value, growing on neutral, acidic and alkaline soil. It needs much light, however, and is therefore found in sunny locations more often than in half-shade. Cultivating niger is easy; it is often grown in Africa and India. Sowing takes place from May to September, and the plant blossoms in October. The harvest takes place about 150–180 days after sowing, i.e. from December to January. Harvesting the seeds is difficult, as they fall out of the flowerheads quite easily. For this reason, workers harvest them manually (Röbbelen et al. 1989, p. 396).

Since the plant ripens late in the year, it only thrives in warm regions. It is noteworthy that the yield is not as high as that of safflower or sunflowers.

Extraction of the Oil

In Ethiopia, the oil is extracted from the seeds of *Guizotia abyssinica Cass.* in a very simple, traditional way. The seeds are ground in mortars and then boiled in water; the oil settles to the surface and can be skimmed off (Von Wiesner 1927, p. 761).

Today, the yellow, drying oil is obtained by boiling the ground achenes (Lieberei and Reisdorff 2007, p. 139).

For industrial use, the oil is extracted with solvents in a Soxhlet extractor (Sarin et al. 2009).

3 Character

Colour

Light yellow to dark yellow (Von Wiesner 1927, p. 761).
Light yellow (Röbbelen et al. 1989, p. 401).
Pale yellow (Axtell 1992, p. 99)

Odour

Weak smell reminiscent of thyme (Von Wiesner 1927, p. 761).

Nutty smell (Hackbarth 1944, p. 345).

Sweetish smell (Röbbelen et al. 1989, p. 401).

Nearly odourless (Salunkhe and Desai 1986, p. 178).

Flavour

Nutty (Röbbelen et al. 1989, p. 401).

Weak, nutty (Salunkhe and Desai 1986, p. 178).

4 Ingredients

Composition of Fatty Acids

Fatty acids	Content in % acc. to Dagne and Jonsson (1997)	Content in % acc. to Schuster (1992), p. 10	Content in % acc. to Dutta et al. (1994)	Content in % acc. to Ramadan and Mörsel (2004)	Content in % acc. to Sarin et al. (2009)	Content in % acc. to Marini et al. (2003)
Lauric acid	–	1–3	–	–	–	–
Myristic acid	–	1–4	–	–	–	–
Palmitic acid	9.6–10.0	5–8	8.0–9.0	17.0	9.2	9.59

Fatty acids	Content in % acc. to Dagne and Jonsson (1997)	Content in % acc. to Schuster (1992), p. 10	Content in % acc. to Dutta et al. (1994)	Content in % acc. to Ramadan and Mörsel (2004)	Content in % acc. to Sarin et al. (2009)	Content in % acc. to Marini et al. (2003)
Palmitoleic acid	0.2	–	–	–	–	0.218
Stearic acid	7.6–8.1	2–5	5.6–8.1	6.52	10.1	7.00
Oleic acid	5.4–7.5	2–6	5.9–11.0	11.2	9.0	7.9
Linoleic acid	65.7–68.7	60–80	71.4–79.2	63.0	71.7	71.0
Linolenic acid	0.2	–	–	–	–	0.376
Arachidic acid	0.5–0.6	0–5	–	–	–	0.459
Eicosenoic acid	0.1	2–3	–	–	–	–
Behenic acid	0.7–0.8	–	–	0.52	–	0.365
Erucic acid	0.4–0.5	–	–	–	–	–
Lignoceric acid	0.5–0.7	–	–	–	–	–

Composition of Triglycerides

Triglycerides	Content in % acc. to Marini et al. (2003)
LLL	37.97
OLL	12.35
PLL	15.01
OOL	2.62
POL	14.89
PPL	1.63
OOO	3.89
Poo+sol	2.51
SOO	1.66

Composition of Sterols

Sterols	Content in µg/g (%) acc. to Dutta et al. (1994)	Content in g/kg acc. to Ramadan and Mörsel (2004)	Content in % acc. to Marini et al. (2003)
Cholesterol	10–31 (0.2–0.8)	–	0.271
Campesterol	490–690 (11.8–13.1)	0.713	11.6
Stigmasterol	560–750 (13.5–14.2)	0.667	13.56
Lanosterol	–	0.113	–
β-Sitosterol	1590–2280 (38.4–43.2)	2.035	44.4
Δ7-Stigmastenol	–	–	11.55
Δ5-Avenasterol	210–350 (5.1–6.6)	0.530	6.00
Δ7-Avenasterol	180–240 (4.1–4.5)	0.164	4.48

N

Composition of Tocopherols

Tocopherols	Content in g/kg acc. to Ramadan and Mörsel (2004)	Content in μg/g, concentration in % acc. to Dutta et al. (1994)	Content in μg/g acc. to Marini et al. (2003)
Total content	–	657–853	797
α-Tocopherol	0.861	94–96	–
β-Tocopherol	0.331	1.0	–
γ-Tocopherol	0.570	3.5	–
δ-Tocopherol	0.185	–	–

Other Ingredients

(Roth and Kormann 2000, p. 140)

Trilinolein 2%
Oelodilinolein 40%
Dioleolinolein 30%
Myristodilinolein 2%
Palmitodilinolein 6%
Palmitooleodilinolein 11%
Stearodilinolein 2%
Stearooleolinolein 4%

β-carotenes: 0.702 g/kg (Ramadan and Mörsel 2004)

Physical Key Figures of Niger Seed Oil

specific weight (15 °C):	0.9272
Point of solidification:	−15 °C
Saponification number:	188.0–192.2

(Hackbarth 1944, p. 345)

density (kg/m^3):	901
Viscosity (mm^2/s) at 40 °C:	28.2
Acid number (mg KOH/gm):	6.03
Iodine number:	131.9
Cloud point:	−2 °C
Flow point:	−9 °C
Oxidative stability (h):	1.41
Moisture content:	0.15%

(Sarin et al. 2009)

Shelf Life

Niger seed oil has a shelf life of 6 months if stored in a cool place (Röbbelen et al. 1989, p. 401).

5 Application

In Pharmacy and Medicine

Essential fatty acids, which are an important part of our diet, can be ontained from niger seed oil. They cannot be produced by the body, but have to be ingested. Niger seed oil contains high amounts of oleic acid and linoleic acid, and is therefore suitable raw material to extract these fatty acids. Since niger seed oil contains a high amount of linoleic acid, it protects against cardiovascular diseases such as coronary heart diseases, arteriosclerosis and high blood pressure. In addition, linoleic acid is a component of phospholipids, which, due to their amphiphile properties, play a significant part in the arrangement of lipid double layers of biological membranes. As a result of its amount of tocopherol, niger seed oil also has an antioxidative effect (Ramadan and Mörsel 2003b).

In Folk Medicine

In Africa, in particular, niger seed oil is applied against rheumatism and syphilis. These applications have not yet been clinically tested (Salunkhe et al. 1992, p. 451).

In Cosmetics

As an ingredient of body-care products, niger seed oil quickly penetrates and moisturises the skin. The moisture balance is restored, tightness of the skin is relieved, and the skin feels smooth and tender again. For this reason, many body lotions, balms and creams contain nigerseed oil. As a result of its high amount of tocopherols, niger seed oil is also added to antiwrinkle creams. The antioxidative effect of tocopherols prevents skin ageing caused by oxidative stress. The skin regains its elasticity, and small wrinkles are reduced (Ramadan and Mörsel 2003b). Thanks to the pleasant smell of niger seed oil and the fact that it does not leave a fatty film on the skin, it is also used as a massage oil (Salunkhe et al.1992, p. 451).

Industrial Uses

In the soap industry, niger seed oil is a frequent substitute for linseed oil, as its properties are similar and it is slightly cheaper (Ramadan and Mörsel 2003b).

Additionally, niger seed oil is regarded as a potential raw material for the production of biodiesel. In the production process, the oil is transesterified using various catalysts. Whether the production of biodiesel is economical depends on the price of the raw material. As niger seed oil is not among the oils traditionally used for this purpose, it has potential in this respect. A study by Sarin et al. tested the fuel properties and stability of niger seed oil and confirmed that the oil is a suitable raw material for biodiesel (Sarin et al. 2009).

As Food

Niger seed oil is a high-quality edible oil. It is rich in linoleic acid and therefore a valuable food. It can also be a substitute for olive oil. Niger seed oil is frequently used to dilute very expensive oils, for example sesame oil. It is also sold mixed with linseed oil (Röbbelen et al. 1989, p. 401).

Guizotia is the most important oil plant in Ethiopia, its share of the native edible oils amounting to 50–60% (Ramadan and Mörsel 2003b).

In Painting

As niger seed oil dries quickly, it is used in the varnish and paint industry. Unfortunately, it has the disadvantage that the paint is dark and remains sticky for a long time. As a result, the use of niger seed oil in painting is limited, and it is hardly used anymore. Today, artists' paints are made from linseed oil or poppyseed oil, which have superior plasticity and durability, as well as less tendency towards yellow than niger seed oil (Von Wiesner 1927, p. 762).

6 Possible Unwanted Side Effects

So far, unwanted side effects have not been detected. One case has been reported in which contact with the seeds of *Guizotia abyssinica* in bird feed led to allergic reactions such as rhinoconjunctivitis, asthma and urticaria (Rodriguez et al. 2003).

Oil of Nutmeg

Oleum Nuciste, Oleum Myristicae Expressum

Synonyms: Muskatbutter (D); Myristica fats (E); huile de noix de muscade (F); grassi di miristica (I); grasas de Myristica (ESP).

S. Krist, *Vegetable Fats and Oils*, https://doi.org/10.1007/978-3-030-30314-3_77

1 Source Plant

Myristica fragrans HOUTT. *(Myristicaceae)*, Nutmeg Tree

Habitat

The nutmeg tree is a native plant of the Molucca Islands Ambon, Banda, Seram, Damar and Nila, which were formerly known as the Spice Islands. Since the twentieth century, it has been widespread in tropical countries, for example in Sumatra, Java, Borneo, New Guinea, Malacca, Penang, Minahasa, the Indian subcontinent and the South-East Asian peninsula (Prentner 2010, p. 202).

From the Molucca Islands, Arabian traders brought the seeds to Europe in medieval times (Lieberei and Reisdorff 2007, p. 329).

Description

The name *Myristica* has its origins in the Greek *myristicos* and means "fragrant", like the Latin *fragrans*. This alludes to the musky odour (Prentner 2010, p. 201).

The nutmeg tree grows up to 20 m tall, but is usually smaller when cultivated. The bark is smooth and brownish-grey; young branches are a green colour. The leaves are alternate, oval-elliptical and entire. *Myristica fragrans* is evergreen and usually dioecious. The male flowers are arranged in racemes and have 9–10 coadunate stamina. The female inflorescences consist of one to three flowers, each of which has only one ovary. After pollination, the ovary develops into a round, fleshy, yellow fruit that looks similar to a peach and is about 5–6 cm in diameter. It is a follicle that opens at a suture. As a result of the fleshy pericarp, it may dehisce at the back as well. This happens about 7–10 months after the flowers have appeared. When the fruit has opened, the dark brown seed becomes visible. It is mistakenly called a nut. The seed is surrounded by a red aril. The kernel consists of a small embryo, much endosperm, and a wrinkly perisperm protruding into the endosperm. On the outside, the kernel is surrounded by a ligneous seed coat (Lieberei and Reisdorff 2007, p. 329–330; Prentner 2010, p. 202).

2 Cultivation and Extraction

Cultivation

Nutmeg trees are cultivated in India, Central America and South America. They need a moist, hot climate (Lieberei and Reisdorff 2007, p. 229).

Water retention should, however, be avoided. *Myristica fragrans* only bears fruit for 5–8 years after planting. The harvest reaches an optimum after 15–20 years and continues for 30–40 years after planting. Each tree bears 1500–2000 fruit annually. The fruit fall from the tree when they are ripe and are collected from the ground (Axtell 1992, p. 103).

The gathered fruit still have their red aril, which is carefully removed and dried. In the process of drying, it becomes hard and its colour changes to a yellowy orange. Subsequently, it is either cut into pieces and ground, or sold under the name "mace". The seeds without the aril are dried until the kernels rattle within. The seed coat is then removed and the seed is limewashed. Formerly, this acted as an insect repellent, but the tradition is still retained and it is regarded as a mark of quality. Indonesia is the main producer (27,000 tons/year) (Lieberei and Reisdorff 2007, p. 330).

Extraction of the Oil

The seeds are ground and subsequently pressed by adding heat (Roth and Kormann 2005, p. 136).

Extraction with solvents is also possible. This is ideally carried out in a Soxhlet extractor (Abdurrasheed and Janardanan 2009).

Fat oil can also be obtained from the aril (Lieberei and Reisdorff 2007, p. 330).

3 Character

Colour

Oil of nutmeg is reddish-brown, lighter in places, and similar to tallow. It has a soft, crumbly texture (Roth and Kormann 2005, p. 136).

Odour

Intensely aromatic (Roth and Kormann 2005, p. 136).

Flavour

Typical taste of nutmeg (Roth and Kormann 2005, p. 136).

4 Ingredients

Composition of Fatty Acids

Fatty acids	Content in % acc. to Collin and Hilditch (1929)	Content in % acc. to Roth and Kormann (2005) p.137	Content in % acc. to Abdurrasheed and Janardanan (2009)	Content in % acc. to Niyas et al. (2003)
Lauric acid	1.5	20.8	8.00	3.0
Myristic acid	76.6	73.4	55.10	90.0
Palmitic acid	10.1	0.3	14.87	6.0
Stearic acid	–	–	7.30	0.5
Petroselinic acid	–	–	–	0.13
Oleic acid	10.5	5.5	–	–
Linoleic acid	1.3	–	–	–

A fat oil can also be extracted from the flowers; its content in % is as follows (Abdurrasheed and Janardanan 2009):

lauric acid:	1.31
myristic acid:	8.11
palmitic acid:	52.26
stearic acid:	7.98

Physical Key Figures: (Roth and Kormann 2005 p. 136)

refractive index: n^{40}_{D}	1.4662–1.4705
density: d^{15}_{4}	0.945–0.960
melting point:	45–51 °C
saponification number:	168–179
acid number:	17–23
iodine number:	40–65
solubility:	completely soluble in hot ethanol, and in cold ethanol to about 55%; soluble in diethyl ether and chloroform (Roth and Kormann 2005 p.136)
unsaponifiable:	17.7%
iodine number:	61.0
(Collin and Hilditch 1929)	
saponification number:	185.27
iodine number:	63.5

(Abdurrasheed and Janardanan 2009).

5 Application

In Pharmacy and Medicine

O

Oil of nutmeg is applied externally as a stimulant (Zachariah et al. 2000).

External use as an ointment is also common (Roth and Kormann 2005, p. 137).

It can also be an ingredient of patches (Lieberei and Reisdorff 2007, p. 330).

In Cosmetics

Since oil of nutmeg has an intense smell, it is used in perfumery (Axtell 1992, p. 103).

6 Possible Unwanted Side Effects

The median lethal dose (LD_{50}) for rats with oral administration is 3640 mg/kg. Humans may suffer from skin irritations when the oil is applied externally. Ingestion may result in allergic reactions (Roth and Kormann 2005, p. 137).

Oilseed Radish Oil

Synonyms: Öl des Ölrettichs (D), Rettichöl (D).

1 Source Plant

Raphanus sativus (L.) *var. oleiferus Pers. (Brassicaceae)*, Daikon

Habitat

The daikon is an old cultivated plant. Cultivation is documented in Egypt from 2000 B.C. and in China from 1100 B.C. onwards. At the beginning of the nineteenth

S. Krist, *Vegetable Fats and Oils*, https://doi.org/10.1007/978-3-030-30314-3_78

century, the plant was brought from China to Central Europe, where it was first cultivated in Silesia (Löw 2003, p. 40).

Description

Raphanus sativus is a herbaceous plant that grows to a height of 100–160 cm. The upright stem is branched, and hollow at the bottom. The root is sturdy, reaching deep into the ground. The leaves are large and slightly lobate, the upper ones pointed and serrate. The inflorescence is a panicle. The colour of the flower can vary from white to blue or violet; different colours on one and the same plant are possible. Pollination is via exogamy, mainly by bees and bumblebees, but also by other insects and by the wind. The fruit is a husk with a long point. It is 3–6 cm long and 1 cm thick. Both flaps have grown together closely, which is why the husk is difficult to open and not prone to falling apart. Within, there is a felty pulp. The seeds are embedded in this soft tissue, 2–3 in each husk. They are either elongated and oval or roundish and flattened, with a length of 4.5 mm and a breadth of 3.5 mm. The surface is like a delicate net and light brown to reddish brown. The thousand-seed weight is between 8 and 14 g (Hackbarth 1944, p. 173; Schuster 1992, p. 42–43).

2 Cultivation and Extraction

Cultivation

In China, Japan and Egypt, daikons are cultivated for their oil. Spain and Romania are the main growing areas in Europe. The daikon is a rather demanding plant. The climate should be mild, during the blooming period especially warm and dry. Later, the plant is usually no longer frost-susceptible. With regard to the soil, sandy, loamy soil is preferable. Sowing should take place as early as possible, ideally from the end of March to mid-April. Since the plant branches out, the sowing quantity can be kept to a minimum (about 12–15 kg/ha). The ideal distance between rows is 30–40 cm. The plant may become infested with pests, especially flea beetles. The harvest takes place from mid-August to the end of August. As the husks do not burst open, they can be left on the plant until they are fully ripe. The harvest should not, however, take place too late in the year, mainly because of bird damage and because the husks may fall off when they are overripe. The yield amounts to between 600 and 1200 kg corns/ha (Hackbarth 1944, p. 173–174; Körber-Grohe 1995, p. 195).

In Central Europe, daikons are mostly used for green manuring, as the plant decomposes the soil with its deep roots. It is also used as a catch crop, especially in the crop rotation of sugar beets, to fight nematodes in the soil (www.landwirtschafts-kammer.de/11.11.2011).

Extraction of the Oil

The seeds are pressed (Löw 2003, p. 40).

3 Character

Colour

Golden yellow (Hackbarth 1944, p. 175).

Odour

Odourless (Hackbarth 1944, p. 175).

Flavour

Mild, pleasant (Hackbarth 1944, p. 175).

4 Ingredients

Composition of Fatty Acids

Fatty acids	Content in % acc. to Schuster (1992), p. 10	Content in % acc. to Downey (1971)	Content in % acc. to Seehuber (1987)	Content in % acc. to Valle et al. (2009)
Palmitic acid	4–6	7.1	–	7.0
Palmitoleic acid	–	2.2	–	–
Stearic acid	1–2	–	–	3.6
Oleic acid	25–32	40.0	17.0–36.4	27.9
Linoleic acid	18–22	16.9	–	7.6
α-Linolenic acid	16–20	14.5	–	4.6
Arachidic acid	–	0.4	–	2.2
Eicosenoic acid	8–12	9.3	–	11.2
Erucic acid	8–24	9.6	8.9–29.6	33.3
Lignoceric acid	–	–	–	0.6
Nervonic acid	–	–	–	2.0

Physical Key Figures

Specific weight at 15 °C:	0.915–0.9175
point of solidification:	−10 to −17 °C
iodine number:	119–124
acid number:	2.77–4.23

Oilseed radish oil is a non-drying oil.
(Hackbarth 1944, p. 175)

5 Application

Industrial Uses

Oilseed radish oil is used as fuel (Lieberei and Reisdorff 2007, p. 138).

A study by Valle et al. proved that oilseed radish oil is excellently suited for the production of biodiesel. Among the oil's many advantages are the rapid growth of daikon, which only takes 90–120 days to develop, the low production costs, the high oil content and the easy production of the oil by cold pressing. In the production of biodiesel, sodium ethoxide can act as a catalyst. This results in reduced water production during the chemical reaction, and transesterification is more efficient and can be carried out in a single step. In this way, the synthesis of biodiesel is very efficient (Valle et al. 2009).

The seed oil contains between 8% and 33% erucic acid, which is why daikon is ebing considered as a possible raw material for the industrial production of erucic acid. A study by Seehuber consluded that daikon is not, however, suitable for the production of oil with a high content of erucic acid. Other species of *Brassica*, like mustard and colewort, are better suited (Seehuber 1987).

In China, oilseed radish oil is traditionally used to produce China ink. The oil is burned, and the soot produced in the process is used to produce the ink (Hackbarth 1944, p. 175).

As Food

If oilseed radish oil is used as an edible oil, its content of erucic acid must be considered. According to the decree on erucic acid (*Erucasäureverordnung*, BGBl. No. 468/1994) it is prohibited to put into circulation edible oils, fats and foodstuffs intended for human consumption that have been produced by adding edible oils and fats and have a total fat content of more than 5%, if they contain more than 5% erucic acid – based on the total content of fatty acids in the fat phase (http://www.ris.bka.gv.at/04.07.2011).

Olive Oil

Oleum Olivarum synonyms: Olivenöl (D); huile d'olive (F)

1 Source Plant

Olea europaea L. (Oleaceae), olive tree

S. Krist, *Vegetable Fats and Oils*, https://doi.org/10.1007/978-3-030-30314-3_79

Habitat

The origins of *Olea europaea* probably lie in the eastern Mediterranean area. Fossil finds dating from the Upper Palaeolithic (35000–8000 B.C.) in France and the Neolithic (8000–2700 B.C.) in Spain and Greece indicate wild olive trees and their possible use. It is certain that by 3000 B.C., the olive tree was a cultivated plant in Syria, Palestine and Crete. It grows in all Mediterranean countries and in those areas near the Black Sea that have no extreme climatic conditions. It prefers porous, calciferous soil that is well-aired and well-drained, so that there is no stagnant moisture. It tolerates heat, but is very frost-susceptible in cold winters. In general, it grows best in a subtropical transition Mediterranean climate with dry, hot summers and cool, wet winters (Kulnik 2003, p. 6; Roth and Kormann 2000, p. 141).

Description

Olea europaea is an evergreen tree that can reach an age of 1000 years. It is 10–16 metres tall. Its knotty, furrowed stem and its dense, silvery crown are characteristic. Young trees have a smooth, grey bark, which becomes darker, more brittle and more crooked with advancing age. The leaves are small and lanceolate; their upper side is greyish green, their lower side silvery grey, shiny and hairy. The small flowers are either white or yellow, with four petals and sepals each, and are arranged in panicles. The fruit is a stone fruit with a thin skin; the fruit pulp is rich in oil. The hard kernel usually contains only one seed, which is also rich in oil. The tree is slow-growing. It bears its first fruit at the age of about 10 years, but it takes two more decades for the harvest to be abundant (Frohn 1998, p. 11).

2 Cultivation and Extraction

Cultivation

All cultivation areas of *Olea europaea* lie between the 30th and 45th degree of northern or southern latitude, with the exception of altitudes near the equator, for example in Peru. The Spanish conquistadors brought the olive tree to Peru in the sixteenth century. From there, it spread to Mexico, California and Hawaii. Today, olive trees are also cultivated in Australia, New Zealand and South Africa. The main growing areas are the European and North African countries around the Mediterranean Sea. Sixty percent of the approx. 750 million olive trees worldwide are to be found in the European Union, which is not only the largest producer of olive oil, but also the main consumer. As demand for the healthy olive oil has been on the increase even in northern countries, more olive trees are being cultivated than ever before. The European Union has been criticised for supporting ecologically problematic large-scale cultivation, which is objectionable because of erosion, water consumption and chemical pest control, and for neglecting small-scale farming. Outside the EU, the main producers of olives and olive oil are Syria, Turkey and Tunisia. In these countries, olive oil production makes up a large amount of the gross domestic product. Many people work in olive cultivation, which constitutes their livelihood. The following table gives an overview of the 10 most important producers (Hänsel et al. 1999, p. 254; Kulnik 2003, p. 12 et seq).

The 10 most important producing countries			
2005	Production (tons)	Yield (kg/ha)	Area (ha)
World	14,762,436	–	7,550,561
Spain	3,919,800	3269.97	1,199,093
Italy	3,715,660	3181.27	1,167,980
Greece	2,583,185	3292.78	784,500
Turkey	850,000	1309.00	649,350
Syria	620,000	1240.00	500,000
Tunisia	600,000	40,000	1,500,000
Morocco	500,000	1041.67	480,000
Algeria	316,489	1322.29	239,350
Egypt	310,000	6326.53	49,000
Portugal	270,000	750.00	360,000

Extraction of the Oil

Olive oil is obtained from the ripe stone fruit of *Olea europaea* by cold pressing or other suitable mechanical methods (Hänsel et al. 1999, p. 254; Frohn 1998, p. 24 et seq; Kulnik 2003, p. 19 et seq; Hunnius 1998, p. 994 et seq; Roth and Kormann 2000, p. 141; Hager 1978, volume 7b, p. 197).

Extraction of Olive Oil in Antiquity

In antiquity, olive oil was extracted using various methods. To obtain oil of the best quality, the olives were not pressed, but rather placed in a basket, and the mixture of oil and water dripping from the basket was collected in a container. Another method was to fill an opening in the rocks with olives. The underside had a hole. Both methods use the pressure produced by the fruits' own weight to squeeze the oil from the fruit. Another method was to pour hot water onto the fruit and then decant the oil from the surface. The methods were improved technically over time. The Romans constructed the *mola olearia*. It consisted of two cylindrical stones attached to an axis. If the axis was turned, the olives were squeezed. After this first cold pressing, hot water was added, and the fruit was pressed a second and third time (Löw 2003, p. 80).

Obtaining Olive Oil of Different Quality

No other oil has so many different kinds and qualities, depending on the methods of harvesting and extraction.

The best oil is the so-called virgin oil (*Oleum virgineum*). The ripe olives are plucked and peeled, and the kernels removed by hand. The fruit pulp is ground and cold pressed under moderate pressure at temperatures of 20–25 °C. The result is a light-yellow to greenish oil that has the typical smell and taste of the fruit pulp (Frohn 1998, p. 24 et seq; Kulnik 2003, p. 19 et seq; Hunnius 1998, p. 994 et seq; Roth and Kormann 2000, p. 141; Hager 1978, volume 7b, p. 197).

The application of slightly higher pressure results in the slightly yellowish *Oleum optimum*, which is of good quality. The oil obtained by this method meets the requirements of the pharmacopoeia (Hunnius 1998, p. 995; Roth and Kormann 2000, p. 141; Hager 1978, volume 7b, p. 197).

By grinding whole olives with their skins and kernels and subsequently cold pressing them, the 1st type of the edible oil is obtained. The press residue is mixed with cold water and pressed again, which results in the 2nd type. The residues now remaining are hot pressed with or without adding hot water; the result of this pressing is oil of a lower quality, which can be used as an edible oil after it has been neutralised, decoloured and steamed, but is mostly sold and used for technical purposes (Hager 1978, volume 7b, p. 197).

The yield is especially good if the olives are left in heaps to ferment for a short time and are then pressed. The oil produced by this method is also an edible oil (Hager 1978, volume 7b, p. 197).

If high pressure and high temperatures are applied, the result is an opalescent, dark yellow to brown oil (*Oleum commune*). This oil smells unpleasant and intense (Roth and Kormann 2000, p. 141).

In addition, there is the so-called white olive oil (*Oleum olivarum album*), which has been decoloured using either animal carbon or sunlight (Hunnius 1998, p. 994 et seq; Roth and Kormann 2000, p. 141).

"Watch oil" is made by squeezing the liquid part from solidified olive oil. As the name suggests, it is used to lubricate delicate watches (Roth and Kormann 2000, p. 141).

Olive kernel oil is obtained by cold pressing or hot pressing cleaned olive kernels. Cold-pressed oil is yellow, whereas hot-pressed oil is green (Frohn 1998, p.34 et seq; Kulnik 2003, p. 19 et seq; Roth and Kormann 2000, p. 141; Löw 2003, p. 107).

Categories of Olive Oil

All the sorts and types of olive oil differ in flavour, odour, colour and their amount of free fatty acids. The current EU Regulation (No. 356/92) differentiates between the following categories of olive oil (Kulnik 2003, p. 25; Roth and Kormann 2000, p. 141):

Virgin Olive Oils

Virgin olive oils are produced exclusively by mechanical or other physical processes under conditions, especially with regard to temperature, that do not result in degradation. They may not be mixed with other kinds of oil. This means that the oil may not undergo any other treatment than washing, decanting, centrifugalising and filtrating. Virgin olive oils are categorised into the following grades and names (Kulnik 2003, p. 25; Roth and Kormann 2000, p. 141):

- extra virgin *olive oil:*
  content of free fatty acids: 0.8 g/100 g oil maximum; organoleptic evaluation: 6.5 minimum

- virgin *olive oil:*
  content of free fatty acids: 2 g/100 g oil maximum; organoleptic evaluation: 5.5 minimum
- ordinarly virgin olive oil:
  content of free fatty acids: 3.3 g/100 g oil; maximum; organoleptic evaluation: 3.5 minimum
- lampante virgin olive oil:
  content of free fatty acids: more than 3.3 g/100 g oil; organoleptic evaluation: less than 3.5

Lampnate oil is virgin olive oil that is not suitable for consumption, one reason being that the content of halogenated solvents is too high. Refining lampante oil produces refined olive oil.

Refined Olive Oil

Content of free fatty acids: 0.5 g/100 g oil maximum (Kulnik 2003, p. 25; Roth and Kormann 2000, p. 141).

Olive Oil

Content of free fatty acids: 1.5 g/100 g oil maximum. "Olive oil" is a mixture of refined and virgin olive oil. The composition of fatty acids is the same as in virgin olive oil (Kulnik 2003, p. 25; Roth and Kormann 2000, p. 141).

Crude Olive-Residue Oil

Crude olive-residue oil is produced by treating olive residue with solvents (Kulnik 2003, p. 25; Roth and Kormann 2000, p. 141).

Refined Olive-Residue Oil

Content of free fatty acids: 0.5 g/100 g oil maximum. Refined olive-residue oil is produced by refining crude olive-residue oil (Kulnik 2003, p. 25; Roth and Kormann 2000, p. 141).

Olive-Residue Oil

Content of free fatty acids: 1.5 g/100 g oil maximum. Olive-residue oil is a mixture of refined olive-residue oil and virgin olive oil (Kulnik 2003, p. 25; Roth and Kormann 2000, p. 141).

The grades are designated and controlled using chemical physical and organoleptic analyses. A chemical physical analysis tests 27 parameters, for example the content of free fatty acids, peroxide number and halogenated solvents, to make sure that they are in the range of the maximum and minimum values. During the organoleptic analysis, 8–12 specially trained test persons examine the oil with regard to "flavour" and "off-flavour" attributes, and their intensity. The International Olive Oil Council (IOOC) has proved and tested this method for years, and has distinct parameters at its disposal. The EU adopted this method on the 1st September 2002. The following illustration gives an overview of the sequence of examination and production of the most important olive oil categories:

3 Character

Colour

Intensely yellow.
Yellow to greenish yellow (Hunnius 1998, p. 995).
Yellow to dark yellow (Kerschbaum and Schweiger 2001, p. 21; Löw 2003, p. 107).
Green.

Odour

Fruity, like pyrazine, oily, of olive, strong.

Volatile Compounds

Analysis of the Volatile Compounds of Olive Oil by Means of Liquid Extraction, Liquid and Gas Chromatography

(Del Mar Caja et al. 2000)

Hexanal	2,4-Heptadienal
1-Acetylcyclohexene	Nonanal
Saturated hydrocarbonate	1-Decene
6-Methyl-hept-5-en-2-one	(E)-2-Decenal
(E)-3-Hexen-1-ol	2,4-Decadienal
3- Hexen-1-ol-acetate	(E,E)-2,4-Decadienal
(E,E)-2,4-Heptadienal	(E,E)-α-Farnesene

Analysis of the Volatile Compounds of Olive Oil Using Flavour Extract-Dilution Analysis

(Reiners 1997, p. 23 et seq)

3-/2-Methylbutanal	Nonanal
1-Penten-3-one	Ethylcyclohexancarboxylate
Ethyl-2-methylpropanoate	(Z)-2-Nonenal
Hexanal	(E)-2-Nonenal
(Z)-3-Hexenal	2-Isobutyl-3-methoxypyrazine
(E)-2-Hexenal	(E,E)-2,4-Nonadienal
Ethyl-2-methylbutanoate	(E,Z)-2,4-Decadienal
1-Octen-3-one	(Z)-3-Nonenal
(Z)-1,5-Octadien-3-one	*(E)-β- d*amascenone
Octanal	(E)-2-Octenal
(Z)-3-Hexenylacetate	Acetous acid
Guaiacol	3-/2-Metyl butyric acid

Aroma Analysis of the Most Important Volatile Compounds in Commercially Available Australian Olive Oil with SPME and GC-MS

(Tura et al. 2003)

Ethylisobutyrate	Nonanal
Hexanal	E,Z-Nona-2,4-dienal
E-eex-2-enal	4-Ethylguaiacol
Hexan-1-ol	E,Z-deca-2,4-dienal
Octanal	E-4,5- epoxy- E-dec-2-enal
Hexylacetate	Damascenone
E-oct-2-enal	

Aroma Analysis of the Volatile Compounds in Olive Oil with SPME and GC-MS

(Torres et al. 2005)

n-Butane	Heptanal
n-Pentane	Octanal
n-Hexane	2-Octenal
n-Heptane	Nonanal
n-Octane	2-Nonenal
Ethanol	Decanal
Cyclobutanol	2-Decenal
1-Octanol	2,4-Decadienal
Pentanal	2-Undecenal
Hexanal	2-Pentylfuran
2-Hexanal	2-Octylfuran

Flavour

Characteristic (Kerschbaum and Schweiger 2001, p. 21; Löw 2003, p. 107).

4 Ingredients

Composition of Fatty Acids

Fatty acids	Content in % acc. to Torres et al. (2005)	Content in % acc. to Roth and Kormann (2000), p. 141	Content in % acc. to Kulnik (2003), p. 31	Content in % acc. to Kerschbaum and Schweiger (2001), p. 21
Palmitic acid	16.72	9.4	7.5–20	12.4
Linoleic acid	12.10	–	3.5–21	11.3
Palmitoleic acid	1.87	–	0.3–3.5	0.9
Oleic acid	66.90	83.5	55–83	67.4
α-Linolenic acid	0.91	4.0	0–1.5	0.5
Stearic acid	1.49	–	0.5–5	2.4
Arachidic acid	–	0.9	–	0.3
Eicosenoic acid	–	–	–	0.3
Behenic acid	–	–	–	0.1
Vaccenic acid	–	–	–	3.8
Myristic acid	–	–	–	0.2

Composition of Triglycerides

Triglycerides	Content in % in Spanish olive oil acc. to Karleskind (1996), p. 228	Content in % in Tunisian olive oil acc. to Karleskind (1996), p. 228
LLL	–	0.8
OLL	0.3	5.8
OLLn	0.9	0.6
OOLn	1.0	1.5
PLL	0.5	2.8
POLn	0.3	1.1
OOL	10.4	18.2 (+PoOO)
PoOO	1.1	–
POL	4.5	12.3
PPoO	0.4	1.2
PPL	0.7	2.1
OOO	43.1	21.8
POO	23.1	20.0
PPO	2.9	5.3
PSPo	0.8	–
PPP	0.5	–
SOO	3.6	3.7
PSO	0.4	1.2
PPS	0.6	0.5

Composition of Sterols

(Concentration in mg/100 g and content in %)

Sterols	Olive oil acc. to Firestone (1999), p. 67	Olive oil acc. to Karleskind (1996), p. 230	Virgin olive oil acc. to Kulnik (2003), p. 31	Refined olive oil acc. to Carstensen (2001), p. 42	Olive oil extra vergine acc. to Carstensen (2001), p. 42
Total content	100	98–184	–	168.0	120.4
Cholesterol	0–0.5	<0.3	–	–	0.1
Brassicasterol	0–0.1	–	–	–	0.1
Campesterol	0–4.0	2–4	2–4	3.2	2.4
Stigmasterol	0–4.0	1–2	–	1.9	0.3
β-Sitosterol	75–80	75–90	65–97	89.0	93.3
Δ5-Avenasterol	4–14	3–14	5–31	3.6	2.6
Δ7-Stigmasterol	0–0.5	<0.8	0–0.8	1.1	0.6
Δ7-Avenasterol	–	<0.7	–	0.2	0.1

Composition of Tocopherols

(Concentration in mg/100 g and content in %)

Tocopherols	Olive oil acc. to Karleskind (1996), p. 230	Refined olive oil acc. to Carstensen (2001), p. 44	Olive oil extra vergine acc. to Carstensen (2001), p. 44
Total content	5–15	19.6	30.5
α-Tocopherol	52–87	68.2	74.3
β-Tocopherol	10–25	4.6	18.2
γ-Tocopherol	7–23	11.2	5.3
δ-Tocopherol	–	16.0	2.2

Other Ingredients

(Kerschbaum and Schweiger 2001, p. 43; Kulnik 2003, p. 31; Roth and Kormann 2000, p. 141; Ulmer 1996, p. 67)

Phytosterols
Chlorophyll
β-Carotene
Vitamin A
Vitamin E
Vitamin K
Polyphenols
Lecithin
Carotenoids
Flavonoids
Squalene
Iron
Copper
Lead

Physical Key Figures of Olive Oil

refractive index: n_D^{20} 1.468–1.471
density: d_{15}^{15} 0.914–0.919
saponification number: 187–196
acid number: 2.0 maximum
iodine number: 78–90
hydroxyl value: 4–12
unsaponifiable: 0.5–1.3%
flash point: 225 °C
ignition point: 343 °C
(Roth and Kormann 2000, p. 141)

Shelf Life

If stored in a cool and dark place, the oil has a shelf life of up to 10 months (Roth and Kormann 2000, p. 141).

5 Application

In Antique Medicine

In antiquity, olive oil was a multi-purpose medicine. It was rubbed in, instilled, administered orally and injected. It was used in combination with (Kulnik 2003, p. 7):

- bread, to treat bloody vomiting;
- wine, to treat inflammation;
- vinegar, for wound healing and to treat cramps;
- cold water, to treat fever;
- warm water, to treat consumption;
- salt, to treat sweating, and
- the ashes of different animals, to treat chilblains and burns.

In Modern Medicine

The applications of *Oleum olivarum* in health-care and as a cure are extensive.

It is a mild laxative, which softens solid faeces and simultaneously acts as a lubricant. After hydrolysis, it has a mildly irritant effect on the intestinal mucosa, however, because of the high amount of free fatty acids (Hänsel et al. 1999, p. 248; Hager 1978; volume 7b, p. 199; Frohn 1998, p. 55).

In addition, it is administered as an ancillary substance in cases of ulcers of the stomach and intestines, to decrease the secretion of gastric acid (Hager 1978, volume 7b, p. 199; Kulnik 2003, p. 73; Frohn 1998, p. 55; Ulmer 1996, p. 67).

In emulsions, olive oil can be administered as part of a nitrogen-free diet during the treatment of kidney complaints (Hager 1978, volume 7b, p. 199).

Oleum olivarum is a good cholagogue because it activates cholecystokinin. Cholecystokinin causes the *Sphincter oddi* to remain open longer, so that a larger amount of bile acid is released (Kulnik 2003, p. 73; Frohn 1998, p. 55; Kiritsakis 1990, p. 159, p. 160; Hunnius 1998, p. 995).

Olive oil has antioxidative, immunomodulatory and anticancerogenic properties due to ingredients such as polyphenols, squalene, vitamin E and phytosterols. These ingredients are not only scavengers, but also increase immune competence, protect the cell nucleus from damage, and inhibit mutagenesis and tumour development.

O

For this reason, *Oleum olivarum is* also effective against the human ageing process (Visioli and Galli 2002; Alarcon De la Lastra et al. 2001; Yaqoob 2002; Fritsche and Rijken 2003; Kushi and Giovannucci 2002; Tuck and Hayball 2002; Kris-Etherton et al. 2002; Kulnik 2003, p. 59 et seq, p. 71; Frohn 1998, p. 52; Kiritsakis 1990, p. 162 et seq).

Olive oil is a dietary oil. Many experimental studies have proved that the substitution of fats containing a high amount of saturated fatty acids with oils rich in oleic acid, especially olive oil, lowers the concentration of cholesterol in the serum as effectively as substitution with fats of polyunsaturated fatty acids. Additionally, it lowers the plasma lipids LDL and VLDL, and increases HDL. As a consequence, consuming olive oil has a positive influence on diseases such as arteriosclerosis, diabetes, arthritis and coronary heart diseases (Visioli and Galli 2002; Alarcon De la Lastra et al. 2001; Harwood and Yaqoob 2002; Fritsche and Rijken 2003; Massaro and De Caterina 2001; Kris-Etherton et al. 2002; Kulnik 2003, p. 59 et seq, p. 70; Frohn 1998, p. 55; Kiritsakis 1990, p. 158; Hänsel et al. 1999, p. 248; Ulmer 1996, p. 67). Olive oil is used topically to soothe burns and abrasions, inflamed skin and rheumatic complaints (Hänsel 1999, p. 248; Hager 1978, volume 7b, p. 199).

In Pharmacy

In pharmacy, olive oil is used in the production of liniments, ointments, patches and enemas. It is also an excipient for peroral and percutaneous application, i.e. for the production of oleaginous solutions and suspensions for injections. It is used locally to soften and remove encrustations of eczema and psoriasis, and to clean the skin of adhering ointments (Roth and Kormann 2000, p. 141; Hager 1978, volume 7b p. 199; Hänsel et al. 1999, p. 255; Hunnius 1998, p. 995; Löw 2003, p. 107).

In Cosmetics

Olive oil does not penetrate the skin well, but is nevertheless used in care cosmetics because of its antimicrobial and antiphlogistic properties. It serves as a base of the following products (Hänsel et al. 1999, p. 248; Roth and Kormann 2000, p. 141):

- skin oils;
- hair oils;
- creams, and
- nail-care products.

Technical Uses

Like many other vegetable oils, olive oil is used to produce lubricants and soap, and as lamp oil (Roth and Kormann 2000, p. 141).

In Folk Medicine

As the effects of olive oil have been known since antiquity, its naturopathic applications are extensive. Olive oil is therefore an all-purpose remedy and is thought to help in the treatment of the following complaints (Kircher 2002, p. 56 et seq):

abscesses, arthritis, arthrosis, bruises, pains in the elbow, frostbite, brittle fingernails, cold feet, inflammation of the joints, gout, shingles, lumbago, itching, varicose veins, tense muscles, contusions, rheumatism, bursitis, phlebitis, sprains, constipation and many more (Ulmer 1996, p. 67 et seqq; Kircher 2002, p. 56 et seq).

As Food

Oleum olivarum is an excellent edible oil. It can be used to season salads, but it is also one of the few vegetable oils that may be heated and is therefore suitable for baking, frying and roasting. The food industry uses it e.g. as a marinade for sardines (Roth and Kormann 2000, p. 141; Hänsel et al. 1999, p. 255; Löw 2003, p. 107).

Oiticica Oil

Oleum Licaniae synonyms: Oiticicaöl (D); huile d'oiticica (F)

1 Source Plant

Licania rigida Benth., (Chrysobalanaceae), oiticica

Habitat

The oiticica originates in Brazil and southern Central America. It is mainly cultivated in Ceará, Rio Grande do Norte, Paraíba and Piauí (Andrews 1972).

Description

The oiticica is about 30 m tall, evergreen and densely leaved, with a very broad crown. The inflorescences are arranged in panicles. The fruit are 3–6 cm long and have a ligneous, brittle shell; they contain a dark brown seed (Salunkhe and Desai 1986, p. 206).

2 Cultivation and Extraction

Cultivation

There are many wild-growing oiticica trees in Brazil, which is why the tree is not often cultivated. It prefers warm climes and much sun. The tree has an extensive root system and thus survives long droughts. The fruit ripen between December and

S. Krist, *Vegetable Fats and Oils*, https://doi.org/10.1007/978-3-030-30314-3_80

March and then fall to the ground. They are gathered by workers (Salunkhe and Desai 1986, p. 206).

Extraction of the Oil

Oiticica oil is obtained from the seeds of *Licania rigida Benth.* by two methods (Salunkhe and Desai 1986, p. 207):

Hot-Water Flotation

The seeds are ground in a mortar. Water is then added, and the mixture is heated. The boiling water causes the oil to rise to the surface, from where it is decanted. This method is primitive and seldom applied today.

Pressing

The shell, which contains chlorophyll, is removed from the seeds. The seeds are either pressed in a hydraulic press or a continuous press, or the oil is extracted from them with a mixture of alcohol and petroleum ether or trichloroethylene.

The oil undergoes autoxidation when exposed to air and then solidifies into a greasy mass. For transport, the oil is heated and therefore stays liquid. At temperatures over 280 °C, it gelatinises (Roth and Kormann 2000, p. 140).

3 Character

Colour

Yellowish (Roth and Kormann 2000, p. 140)
Fresh oil: yellowish.
After oxidation: brownish.

Odour

Unpleasant smell (Roth and Kormann 2000, p. 140).

Flavour

Unpleasant taste (Roth and Kormann 2000, p. 140).

4 Ingredients

Composition of Fatty Acids

Fatty acids	Content in % acc. to Roth and Kormann (2000), p. 140	Content in % acc. to List and Hörhammer (1976), 4th edition, 5th volume, p. 507
Licanoic acid	73–83	70–82
Oleic acid	4–6	4–12
Palmitic and stearic acid	10–11	10–11
Isolicanoic acid	–	Traces
Myricetin	–	Traces
Linoleic acid	–	4

Physical Key Figures of Oiticica Oil

refractive index: n_D^{40} 1.5050–1.5140
density: d_{20}^{20} 0.95–0.97
melting point: 15 °C
saponification number: 186–195
iodine number: 139–185
unsaponifiable: 0.5–1%
(Roth and Kormann 2000, p. 140)

5 Application

In Folk Medicine

In Brazil, oiticica oil is used externally as a liniment to treat rheumatism. So far, this application has not been clinically examined, and there are no studies available (Franke 1989, 4th edition, p. 168).

In Cosmetics

Oiticica oil would be suitable for soap production, but it is not used for this purpose because of its unpleasant smell (Roth and Kormann 2000, p. 140).

As Food

Oiticica oil does not have a pleasant taste and is not used as an edible oil (Roth and Kormann 2000, p. 140).

Industrial Uses

As oiticica oil is a matt oil that dries quickly, it is used in the paint and varnish industry. It is often mixed with linseed oil and sesame oil to achieve even better results. These coatings dry very quickly, leaving a resistant, insoluble, waterproof, elastic film. As the colour of oiticica oil changes to brown when the oil comes into contact with air, the oil is no substitute for colourless, lightfast or white paint (Andrews 1972).

Mixed with castor oil, oiticica oil is used to seal floors. The floors are made shiny and resistant to wear caused by friction. As oiticica oil is waterproof, it prevents moisture expansion of parquet floors (Andrews 1972).

The waterproof properties of oiticaca oils are also used in the production of wood fibreboard, to protect wood fibreboard and particle board from water. Wood fibreboard that has been impregnated with oiticica oil is water-resistant and does not need to be impregnated frequently. In addition, the wood fibreboard becomes more flexible without its rigidity being impaired (Andrews 1972).

6 Possible Unwanted Side Effects

Oiticica oil is not ingested because of its unpleasant taste; no studies are therefore available on possible side effects. Deaths of animals that have eaten the press cake of oiticica oil as fodder have been reported. Analyses testing for a content of alkaloids, derivates of hydrocyanic acid and toxic protein had negative results. It was thus assumed that the toxic effects were due to the seeds having been erroneously mixed with seeds of *Ricinus communis* (De Souza 1946).

Ootanga Oil

Synonyms: Oontangaöl, Wassermelonenkernöl (D); karingda oil, watermelon seed oil (E); huile de graine de pastèque (F)

1 Source Plant

Citrulluslanatus var. vulgaris Schrad. Mansf. (Cucurbitaceae), *Citrullus lanatus* (Thunb.) Matsum. et Nakai *var. lanatus., watermelon*

S. Krist, *Vegetable Fats and Oils*, https://doi.org/10.1007/978-3-030-30314-3_81

Habitat

Three species of *Citrullus* are native plants of Africa and Asia, but only *Citrullus vulgaris* and *Citrullus lanatus* are used for oil production.

2 Description

Citrullus lanatus is an annual plant with stems that can reach a length of 10 m; they are either creeping or they lie on the ground, and have tendrils. The hairy leaves are 5–20 × 3–19 cm large; they are palmately compound, with 3–5 lobes on stems that are 2–19 cm long. From December to May, the yellow female and male flowers appear. The fruit are round melons with a diameter of up to 20 cm. Their peel is greenish yellow, with longitudinal stripes of a darker green colour. The fruit pulp is usually light yellow and contains many seeds, which may be blue, light red, grey or white.

3 Cultivation and Extraction

Cultivation

Watermelons are cultivated in Asia, the former USSR and Europe mainly because of their fruit. The kernels contain 52–57% oil and 34% protein (Salunkhe et al. 2003, p. 460).

Extraction of the Oil

To extract the oil, the fruit pulp is removed; the seeds are dried, unshelled, reduced to small pieces and extracted with hexane (Das et al. 2002).

4 Character

Colour

Yellowish to greenish semi-drying oil (Salunkhe et al. 2003, p. 460).
Light-yellow oil (Das et al. 2002).
Golden-yellow oil, liquid at room temperature (Onyeike and Acheru 2002).

Odour

Pleasant (Salunkhe et al. 2003, p. 460).

Flavour

Mild, fruity, seed-like.

5 Ingredients

The seed oil of *Citrullus lanatus* consists of about 0.27% hydrocarbonates, 94.90% triglycerides, 1.41% free fatty acids, 1.12% sterols, 0.35% diglycerides, 0.98% monoglycerides and 0.96% phospholipids (El-Adawy and Taha 2001).

Composition of Fatty Acids

Fatty acids	Content in % acc. to El-Adawy and Taha (2001)	Content in % acc. to Badifu (1993)	Content in % acc. to Salunkhe et al. (2003)	Content in % acc. to Firestone (1999), p. 108
Myristic acid	0.11	–	–	–
Palmitic acid	11.30	10.06	11–19	9–11
Palmitoleic acid	0.29	–	–	0.5–1
Stearic acid	10.24	9.37	13–17	5–10
Oleic acid	18.07	15.55	14–22	13–19
Linoleic acid	59.64	65.03	52–58	62–71
α-Linoleic acid	0.35	–	–	–
Arachidic acid	–	–	–	0–1
Unsat. fatty acids in total	78.35	–	–	–
Sat. fatty acids in total	21.65	–	–	–
Monounsat. fatty acids	18.36	–	–	–
Polyunsat. fatty acids	59.99	–	–	–

Composition of Tocopherols

Ootanga oil has a vitamin E content of 0.1 mg/100 g oil.

Other Ingredients

Vitamin B1: 0.05 mg/100 g
Vitamin B2: 0.03 mg/100 g
Vitamin C: 25 mg/100 g
Iron: 0.352 mg/100 g
Copper: 0.158 mg/100 g
Chloride: 197 mg/100 g

Zinc: 0.474 mg/100 g
Potassium: 16.5 mg/100 g
Sodium: 1.2 mg/100 g
Phosphate: 42.61 mg/100 g
Sulphate: 0.018 mg/100 g

Physical Key Figures of Ootanga Oil

density: d^{25} 0.919 ± 0.007
refractive index: n_D^{25} 1.4696 ± 0.001
saponification number: 201 ± 2.25
iodine number: 115 ± 3.36
acid number: 2.82 ± 0.15
peroxide number: 3.40 ± 0.38
ester value: 193.58 ± 3.15
free fatty acids (% as oleic acid): 1.41 ± 0.11 (El-Adawy and Taha 2001)

6 Application

In Cosmetics

The oil can be added to sunscreens and after sun lotions as well as to products for damaged, irritated and sensitive skin.

To examine the stability of cosmetic emulsions, cosmetic oil-in-water emulsions with ootanga oil were produced. After determining the physicochemical parameters, the researchers carried out stability tests. The emulsions were stable in freeze-thaw tests, centrifugation at 5000 r/min over a period of 30 min, cyclic changes of temperature and storage at room temperature for 36 months. The results show that emulsions from ootanga oil are as stable as emulsions from paraffin oil (Oyedeji and Oderinde 2004).

As Food

Since ootanga oil contains a high amount of mineral nutrients, it is recommended as a salad oil and as an edible oil in a diet for patients with a low level of mineral nutrients (Onyeike and Acheru 2002).

Other Uses

O

Ootanga oil is used as a lamp oil (Salunkhe et al. 2003), for soap production and as a lubricant.

Palm Kernel Oil/Palm Oil

Oleum guineensis synonyms: palm kernel oil: Palmkernöl (D); huile de palmiste (F). Palm oil: Palmöl (D); huile de palme (F)

1 Source Plant

Elaeis guineensis Jacq. *(Arecaceae)*, African oil palm

S. Krist, *Vegetable Fats and Oils*, https://doi.org/10.1007/978-3-030-30314-3_82

Habitat

The African oil palm and its fruit are the source of two different oils: palm oil, which is obtained from the mesocarp of the fruit, and palm kernel oil, which is extracted from the seeds (Salunkhe et al. 1992, p. 217).

The African oil palm has its origins in tropical Africa and Angola (Roth and Kormann 2000, p. 61). Increasing use of vegetable fats and increasing demand led to an increase in cultivation, so that today, the plant is cultivated in tropical zones all over the world. The main growing areas are in South-East Asia, Africa and South America. The most important producers are Malaysia and Indonesia with a share of about 50% and 30% respectively of the global production.

Description

The African oil palm can grow to a height of 30 m. It has enormous leaves, a dense crown and a stem that can measure up to 50 cm in thickness. The crown consists of about 40 leaves that are up to 7.5 m long. The leaf stem measures up to 1 m and is thorned on its upper end; the thorns are about 4 cm long. The male and female inflorescences develop in periodical alternation. The fruit are arranged in infructescences that can weigh up to 25 kg and consist of 1000–4000 fruit (Roth and Kormann 2000, p. 62). The individual fruit are 2.5–5 cm long and 2.5 cm in diameter; they weigh between 3 and 30 g. The epicarp is thin and reddish-orange in colour. Depending on the species, the colour can vary from yellow to orange, red, brown or black. The fibrous, oily mesocarp is orange or reddish brown. The seed of the African oil palm is a nut, consisting of a shell or endocarp, and one or more kernels. The kernels contain layers of hard, oily endosperm and an embryo (Salunkhe et al. 1992, p. 219). The fruit's mesocarp contains 45–60% oil (palm oil), and the seeds 43–52% oil (palm kernel oil) (Roth and Kormann 2000, p. 62).

2 Cultivation and Extraction

Cultivation

The ideal growing areas lie between the 10th degrees of northern and southern latitude at altitudes of up to 700 m (Roth and Kormann 2000, p. 62). The annual mean temperature should lie between 24 °C and 27 °C, and the annual precipitation should amount to 2000–3000 mm. Continuous changes between sunny and rainy weather are important for the development of the flowers and fruit. Deep, slightly acidic, friable soil rich in humus is ideal for the development of the root. For propagation, the pulp is removed from the seeds, which are kept in boxes at temperatures between 37 °C and 40 °C. As soon as the plugs have developed 2–3 green leaves, they are planted into beds in sunny locations. The plants are bedded out after 10–14 months, and after about 4 years, the palm begins its vertical growth and bears fruit for the first time. Depending on the geographical position, the fruit develop within 5–8 months. The greatest yields occur between the 12th and 20th year. The average annual yield amounts to about 120 kg per plant (Roth and Kormann 2000, p. 62). The large corporations producing palm oil employ practices harmful to the environment, which have been criticised repeatedly by environmental organisations. Meanwhile, there has been a partial conversion to more environment-friendly, sustainable palm oil production, for example in southern Thailand (Stadlober 2012, p. 9)

Extraction of the Oil

Palm kernel oil is a by-product of the production of palm oil. The fruit are pressed in mills. Palm oil is obtained from the fleshy mesocarp, but the nuts endure the pressure without breaking. They are cleaned to remove the fibres of the mesocarp, unshelled and dried until their moisture content is reduced to 8%; this last step is necessary to prevent mould growth. The palm kernels are then pressed in screw presses to obtain palm kernel oil and palm kernel meal. To increase the oil yield, the press cake can be extracted with solvents afterwards. This process is, however, becoming increasingly obsolete, as it does not appear to be economically worthwhile (Pantzaris and Basiron 2002, p. 174).

Unrefined palm oil is also called red palm oil. The red colour is caused by an unusually high amount of carotinoids.

3 Character

Colour

Palm Kernel Oil

White, slightly yellowish.
Nearly colourless (Salunkhe et al. 1992, p. 229).

Palm Oil

Buttery, dark-yellow to reddish-yellow fat (Roth and Kormann 2000, p. 142).

Odour

Palm Kernel Oil

Herbaceous, similar to eucalyptus, slightly green, aromatic, slightly fruity, sweet, touch of citrus.
Nearly odourless (Roth and Kormann 2000, p. 142).

P

Palm Oil

Weak, fresh, slightly green.

4 Ingredients

Composition of Fatty Acids

Palm Kernel Oil

Fatty acids	Content in % acc. to Karleskind (1996), p. 203	Content in % acc. to Firestone (1999), p. 69	Content in % acc. to Hamilton (1995), p. 165	Content in % acc. to nature.de/ 24.04.06
Caproic acid	<0.8	0–0.8	0.3	–
Caprylic acid	2–5	1.9–6.2	4.4	2.7
Capric acid	3–5	2.6–5.0	3.7	3.1
Lauric acid	44–51	40–55	48.3	46.9
Myristic acid	15–17	14–18	15.6	16.4
Pentadecanoic acid	–	–	–	–
Palmitic acid	7–10	6.5–10.3	7.8	8.9
Palmitoleic acid	<0.1	–	–	–
Margaric acid	–	–	–	–
Stearic acid	2–3	1.3–3	2.0	2.1
Oleic acid	12–18	12–21	15.1	16.6
Linoleic acid	1–4	1–3.5	2.7	2.9
α-Linolenic acid	<0.7	0–0.7	–	–
Arachidic acid	<0.3	0–0.3	–	–
Eicosanoic acid	<0.5	0–0.5	–	–

The table shows that palm kernel oil is highly saturated, and lauric acid has an average share of 50%.

Palm Oil

Fatty acids	Content in % acc. to Hui (1996), p. 24 et seq	Content in % of Malaysian palm oil acc. to Hamilton (1995), p. 161
Lauric acid	0.3	0.1–1.0
Myristic acid	1.1	0.9–1.5
Palmitic acid	45.1	41.8–46.8
Palmitoleic acid	0.1	0.1–0.3
Stearic acid	4.7	4.2–5.1
Oleic acid	38.8	37.3–40.8
Linoleic acid	9.4	9.1–11
α-Linolenic acid	0.3	0–0.6
Arachidic acid	0.2	0.2–0.7

The table above illustrates that palm oil is especially rich in palmitic acid (16:0) and oleic acid (18:1).

Composition of Triglycerides

Palm Kernel Oil

Chain length of triglycerides	Content in % acc. to Siew (2001)	Chain length of triglycerides	Content in % acc. to Siew (2001)
C26	1.07	C42:2	0.84
C28	0.64	C42:1	5.12
C30	1.41	C42	4.11
C32	6.11	C44:1	3.44
C34:1	0.54	C44	4,39
C34	8.36	C46:2	2.38
C36	21.35	C46:1	1.78
C38:2	0.62	C46	2.01
C38:1	0.8	C48:3	1.92
C38	15.23	C48:2	1.78
C40:2	0.88	C48:1	0.97
C40:1	4.55	C48	0.24
C40	8.64	C50:2	0.47
		C50:1	0.34

Palm Oil

Chain length of triglycerides	Content in % acc. to Karleskind (1996), p. 239
C46	<1
C48	4–11
C50	40–4
C52	38–43
C54	6–11

Composition of Sterols

Palm Kernel Oil

(concentration in mg/100 g and content in %)

Sterols	Palm kernel oil acc. to Karleskind (1996), p. 203	Palm kernel oil acc. to Firestone (1999), p. 69
Total content	60–120	79–141
Cholesterol	1–2	0.6–3.7
Brassicasterol	–	0–0.8
Campesterol	9–10	8.4–12.7
Stigmasterol	12–16	12.0–16.6
β-Sitosterol	65–73	62.6–73.1
Δ5-Avenasterol	4–7	1.4–9.0
Δ7-Stigmasterol	<0.2	0–2.1
Δ7-Avenasterol	<0.2	0–1.4

P

Palm Oil

(concentration in mg/100 g and content in %)

Sterols	Palm oil acc. to Karleskind (1996), p. 240	Palm oil acc. to Firestone (1999), p. 72
Total content	40–90	36.2–62.7
Cholesterol	<3	2.6–6.7
Campesterol	19–21	18.7–27.5
Stigmasterol	11–13	8.5–13.9
β-Sitosterol	56–59	50.2–62.1
Δ5-Avenasterol	2–3	0–2.8
Δ7-Stigmasterol	1–2	0.2–2.4
Δ7-Avenasterol	1–2	0–5.1
Ergosterol	1–2	–
Fucosterol	<1	–

Composition of Tocopherols and Tocotrienols

Palm Kernel Oil

(concentration in mg/100 g and content in %)

Tocopherols/ tocotrienols	Palm kernel oil acc. to Karleskind (1996), p. 203	Palm kernel oil acc. to Pantzaris and Basiron (2002), p. 179	Palm kernel oil acc. to Firestone (1999), p. 69
Total content	8–10	31.9	0–26
α-Tocopherol	1–10	2.2	0–4
β-Tocopherol	1–30	21.0	0–25
γ-Tocopherol	1–10	8.7	0–26
Tocotrienols	1–40	–	–
α-Tocotrienol	–	–	0–6
γ-Tocotrienol	–	2.3	–

Palm Oil

(concentration in mg/100 g and content in %)

Tocopherols/ tocotrienols	Palm oil acc. to Karleskind (1996), p. 240	Palm oil acc. to Firestone (1999), p. 72
Total content	32–100	–
α-Tocopherol	16–20	0.4–19.3
β-Tocopherol	<1	0–23.4
γ-Tocopherol	<1	0–52.6
Δ-Tocopherol	<1	0–12.3
Tocotrienols	80–83	9.8–150
α-Tocotrienol	–	0.4–33.6
γ-Tocotrienol	–	1.4–71
Δ-Tocotrienol	42–45	0–37.7

Other Ingredients

Palm Oil

Carbon hydrides: 100–110 mg/100 g
Carotene: 50–70 mg/100 g

Physical Key Figures of the Oil

Palm Kernel Oil

refractive index: 1.448–1.452
density (40/20 °C): 0.899–0.94
iodine number: 14.1–21
saponification number: 230–234
melting point: 23–30 °C
unsaponifiable: 0.1–0.8%
acid number: 0.14
peroxide number: 0.17

Palm Oil

refractive index: 1.449–1.456
density (50/20 °C): 0.891–0.899
iodine number: 49–55
saponification number: 190–209
melting point: 33–40 °C
unsaponifiable: 0–1.2%
(Firestone 1999, p. 71)

5 Application

Palm kernel oil:

In Pharmacy and Medicine

In the pharmaceutical industry, palm kernel oil serves as a base for various creams (Salunkhe et al. 1992, p. 241).

Industrial Uses

In Food Production

In the food industry, palm kernel oil is often used in margarine production. The relatively high amount of short-chain fatty acids contributes to a taste similar to butter (Pantzaris and Basiron 2002, p. 196). At body temperature, palm kernel oil melts quickly and has a pleasant cooling effect. It is therefore often added to chocolate icing, ice cups, cream coatings and cool-melting chocolate fillings. The oil is also used for the production of toffees and caramels, since it contributes to hardness and flavour (Pantzaris and Basiron 2002, p. 196 et seq). By modifying palm kernel oil, various high-quality special fats are produced for the sweets industry.

In Oleochemistry

Industrially, palm kernel oil is used similarly to palm oil. To a great extent, it is used to produce oleochemical interstage products for the cosmetics industry and the detergent industry.

In the Aluminium Industry

Palm kernel oil is used as grease during rolling and deep-drawing processes for aluminium sheeting.

Palm oil:

Industrial Uses

In industry, palm oil is used to produce soap and candles and to obtain lauric acid, which is a raw material for various tensides (sodium lauryl sulphate, sorbitan monolaureate, etc.) (Roth and Kormann 2000, p. 143).

Recently, palm oil has also been used in the production of bio fuel. The production of this allegedly environment-friendly product has, however, led to massive ecological destruction in Indonesia.

As Food

As a result of its excellent stability against heat and oxidation, palm oil as an edible oil is well-suited for frying, cooking and roasting.

It is also used in the food industry for baking and in the production of margarine and confectionery (Karleskind 1996, p. 242).

Possible Unwanted Side Effects

Cultivation of oil palms on an industrial scale not only poses a major problem for the environment in the main countries of production, the consumption of palm oil also gives rise to health concerns. The European Food Safety Authority (EFSA) warns against so-called process contaminants in food containing refined palm oil. According to EFSA, process contaminants on a glycerine base, which occur in palm oil, but also in other vegetable oils, margarines and processed food, cause health concerns for young consumers (baby food) who partake in average amounts of these foodstuffs, as well as for consumers of all ages if large amounts are consumed (www.efsa.europa.eu/de/press/news/160503-0, consulted 14.4.2019).

Process contaminants, such as glycidyl-fatty acid ester (GE), 3-monochloropropanediol (3-MCPD) and 2-monochloropropanediol (2-MCPD) as well as the esters of their fatty acids are created during food processing, especially during the process of refining vegetable oils at high temperatures (about 200 °C). In the body, glycidyl-fatty acid esters are broken up, and glycidol is released. This substance is probably carcinogenic and mutagenic (www.ages.at/themen/rueckstaende-kontaminanten/mcpds-glycidyl-ester/, consulted 14.4.2019). According to a risk study of EFSA, palm oil/palm fat contains a medium content of 3.955 μg/kg glycidol (in comparison, sunflower oil contains 269 μg/kg) (EFSA Panel on Contaminants in the Food Chain (CONTAM) 2016).

An examination of the EFSA panel showed that GE content in palm oil and palm fat was reduced by half between 2010 and 2015, which is the result of measures taken voluntarily by the producers. This contributed to a reduction of consumer exposure to these substances to a considerable degree (http://www.efsa.eurpoa.eu/de/press/news/160503-0, consulted 14.4.2019). The EU passed a regulation for the step-by-step reduction of contamination; the maximum values for glycidyl-fatty acid esters are included in the regulation (EU) 2018/290 of the commission of February 26th, 2018, amending the regulation (EC) No. 1881/2006 regarding the maximum content of glycidyl-fatty acid esters in vegetable oils and fats, formulations for babies and food for special medical puropses for babies and small children (eur.lex.europa.eu/legal-content/DE/TXT/?uri=CELEX:32018R0290, consulted 14.4.2019).

Papaya Oil

synonyms: Papayasamenöl (D); huile de papye (F), olio (di semi) di papaya (I), aceite de papaya (ESP)

1 Source Plant

Carica Papaya L. (Caricaceae), Papaya

S. Krist, *Vegetable Fats and Oils*, https://doi.org/10.1007/978-3-030-30314-3_83

P

Habitat

Carica papaya originates in Central America, where it is still widespread from the Yucatán peninsula to Costa Rica. Via Malaysia and India, the papaya reached the southern Pacific in the eighteenth century (Lieberei and Reisdorff 2007, p. 172–173).

Description

The papaya's appearance is reminiscient of a palm, but the plant actually belongs to the *Caricaceae* family. The stem is up to 9 m tall and up to 30 cm broad at the base, tapering towards the top, where the deeply indented, palmate leaves sit. They are about 1 m in breadth and short-lived, leaving a triangular scar on the shoot after they have fallen off. The papaya is not a tree, as the sprout does not lignify except for the tracheae. It is a perennial plant that could be described as an arborescent herb. Papayas are dioecious plants that are pollinated by insects. The flowers are white or yellow and appear at the axillary buds. The ovary is superior and consists of 3–5 conjoined carpels. The fruit is a pear-shaped berry; several fruit are arranged in clusters underneath the leaves. Beneath the leathery skin, there is 1.5–4 cm of fruit pulp of an orange or yellow colour. In the middle lies an amniotic cavity with up to 1000 black seeds (Lieberei and Reisdorff 2007, p. 173).

2 Cultivation and Extraction

Cultivation

The native people of Central and South America already cultivated *Carica papaya* before the arrival of Columbus; it is now planted in tropical and subtropical regions worldwide in many different varieties. As *Carica papaya* is a tropical plant, it needs temperatures of 21–33 °C for growth. Even a few hours of temperatures below 12 °C cause damage. On the other hand, the papaya's water requirements are not high for a tropical plant; about 100 mm precipitation per month is sufficient. Neither does the papaya make any great demands on the soil; only stagnant moisture is a problem. Papayas live for about 15 years, but cultivated plants are removed after 4–5 years because the yield drops and the harvest becomes too costly. Papayas develop fruit throughout the year, the yield amounting to about 4–40 t/ha annually (Lieberei and Reisdorff 2007, p. 173–174).

Most important countries of production (Lieberei and Reisdorff (2007), p. 174)

2004	Yield (tons)
World	6,800,000
Brazil	1,612,000
Mexico	955,000

Most important countries of production (Lieberei and Reisdorff (2007), p. 174)	
2004	Yield (tons)
Nigeria	755,000
Indonesia	732,000
India	700,000
Ethiopia	230,000
Peru	195,000
China	154,000
Philippines	132,000
Venezuela	132,000

Extraction of the Oil

To extract the oil, the seeds are washed and air-dried for 10 days. In order to remove the rough shell (sarcotesta) from the endosperm, they are soaked in water overnight. Subsequently, they are dried again, then ground, and the oil is finally extracted using a Soxhlet extractor. If the seeds, sarcotesta and endosperm are separated into individual fractions, the oil yield is as follows:

seeds:	28% oil
endosperm:	54% oil
sarcotesta:	7% oil

(Nguyen and Tarandjiiska 1995)
The seeds can also be pressed (Lee et al. 2011).

3 Character

Colour

Yellow (Lee et al. 2011).
Slightly green if extracted with ether (Chan and Heu 1978).

Odour

Papaya oil is pungently caustic, similar to cress (Lee et al. 2011).

4 Ingredients

Composition of Fatty Acids

Fatty acids	Content in % acc. to Badami and Daulatabad (1967)	Content in % acc. to Passera and Spettoli (1981)	Content in % acc. to Subrahmanyam and Achaya (1957)	Content in % acc. to Chan and Heu (1978)	Content in % acc. to Puangsri et al. (2005)
Lauric acid	0.4	–	–	0.13	–
Myristic acid	0.4	–	0.2–4.6	0.16	0.1–0.2
Palmitic acid	16.2	15.10	12.0–17.2	15.13	12.8–13.9
Palmitoleic acid	0.8	–	1.3	–	0.2–2.1
Stearic acid	5.0	5.00	1.8–5.5	3.61	4.4–4.9
Oleic acid	74.3	76.30	77.3–80.6	71.60	75.9–76.8
Linoleic acid	0.4	3.40	0.4–2.2	7.68	3.0–3.3
Linolenic acid	–	–	–	0.60	0.1–0.2
Arachidic acid	0.9	–	0.3	0.87	0.4
20:1	–	–	–	–	0.3
Behenic acid	1.6	–	–	0.22	–

The composition of fatty acids of papaya oil is comparable to that of olive oil. The content of oleic acid is similarly high, but papaya oil contains less linoleic acid and linolenic acid than olive oil. This results in higher stability against oxidation processes (Nguyen and Tarandjiiska 1995).

Composition of Triglycerides

Triglycerides	Content in % acc. to Puangsri et al. (2005)	Content in % acc. to Lee et al. (2011)
LLL	–	0.02
OLL	00.3–0.7	0.03
PLL	–	0.03
OLO	–	2.54
PLO	–	1.72
PLP	–	0.18
OOL	3.7–4.3	–
POL	2.3–2.7	–
OOO	43.6–44.6	43.77
POO+SOL	30.0–30.5	–
POO+SLO	–	33.83
POP	–	6.19
OGO	–	0.42
PPO	5.1–5.6	–
SOO	9.8–11.0	8.37
POS	–	2.41
SOS	–	0.20
SSP	–	0.27
PSO	3.1–3.8	–

Triglyceride groups	Content in % acc. to Nguyen and Tarandjiiska (1995)
S_2M	16.1
SM_2	38.0
S_2D	34.5
M_3	34.5
SMD	4.3
M_2D	5.2
SD_2	0.4
MD_2	0.6
SMT	0.3
M_2T	0.5

Composition of Sterols

Sterols	Papaya oil in g/kg^{-1} acc. to Lee et al. (2011)
Total content	6.5–6.8
Campesterol	0.8–1.0
Stigmasterol	0.4–0.6
β-Sitosterol	5.1–5.4

Physical Key Figures: (Chan and Heu 1978)

refractive index (40 °C):	1.4627
density (25 °C):	0.9130
unsaponifiable:	2.11%
saponification number:	193.4
iodine number:	74.77
viscosity (centipoise):	339.41

5 Application

In Pharmacy and Medicine

A current study by Afolabi et al. examined the effect of papaya oil on rats. Rats that received a diet containing 10% papaya oil showed an increase in brain weight but a decrease in body weight. In addition, there was an increase in reduced glutathione in the erythrocytes. A decrease in cholesterol and phospholipids was registered in the brain. According to the study by Afolabi et al., the oil may have a damaging effect on the brain. It should not therefore be used as food, but as a medicine, so that the intake can be controlled (Afolabi et al. 2011).

Papaya oil may contain 4.0–23.3 mg/g^{-1} benzylisothiocyanate, which is anticarcinogenic. According to Lee et al., it is a potential inhibitor of carcinogens of the breast, lungs and liver. Further studies are necessary to discover possible potential risks of using papaya oil (Lee et al. 2011).

P

6 Possible Unwanted Side Effects

Papaya seeds contain about 0.65% benzylisothiocyanate in the oil and about 1.08% benzylglucosinolates in the defatted flour (Nguyen and Tarandjiiska 1995). Subacute oral toxicity of benzylisothiocyanate is under discussion (Lewerenz et al. 1992).

The benzylisothiocyanate contained in papaya oil is produced by hydrolysing benzylglucosinolate, which is found in the endosperm. This is carried out using the enzyme thioglucosidase, which, according to Chan et al., may be removed or inactivated by thorough washing or heating (Chan and Heu 1978).

Parsley Seed Oil

Oleum Petroselini synonym: Petersiliensamenöl (D)

1 Source Plant

Petroselinum crispum (Mill.) *Nyman ex* A.W. Hill *(Apiaceae)*, parsley

S. Krist, *Vegetable Fats and Oils*, https://doi.org/10.1007/978-3-030-30314-3_84

Habitat

Parsley grows in the whole Mediterranean area, especially on stony ground near water. In its wild form, it is common throughout Europe. It is also found in Iceland, Norway and Greenland. Parsley was imported to North and South America, South Africa, India, Japan, Australia and New Zealand, where it also grows wild. The plant has been a valued medical plant since antiquity. The Romans brought it to Europe. It was common in monastery gardens throughout Europe, from where it spread, but it is rarely cultivated on a large scale. It was only in the twentieth century that larger areas were planted with parsley to obtain the seeds (Schuster 1992, p. 181).

Description

Parsley is a biennial or perennial herbaceous plant. The flowers and fruit only appear in the second year, and the plant may then continue to develop flowers and fruit for several years. Parsley has a long, spindle-shaped, fairly branched taproot, which is either thin and hard, or fleshy and thickened. The shoot appears in the second year and usually consists of several stems. The stem is round, narrowly grooved and often hollow; it can reach a height of 30 to 120 cm and is branched from the middle. There are three main varieties of leaves: flat-leaf, curly-leaf and fine-leaf parsley. The leaves are dark green, shiny and triangular, either divided several times to form three-toothed segments or pinnate. The basal leaves and lower leaves of the stem are stalked; the upper leaves are stalkless. Heterophila is quite common. The flowers are arranged in 10–20-rayed, terminal umbels on long stems. The plant develops 1–3 lanceolate bracts. The individual flowers are either entire or male. The petals are small, approximately 0.7 mm long and greenish yellow. The fruit is broad, oval to heart-shaped, with thin longitudinal ribs. It is greyish brown, 2.5–3.0 mm long and 2 mm thick. The fruitlets are flat on the inside, with a heavily bent back (Schuster 1992, p. 180).

2 Cultivation and Extraction

Cultivation

Parsley is undemanding as far as the soil is concerned: It grows well in moist soil and sunny locations, but does not tolerate acidic soil. The seeds should be sown between April and July. They germinate about 3 weeks later. The flowers appear from the June to the August of the following year; the fruit ripen between July and September.

Extraction of the Oil

Parsley seed oil is obtained from the fruit of *Petroselinum crispum* either by cold pressing or by extraction with petroleum ether (Roth and Kormann 2000, p. 144; Young-Hoi et al. 1990).

3 Character

Colour

Reddish-brown to greenish-brown (Roth and Kormann 2000, p. 144).

Odour

Characteristic smell of parsley (Young-Hoi et al. 1990).

Volatile Compounds

(Young-Hoi et al. 1990)

Components	Content in %	Components	Content in %
α-Thujene	0.19	β-Phellandrene	7.52
α-Pinene	22.28	Limonene	0.10
Camphene	0.13	γ-Terpinene	0.10
Sabinene	0.87	Terpinolene	0.14
β-Pinene	16.20	δ-Cadinene	0.35
Myrcene	0.54	p-1,3,8-Menthatriene	0.28
α-Phellandrene	0.28	p-*M*ethylacetophenone	0.10
p-Cymene	0.11	β-Cymen-8-ol	0.09
α-Terpineol	0.07	β-Farnesene	0.08
Myristicin	47.54	β-Bisabolene	0.07
Elemicin	2.64	Apiols	0.07

Flavour

Characteristic taste of parsley (Young-Hoi et al. 1990).

P

4 Ingredients

Composition of Fatty Acids

Fatty acids	Content in % acc. to Roth and Kormann (2000), p. 144	Content in % acc. to Firestone (1999), p. 73 et seq	Content in % acc. to Schuster (1992), p. 180
Palmitic acid	3	2	–
Stearic acid	–	1	–
Oleic acid	15	12–15	8–10
Petroselinic acid	76	69–76	68–72
Linoleic acid	6	6–14	16–20
Others	1	–	4–8

Physical Key Figures of Parsley Seed Oil

refractive index: n_D^{40} 1.4778–1.4800
density: d_{15}^{15} 0.924
melting point: −15 °C
saponification number: 170–177
iodine number: 109
unsaponifiable: 2.2%

(Roth and Kormann 2000, p. 144)

Characteristics of the Oil

Apiol is a natural ingredient of the essential oil of *Petroselinum crispum*. The plant has different chemical races; there is a myristicin variety with a myristicin content of up to 77% (apiol content 0–3%), an apiol variety with an apiol content of 58–80% (myristicin content 9–30%), and an allyltetramethoxybenzole variety that only contains traces of parsley apiol. Apiol and myristicin are responsible for the spasmolytic, mildly diuretic, strong uterus-stimulating effect. An allergenic effect has been ascribed to the substance.

5 Application

In Folk Medicine

As a Dehydrating Agent

In folk medicine, parsley seed oil is used as a dehydrating agent in cases of oedema as well as to treat kidney complaints and complaints of the urinary tract collection system. Today, this application is obsolete, as the treatment is unlikely to achieve any therapeutic benefit. A study on white mice, which were administered a saturated solution of the oil in physiological saline solution subcutaneously and perorally, could only prove increased diuresis within a narrow dosage range. Increasing the dosage led to a decrease in diuresis (Hänsel and Sticher 1992, 5th edition, 6th volume, p. 112).

Other Uses

In folk medicine, parsley seed oil is also used to treat complaints of the gastrointestinal tract, to promote digestion, and to treat dysmenorrhoea and period pain. These applications have not yet been scientifically examined (Hänsel and Sticher 1992, 5th edition, 6th volume, p. 112).

In Ayurvedic Medicine

Parsley seed oil is used as a liniment to treat arthritis and rheumatism, and is also used against haemorrhoids, spider veins and cellulitis.

In Cosmetics

Since parsley seed oil has a pleasant smell and a high content of volatile compounds, it can be added to the following cosmetic products (Young-Hoi et al. 1990):

- soap;
- perfume, and
- creams.

As Food

Parsley seed oil is an edible oil for the seasoning of meat, industrially produced soups and gravy, non-alcoholic drinks, bakery products and other food (Young-Hoi et al. 1990).

6 Possible Unwanted Side Effects

There are several reports dating from the 1930s and 1940s on abortion attempts and cases of poisoning with parsley seed oil. As a result of the apiol content, in particular, the contraction of the smooth muscles of the bladder, bowels and uterus is increased. Cases of poisoning are accompanied by headaches, gastroenteritis, an increase in the pulse rate, shock and coma. Other symptoms may include irritation of the kidneys, which may lead to haematuria and liver damage. External application may lead to skin irritation and allergies. Parsley seed oil should not be used during pregnancy or in cases of nephritis (Hänsel and Sticher 1992, 5th edition, 6th volume, p. 112).

Passion Fruit Seed Oil

Oleum Passiflorae – synonyms: Passionsfruchtsamenöl (D)

1 Source Plant

Passiflora incarnata L. (Passifloraceae), passion flower

S. Krist, *Vegetable Fats and Oils*, https://doi.org/10.1007/978-3-030-30314-3_85

Habitat

Passiflora incarnata L. is a native plant of the southeastern states of North America, for example Florida, Texas, Virginia, Oklahoma, Missouri and North Carolina. It is also found on the Bermuda Islands, the Antilles, in Central America – especially in Mexico – and in Brazil, Argentina and India. The passion flower's splendid blossoms already fascinated the Spanish explorers of the sixteenth century, who came across this climbing plant in Peru. In 1609, it was imported to Europe. In Central Europe, it can only be cultivated outside in warmer regions. The main growing areas lie in India and North America, and less important ones in Spain and Italy. The passion flower received its name in the seventeenth century, when monks interpreted the individual parts of the flower as symbols of Christ's Way of the Cross (Hänsel and Sticher 1992, 5th edition, 6th volume, p. 35; Krenn 2002).

Description

The flesh-coloured passion flower is a climbing shrub that can grow to a height of 10 m. It has thin, green, lignified stems, which are usually angular on young parts of the plant, and later grow to be slightly grey and round, with a longitudinally striped bark. The leaves are alternate, stalked and three-lobed; their blades have netted venation. The leaves are 6–15 cm long, with dentate edges. The leaf stems and leaf veins are covered with fine hair, which are denser on the lower side of the leaf than on the upper side. On the leaf blade and stem, there are extrafloral nectaries forming protuberances. From the axils, stipules and cirri emerge; the latter are formed like corkscrews. The flower stems are up to 8 cm long and axillary; they only bear one flower each. The flower is perfect, radial, and measures 5–9 cm in diameter. The five sepals are compact, green on the outside and white on the inside. The five delicate petals are white to pale lavender. Within the petals, there is a corolla appendage formed from four radiate corona filaments, which are arranged concentrically around the axis core; they are white on the inside and purple on the outside. The trifoliate ovary has three styli ending in thickened stigmata. The five stamina are joined at the base and have grown together with the androgynophore. The fruit is an oval, edible berry about 6 cm long. It is a yellowish or pale-orange colour; it is known as a passion fruit. It contains many yellow to brownish-yellow seeds 5–8 mm in size (Hänsel and Sticher 1992, 5th edition, 6th volume, p. 35).

2 Cultivation and Extraction

Cultivation

The passion flower prefers sunny locations and dry, not too fertile soil. As a climbing plant, it grows in dry hedges and on road embankments. The plant can hibernate outside in mild, frost-protected locations. It blooms from June to July; the fruit ripen between September and November (Hänsel and Sticher 1992, 5th edition, 6th volume, p. 36).

Extraction of the Oil

Passion fruit seed oil is extracted from the ground seeds of *Passiflora incarnata L.* with n-hexane, and subsequently refined (Quiroga et al. 2000).

The seeds can also be cold pressed in mechanical presses (Pruthi 1962).

A study examined the method of extracting the oil via microwaves. A slightly altered microwave oven was used, and the seeds were soaked in water 24 hours prior to extraction. This method is very quick: It takes only 120 seconds to extract the oil. It also results in a higher yield, and the oil smells and tastes better (Hao et al. 2001).

3 Character

Colour

Light yellow (Hao et al. 2001).

Odour

Pleasant smell (Hao et al. 2001).

Volatile Compounds

Analysis using GC-MS (Huang et al. 2003)

Linalool	Limonene	m-Cymene
β-Ionon	Nerolidol	2-Hexenal
α-Terpineol	β-Ocimene	2-Propylfurane
γ-Terpinene	2-Carene	Hexanol
Benzaldehyde	α-Ocimene	Octanol
2-Furancarboxaldehyde	Methylbenzene	Ethyl butyrate
Citronellol	Benzenacetaldehyde	Ethyl hexanoate
Geraniol	Citronellyl butanoate	2-Methylpropyl butanoate
β-Myrcene	Hexylacetate	Pentyl hexanoate

Flavour

Pleasant taste (Hao et al. 2001).

4 Ingredients

Composition of Fatty Acids

Fatty acids	Content in % acc. to Lopez (1980)	Content in % acc. to oilsbynature.com/10.08.05
Myristic acid	Traces	<0.2
Palmitic acid	6.78	5–15
Stearic acid	1.76	<5
Oleic acid	19.0	10–22
Linoleic acid	59.9	55–80
α-Linolenic acid	5.4	<1.5
Arachidic acid	0.34	<0.7
Lauric acid	–	<0.2
Palmitoleic acid	–	<1.0
Eicosanoic acid	–	<0.5
Behenic acid	–	<0.3
Erucic acid	–	Traces
Lignoceric acid	–	<0.5

Physical Key Figures of Passion Fruit Seed Oil

refractive index: d_D^{25} 1.4727
density: d_{25}^{25} 0.9214
saponification number: 190.3
iodine number: 142.10
acid number: 0.2
unsaponifiable: 0.65%

(Pruthi 1962)

Shelf Life

Passion fruit seed oil becomes rancid quickly and does not have a long shelf life (Pruthi 1962).

5 Application

In Pharmacy and Medicine

Passion fruit seed oil can be applied in minor cases of cardiovascular diseases and to treat nerve-related cardiac complaints. The oil also lowers blood pressure and, for this reason, is recommended for persons with high blood pressure as an auxiliary measure to drug therapy (Zhang et al. 2000).

Passion fruit seed oil has an anti-inflammatory and pain-soothing effect and is added to ointments to treat irritated skin.

In Cosmetics

As a result of its high vitamin A content, which has a protective and antioxidative effect on the skin, and its relaxing effect on the muscles, passion fruit seed oil is used in the following products:

- sunscreens;
- bath oils and massage oils;
- day creams;
- body lotions;
- balms, and
- soap.

As Food

Passion fruit seed oil is used as an edible oil especially in Brazil (Lopez 1980).

In the Food Industry

Since passion fruit seed oil has a pleasant taste, it is added to sorbets, drinks, jam and confectionery.

Peach Kernel Oil

Oleum Persicarum synonyms: Pfirsichkernöl (D); huile de noyaux de pêche (F)

1 Source Plant

Prunus persica (L.) Batsch (Rosaceae), peach

S. Krist, *Vegetable Fats and Oils*, https://doi.org/10.1007/978-3-030-30314-3_86

Habitat

Although the Latin name *Prunus persica* suggests otherwise, the peach originates in China, from where it came to Europe via Persia (Dassler and Heitmann 1991, p. 174). Today, it is grown especially in warm, temperate or subtropical climates.

Description

In cultivation, the tree can grow 3–4 m tall. The flowers are pink or purple; they sit on the branches either aloney or in pairs, usually without leaves. There are many fruit varieties. The fruit pulp is either white or yellow, and there are early- and late-ripening fruit, as well as freestone and clingstone varieties. The nectarine *(Prunus persica var. Nucipersica) is a variety of the peach* (Roth and Kormann 2000, p. 99) and has a smooth skin, in contrast to the peach's skin, which is covered with velvety hair (Dassler and Heitmann 1991, p. 174). The light brown stone kernels are 3.5–4 cm long and 2.5 cm broad. They have an oval, pointed form, and are deeply furrowed and sharp-edged on both sides. Within the thick stone layer lie cinnamon-brown oval seeds that are about 1.5 cm long, more than 1 cm broad and about 2 mm thick; their edges are sharp. The kernels contain 30–45% fat oil (Roth and Kormann 2000, p. 99).

2 Cultivation and Extraction

Cultivation

Peach trees can be cultivated in all warm temperate climes without late or early frosts (Dassler and Heitmann 1991, p. 174). They need light and a warm, protected location. Rich, loose, sandy, neutral soil is best. It is better to plant the trees in spring than in autumn. Peaches are self-fertile, and a pollinating cultivar is therefore not necessary (blumenboerse.ch/24.05.06). The trees bloom from March to April; the harvest takes place from August to September (Roth and Kormann 2000, p. 99) and is ideally carried out in stages and by hand, as the fruit are pressure- and shock-sensitive. The most important producers are Italy, Greece, France and Spain (Dassler and Heitmann 1991, p. 174 et seqq).

Extraction of the Oil

To obtain the oil, the ligneous shells of the fruit are removed, the kernels cut into pieces and the oil extracted using a solvent (i.e. petroleum ether) (Lawson 1995, p. 124).

3 Character

Colour

Weakly yellow oil (Dhar and Chauhan 1963).

Odour

Weakly nutty and pithy, fresh, delicate smell of almonds.

Flavour

Weakly nutty.

4 Ingredients

Composition of Fatty Acids

Fatty acids	Content in % acc. to Firestone (1999), p. 74	Content in % acc. to nature.de/16.01.07
Myristic acid	0–1	–
Palmitic acid	6–13	5.4
Palmitoleic acid	0.2–0.5	0.6
Stearic acid	2–6	2.7
Oleic acid	61–70	63.9
Linoleic acid	15–29	25.3
α-Linolenic acid	–	0.1
Arachidic acid	0.6	–

Composition of Triglycerides

(Firestone 1999, p. 74)

Triglycerides	Content in % acc. to Firestone (1999), p. 74
POO	7
SOO	2
PPL	0.3
OOO	31
POL	9
OOL	28
PLL	2
OLL	17
LLL	4

Composition of Sterols

Sterols	Content in % acc. to Firestone (1999), p. 74	Content in % acc. to Lawson (1995), p. 126
Cholesterol	–	12.94
Campesterol	3.5–5	162.15
Stigmasterol	0.4–1.0	–
β-Sitosterol	90–93	3265.75
Δ5-Avenasterol	2–3	210.48
Δ7-Stigmasterol	0–0.5	187.35
Δ7-Avenasterol	–	31.95

Composition of Tocopherols

(concentration in mg/100 g and content in %)

Tocopherols	Peach kernel oil acc. to Firestone (1999), p. 74	Peach kernel oil acc. to Belitz and Grosch (1987), p. 223
Total content	8.8–12.3	–
α-Tocopherol	6.2–7.4	6.4
β-Tocopherol	1.5–3.7	1.3
γ-Tocopherol	0.3–1.1	1.0

Physical Key Figures of Peach Kernel Oil

refractive index: 25 °C 1.468–1.470
40 °C 1.459–1.462
density: 25 °C 0.913–0.918
iodine number: 94–110
saponification number: 189–194
acid number: 0.14
unsaponifiable: 0.7–1.0%

(Firestone 1999, p. 74 – nature.de/24.04.06)

Shelf Life

Peach kernel oil has a shelf life of about 6–8 months and should be stored in a dark place, and not above room temperature.

5 Application

In Research

A study examined the effects of peach kernel oil on the physical development and sexual maturity of female rats. The oil was injected either once intramuscularly in prepuberty (0.2 ml at an age of 35 days), or five times at an age of 31–35 days. The result was that, depending on the dosage, peach kernel oil inhibited the physical development of rats. It inhibited the ovarian endocrine function, significantly delayed the beginning (opening of the vagina) and end phase (ovulation) of sexual maturity, or inhibited sexual maturity completely in 30% of the cases (Litvinova 1998).

Industrial Uses

In comparison to other oils, peach kernel oil is healthy in cases of blood pressure complaints and coronary diseases, and can therefore be used in the food industry (Lawson 1995, p. 126).

The oil is used to adulterate almond oil, as an identification test is rarely unambiguous; the adulteration is only detectable with a considerable amount of analytical work (Carstensen 2001, p. 20).

In Cosmetics

Peach kernel oil is often used in the cosmetics industry for care cosmetics (Carstensen 2001, p. 3) as well as massage products for the face and body. It can be added to all cosmetic recipes and is effective in cases of dry, ageing, flaky and chapped skin as well as in treating skin poorly supplied with blood.

Peanut Oil/African Peanut Oil

Oleum Arachidis, Oleum Arachidis african. synonyms: Erdnussöl (D); arachis seed oil, groundnut oil (E); huile d'arachide (F)

1 Source Plant

Arachis hypogaea L. *(Fabaceae)*, peanut

S. Krist, *Vegetable Fats and Oils*, https://doi.org/10.1007/978-3-030-30314-3_87

Habitat

According to botanical nomenclature, the peanut is a leguminous plant. Its origins probably lie in Bolivia. It is an important food plant of the tropics and subtropics and is cultivated in these climes on all continents. India and China are the main producers; peanuts were already cultivated in these countries in pre-Christian times. Today, African countries like Nigeria and Senegal supply about 20% of the global production, followed by the USA with a share of 9% (Dassler and Heitmann 1991, p. 192). The peanut harvest worldwide amounts to 33 million tons per year.

Description

Arachis hypogaea is an annual plant up to 60 cm tall. It is branched from the base, grows close to the ground and is densely covered with oval, paripinnate leaves. The flowers are golden yellow and 1–2 cm long; the inflorescences are short-stemmed and axial. After pollination, a carpophore formed like a nail develops at the base of the ovary; it bows down and penetrates the ground. When the tip of the carpophore has reached a depth of 5–10 cm, the fruit starts to develop. No fruit develop above ground. The fruit has a yellow, wrinkly shell, which usually contains two seeds in a thin, reddish-brown peel. All varieties of peanut can be divided into two groups: those with upright stems and those with creeping stems. All pollinated ovaries of the creeping varieties reach the ground, which is not always the case for the upright varieties (Roth and Kormann 2000, p. 40 et seq.).

2 Cultivation and Extraction

Cultivation

The peanut prefers loamy, sandy soil that retains water and is neutral to slightly alkaline. It is a thermophilic plant, but must be cultivated under irrigation in dry conditions. The plant does not tolerate long periods of wetness, which cause it to die back. Crust-forming soil is not suitable for the cultivation of peanuts, as it prevents the carpophore from penetrating the surface of the ground. Upright varieties require loosening of the ground and earthing up to support fructification. Peanut plants do not tolerate frost (Roth and Kormann 2000, p. 41).

Extraction of the Oil

Peanuts are harvested by cutting the taproot and lifting the shoots out of the ground. The fruit are dried for 2–4 weeks. The seeds contain about 45% oil (nature.de/08.02.06). To extract the oil, the shell is removed using crimpers or disc mills. The kernels are then cleaned above sieves, reduced to small pieces in roller mills and prepressed in screw compressors. The remaining oil is extracted with hexane (Hager 1993, p. 317). Sometimes the peanuts are freeze-dried before pressing (Tseng et al. 1993). Cold pressed and extracted oil can easily be neutralised with bases, bleached with activated carbon and deodorised. Traces of aflatoxins[1] ending up in the oil are inactivated by alkaline treatment in the process of neutralisation (Hager 1993, p. 317). From 100 kg peanuts containing about 30 kg shells and 70 kg seeds, about 34 kg oil can be extracted (Karleskind 1996, p. 186).

By partial hydrogenation, peanut oil can be further processed to *Arachidis oleum hydrogenatum* (hardened peanut oil) (Hager 1993, p. 319), which is used as an ointment base (Frohne 2002).

3 Character

Colour

Peanut oil: clear, yellowish, viscous liquid.
Hardened peanut oil: white mass, fatty to the touch (Hager 1993, p. 319).
African peanut oil: colourless, slight yellow tint.

Odour

Peanut oil: a slight to pronounced smell of peanuts, slightly fruity, roasted (if roasted before pressing).
Hardened peanut oil: almost odourless (Hager 1993, p. 319).
African peanut oil: touch of peanuts.

Volatile Compounds

Examinations using **aroma extract dilution analysis** (AEDA) and gas chromatography olfactometry of headspace samples (GCOH) detected the following potent odorous substances in peanut oil (Matsui et al. 1998):

Odorous substance	Quality of the odour
Ethyl-2-methylbutanoate	Fruity
2-Ethyl-3,5-dimethylpyrazine	Roasted
2,3-Diethyl-5-methylpyrazine	Roasted
(Z)-2-Nonenal	Fatty, green
(E,E)-2,4-decadienal	Fatty, fried
*(E)-β-d*amascenone	Like baked apples
Ethyl-isobutyrate	Fruity

[1] Aflatoxins are a group of structurally similar mycotoxins produced by mould fungi (*aspergillus* and *penicillium* species). Aflatoxin B1 is acutely toxic and is among the most carcinogenic substances. Acute poisoning causes liver damage, in particular (Hunnius 2004, p. 35).

Odorous substance	Quality of the odour
3-Mercapto-2-butanone	Like cooked meat
2-Acetyl-1-pyrroline	Roasted, sweet
Dimethyltrisulfide	Like sulphide
2-Propionyl-1-pyrroline	Roasted
3-Ethyl-2,5-dimethylpyrazine	Roasted
2-Furfurylthiol	Sweet, smoky
2-Ethenyl-3,5-dimethylpyrazine	Roasted
(E,Z)-2,4-Nonadienal	Green
Unknown	Like sulphide
(E,Z)-2,4-Decadienal	Fatty, green
2-Methoxyphenol (guaiacol)	Burnt
2-Phenyl-2-butenal	Green, phenolic
Vanillin	Like vanilla
3-Methylbutanal	Malty
d-Limonene	Like lime
Octanal	Fatty
2,5-(or 2,6-) Diethyl pyrazine	Sweet
(E)-2-Nonenal	Fatty, green
Phenyl acetaldehyde	Sweet, like honey
2-Methyl-5-((E)-1-propenyl) pyrazine	Sweet, earthy
(E,E)-2,4-Nonadienal	Fatty, fried
(E)-2-Undecenal	Fatty
Hexanoic acid	Sweetish
2-Methoxy-4-vinylphenol-(4-vinylguaiacol)	Pungent, phenolic
∆-Dodecalactone	Sweet
Hexanal	Green
(E)-2-heptenal	Fatty, green
2-Ethyl-5-methylpyrazine	Sweet
2-Ethyl-3-methylpyrazine	Roasted
(E,Z)-2,6-Nonadienal	Like cucumber
3-Methylbutyric acid	Sweetish
2-Methylpropanal	Malty
1-Penten-3-on	Malty
Propanal	Sweet
2-Methylbutanal	Green
2,3-Butandione (diacetyl)	Buttery

According to the relevant literature, the total content of odorous substances in peanut oil amounts to between 19.77 mg/kg (Cheng et al. 1989) and 21.32 mg/kg (Tseng et al. 1993). Pyrazines are the most important group with a total amount of 12.33 mg/kg peanut oil (Cheng et al. 1989).

The proportion of volatile compounds increases further if the temperature of the oil is increased (Chung et al. 1993) or if the peanuts are roasted before pressing. The amount of volatile compounds increases in proportion to the period of roasting (Tseng et al. 1993).

Flavour

Peanut oil: The taste of peanut oil from the first cold pressing is very pleasant (Hänsel et al. 1999, p. 253), mild.

Hardened peanut oil is almost tasteless (Hager 1993, p. 319).

African peanut oil: mild, nutty.

4 Ingredients

Composition of Fatty Acids

The composition of fatty acids in peanut oil varies depending on the producing area. The following table shows the difference in the composition of fatty acids between peanut oil from Africa and South America, as well as of hardened peanut oil (monograph European Pharmacopeia IV 2002, p. 1777) with a comparatively high proportion of oleic acid. For purposes of comparison, two known mean values have also been included.

Fatty acids	Content in %, peanut oil from Africa acc. to Karleskind (1996), p. 188	Content in %, peanut oil from South America acc. to Karleskind (1996), p. 188	Content in % acc. to DGF (2003)	Content in % acc. to Firestone (1999), p. 75
Lauric acid	–	–	0–0.1	0–0.1
Myristic acid	–	–	0–0.1	0–0.1
Palmitic acid	8–13	10–13	8.0–14.0	8.3–14.0
Palmitoleic acid	0–0.3	0–0.1	0–0.2	0–0.2
Stearic acid	3–4	1–4	1–4.5	1.9–4.4
Oleic acid	48–66	35–41	35–69	36.4–67.1
Linoleic acid	14–28	35–41	12–43	14.0–43.0
α-Linolenic acid	0–0.3	0–0.3	0–0.3	0–0.1
Arachidic acid	1–2	1–2	1.0–2.0	1.1–1.7
Eicosenoic acid	1–2	1–2	0.7–1.7	0.7–1.7
Behenic acid	2–4	4–5	1.5–4.5	2.1–4.4
Cetoleic acid	–	–	0–0.3	0–0.3
Lignoceric acid	1–2	1–2	0.5–2.5	1.1–2.2

It is noteworthy that oleic acid is the main component in African peanut oil, whereas in peanut oil from South America, the amounts of linoleic acid and oleic acid are balanced.

In peanut oil with a high content of oleic acid, the amount of oleic acid is increased at the expense of the amount of linoleic acid by hydrogenation of the double bond.

P

Composition of Triglycerides in Peanut Oil

Triglycerides	Content in % acc. to Firestone (1999), p. 140	Content in % acc. to Bockisch (1998), p. 247	Content in % acc. to Karleskind (1996), p. 187
PSO	0.6	2.2	2.2
PPO	1	–	2.3
POO	6	6.7	6.7
SOO	4	1.5	1.5
PPL	2	2.9	2.9
OOO	5	11.8	11.8
POL	13	12.9	12.9
OOL	22	19.4	19.4
PLL	8	5.1	5.1
OLL	26	18.3	18.3
LLL	6	2	2
AOO	4	–	–
POP	–	2.3	–
PLP	–	5.3	–
SOL	–	3.9	3.0
SLL	–	1.4	1.4
LOL	–	2.2	–
LLG	–	–	0.4

Composition of Sterols

(concentration in mg/100 g and content in %)

Like the composition of fatty acids, the composition of sterols varies depending on the producing area (Karleskind 1996, p. 188). The following table also shows the difference in the composition of sterols of refined and raw peanut oil (Carstensen 2001, p. 42):

Sterols	African peanut oil	South American peanut oil	Refined peanut oil	Raw peanut oil
Total content	190–330	260–320	329.6	318.7
Cholesterol	0–0.7	0–0.4	0.6	1.1
Campesterol	12–15	16–19	11.2	8.8
Stigmasterol	8–11	6–9	11.3	7.3
β-Sitosterol	58–66	61–67	66.2	69.2
Δ5-Avenasterol	9–14	7–9	6.8	8.4
Δ7-Stigmasterol	0–1	0–1	0.2	1.1
Δ7-Avenasterol	1–2	0–1	–	2.7

Composition of Tocopherols

(concentration in mg/100 g and content in %)

Again, the contents of African and South American peanut oil (Karleskind 1996, p. 188), as well as that of refined and raw peanut oil, differ (Carstensen 2001, p. 44):

Tocopherols	African peanut oil	South American peanut oil	Refined peanut oil	Raw peanut oil
Total content	28–73	30–100	38.8	17.3
α-Tocopherol	42–59	47–65	42.8	30.4
β-Tocopherol	2–5	1–3	2.1	2.9
γ-Tocopherol	32–52	30–50	49.3	42.8
δ-Tocopherol	2–4	2–3	5.8	23.9

Other Ingredients

(Karleskind 1996, p. 188)

Hydrocarbons
Squalene
Glycerine
Lipoids
Sodium
Potassium
Calcium
Magnesium

Physical Key Figures

Peanut Oil

refractive index: 40 °C 1.460–1.465
density: 0.912–0.918
iodine number: 83–107 (63–75 if hardened)
saponification number: 187–196
melting point: −2 °C
unsaponifiable: 0–1.0%

(Firestone 1999, p. 75; Hager 1993)

African Peanut Oil

refractive index: n_D^{20} 1.470–1.472
density: d^{20} 0.914–0.917
iodine number: 85–90
saponification number: 189–196

(Karleskind 1996, p. 187)

Shelf Life

Peanut oil has a shelf life of 2 years.

5 Application

In Pharmacy and Medicine

As a result of its proportion of unsaturated fatty acids in the triacylglycerides, peanut oil is part of the group of oils that lowers the cholesterol level of human blood. It is also used as an enema to soften solid faeces in the rectum (Hager 1993, p. 319). Other important applications include:

In Pharmaceutical Technology
Peanut oil serves as a vehicle for liposoluble agents for external, enteral and parenteral use, especially for sex hormones with a repository effect, and as an oil for eye drops (Hager 1993, p. 319; Hänsel et al. 1999, p. 247 et seq).

In Dermatology
Peanut oil is used to treat encrustations and dandruff on the scalp. It can also be used in baby care, as a bath essence, in the treatment of subacute and chronic eczema, for atopic eczema and ichthyosis. The recommended dosage for oil baths is 4 ml in 10 litres of water, 2–3 times a week (Hager 1993).

In Magistral Formulae
As a result of its water absorption capacity, hardened peanut oil is used as an ointment base (Frohne 2002), but only rarely, as despite partial hydrogenation of the double bonds in the portion of fatty acids in the triacylglycerides, it becomes rancid easily. It is, however, part of old recipes like camphor ointments (Pharmacopoea Helvetica VII), *unguentum aromaticum* and *unguentum leniens* (Austria Pharmacopoeia 81) (Hager 1993, p. 319).

In Cosmetics

As a Massage Oil
Peanut oil is only absorbed into the skin slowly, as it almost lacks the ability to penetrate the intercellular space of the *stratum corneum* (Hänsel et al. 1999, p. 248). For this reason, it is mainly suitable as a base oil for massage in cases of tension in the back, circulatory disorders and for rheumatism prevention.

As Protection for the Skin
In cosmetics, peanut oil is used as a skin oil and suntan oil because of its high fat content (Carstensen 2001, p. 3; Hager 1993, p. 319). It is also suitable for baby care and childcare and as a cosmetic care product for the skin and scalp. Peanut oil softens up crusts and dandruff and is thus an excellent bath essence for persons suffering from eczema and dry skin (Roth and Kormann 2000, p. 29).

As Food

The oil is excellent for roasting and frying, as it can be heated to 230 °C. It is only of limited suitability as a salad oil, however (Kircher 2002, p. 49).

Industrial Uses

In the food industry, peanut oil is often added to soups and spices (Carstensen 2001, p. 4). It is also used in the production of margarine, soap and paint.

In Folk Medicine

Oleum Arachidis is used in folk medicine to treat the following complaints (Kircher 2002 p. 29 et seq):

- arteriosclerosis;
- cystitis;
- dimpled skin after sunbathing;
- sunburn;
- stress;
- constipation, and
- digestive weakness.

6 Possible Unwanted Side Effects

Skin Irritations

Peanut oil should not be applied to weeping, acutely inflamed, irritated or oily skin. Oil baths containing peanut oil should not be used in cases of fresh *Psoriasis pustulosa*.

Allergic Reactions

Peanuts contain dangerous allergens, the most important of which is the peanut allergen Ara h2, a protein. Even the oil may contain tiny amounts of the protein and thus cause allergic reactions. In an English study dating from 1997, 60 persons with a peanut allergy were given refined and virgin peanut oil. None of the test persons showed a reaction after ingesting the refined oil, but 6 of them reacted after ingesting virgin peanut oil (Hourihane et al. 1997).

About 1% of the population of the USA has a peanut allergy; this is a significant problem, as regular consumption of peanut butter and the addition of peanuts to sweets may already lead to allergisation in early childhood. Peanuts and peanut oil are hidden in many foods, which makes it difficult for patients to avoid them. This situation is particularly dangerous, since 24% of all life-threatening anaphylactic reactions to food can be traced back to peanuts (Senti et al. 2000).

Carcinogen Potential

Since the amount of oleic acid in African peanut oil is considerable, it should be noted that oleic acid is classified in category 4 of products that are suspected of causing cancer, and it is in water pollution class 1 (Roth and Kormann 2000 p. 193).

Pear Seed Oil

synonyms: Birnenkernöl (D); Pear oil (E)

S. Krist, *Vegetable Fats and Oils*, https://doi.org/10.1007/978-3-030-30314-3_88

P

1 Source Plant

Pyrus communis L. *(Rosaceae)*, Pear

Habitat

The pear is thought to originate in Asia Minor, between the Caucasus mountains and the western coast of the Black Sea (Lieberei and Reisdorff 2007, p. 198).

Description

The source plant of *Pyrus communis* is *Pyrus pyraster*, the wild pear, with the subspecies *pyraster* and *achras*. The pear is a tree with round, hairless leaves. The flowers are white, often with red anthers, and arranged in corymbs. The tree blossoms in April. The ovary is epigynous and develops into a follicle. The ripe fruit is very juicy, but it does not keep long (Lieberei and Reisdorff 2007, p. 198; Guimpel et al. 1815, p. 99).

2 Cultivation and Extraction

Cultivation

The cultivation of pears can be traced back to as early as 900 B.C. in Greece. The Romans cultivated pears from about 100 B.C. onwards. In the sixteenth century, there were already 50 different cultivars of pears in Germany. Pears need warm habitats and loamy or clayey soil with plenty of sand. Sowing takes place in the autumn, and sometimes also in spring. The global production of pears was 18.7 million tons in 2004; the main producer was China with 10,767,000 tons, followed by Italy with 877,000 tons and the USA with 796,000 tons (Lieberei and Reisdorff 2007, p. 198; Guimpel et al. 1815, p. 99–100). More than half of the global production therefore takes place in China. As the seeds accumulate as by-products in production as well as after the consumption of fresh fruit, oil extraction has a commercial use (Yukui et al. 2009).

Extraction of the Oil

The dried seeds are first ground, and the oil is then extracted with solvents. The seeds contain 179 g/kg oil (Yukui et al. 2009).

The seeds can also be pressed in screw compressors or stamping presses (Reinberg et al. 2009, p. 29).

3 Character

Odour

The odour is similar to apricot oil (Reinberg et al. 2009, p. 47).

4 Ingredients

Composition of Fatty Acids

Fatty acids	Content in % acc. to Schuster (1992), p. 215	Content in g/100 g oil acc. to Yukui et al. (2009)	Content in % acc. to Reinberg et al. (2009), S. 41	Content in % acc. to Bizal (2006)
Lauric acid	0.5–1.0	–	–	–
Palmitic acid	2–4	6.388	13.5	8.6–10.46
Palmitoleic acid	–	0.119	–	–
Stearic acid	–	1.746	3.0	1.52–2.01
Oleic acid	60–70	20.281	27.2	24.92–38.74
Linoleic acid	4–6	56.801	51.5	51.12–62.99
Linolenic acid	4–8	0.320	1.1	–
Arachidic acid	–	1.251	2.6	–
Eicosanoic acid	–	0.275	–	–
Eicosadienoic acid	–	0.050	–	–
Behenic acid	–	0.238	0.8	–
Lignoceric acid	–	0.109	0.3	–

According to Yukui et al. (2009), pear seed oil contains 9.732 g/100 g saturated fatty acids, 77.846 g/100 g unsaturated fatty acids and 20.675 g/100 g monounsaturated fatty acids.

Composition of Triglycerides

Triglycerides	Content in atomic percent acc. to Deineka and Deineka (2004)
$LnL_2 + Ln_2O$	0.4
$L_3 + LnLO$	20.3
$LnLP + Ln_2S$	0.8
$L_2O + LnO_2$	27.6
$L_2P + LnLS + LnOP$	2.3
LO_2	18.5
$L_2S + LOP + LnOS$	11.3
$LP_2 + LnPS$	1.2
O_3	11.2
$LOS + O_2P$	4.5
$LPS + OP_2$	0.1
O_2S	1.8

P

Phospholipids

Phospholipids	Content in % acc. to Zlatanov et al. (1997)
Total content	9 g/kg
Phosphatidylcholine	30.1
Phosphatidylethanolamine	13.8
Phosphatidylinositol	29.3
Phosphatidic acid	5.5

Physical Key Figures

refractive index: n_D^{20}	1.474
density: d_{15}^{15}	0.9168
saponification number:	180
iodine number:	109
unsaponifiable:	1.7%

(Roth and Kormann 2005, p. 159)

specific weight:	0.9177 (20 °C)
refractive index:	1.4717 (21 °C)
saponification number:	181.7–186
iodine number:	113–120

(Hackbarth 1944, p. 343)

5 Application

In Cosmetics

Since the oil has a pleasant smell, it is added to soaps, peeling products and bath teas. Pear seed oil is often substituted for apricot oil, however, as the latter has a similar smell and is less expensive (Reinberg et al. 2009, p. 47).

As Food

The pressed oil is an edible oil (Reinberg et al. 2009).

Pecan Oil

Oleum Pecan synonyms: Pekannussöl, Hickorynussöl (D); pecan nut oil (E)

1 Source Plant

Carya illinoinensis (Wangenh.) K. Koch *(Juglandaceae)*, pecan

Habitat

The pecan is native between the Gulf of Mexico and the Atlantic Coast, as well as between Iowa and New York state. It is not clear where the pecan came from. The

S. Krist, *Vegetable Fats and Oils*, https://doi.org/10.1007/978-3-030-30314-3_89

Mississippi region as well as Texas have been named as possible regions of origin. Today, the tree grows mainly in Kentucky, Indiana, Iowa, Nebraska, but also in Alabama, Louisiana and Texas. Additionally, it is found in Mexico, India and China, where it is also cultivated. Cultivation began as late as 1800, and since 1960, the tree has also been grown in Israel, Australia and South Africa. The USA is the largest producer, followed by Mexico and Australia; other countries only have a small share in production (Roth and Kormann 2000, p. 51).

Description

The pecan is a deciduous tree 30–50 m tall, and can reach an age of up to 300 years. The leaves are imparipinnate, and the leaflets crescent-shaped and slightly asymmetrical. Pecan wood is very hard and robust. The flowers are monoecious and are pollinated by the wind. The fruit resemble stone fruit, but are actually nuts surrounded by conjoined sepals. They look like large acorns. Beneath the ligneous, light- to dark-brown shell, there are two nuts. Their form resembles brains or walnuts, but the pecan nuts are longer than walnuts, and their shell is thinner. The form of the nuts can vary slightly. The cotyledons of the seeds contain about 70% fat rich in oil and linoleic acid, 9% protein, 4% carbohydrates and nearly 10% fibre.

2 Cultivation and Extraction

Cultivation

The pecan prefers cold winters, but tolerates high temperatures in summer. The soil should be loose and moist. Pecans need much sun; they do not grow in the shade. Most varieties produce nuts after 4–7 years of cultivation in the same place; harvests of commercial value start after 8–12 years. The period of rising yields can be shortened by employing tending strategies such as irrigation, good management and dense planting. The final distance between the plants usually amounts to 15–20 m or more; it depends on the water supply and soil fertility. It is not easy to grow pecans, as mice often eat the seeds. The harvest takes place in the autumn, when the nuts fall from the trees. The nuts that do not fall from the trees by themselves are shaken off. Afterwards, they are dried in dry, dark, well-aired rooms for about 3 weeks.

Extraction of the Oil

Pecan oil is obtained from the seeds of *Carya illinoinensis* by cold pressing or extraction with ether (Toro-Vazquez and Perez-Briceno 1998).

3 Character

Colour

Clear and straw-coloured (Heller and Heston 1932; Deiler and Fraps 1910).

Odour

Pleasantly nutty smell (Deiler and Fraps 1910).

Flavour

Mild and pleasant taste (Roth and Kormann 2000, p. 143).
Pleasant, nutty taste (Deiler and Fraps 1910).

4 Ingredients

Composition of Fatty Acids

Fatty acids	Content in % acc. to Roth and Kormann (2000), p. 143	Content in % acc. to Firestone (1999), p. 75 et seq	Content in % acc. to libertyvegetableoil.com/10.08.05
Myristic acid	Traces	–	0–0.1
Palmitic acid	3.3	5–11	6.7–7.7
Palmitoleic acid	–	0.1–0.2	0.2–0.8
Stearic acid	1.9	1–6	1.9–2.3
Oleic acid	77.8	49–69	43.0–51.0
Linoleic acid	15.8	19–40	37.0–45.0
α-Linolenic acid	–	0–3	1.4–2.4
Arachidic acid	0.1	0.1–0.2	0–0.5
Eicosanoic acid	–	0.2–0.3	0.2–0.6
Behenic acid	–	0.1–0.2	0–0.1
Heptadecanoic acid	–	–	0–0.1
9-Heptadecenoic acid	–	–	0–0.1

Composition of Triglycerides

Triglycerides	Content in % acc. to Firestone (1999), p. 140
POO	3–5
SOO	0.1–0.7
OOO	4–10
POL + PoOO	8–10
SOL	0.1–1
OOL	24–29
SLL	0.3–1
OLL	24–29
LLL	12–17

Triglycerides	Content in % acc. to Firestone (1999), p. 140
LnLO	0.5–1
LnOO	6–9
LnLL	1–3
LnLnL	0.1–1
LnLnLn	0.3–1

Composition of Sterols

(total content in mg/kg and composition in %)

Sterols	Pecan oil acc. to Firestone (1999), p. 75 et seq
Total content	1000–2900
Cholesterol	–
Brassicasterol	–
Campesterol	3.5–4.5
Stigmasterol	1–2
β-Sitosterol	81–93
Δ5-Avenasterol	1–8
Δ7-Stigmasterol	0.2–0.6
Δ7-Avenasterol	0.2–0.4

Composition of Tocopherols

(total content in mg/kg and composition in %)

Tocopherols	Pecan oil acc. to Firestone (1999), p. 75 et seq
Total content	88–420
α-Tocopherol	50–370
β-Tocopherol	–
γ-Tocopherol	20–125
Δ-Tocopherol	0–40

Physical Key Figures of Pecan Oil

refractive index: n_D^{20} 1.470
density: d_{15}^{15} 0.917
saponification number: 189–198
iodine number: 97–107
unsaponifiable: 0.4%

(Roth and Kormann 2000, p. 143)

Shelf Life

Pecan oil becomes rancid relatively quickly and does not keep long (Heller and Heston 1932).

5 Application

In Pharmacy and Medicine

Pecan oil contains a high amount of monounsaturated fatty acids. These fatty acids have a positive influence on the heart, blood vessels and cholesterol metabolism. Food rich in mono- and polyunsaturated fatty acids can lower high LDL cholesterol levels, which are jointly responsible for arteriosclerosis. On the other hand, the HDL cholesterol level increases. The risk of developing arteriosclerosis or myocardial infarction is thus significantly lower, which is why pecan oil is recommended for preventing these diseases (Toro-Vazquez and Perez-Briceno 1998).

Pecan oil is a source of natural vitamin E. It contains high amounts of tocopherol and has antioxidant properties. Tocopherol protects polyunsaturated fatty acids in the lipids of biological membranes from the destructive influences of molecular oxygen (Toro-Vazquez and Perez-Briceno 1998).

In Cosmetics

Since it has a pleasant smell, pecan oil is used as a massage oil, even though it does not penetrate the skin quickly and thus leaves a fatty film. For this reason, the oil is not often used in cosmetics.

As Food

Pecan oil has a very pleasant, nutty taste and is therefore ideal for salads and sweet dishes such as cakes. In addition, it has a positive influence on fat metabolism, as mentioned above (Heller and Heston 1932).

6 Possible Unwanted Side Effects

Deficiency Syndromes

A study on rats examined the content of vitamins A and D in pecan oil and their effect. Apart from pecan oil, the rats were given a diet free of vitamins A and D. After 4 weeks, the bones showed rachitic symptoms, which could be traced back to the lack of vitamin D in pecan oil. Rickets is a vitamin D deficiency syndrome;

the bones only absorb and store insufficient amounts of calcium, and soften as a result. As pecan oil does not contain much vitamin A either, the rats showed the first symptoms of xerophthalmia after 4 weeks. Xerophthalmia is a vitamin A deficiency syndrome leading to a drying-up of the ocular conjunctiva and cornea. Vitamin A has an important influence on sight and plays a role in the development of rhodopsin. Other deficiency syndromes are night blindness (hemeralopia) and keratomalacia, which results in a softening of the cornea. Other complaints that may develop include skin diseases such as hyperkeratosis and parakeratosis, changes to the mucous membranes of the respiratory passages, and impaired growth (Heller and Heston 1932; Burger and Wachter 1993, 7th edition, p. 1485, 1490).

Allergenic Potential

Pecan nuts frequently trigger allergies; even life-threatening reactions have been described. Oil from pecan nuts is a potential source of allergies.

Pepper Oil

Oleum Piperis nigris – synonym: Pfefferöl (D)

1 Source Plant

Piper nigrum L. (Piperaceae), black pepper

S. Krist, *Vegetable Fats and Oils*, https://doi.org/10.1007/978-3-030-30314-3_90

Habitat

Pepper is a native plant of the Malabar Coast in India. It prefers soil rich in humus and locations near the coast of tropical or subtropical regions, in warm half-shade. The wild form of *Piper nigrum* has not yet been identified beyond a doubt. There are, however, very closely related species in southern India and Myanmar (Hager 1978, volume 6a, p. 697; Jagella 1999, p. 1).

Description

Piper nigrum is a perennial climbing shrub that can grow up to 10 m tall. It climbs using aerial roots. The leaves are leathery, dark green, roundish-ovate or heart-shaped, and have seven longitudinal leaf veins. The usually dioecious flowers are arranged in loose, pendant spikes of about 10 cm length. The 20–30 stone fruit resemble berries and form a panicle similar to the blackcurrant. It is red when ripe. Within lies the seed, the peppercorn. Black pepper is obtained by harvesting the fruit when they are still unripe. The pre-dried berries are taken from the panicle and dried again in the sun. Due to an enzymatic reaction, the previously green fruit turn black. Green pepper is also harvested in an unripe state, but subsequently preserved in salt water or freeze-dried. In this way, the green pepper-corns are prevented from turning black by inhibition of the enzyme o-phenyloxidase. To obtain white pepper, the harvest must not be carried out until the fruit are fully ripe. After removing the outer, soft layers of the pericarp, the remaining grey fruit is dried in the sun until its colour changes to yellowish white (Hager 1978, volume 6a, p. 697; Jagella 1999, p. 1).

2 Cultivation and Extraction

Cultivation

Black pepper has always been highly valued throughout the world. When the commercial routes from Europe to Central Asia and India opened in the wake of Alexander the Great's Asian campaign, pepper, which was cultivated in India, Malaysia and Indonesia, was imported to Europe for the first time. The new spice grew immensely popular in a very short time and became commercially important merchandise. Arab merchants had a monopoly on the pepper trade since antiquity and brought pepper to their European trade partners via Arabia and Egypt, without giving them information on its country of origin. Prices were astronomically high, which today seems almost unthinkable. At the end of the fifteenth century, the Portuguese explorer Vasco da Gama reached India and founded several Portuguese trading posts. Of course, this did not reduce the price, but only shifted the profit to another country. It was only in times of modern shipping that pepper became affordable for everyone. The last decades of the twentieth century saw an increase in the global production of pepper because new plantations appeared in Thailand, Vietnam, China, Sri Lanka and Brazil. The most important growing areas today are still in India, followed by Indonesia: The two countries together produce about 50% of the global harvest. By far the largest importer is the USA with about 40,000 tons, followed by Germany (Jagella 1999 p. 2).

Extraction of the Oil

The fat oil of *Piper nigrum* is cold pressed or extracted from the seeds (Daulatabad et al. 1995).

3 Character

Colour

Yellow.

Odour

Peppery, pungent, smell of cloves and mango peel.

4 Ingredients

Composition of Fatty Acids

Fatty acids	Content in % acc. to Daulatabad et al. (1995)	Content in % acc. to uni-hamburg.de/12.02.04
Capric acid	4.1	–
Lauric acid	2.5	1
Myristic acid	3.1	1.6
Palmitic acid	27.2	28.8
Stearic acid	7.3	–
Oleic acid	29.9	32.4
Linoleic acid	7.7	31
Linolenic acid	–	5.1

Other Ingredients

(Dasgupta and Datta 1980)

Whether and to what amount the spicy substance piperine is contained in the fat oil could not be determined. As pungency is distinctly detectable in odour tests, it must be assumed that piperine is to be found in fat pepper oil, at least in small quantities.

Physical Key Figures of Pepper Oil

refractive index: n_D^{16} 1.4735
saponification number: 185
iodine number: 129
unsaponifiable: 2.7%
point of solidification: 12 °C
(Roth and Kormann 2000, p. 163)

5 Application

Black pepper is a spice used worldwide. In pharmacy and medicine, it is used as a stomachicum and a skin irritant. In folk medicine, it is used to treat scabies and neuralgia. To a greater part, all these applications can be traced back to the spicy substance piperine. Whether this is also applicable for the fat oil still needs to be examined scientifically (Hager 1978 volume 6a, p. 697).

Pequi Oil

synonyms: Pequiöl (D); pequi butter (E); huile de Péqui (F); Óleo de Pequil (Pt).

1 Source Plant

Caryocar brasiliense Camb. *(Caryocaraceae)*, Pequi

Habitat

Caryocar brasiliense is found from Costa Rica to southern Brazil and Paraguay, especially in the Brazil savannahs (Dantas de Araujo 1995).

S. Krist, *Vegetable Fats and Oils*, https://doi.org/10.1007/978-3-030-30314-3_91

Description

Caryocar brasiliense is a tree with a voluminous, broad crown and grows up to 15 m tall. The branches are alternate, often horizontal, which is why it often looks like a bush. The wood is yellowish, with distinct growth rings. The ligneous rhizome annually produces not woodened sprouts above ground. The fruit are stone fruit containing two seeds each. They weigh about 61–73 g and measure about 12 cm in diameter. The mesocarp is thick, fatty or fleshy, and has an intense smell and taste. It consists of an outer, whitish layer and an inner, yellowish layer. The endocarp is ligneous, with spikes on the outside. The seeds are kidney-shaped and weigh approximately 1.5 g. The seeds as well as the mesocarp contain oil (Dantas de Araujo 1995).

2 Cultivation and Extraction

Cultivation

For growth, the pequi needs tropical climates. The main growing areas are Brazil and Guyana. The blooming period is before and at the beginning of the rainy season, i.e. from August to September. The fruit develop and ripen after the rains. The harvest takes place from January to March. The fruit should be processed quickly; long storage activates enzymes that result in an oil that becomes rancid quickly (Axtell 1992, p. 37, 38; Dantas de Araujo 1995).

Extraction of the Oil

The native population of South America employs a simple method to obtain the oil contained in the fruit pulp. They cook the nuts in water simmering at a low heat and stir carefully until the oil is extracted. The oil rises to the surface and is removed with a spoon; further processing is not necessary: The oil can be used directly (Dantas de Araujo 1995).

On an industrial scale, the oil can be extracted with various solvents. Hexane is the most common, as it has a strong affinity to oils and fats and is easy to remove. Water can also be used as a solvent, mostly in oil production on a small scale. It has the disadvantage of a low yield and a high risk of microbial contamination. In addition, high temperatures are necessary, which destroy components of functional importance, for example carotenoids. A study by Aquino et al. tested several solvents (hexane, acetone, ethyl alcohol and their mixtures) for the extraction of pequi oil. Acetone and hexane achieved the best yield, especially their mixtures with ethyl alcohol. The various solvents had no influence on the composition of fatty acids, the iodine number, the saponification number or the refractive index. Extraction with acetone and ethyl alcohol as pure solvents resulted in a higher amount of carotenoids in the oil (Aquino et al. 2011).

3 Character

Colour

Intensely yellow to orange (Segall et al. 2006).

Odour

Weak (Dantas de Araujo 1995).

4 Ingredients

Composition of Fatty Acids

Fatty acids in the fruit pulp	Content in % acc. to Garcia et al. (2007)	Content in % acc. to Aquino et al. (2011)	Content in % acc. to Miranda-Vilela et al. (2009b)	Content in % acc. to Dantas de Araujo (1995)	Content in g/kg^{-1} acc. to Segall et al. (2006)
Caproic acid	–	–	–	0.8	–
Caprylic acid	–	0.03–0.05	–	0.1	–
Capric acid	–	–	–	0.1	–
Lauric acid	–	0.04	–	0.1	–
Myristic acid	0.2	0.08–0.18	–	0.3	–
Palmitic acid	41.1	39.64–41.55	41.78	39.0	442.8
Palmitoleic acid	0.5	0.96–1.11	0.67	1.6–1.7	–
Margaric acid	–	0.09–0.50	–	–	–
Stearic acid	1.9	1.73–2.58	1.28	0.7–1.2	25.8
Oleic acid	54.0	52.90–54.79	54.28	51.7–54.0	487.1
Vaccenic acid	0.3	–	–	–	–
Linoleic acid	0.9	1.10–1.32	1.36	2.0–3.5	44.3
Linolenic acid	–	0.21–0.29	0.51	1.0–1.2	–
Arachidic acid	0.2	0.16–0.20	0.12	–	–
Gondoic acid	0.7	0.15–0.22	–	–	–
Eicosenoic acid	0.2	–	–	–	–

Fatty acids in the seed	Content in % acc. to Dantas de Araujo (1995)
Myristic acid	0.4–0.5
Palmitic acid	32.0–33.0
Palmitoleic acid	0.4–0.7
Stearic acid	2.6–3.0
Oleic acid	44.0–47.0
Linoleic acid	15.0–19.0
Linolenic acid	0.3–0.6
Arachidic acid	0.5–0.8

P

Composition of Triglycerides

Triglycerides	Content in g/kg acc. to Segall et al. (2006)
OOO	56
POO	466
POP	452
OOS	5.2

Other Ingredients

Carotenoids	Content in mg/100 g acc. to Miranda-Vilela et al. (2009b)
Total	6.75–28.66
β-Carotene	6.26–11.5
Lycopene	1.12–2.08

Physical Key Figures

iodine number:	49.19–51.34
saponification number:	197.20–197.49
acid number:	1.54–4.00
refractive index:	1.457–1.459

(Aquino et al. 2011)

Shelf Life

A study by Garcia et al. examined the thermal stability of pequi oil in comparison to other oils with a high content of oleic acid. Stability depends heavily on the chemical structure of the oil. Oils with a high amount of unsaturated fatty acids are less stable than oils with saturated fatty acids. The result of the study was that pequi oil was most stable in a nitrogen atmosphere; in synthetic air, it produced poor results. The reason is probably that of all the oils examined in the study, pequi oil contains the highest amount of short-chain fatty acids (Garcia et al. 2007).

5 Application

In Pharmacy and Medicine

Two studies by Miranda-Vilela et al. tested the effect of pequi oil on athletes when it is administered orally. The test persons ran distances of their own choice and were examined both before and after they had ingested pequi oil. The races took place in the same conditions and at the same intensity. Pequi oil was administered in capsules over a period of 14 days. Each capsule contained 400 mg pequi oil. Study (a) showed a reduction of damage to the DNA and a reduction of tissue lesions, muscle damage and lipid peroxidation after ingestion of pequi oil. It protected the runners from oxidative stress induced by sports and can thus be regarded as a good antioxidative dietary supplement. Study (b) proved anti-inflammatory effects and a reduction of the total amount of cholesterol as well as of LDL. There was also a tendency towards a decrease in arterial blood pressure (Miranda-Vilela et al. 2009a).

As Food

Pequi oil is ideal for cooking and can be a substitute for butter (Dantas de Araujo 1995).

In Cosmetics

Due to its pleasant smell, the oil obtained from the seeds of pequi is often used in cosmetic products (Dantas de Araujo 1995) and as a massage oil.

Perilla Oil

Oleum Perillae synonyms: Perillaöl (D); perilla seed oil (E); huile de perilla (F)

1 Source Plant

Perilla frutescens (L.) Britt. *(Lamiaceae) syn. Perilla ocymoides* L. *Dentidia nankinensis*, perilla, shiso, Chinese basil

S. Krist, *Vegetable Fats and Oils*, https://doi.org/10.1007/978-3-030-30314-3_92

P

Habitat

Perilla is a native plant of East and South-East Asia, from the mountainous regions of India and China. It grows up to 1200 m above sea level (Yu, 1997). Today, the main growing areas are in the USA, China, Japan, Korea, Vietnam, Myanmar, India and Iran. As a result of its oil, the plant is also cultivated in the former Soviet Republics, Cyprus, South Africa and Austria (Hager 1998, p. 333).

Description

Perilla is an annual plant that grows 0.5–1 m tall. Its leaves sit on long stems and are broad and oval, with curly edges. The small white flowers stand in false whorls. They are bell-shaped, bilabiate, and 3–10 mm long. The fruit are either oval or round, and measure about 1.5 mm in diameter (Hager 1998, p. 328 et seqq). The fruit is a capsule, and open at the top (Löw 2003, p. 60).

2 Cultivation and Extraction

Cultivation

The plant is relatively undemanding; it is an ornamental plant, and prefers sunny, warm locations with rich, permeable soil, regular irrigation and occasional fertilisation (Roth and Kormann 2000, p. 90).

Extraction of the Oil

The seeds are cold pressed to obtain the oil (Roth and Kormann 2000, p. 90).

3 Character

Colour

Medium-yellow oil, liquid at room temperature.

Yellow, drying oil (Roth and Kormann 2000, p. 144).

Odour

Intensely fatty, fusty, tart, acetous, slightly green.

Volatile Compounds

Olfactory compounds in perilla oil dependent on the roasting temperature of the seeds before extraction (Kim et al. 2000):

Compounds	Perilla oil (peak area ratio)					
	Unroasted	150 °C (roasting temperature of the seeds)	160 °C (roasting temperature of the seeds)	170 °C (roasting temperature of the seeds)	180 °C (roasting temperature of the seeds)	190 °C (roasting temperature of the seeds)
Aldehydes:						
2-Methylpropanal	–	1.09	1.67	2.86	2.61	2.81
3-Methylbutanal	–	1.82	2.53	2.62	1.72	2.58
2-Methylbutanal	–	1.29	2.42	4.60	5.35	7.73
Hexanal	Traces	Traces	Traces	Traces	Traces	Traces
2-Heptenal	–	–	–	–	–	0.37
2,4-Heptadienal	–	–	–	–	0.17	0.40
Perillaldehyde	–	0.16	0.15	0.19	0.19	0.19
Ketones:						
2-Propanone	–	0.30	0.30	0.52	1.07	1.72
2-Butanone	–	1.15	1.12	1.68	3.57	4.83
2-Pentanone	–	0.22	0.27	0.18	0.16	0.52
Perilla ketone	0.06	0.16	0.14	0.14	0.14	0.18
Alcohol:						
1-Penten-3-ol	–	0.73	0.80	1.20	1.63	2.48
Furans/oxazols:						
2,5-Dimethylfuran	–	0.22	0.15	0.19	0.25	0.37
2-Furfural	–	–	–	–	0.45	2.74
Furfuryl alcohol	–	–	–	–	0.58	5.24
5-Methyl-2-furfural	–	–	–	–	–	0.61
Trimethyl oxazol	–	–	–	–	0.13	0.19
3,5-Dimethyl	–	–	–	–	–	0.24
Nitrogenous components:						
Pyrazine	–	–	–	–	0.51	2.75
2-Methylpyrazine	–	–	0.29	1.10	4.62	15.86
2,5(6)-Dimethylpyrazine	–	0.26	0.60	1.92	3.59	6.70
Ethylpyrazine	–	–	–	0.17	0.45	1.35
2,3-Dimethylpyrazine	–	–	–	0.18	0.45	0.92
2-Ethyl-6-methylpyrazine	–	–	–	0.12	0.25	0.58
2-Ethyl-5-methylpyrazine	–	–	–	0.16	0.24	0.38
Trimethylpyrazine	–	0.12	0.18	0.48	0.74	1.08
2-Ethyl-3-methylpyrazine	–	–	–	–	0.11	0.33
2-Acetylpyrazine	–	–	–	–	–	0.24
2-Ethyl-2,5-dimethylpyrazine	–	–	–	0.20	0.29	0.42
Pyridine	–	–	–	0.12	0.27	0.84

P

Compounds	Perilla oil (peak area ratio)					
	Unroasted	150 °C (roasting temperature of the seeds)	160 °C (roasting temperature of the seeds)	170 °C (roasting temperature of the seeds)	180 °C (roasting temperature of the seeds)	190 °C (roasting temperature of the seeds)
1H-pyrrole	–	0.15	0.16	0.22	0.68	2.44
Sulphurous components:						
Dimethyldisulfide	–	0.16	0.20	0.29	0.51	0.82
3-Methylthiophene	–	–	–	–	0.14	0.23
4-Methylthiazole	–	–	–	–	0.23	0.90
5-Methylisothiazole	–	–	–	–	–	0.14
3,4-Dimethylisothiazole	–	–	–	–	0.15	0.52
4,5-Dihydro-2-methylthiazole	–	–	–	–	–	0.20
Others:						
Acetic acid methyl ester	–	–	–	0.14	0.22	0.40
Chloroform	–	–	–	–	–	1.45
Acetic acid	–	0.15	0.16	0.18	1.09	5.98
Phenol	–	0.29	0.29	0.24	0.25	0.32
Octane	Traces	0.87	0.84	0.38	0.33	0.55
2-Octene	–	–	–	–	0.25	0.72
Trans-caryophyllene	Traces	Traces	Traces	Traces	Traces	Traces

Flavour

Similar taste to linseed oil (Roth and Kormann 2000, p. 144). Pleasant (Löw 2003, p. 60).

Aromatic taste, faintly resembling cinnamon, aniseed or liquorice, tart.

4 Ingredients

Examinations of 5 different varieties of *Perilla frutescens* seeds showed an oil content of 38.6–47.8% of the dry weight, composed of 91.2–93.9% neutral lipids, 3.9–5.8% glycolipids and 2.0–3.0% phospholipids. The neutral lipids consist, to a great extent, of triglycerides (88.1–91.0%), and, to a small extent, of sterol esters, hydrocarbons, free fatty acids, free sterols and partial glycerides (Yu 1997, p. 93 et seq).

Composition of Fatty Acids

Fatty acids	Content in % acc. to Firestone (1999), p. 76	Content in % acc. to Hager (1998), p. 333	Content in % acc. to Rossell Pritchard (1991)
Palmitic acid	6–7	–	7
Stearic acid	1–2	–	2
Oleic acid	13–15	21	13
Linoleic acid	14–17	11	14
α-Linolenic acid	44–64	55	64
Saturated FA	–	9	–

Composition of Sterols

Sterols	Content in mg/100 g acc. to Yu (1997), p. 97
Campesterol	9.4–10.4
Stigmasterol	3.1–10.0
β-Sitosterol	54.7–72.6
Δ5-Avenasterol	5.4–32.8

Composition of Tocopherols

Tocopherols	Content in mg/100 g acc. to Matthaus et al. (2003)
α-Tocopherol	57
β-Tocopherol	37
γ-Tocopherol	538
Δ-Tocopherol	40

Physical Key Figures of Perilla Oil

density: d_4^{15} 0.927–0.933
refractive index: n_D^{15} 1.4825–1.4850
saponification number: 187–197
iodine number: 196–206
unsaponifiable: 0.4–1.5%

(Roth and Kormann 2000, p. 144)

5 Application

In Pharmacy and Medicine

Adjuvant Treatment of Diabetes Mellitus

A study in which 101 patients suffering from type 2 diabetes participated tested the effects of perilla oil. The test persons were given perilla oil in capsules over a period of 12 weeks; the oil contained a propoertion of about 58% α-linolenic acid and contained the antioxidant vitamins C and E. The result was a significant decrease in triglycerides in the serum, the total cholesterol level and LDL cholesterol in comparison to the initial values, whereas HDL cholesterol increased. For 63% of the patients, the effect was rated very good or good, and tolerability was rated good to very good by 95% of the test persons (Schwedes et al. 2004).

As a Dietary Supplement

A long-term study on 20 elderly Japanese patients helped to explain on the long-term effects of dietary α-linolenic acid from perilla oil on fatty acids in the serum and risk factors for coronary diseases. For a duration of 10 months, perilla oil was used instead of soy bean oil for cooking, which resulted in an increase of 3 g dietary α-linolenic acid per day. After 3 months, this increased the level of α-linolenic acid from 0.8% to 1.6%. After 10 months, the value of DHA (docosahexaenoic acid) had increased from 5.3% to 6.4%, and the value of EPA (eicosopentaenoic acid) had increased from 2.5% to 3.6% in the total serum lipid, without any significant effects on body weight, total cholesterol in the serum, triacylglycerol, glucose, insulin, HbA1c-concentrations, the number of thrombocytes and aggregation, prothrombin time, fibrinogen or the normal blood test results. The results showed that perilla oil as a dietary supplement has no serious contraindications, even for elderly people (Ezaki et al. 1999).

To Treat Allergies

Several studies tested the oil's effect on allergies such as hay fever and atopic dermatitis. The results can be traced back to the α-linolenate/linolate ratio and the inhibition of leukotriene production (Yu 1997, p. 5 et seqq).

Positive Effects on Lipid Metabolism

A study on rats showed that α-linolenic acid, which is contained in perilla oil with a share of 58.4%, lowers the cholesterol level in the serum more effectively than linoleic acid, independently of age (Ihara-Watanabe et al. 2000).

Another study on hyperlipidemic rats showed a significant decrease in the total cholesterol level in the serum and LDL cholesterol, as well as an increase in the ratio of HDL cholesterol/total cholesterol and HDL cholesterol/LDL cholesterol. These effects were more pronounced than with administration of Clofibrat. The results show that perilla oil can be used to regulate the lipid metabolism and for the prevention as well as the treatment of arteriosclerosis (Yan et al. 1993).

To examine the growth of visceral fat tissue under the influence of a diet of perilla oil, rats were fed a diet containing 12 g/100 g dietary perilla oil for a period of 4 months. The results showed that excessive growth of visceral fat tissue was significantly inhibited (Okuno et al. 1997).

Perilla Oil as Osteoporosis Prophylaxis

In an examination of the effects of DHA (docosahexaenoic acid), EPA (eicosopentaenoic acid) and perilla oil in combination with food ingestion reduced by 50%, an increase in breaking force and bone density as well as a decrease in the levels of desoxypiridinoline and calcium in the urine were observed. This indicates that DHA, EPA and perilla oil lower the excretion of calcium via the urine, and thus inhibit osteoporosis by way of a reduced diet (Sun et al. 2004).

Advantages of Linolenic Acid

Tests on rats showed that in comparison to animal fats, a diet containing perilla oil lowered plasma cholesterol, the triglyceride level, TXB_2 in the colon mucosa, the PGE_2 level and the faecal excretion of bile acid. Secondary bile acids like deoxycholinic and lithocholinic acids are tumour promoters in the mucosa of the colon.

Perilla oil reduced the risk factors of bowel cancer most effectively, but induced an increase in lipid peroxidation in the tissue. It is therefore recommended for perilla oil to be used in combination with other oils, for example fish oil, to reduce the risk of bowel cancer and cardiovascular diseases (Park et al. 1995).

In Traditional Chinese Medicine

In traditional Chinese medicine, coughs, breathing difficulties caused by mucus formation in the bronchial tubes, and constipation are treated with perilla oil. These uses have not yet been proven medically (Hager 1998, p. 335).

As Food

In order to test the suitability of perilla oil as a substitute for margarine and butter, the oil was esterified with lard and compared to butter, lard, margarine and hardened soya bean oil in a study on SHRSP[1] rats. The rats fed with butter lived longest, followed by those fed with perilla lard, lard, margarine and hardened soya bean oil. Further studies need to clarify whether perilla oil is a suitable substitute fat in human nutrition (Tatematsu et al. 2003).

Perilla frutescens is a traditional oil fruit; the native population of north-eastern India have been eating it for years without any side effects. A study was carried out to evaluate any possible toxicity. Twenty-four rats were fed on a diet of 10% perilla oil for 18 weeks. In comparison with peanut oil, no significant differences were observed. The oil can therefore be used as an edible oil in combination with other vegetable oils (Longvah et al. 1999).

Industrial Uses

In China, perilla oil is used in the dye industry and for the impregnation of paper (Roth and Kormann 2000, p. 144); this is because the oil has the highest known iodine number and thus dries more quickly than linseed oil (Löw 2003, p. 60). The oil is ideal for producing quick-drying paint. In combination with paint resin, it is a bright, transparent varnish. It can also be a substitute for linseed oil or tung oil in paints, varnish, linoleum, oilskin, watertight paper, printing ink and similar products. It is stable and of high quality (Yu 1997, p. 5).

[1] SHRSP (stroke-prone spontaneously hypertensive rats), a strain of rats that develops high blood pressure and cerebral haemorrhage, especially if the drinking water is substituted with a saline solution.

P

Possible Unwanted Side Effects

A study on animals commissioned by EFSA (European Food Safety Authority) concluded that perilla aldehyde, a flavouring agent that appears in perilla oil in small quantities, induces DNA damage in the liver (www.efsa.eurpoa.eu; 23.8.2015).

Pine Kernel Oil

synonyms: Pinienöl, Pinienkernöl (D); stone pine oil (E); huile de pignon (F)

1 Source Plant

Plants of the *Pinus* species, especially *Pinus pinea* L. (Pinaceae), but also *Pinus pinaster Aiton* and *Pinus sylvestris* L., as well as other species of pine, pine

S. Krist, *Vegetable Fats and Oils*, https://doi.org/10.1007/978-3-030-30314-3_93

Habitat

Pine kernel oil can be obtained from several pine species. Most common on the global market is *Pinus pinea* (Nergiz and Dönmez 2004).

The plant has its origins in the coastal regions of the western Mediterranean area and is also cultivated in the eastern Mediterranean region (Roth and Kormann 2000, p. 94). The tree grows along coasts from Portugal to Turkey as well as in North Africa. It also finds ideal conditions on Madeira and the Canary Islands.

Description

The stone pine *Pinus pinea* is a tall, evergreen conifer with a broad, umbrella-shaped crown; it can reach a height of 20–30 m. The dark-green to grey-green needles are 10–20 cm long, about 2 mm broad, and are arranged in pairs on the spurs. The male cones are oval and about 1 cm in size; the female cones are brown to reddish-brown and 8–14 cm long. They only ripen in the third year after the flower and contain seeds of a dull reddish-brown colour. The seeds are about 2 cm in length and rounded on both sides. The kernel is cylindrical or spindle-shaped, surrounded by a reddish-brown pellicle, which in turn is surrounded by a hard shell (Roth and Kormann 2000, p. 94).

2 Cultivation and Extraction

Cultivation

As pines are frost-susceptible, they are cultivated almost exclusively in the Mediterranean area on loose, dry, warm, sandy soil. They tolerate dry periods and have been planted as shade trees along streets and alleys since antiquity (Roth and Kormann 2000, p. 95).

Extraction of the Oil

The oil is obtained by pressing the seeds that are rich in fat (approx. 50%) (Roth and Kormann 2000, p. 145).

3 Character

Colour

oil from *Pinus pinea*: liquid, clear, brownish-yellow oil (Roth and Kormann 2000, p. 145).

oil from *Pinus sylvestris*: brownish-yellow (Löw 2003, p. 63).

Odour

oil from *Pinus pinea*: nearly odourless (Roth and Kormann 2000, p. 145).

Flavour

oil from *Pinus pinea*: sweetish taste (Roth and Kormann 2000, p. 145).

4 Ingredients

Pine kernels contain 31 to 68% oil and some unusual fatty acids with a cis-5-ethylene bond, containing polymethylene (Wolff and Bayard 1995).

It is generally assumed that methyl-substituted saturated fatty acids are characteristic components of animal lipids and micro-organic lipids that do not usually occur in plants. 14-MHD (14-methylhexadecanoic acid) was, however, found in all species of *Pinaceae*, in a quantity of about 0.86–1.40% (Wolff et al. 1997).

Composition of Fatty Acids

Pinus pinea

The seeds of *Pinus pinea* contain 45–50% oil (Löw 2003, p. 76), of which the main components are oleic acid (48%), palmitic acid (5.4%) and stearic acid (0.6%) (Roth and Kormann 2000, p. 145).

Fatty acids	Content in % acc. to Firestone (1999), p. 77	Content in % acc.to Wolff and Bayard (1995)	Content in % acc. to Nergiz and Dönmez (2004)
Myristic acid	–	–	0.05 ± 0.004
Palmitic acid	6–8	5.55	6.49 ± 0.078
Palmitoleic acid	0.1–0.2	0.16	0.224 ± 0.03
Margaric acid	–	0.05	Traces
Stearic acid	2–3	3.20	3.47 ± 0.1
Oleic acid	36–39	36.34	38.60 ± 0.59
Linoleic acid	47–51	47.19	47.6 ± 0.3
α-Linolenic acid	0.6	0.63	0.68 ± 0.001
Pinolenic acid	0.4	0.35	–
Arachidic acid	0.5	–	0.54 ± 0.03
Eicosenoic acid	0.7	0.74	0.79 ± 0.06
Eicosadienoic acid	0.5	–	–
Eicosatrienoic acid	2.5	2.47	–
Behenic acid	–	–	0.13 ± 0.01
Lignoceric acid	–	–	3.02 ± 0.02

Pinus sylvestris

The kernels contain 25–30% oil, of which the main components are linoleic acid (57.9%), linolenic acid (25.4%), oleic acid (9.5%), palmitic acid (4.3%) and stearic acid (3.1%) (Löw 2003, p. 63).

Fatty acids	Content in % acc. to Wolff and Bayard (1995)
Palmitic acid	3.36
Palmitoleic acid	0.35
Margaric acid	0.05
Stearic acid	1.78
Oleic acid	14.36
Linoleic acid	44.84
α-Linolenic acid	0.38
Pinolenic acid	21.65
Eicosenoic acid	1.14
Eicosadienoic acid 5c,11c	0.50
Eicosadienoic acid 11c,14c	0.99
Eicosatrienoic acid	5.46

Pinus pinaster

Fatty acids	Content in % acc. to Pasquier et al. (2001)	Content in % acc. to Asset et al. (1999)	Content in % acc. to Gresti et al. (1996)
Palmitic acid	5.3	3.6	3.8
Palmitoleic acid	–	0.2	0.2
Margaric acid	–	0.1	–
Stearic acid	2.8	2.4	2.8
Oleic acid	27.9	18.1	18.8
Linoleic acid	45.8	55.9	52.2
α-Linolenic acid	1.2	1.3	1.3
Pinolenic acid	6.6	7.1	8.6
Arachidic acid	–	–	0.3
Eicosenoic acid	–	1.0	1.1
Eicosadienoic acid 5c,11c	0.7	0.8	0.8
Eicosadienoic acid 1 c,14c	–	0.8	0.9
Eicosatrienoic acid	6.0	7.1	8.1

Pinus koraiensis

Fatty acids	Content in % acc. to Gresti et al. (1996)	Content in % acc. to Wolff and Bayard (1995)
Palmitic acid	4.4	4.20
Palmitoleic acid	0.1	0.10
Margaric acid	–	0.03
Stearic acid	2.3	1.82
Oleic acid	29.4	24.06
Vaccenic acid	–	1.46
Linoleic acid	43.9	43.38
Pinolenic acid	14.2	14.92
α-Linolenic acid	0.2	0.17
Arachidic acid	0.3	–

Fatty acids	Content in % acc. to Gresti et al. (1996)	Content in % acc. to Wolff and Bayard (1995)
Eicosenoic acid	1.2	1.03
Eicosadienoic acid 5c,11c	0.1	0.10
Eicosadienoic acid 11c,14c	0.6	0.49
Eicosatrienoic acid	1.1	0.90

Composition of Triglycerides in Pinus pinea Seed Oil

Triacylglyceroles	Content in % acc. to Nergiz and Dönmez (2004)
LLL	10.8
OLnL	2.23
PLnL	0.83
OLnO	23.5
PLL	5.32
PLnO	1.23
LOL	18.6
PLO	10.6
LPP	0.87
OOO	10.3
SLO	3.39
POO	5.38
OPP	0.77
SOO	1.87
SLS	0.25
POS	0.46
OSS	0.28
PSS	0.14

Physical Key Figures of Pine Kernel Oil

Pinus pinea

density: d_{15}^{15} 0.920–0.932
refractive index: n_D^{40} 1.467–1.485
saponification number: 192–198
iodine number: 118–125
unsaponifiable: 0.5–2%
point of solidification: −21 to −27 °C
(Roth and Kormann 2000, p. 145)

Pinus sylvestris

saponification number: 194
iodine number: 184
unsaponifiable: 1.2%
point of solidification: −30 °C
(Löw 2003, p. 63)

P

5 Application

In Pharmacy and Medicine

Influence on the Lipid Level

A study by a Korean university examined dietary pine kernel oil from *Pinus koraiensis* for the purpose of obtaining concentrated pinolenic acid by crystallisation using urea and to examine the cholesterol-lowering effect of pinolenic acid on the activity of the LDL receptors in human liver cells (HepG2 cells). Treatment of HepG2 cells with an alcoholic extract of fatty acids with a high amount of pinolenic acid resulted in an increased intake of 3,3′-dioctadecylindocarbocyanine-LDL into the cell. This result indicates properties lowering the LDL cholesterol level and an increase in the hepatic LDL absorption of the extract (Lee et al. 2004a).

A study on mice with a deficiency in ApoE examined the effect of an oil made from the seeds of *Pinus pinaster* on the lipoprotein metabolism. Sunflower oil and lard were used as comparison material. Compared with these, the levels of cholesterol, VLDL+IDL cholesterol and VLDL+IDL phospholipids were lower if pine seed oil was used, but the levels of triglycerides and VLDL+IDL triglycerides were higher. Despite these changes, *Pinus pinaster* seed oil did not prevent the development of arteriosclerosis in mice with a deficiency in ApoE. However, the VLDL+IDL-lowering potential of the oil in the study on mice with a deficiency in ApoE appears promising (Asset et al. 1999).

A study illustrated the effect of *Pinus koraiensis* seed oil on the composition of serum lipids of rats and the immune reaction of mice. Rats fed a diet rich in cholesterol-containing pine kernel oil did not show any differences from the control group with regard to increase in body weight and feeding habits. The total plasma cholesterol, triacylglycerol and phospholipid level were significantly reduced, the HDL cholesterol level was increased, and the VLDL and LDL cholesterol levels were decreased. The diet fed to mice showed effects on the intestinal immune system. Administration of 5% pine kernel oil over a period of 20 days resulted in stimulation of the lysosomal enzyme activity in macrophages. These results document the oil's hypocholesterolemic effect on rats, as well as the stimulation of cell proliferation in the spinal bone marrow and the heightened reaction of macrophages in mice (Lee et al. 2004b).

Another study examined the influence of *Pinus koraiensis* seed oil on rats compared to that of linseed oil, thistle oil and evening primrose oil. The hypocholesterolemic effect of pine kernel oil was lower than that of linseed oil, but higher than that of thistle oil. However, the dietary oils had no influence on the concentration of liver lipids. The amount of ADP-induced platelet aggregation and the production of prostacyclin remained the same; neither did the oil have any effect on the rats' growth. Pine kernel oil reduced the age-related increase in blood pressure of rats after 5 weeks, an effect that remained constant until the end of the study after 8 weeks. The oil had a positive effect on physiological functions preventing hypercholesterolemia, thromboses and hypertension (Sugano et al. 1994).

Various tests on animals proved the lipid lowering potential of pine kernel oil. Future studies should examine whether the oil is a suitable dietary supplement and should ensure that no unwanted side effects are involved (Pasquier et al. 2001).

In Dietetics

Pinolenic acid (5,9,12–18:3), which is contained in the seeds of members of the *Pinaceae* family, has a function in the biosynthesis of eicosanoids, lowers blood pressure and is employed in dietetic nutrition (Imbs and Pham 1996).

Pistachio Oil

Oleum Pistaciae synonyms: Pistazien(kern)öl (D); huile de pistache (F)

1 Source Plant

Pistacia vera L. (Anacardiaceae), pistachio

S. Krist, *Vegetable Fats and Oils*, https://doi.org/10.1007/978-3-030-30314-3_94

Habitat

Pistacia vera originates in Central Asia and the Mediterranean area. Alexander the Great brought it to Greece, and the Romans brought to to Sicily. Even today, pistachios grow wild in several countries, for example in Afghanistan and India. They prefer dry, desert-like regions and are very frost-susceptible. *Pistacia vera* has a biennial crop sequence, which is why there are in turn small and large amounts of crop (Hager 1978, volume 6a, p. 730; Roth and Kormann 2000, p. 145).

Description

Pistacia vera is an evergreen, deciduous tree that can reach a height of 8–12 m. The crown is spreading and forms a dense canopy of leaves. The leaves are pinnate or bipinnate, greyish green, with stalkless, ovate leaflets. The panicle is short; the inconspicuous, axillary flowers are a reddish colour. They develop into elongated, oval stone fruit that are about 2–3 cm long. The fruit are brownish red and wrinkly, with a thin layer of fruit pulp that tastes of turpentine. They contain the seeds that are sold as pistachios or "green almonds". The seed is usually triangular, a green, brownish or violet colour and 20 mm long. It is slightly compressed at the sides and protected by a whitish, hard shell (Hager 1978, volume 6a, p. 730).

2 Cultivation and Extraction

Cultivation

Pistacia vera is cultivated for commercial purposes in Iran, Turkey, Syria, Greece and Italy. In the 1970s, large plantations were planted in the USA, especially in California. Today, the USA is one of the main producders, in addition to Turkey and Iran (Hager 1978, volume 6a, p. 730; Yildiz et al. 1998).

Extraction of the Oil

Pistachio oil is obtained by cold pressing or extracting the pistachio seeds (Roth and Kormann 2000, p. 145).

3 Character

Colour

Depending on the mode of extraction golden yellow to dark green (Roth and Kormann 2000, p. 145).

Odour

Sweet, fruity, smell of pistachio seeds, slightly green.

Nearly odourless (Roth and Kormann 2000, p. 145)

Volatile Compounds

SPME-GC-MS analysis (Bail et al. 2009b):

Acetous acid	2-Methylpyrazine
3-Methylpropanal	Furfural
2-Methylbutanal	Gamma-butyrolactone
3-Hydroxy-2-butanone	Trans-2-heptanal
Hexanal	Limonene
Octane	2-Ethyl-3-dimethylpyrazine

Flavour

Delicately nutty.

Nearly flavourless (Roth and Kormann 2000, p. 145).

4 Ingredients

Composition of Fatty Acids

Fatty acids	Content in % acc. to Roth and Kormann (2000), p. 146	Content in % acc. to Yildiz et al. (1998)	Content in % acc. to Kerschbaum and Schweiger (2001), p. 19
Palmitic acid	8.2	11.7	11.5
Linoleic acid	20.0	18.7	29.2
Palmitoleic acid	–	1.3	0.8
Oleic acid	69.6	65.15	52.2
α-Linolenic acid	–	–	0.6
Stearic acid	1.4	3.15	0.9
Arachidic acid	–	–	0.1
Behenic acid	–	–	0.1
Vaccenic acid	–	–	4.5
Myristic acid	0.6	–	0.1
Eicosenoic acid	–	–	0.2

Composition of Sterols

(concentration in mg/100 g and content in %)

Sterols	Pistachio oil acc. to Firestone (1999), p. 80
Total content	201
Campesterol	5
Stigmasterol	2
β-Sitosterol	77
Δ5-Avenasterol	7
Δ7-Stigmasterol	1

Composition of Tocopherols

Tocopherols	Content in mg/100 g acc. to Kerschbaum and Schweiger (2001), p. 36
Total content	21.6
α-Tocopherol	8.4
γ-Tocopherol	13.2

Other Ingredients

(Kerschbaum and Schweiger (2001), p. 20, 43; Yildiz et al. 1998):

Vitamin E
Mineral nutrients
Vitamin A
Vitamin B_1
Vitamin B_2

Physical Key Figures of Pistachio Oil

refractive index: n_{25}^{40} 1.467
density: d_4^{15} 0.914–0.920
saponification number: 187–196
iodine number: 86–98
unsaponifiable: 0.5–3%
point of solidification: −5 to −11 °C

(Roth and Kormann 2000, p. 146)

Shelf Life

Pistachio oil is relatively stable against oxidation. It thus has a longer storage life than other oils made from nuts (Yildiz et al. 1998).

Characteristics of the Oil

Pistachio oil is comparable to almond oil with regard to its ingredients and composition of fatty acids (Kerschbaum and Schweiger (2001), p. 20).

5 Application

In Pharmacy and Medicine

Pistachio oil contains large amounts of β-sitosterol, which competitively inhibits cholesterol in the organism. Its application in cases of hypercholesterolaemia is conceivable, but no cases of actual application have been mentioned in literature.

In Cosmetics

Oleum pistaciae has excellent moisturising properties and is therefore added to lotions for dry skin, sunblock and after-sun products, as well as nail-care products. It is also used as an emollient (Kerschbaum and Schweiger (2001), p. 45; Hannon 1997).

As Food

Pistachio oil is mainly used as a salad oil. As a result of its delicate nutty flavour, which is comparable to almond oil, it is also used in the sweets industry (Kerschbaum and Schweiger 2001, p. 45).

Plum Kernel Oil

synonyms: Zwetschkenkernöl, Pflaumenkernöl (D); prune kernel oil (E); huile de noyaux de prune (F); olio di nocciolo di prugna (I); aceite del hueso de ciruela (ESP)

S. Krist, *Vegetable Fats and Oils*, https://doi.org/10.1007/978-3-030-30314-3_95

P

1 Source Plant

Prunus domestica L. *(Rosaceae)*, Plum

Habitat

The origins of the plum are not known. They presumably lie in the southern Black Sea area, like those of the cherry; *Prunus cerasifera*, the cherry plum, is probably one of its ancestors. The Romans imported it to northern Central Europe (Lieberei and Reisdorff 2007, p. 188).

Description

The plum tree can reach an age of 20–30 years. Its leaves are ovate, hairless on both sides, with bluntly serrated edges (Guimpel et al. 1815, p. 85).

Within the bud, the leaves are rolled. The hypanthium is flat-patelliform to tubular (Danert 1993, p. 203).

A stone fruit develops from the half-inferior ovary; its size, colour and taste differ (Lieberei and Reisdorff 2007, p. 188).

The one-seeded stone fruit have a hard endocarp and a juicy fruit pulp. The skin is usually bluish violet (Danert 1993, p. 203).

The kernels are a light brown colour (Kaßner 1918).

2 Cultivation and Extraction

Cultivation

The main growing areas of the plum are China, Europe and South America. Loamy, sandy soil is best. If the trees are cultivated in unfavourable locations, the fruit are smaller. Depending on the variety, the harvest takes place between August and September (Guimpel et al. 1815, p. 85, 86, Hackbarth 1944, p. 342).

Plum production in 1000 t (Lieberei and Reisdorff 2007, p. 433)	
	2005
Globally	9863
China	4636
Serbia and Montenegro	580
Germany	568
Romania	409
USA	300
France	283

Extraction of the Oil

The oil is cold pressed (Wimmer et al. 2003).

First, the fruit pulp is removed to isolate the kernels. The seeds are then ground and pressed in a hydraulic press under pressure of up to 350 atm (Kaßner 1918).

Extraction with solvents is also possible (Kaßner and Eckelmann 1914).

3 Character

Colour

Yellowish brown (Hackbarth 1944, p. 342, Wimmer et al. 2003)

Golden yellow (Kaßner 1918).

Light yellow, clear (Kaßner and Eckelmann 1914)

Odour

Like bitter almonds (Hackbarth 1944, p. 342, Wimmer et al. 2003)

Flavour

Pleasant, with a touch of bitter almonds (Wimmer et al. 2003)

4 Ingredients

Composition of Fatty Acids

Fatty acids	Content in % acc. to Matthäus and Özcan (2009)	Content in % acc. to Kamel and Kakuda (1992)	Content in % acc. to Wimmer et al. (2003)	Content in % acc. to Schuster (1992), p. 215
Palmitic acid	5.4–7.3	6.3	6–12	6–8
Palmitoleic acid	–	0.5	–	–
Stearic acid	1.3–1.4	1.4	4–9	4–6
Oleic acid	63.9–78.5	62.0	55–65	55–60
Vaccenic acid	1.1–1.2	–	–	–
Linoleic acid	9.7–26.9	29.6	15–35	25–35
α-Linolenic acid	0.1–0.2	–	–	–
Arachidic acid	0.1	0.3	–	–

P

Composition of Triglycerides

Triglycerides	Content in % acc. to Deineka (2003)	Content in % acc. to Deineka et al. (2002)
LLL	8.05	7.2
LLO	17.6	16.2
LLP	–	2.1
LOO	33.5	26.9
LLS	–	4.5
OOO	40.9	33.3
LOS	–	8.0
OOS	–	2.7

Composition of Tocopherols

Tocopherols	Content in mg/kg acc. to Matthäus and Özcan (2009)
Total content	222.7–374.4
α-Tocopherol	24.1–27.1
α-Tocotrienol	31.4–41.6
β-Tocopherol	2.3–4.0
γ-Tocopherol	133.1–302.1
δ-Tocopherol	11.4–18.9

Physical Key Figures of Plum Kernel Oil

refractive index: n_D^{25}	1.468–1.470
saponification number:	188–190
iodine number:	91–104

(Roth and Kormann 2005, p. 159)

acid number:	1.3
saponification number:	192
iodine number:	108
hydroxyl number:	6.5
unsaponifiable:	0.60%

(Kamel and Kakuda 1992)

specific weight:	0.916–0.9195
light refraction:	1.4705–1.4715 (20 °C)
saponification number:	188.1–198.5
iodine number:	104–121
Plum kernel oil is a non drying oil.	

(Hackbarth 1944, p. 343)

Shelf Life

The oil has a shelf life of 6–8 months (Wimmer et al. 2003).

5 Application

In Cosmetics

Plum kernel oil is often added to cosmetic products, as it is suitable for all skin types. It penetrates the skin quickly and forms part of many massage oils and bath oils (Wimmer et al. 2003).

As Food

Plum kernel oil is a good edible oil, especially for desserts and salads (Wimmer et al. 2003).

Pomegranate Seed Oil

synonyms: Granatapfelkernöl (D)

S. Krist, *Vegetable Fats and Oils*, https://doi.org/10.1007/978-3-030-30314-3_96

1 Source Plant

Punica granatum L. (Punicaceae), pomegranate

Habitat

The pomegranate originates in South-West Asia, especially Iran and Afghanistan. There are still wild and half-cultivated forms in Syria. Today, pomegranates are found in the Mediterranean area, the Middle East, South Africa, South Asia, China, Australia, the south-western USA, and South America (Hager 1994, p. 327 et seq).

Description

The pomegranate can reach a maximum height of 5 m and is thus a rather small tree. It has a reddish-brown to grey bark and is often cultivated as a bush. Its robust appearance is heightened by its hard leaves and the fact that it usually has thorns. The leathery leaves are lanceolate, about 5–8 cm long and usually arranged alternately in pairs on the tetragonal branches. *Punica* flowers are funnel-shaped, about 3 cm in diameter, and in their wild form are a fluorescent red colour. The fruit is called pomegranate; it is a fanned, round mock berry up to 12 cm in size, with a rough, leathery, reddish-brown to leather-brown peel. It contains many seeds, which have a fluorescent red skin when the fruit is ripe (Hager 1994, p. 326). The seeds make up about 3% of the fruit's weight and contain 12–20% oil (Lansky and Newman 2007). Since the pomegranate has many seeds, it has been regarded as a symbol of fertility.

2 Cultivation and Extraction

Cultivation

Punica granatum prefers hot, dry climes and does not make great demands on the quality of the soil (Hager 1994, p. 328). It can be grown in half-shade if necessary, although this has a negative effect on fructification.

Extraction of the Oil

The seeds are cold pressed under exclusion of heat, light and oxygen.

3 Character

Colour

Light-yellow oil (Umarov et al. 1969).

Odour

Oily, sweetish-mellow, green, pungent. Odourless (Umarov et al. 1969).

4 Ingredients

Composition of Fatty Acids

Fatty acids	Content in % acc. to Tsuzuki et al. (2006)	Content in % acc. to Chen et al. (2003)	Content in % acc. to Schubert et al. (1999)	Content in % acc. to Umarov et al. (1969)
Lauric acid	–	–	–	0.46
Palmitic acid	3.9	2.47	4.8	6.27
Palmitoleic acid	–	–	–	1.66
Margaric acid	–	–	–	1.16
Stearic acid	2.6	1.66	2.3	3.87
Oleic acid	6.6	4.07	6.3	10.86
Linoleic acid	6.9	4.85	6.6	13.66
Punicic acid	74.5	81.62	65	59.14
Eicosanoic acid	–	0.53	–	–
Arachidic acid	–	0.44	–	2.92

The predominant fatty acid in pomegranate oil is punicic acid, a conjugated cis-9, cis-11, trans-13-octadecatrienoic acid (Hager 1994, p. 327; Umarov et al. 1969).

Physical Key Figures of Pomegranate Oil

relative density: 0.9362
refractive index: 1.5160 (at 20 °C)
acid number: 1.83
unsaponifiable: 2.26%

(Umarov et al. 1969)

Shelf Life

The oil should be stored in a cool (10–20 °C), dry and dark place in order to guarantee a shelf life of at least 12 months from the production date.

5 Application

In Pharmacy and Medicine

Effect

The pucinic acid contained in pomegranate oil may inhibit the biosynthesis of prostaglandins. A study established that an extract of pomegranate oil rich in phenols has a strong inhibiting effect on lipoxygenase and cyclooxygenase (COX), and thus stops inflammatory processes in the body. If COX is inhibited by conventional, nonsteroidal antiphlogistics, this may have a negative effect on cardiovascular functions, as PGI_2 (prostacyclin) is also inhibited. PGI_2 is responsible for the cardiovascular balance; it inhibits platelet aggregation, is a vasodilator and regulates the expression of endothelial cell adhesion molecules (Lansky and Newman 2007).

Tumour Prophylaxis

In an ex vivo study that exposed cells to the chemical carcinogen 7,12-dimethylbenz[a]anthracene (DMBA), breast tumours could be prevented from appearing by treating mice with cold pressed pomegranate oil; the success rate was 87%. Equally, skin tumours of female mice appeared less often if pomegranate oil was applied externally. In addition, rats exposed to carcinogenic azoxymethane developed bowel cancer less often if the oil was applied (Lansky and Newman 2007).

Inhibition of Enzymes

The enzyme aromatase catalyses the generation of oestrone and oestradiol from androstenedione and testosterone. As a limiting factor of endogenous oestrogen synthesis, aromatase may promote hormone-dependent tumours. Inhibiting the aromatase enzyme complex is an important therapeutic goal in the treatment of many hormone-dependent diseases, for example breast cancer. Pomegranate oil has a weak inhibitory effect on this enzyme. However, the oil has a much more potent inhibitory effect on 17β-hydroxysteroid-dehydrogenase of type 1. It catalyses the transformation of oestrone into the much more effective 17β-oestradiol, which also supports the growth of hormone-dependent breast tumours (Lansky and Newman 2007).

As an Antioxidant

According to a study, cold-pressed pomegranate oil has strong antioxidant properties, comparable to that of butylated hydroxyanisole or green tea made from *Thea sinensis* (Schubert et al. 1999).

Oestrogen Properties

In addition to its unique composition of fatty acids, cold-pressed pomegranate oil has another outstanding characteristic: It contains a number of different oestrogen components, for example phytoestrogens with its important representative coumestrol, and another steroid called oestrone (Schubert et al. 1999).

In Cosmetics

The oil is used as a facial oil for dry, ageing and stressed skin. Pomegranate oil is primarily applied as an active ingredient in combination with base oils. It supports the formation of cells and regenerates tissue. It also has an anti-wrinkle effect, smooths the skin and is thus recommended for ageing skin.

Poppyseed Oil

Oleum Papaveris synonyms: Mohnöl (D); huile d'œillette (F)

1 Source Plant

Papaver somniferum L. (Papaveraceae), opium poppy

S. Krist, *Vegetable Fats and Oils*, https://doi.org/10.1007/978-3-030-30314-3_97

Habitat

The opium poppy is a native plant of the eastern Mediterranean area and West Asia. It is one of the oldest cultivated plants; seeds dating from the New Stone Age (4600–3800 B.C.) were found in several regions of Central Europe. In Poland and south-eastern Spain, poppy-seeds and capsules dating from the Neolithic period (about 2500 B.C.) were dug up. There were also findings in Greece, Crete and Cyprus, which indicate the cultivation of poppy plants as well as the wild growth of poppy plants as weeds. Written records from the eighth century B.C. mention poppy cultivation in Greece and the use of poppies as a medicine. Roman soldiers brought the opium poppy as far as England and Scotland. In the Middle East, poppies are mentioned for the first time in the seventh and sixth century B.C., in Assyrian and Persian sources. In the Far East, China, Japan and India, the opium poppy has only been known since medieval times. Today, the main growing areas are in India, Asia Minor, South-East and South-West Asia, Central and Eastern Europe, Turkey, Iran and the Balkans (Schuster 1992, p. 124).

Description

The opium poppy is an annual plant 30–150 cm tall. It has a sturdy taproot with strong lateral roots. The upright stem is unbranched or branched, usually hairless, more rarely covered with stiff hair, and frosted bluish green. The whole plant contains white chyle, which is obtained by slitting the capsules slightly. The leaking chyle is then dried and forms raw opium, which contains about 45 different alkaloids, including morphine, codeine and papaverine. The leaves are cauline, stalkless, and elongated-oval in form; their edges are either crenate or serrate. The opium poppy is uniflorous. The fairly upright flowers measure up to 10 cm in diameter. The two sepals are green, the four petals violet to white or red, roundish to obovate, and usually have a dark spot at the base. The petals are either entire or wavy, and to a certain extent notched or fringed. The numerous stamina have filaments broadening at the top, and elongated, bluish-green anthers. The fruit is a capsule or poricidal capsule consisting of 8–12 carpels. The kidney-shaped seeds are different colours depending on the plant variety; they are generally bluish grey, white, yellow, greyish red or black. For this reason, a distinction is made between the varieties grey poppy, white poppy and blue poppy. The oil content of the seed amounts to 33–49% (List and Hörhammer 1978, fourth edition, sixth volume, p. 404; Bernäth 1998, p. 337; Singh et al. 1990; Eklund and Agren 1975, p. 188).

2 Cultivation and Extraction

Cultivation

Cultivation is labour-intensive and involves high amounts of manual labour.

The opium poppy grows best in warm, protected, moderately moist locations. The soil should be warm, rich and especially calciferous, as the opium poppy has high calcium requirements. Sandy, loamy or clayey soil as well as soil wetness are not suitable for poppy cultivation. A supply of too much nitrogen has a negative influence on growth. Despite its relatively high water requirements, the opium poppy is susceptible to too much moisture. Wet and cold locations with strong winds are unsuitable for cultivation. Monoculture as well as crop combinations are possible. The seeds should be sown as early in the year as possible, in order to use the remaining winter moisture to advantage for the growth of the young plants. Opium poppies are not susceptible to late frosts. The harvest takes place by cutting or threshing when the seeds rustle in the capsules, which begin to turn brown around that time (Hackbarth 1944, p. 103).

In Germany, cultivation of opium poppies used to be illegal, since opium can be produced from the chyle of the unripe capsules. Various intoxicants can be produced from the chyle, for example opium, morphine and heroin. By subsuming poppy cultivation under the law of narcotics, it was hoped that cultivation for the purposes of drug abuse would be stopped. Meanwhile, varieties containing no or only small amounts of alkaloids have been developed. The blue opium poppy variety "Przemko" is currently the only variety with only small amounts of morphine that is permitted to be cultivated in Germany (Schuster 1992, p. 124).

Extraction of the Oil

The seeds of *Papaver somniferum* are either cold pressed or pressed by adding heat. Cold pressing results in an oil of higher quality, which is especially suitable as an edible oil. Oil pressed at temperatures of 60–70 °C is of inferior quality and is mainly used for technical purposes to produce paint (Bauer 1928, p. 207). Another method is extraction with supercritical CO_2 (Bozan and Temelli 2003).

The blending of high-quality poppy seed oil with much cheaper sunflower oil may pose a problem. Blending of this kind is hard to detect using conventional methods, as the compositions of the oils' fatty acids are nearly identical. A new method, which analyses the volatile compounds of the oil using SPME-GC-MS analysis, makes it possible to detect even the smallest amounts of sunflower oil in poppyseed oil (Krist et al. 2006c).

3 Character

Colour

Cold-pressed poppy seed oil: light yellow to light golden-yellow (Bauer 1928, p. 207).

Hot pressed poppy seed oil:

P

Light yellow with discreet green reflexes.
Dark yellow to reddish (Von Wiesner 1927, p. 757).
Light yellow (Roth and Kormann 2000, p. 137).

Odour

Cold-pressed oil from grey poppy seeds: slightly nutty, sweetish (Krist 2002, p. 148).

Cold-pressed oil from white poppy seeds: slightly nutty, smell of poppy, touch of hazelnut and peanut (Krist 2002, p. 149).

Cold-pressed oil from blue poppy seeds: pleasantly weak, slightly nutty, mild smell (Krist 2002, p. 150).

Volatile Compounds

(Krist et al. 2005)

α-Pinene	2-Methylbutanal
Camphene	γ-Butyrolactone
2-Pentylfurane	γ-Hexalactone
3-Carene	2-Methylpyrazine
ρ-Cymene	Pentanal
Limonene	Hexanal

Flavour

Cold-pressed oil from grey poppy seeds: weak, fatty, slight peanut note, nutty, sweet (Krist 2002, p. 148).

Cold-pressed oil from white poppy seeds: fatty, rancid, soft and pleasant, slightly nutty (Krist 2002, p. 149).

Cold-pressed oil from blue poppy seeds: fatty, touch of peanut skin, strong and full taste (Krist 2002, p. 150).

Hot-pressed poppyseed oil: scratchy taste (Bauer 1928, p. 207; Wiesner 1927, p 757).

4 Ingredients

Composition of Fatty Acids

Fatty acids	Content in % acc. to Roth and Kormann (2000), p. 137	Content in % acc. to Firestone (1999), p. 81	Content in % acc. to Krist et al. (2005)
Myristic acid	–	0.1–0.7	0–1.1
Palmitic acid	Traces	7–11	7.8–30.66
Palmitoleic acid	–	0.8–1.6	–

Fatty acids	Content in % acc. to Roth and Kormann (2000), p. 137	Content in % acc. to Firestone (1999), p. 81	Content in % acc. to Krist et al. (2005)
Stearic acid	Traces	1–4	1.4–10.9
Oleic acid	30	16–30	13.2–36.8
Linoleic acid	60	62–73	18.4–80
α-Linolenic acid	5	–	0–9.4
Lauric acid	–	–	0–13.4

Composition of Triglycerides

(content in %)

Triglycerides	White poppyseed oil acc. to Krist et al. (2006c)	Grey poppyseed oil acc. to Krist et al. (2006c)	Blue poppyseed oil acc. to Krist et al. (2006c)
LnPP	–	0.1 ± 0.2	0.2 ± 0.2
LPP	1.0 ± 0.1	0.8 ± 0.1	1.3 ± 0.2
OPP	0.1 ± 0.0	–	0.2 ± 0.0
LnLP	1.1 ± 0.1	0.6 ± 0.6	1.0 ± 0.1
LLP	24.2 ± 1.5	23.5 ± 0.7	27.1 ± 0.8
LOP	3.8 ± 0.4	3.5 ± 0.6	5.1 ± 0.3
LSP	0.3 ± 0.3	0.4 ± 0.3	–
SSP	1.1 ± 0.4	2.0 ± 1.2	–
LnLnLn	0.4 ± 0.3	0.3 ± 0.3	0.3 ± 0.0
LnLnL	–	–	0.1 ± 0.1
LnLL	1.3 ± 0.1	1.2 ± 0.1	1.2 ± 0.2
LLL	43.5 ± 1.0	41.1 ± 1.0	39.5 ± 0.5
LLO	12.5 ± 0.6	15.1 ± 0.3	14.0 ± 0.1
LOO	7.3 ± 0.1	7.4 ± 0.2	7.0 ± 0.3
LOS	0.1 ± 0.2	0.5 ± 0.3	0.4 ± 0.1
LSS	0.5 ± 0.2	0.4 ± 0.2	–
LLG	0.2 ± 0.2	0.1 ± 0.2	–
LLA	0.6 ± 0.1	0.2 ± 0.2	0.4 ± 0.1
LOA	0.1 ± 0.1	0.1 ± 0.2	–
OOA	0.1 0.3	0.1 ± 0.2	–

Composition of Sterols

Sterols	Content in % acc. to Firestone (1999), p. 81
Cholesterol	–
Brassicasterol	–
Campesterol	22
Stigmasterol	3
β-Sitosterol	68
Δ 5-Avenasterol	2
Δ7-Stigmasterol	2
Δ7-Avenasterol	–

P

Other Ingredients

(Johansson 1979)

Calcium
Magnesium
Iron
Copper
Zinc
Manganese
Vitamin B1

Physical Key Figures of Poppyseed Oil

refractive index: n_D^{20} 1.4750–1.4774
density: d_4^{20} 0.919–0.927
point of solidification: −17 to −27 °C
saponification number: 189–197
iodine number: 133–197
acid number: not over 8
unsaponifiable: 0.4–1.2%

(Roth and Kormann 2000, p. 137)

Shelf Life

Poppyseed oil has a shelf life of about 9 months if it is kept in a closed bottle in a cool and dark place. After opening, the oil should be stored in the refrigerator and be used within 8 weeks.

5 Application

In Pharmacy and Medicine

As poppyseed oil has a mild smell and penetrates the skin quickly, it is added to liniments, ointments and emulsions (Hackbarth 1944, p. 120).

It is also used in chemotherapy: When injected into the hepatic artery, iodised poppyseed oil accumulates selectively as micro droplets only in the tissue of hepatocellular carcinoma, where it remains for over 3 weeks and attacks tumour cells (Bhattacharya et al. 1994; Krist et al. 2005; Higashi and Setoguchi 2000).

Iodine addition compounds of the ethyl esters of fatty acids of poppyseed oil are used as a radio-opaque substance for sonography, lymph-angiography and for medical imaging of the paranasal sinuses (Choi et al. 1989; Krist et al. 2005; Burger and Wachter 1993, seventh edition, p. 1045).

Iodised poppyseed oil can be administered in cases of iodine deficiency, but also as a diagnostic additive in sonography, hystero contrast sonography and angiography (Benmiloud et al. 1994; Furnee et al. 1998).

In Cosmetics

Soap is produced from hot-pressed poppyseed oil. It is easy to use hot-pressed poppyseed oil to produce hard soap (Janystin 1978, p. 622; Hackbarth 1944, p. 120).

Poppyseed oil penetrates the skin quickly; it is moisturising and improves skin elasticity. It therefore forms part of care products such as anti-wrinkle creams, body lotions and balms, and is well suited for dry skin.

Poppyseed oil is also a massage oil (Martindale 1993, p. 1404).

As Food

Since it has an interesting, aromatic taste, cold-pressed poppyseed oil is a high-quality edible oil. Its high content of unsaturated fatty acids can lower high cholesterol levels. Poppyseed oil should not, however, be heated above temperatures of 170 °C, and is thus only suitable for frying and baking to a limited extent. It is mostly used for cold meals like salads, uncooked vegetarian food, sweets and muesli. Since it has a delicate nutty taste, it enhances sweet dishes like *Mohnnudeln* (poppyseed noodles) and poppyseed cakes. The oil is a flavour enhancer, bringing out the food's flavour (Hackbarth 1944, p. 120).

In Painting

As a result of its high drying capacity, poppyseed oil is a suitable raw material for the production of artists' paints. Its disadvantages are that it decomposes at high temperatures and resoftens, and that it can be dissolved by chemical reagents. These drawbacks can be avoided by using so-called stand oil. This is produced by heating poppyseed oil in a carbonated current to 250–260 °C for 60 hours. Today, the oils used most often for the production of artists' paints are poppyseed oil and linseed oil. In comparison with linseed oil, poppyseed oil has less tendency towards yellow and has a more stable plasticity, but it is more sensitive in processing. The poppyseed oil best suited for painting is obtained from white poppyseeds. Poppyseed oil paints should not be used on fat, non-absorbent picture supports (Hackbarth 1944, p. 120).

Industrial Uses

As poppyseed oil has similar properties to linseed oil and penetrates surfaces quickly, it is used in furniture care for wood and leather materials.

P

6 Possible Unwanted Side Effects

Poppyseed oil may contain traces of opium alkaloids if the seeds were contaminated with chyle during harvesting. In poppyseeds, a content of 2–964 mg/kg morphine and 0.1–79 mg/kg codeine was validated. False positive drug screenings may be the result. Ingestion of a few grams of poppy seeds can lead to a positive morphine result in urine tests according to US standards. The threshold values of the United States Department of Defense are 4000 ng/ml morphine and 2000 ng/ml codeine in urine, and those of the National Institute on Drug Abuse (NIDA) are 300 ng/ml for both morphine and codeine. This is a serious problem for some occupational groups as well as for drivers, who have to abstain from drug use completely, since suspicions of drug abuse are quickly entertained (Leichtfried 2002, p. 42).

Poppyseed can cause allergies. Anaphylactic reactions have been reported; among the symptoms are gastrointestinal disorders, breathing difficulties and breathlessness, skin irritations, swelling of the oral mucosa, vomiting and urticaria. Cross-reactions with kiwi seeds or sesame are possible (Keskin and Sekerel 2006; Oppel et al. 2006).

Pumpkin Seed Oil

Oleum Cucurbitae synonyms: Kürbiskernöl (D); huile de pépins de courge (F)

1 Source Plant

Cucurbita pepo L. (Cucurbitaceae), pumpkin

S. Krist, *Vegetable Fats and Oils*, https://doi.org/10.1007/978-3-030-30314-3_98

P

Habitat

The pumpkin has its origins in the states of Alabama, Arkansas, Illinois and Missouri and in Mexico, from where the Spanish brought it to Europe about 400 years ago. Wild forms of *Cucurbita pepo* are also found in Africa and Asia; it is assumed that they were carried by the ocean, on the tide. Today, pumpkins are cultivated throughout the world, especially in Austria, where three species of *Cucurbita pepo styriaca* have been registered in the breeding book of cultivated plants todate. Pumpkins do not make many demands on the soil. Apart from pure sandy soil, they can be grown anywhere. Well fertilised, warm soil with a sufficient content of fine earth and lime is best. Pumpkins do, however, need specific climatic conditions. They are frost-susceptible and need sufficient precipitation to thrive (Waniorek 1997, p. 9; Boucek 1997, p. 3; Löw 2003, p. 45 et seq).

Description

The pumpkin is an annual plant. The shoot is procumbent or climbing by means of its branched tendrils, which are up to 10 m long, angular, knobbly and covered with stiff hair. The leaves stem from a heart-shaped base; they are five-lobed, with pointed lobes separated from each other by distinct recesses, bristly and covered with stiff hair. Male flowers stand in bunches in the axils of the leaves, while female flowers stand alone. The golden-yellow corolla has a breadth of 7–10 cm. The fruit is a berry, which is large, either round or elongated, and 15–40 cm in diameter. It can vary in form and colour. It contains fibrous pulp; the placenta liquefies when the fruit is ripe. *Cucurbita pepo var styriaca* differs from *Cucurbita pepo* inasmuch as the four outer cell layers are not lignified. This oil pumpkin without a shell has only been cultivated for about 100 years (Hager 1978, volume 4, p. 360; Boucek 1997, p. 2 et seq).

2 Cultivation and Extraction

Cultivation

Today, pumpkins are grown everywhere. The following table gives an overview of the main countries of production.

The most important countries of production with pumpkin production given in 1000 kg (fao.org/29.04.04):

Country	2000
China	3,595,117
India	3,400,000
Ukraine	891,000
Egypt	649,640
Iran	450,000

Country	2000
Mexico	440,000
Italy	366,392
South Africa	338,000
Argentina	323,079
Turkey	300,000
Spain	300,000
Total production worldwide	15,934,145

Austria does not feature in the list, yet pumpkins are an important agricultural product, especially in Styria. In 2002, 18,000 ha were used for the cultivation of oil pumpkins in Austria, 13,000 ha of which were in Styria. On a value basis, 11% of the province's vegetable products are pumpkins. Austria produces pumpkinseeds amounting to 16 million euros and about 2 million litres of pumpkin seed oil.

Extraction of the Oil

Pumpkinseed oil is obtained from the seeds of *Cucurbita pepo L.* by cold pressing or pressing by adding heat. The seeds may, but do not need to be peeled before they are pressed. In Styria, they are roasted before they are peeled: The seeds are ground and kneaded to a pulp by adding water and salt. This mass is stirred continuously and roasted for about half an hour, until the water has evaporated. The typical nutty aroma is already noticeable at this stage and can be traced back to the Maillard reaction that takes place during the process of roasting. The temperature of the pan is 60–80 °C. Finally, the mass is poured into the press, into which steel plates are inserted at intervals. The pressure applied is 300–350 bar (Boucek 1997, p. 6; Löw 2003, p. 47; Roth and Kormann 2000, p. 131; Hager 1978, volume 7b, p. 175).

3 Character

Colour

Dark green to black.

Cold pressed oil: greenish (Roth and Kormann 2000, p. 131).
Hot pressed oil: fluorescent red (Roth and Kormann 2000, p. 131).

Odour

Intensely nutty, of pyrazine.

Nutty, green, herbaceous, roasted (Boucek 1997, p. 78).

Very nutty, touch of peanuts, herbaceous, mildly aromatic (Buchbauer et al. 1998).

Volatile Compounds

Volatile Compounds in Pumpkinseed Oil via Aroma Extract Dilution Analysis and Gas Chromatography Olfactory Methods

(Matsui et al. 1998)

Ethyl-2-methylbutanoate	3-Methylbutanal
2-Ethyl-3,5-dimethylpyrazine	(E)-2-Nonenal
2,3-Diethyl-5-methylpyrazine	Phenyl acetaldehyde
(Z)-2-Nonenal	(E,E)-2,4-Nonadienal
(E,E)-2,4-Decadienal	2-Methoxy-4-vinylphenole
(E)-ß- damascenone	Hexanal
2-Acetyl-1-pyrroline	2-Ethyl-5-methylpyrazine
Dimethyltrisulfide	3-Methylbutanoic acid
3-Ethyl-2,5-dimethylpyrazine	Trans-4,5-Epoxy-(E)-2-decenal
2-Ethenyl-3,5-dimethylpyrazine	4-Hydroxy-2,5-dimethyl-3(2H)-furanon
Benzaldehyde	4-Ethyl-2-methoxyphenol
2-Methyl-6-isopropylpyrazine	

Analysis of the Volatile Compounds in Pumpkinseed Oil via HS-SPME GC-MS

(Siegmund and Murkovic 2004)

2-Methylpropanal	2-Methyl-1-butanol
2-Butenal	1-Pentanol
3-Methylbutanal	1-Hexanol
2-Methylbutanal	Phenyl methanol
Pentanal	Phenyl ethanol
Hexanal	2-Methylfuran
2-Hexenal	2-Pentylfuran
2-Heptenal	2-Furancarboxaldehyde
Nonanal	2-Furanmethanol
Benzaldehyde	Dimethylsulfide
Phenyl acetaldehyde	3-(methylthio)-propanal
2,3-Butanedione	2-Methylpyrazine
2-Butanone	2,5-Dimethylpyrazine
2-Bentanone	2-Ethylpyrazine
2-Heptanone	2-Ethyl-5,(6)-methyl-pyrazine
1-Penten-3-ol	2-Acetylpyrrole
3-Methyl-1-butanol	2-Ethyl-3,6-dimethyl-pyrazine

Flavour

Nutty (Hänsel et al. 1999, p. 261; Hager, 1978 volume 7b p. 175; Löw 2003 p. 47).

Highly aromatic (Ulmer 1996, p. 63).

Intense, unique taste (Kerschbaum and Schweiger 2001, p. 17).

4 Ingredients

Composition of Fatty Acids

Fatty acids	Content in % acc. to Boucek (1997), p. 15	Content in % acc. to Nature certificate of analysis/28.08.03	Content in % acc. to Kerschbaum and Schweiger (2001), p. 16
Palmitic acid	9.49–14.5	11.4	12
Linoleic acid	36.6–60.8	49.5	48.6
Palmitoleic aid	0.1	0.1	0.1
Oleic acid	21–46.9	31.8	31.8
Linolenic acid	0.1	0.3	0.1
Stearic acid	3.11–7.42	5.4	5.1
Arachidic acid	0.3	0.4	0.4
Gondoic acid	0.1	0.1	–
Behenic acid	0.1	0.2	0.3
Myristic acid	0.1	0.1	0.1
Cetoleic acid	0.1	–	–
Lignoceric acid	0.2	–	–
Vaccenic acid	–	–	1.3

Composition of Triglycerides

Triglycerides	Content in % acc. to Karleskind (1996), p. 180
LLL	27.8
OLL	24.2
PLL	19.9
OOL	7.3
SLL	4.5
POL	9.4
PPL	1.7
OOO	0.9
SOL	1.5
POO	1.5
PPO	1.3
PPP	–

Composition of Sterols

(concentration in mg/100 g and content in %)

Sterols	Pumpkin seed oil acc. to Firestone (1999), p. 83	Pumpkin seed oil acc. to Karleskind (1996), p. 181	Pumpkin seed oil acc. to Carstensen (2001), p. 42
Total content	–	360	440.9
Cholesterol	–	–	–
Brassicasterol	–	–	0.2
Campesterol	–	0.1	1.7
Stigmasterol	1	0.1	2.2

Sterols	Pumpkin seed oil acc. to Firestone (1999), p. 83	Pumpkin seed oil acc. to Karleskind (1996), p. 181	Pumpkin seed oil acc. to Carstensen (2001), p. 42
β-Sitosterol	–	1.0	–
Δ5-Avenasterol	–	–	–
Δ7-Stigmasterol	4	–	7.0
Δ7-Avenasterol	10	–	15.5
24-Methyl-cholest-7-enol	6	–	
Δ7,25-Stigma-stadienol	22	19.5	1.7
Δ7,22,25-Stigmastatrienol	29	22.4	28.8
Spinasterol	27	33.2	53.1

Composition of Tocopherols

(concentration in mg/100 g and content in %)

Tocopherols	Pumpkin seed oil acc. to Karleskind (1996), p. 181	Pumpkin seed oil acc. to Carstensen (2001), p. 44
Total content of tocopherols	84	56.6
α-Tocopherol	7.3	13.7
β-Tocopherol	–	0.5
γ-Tocopherol	89.3	79.1
Δ-Tocopherol	3.2	6.7

Other Ingredients

(Kerschbaum and Schweiger 2001, p. 43)

Phytosterols
Vitamin E
Vitamin A
Vitamin B
Carotenoids
Magnesium
Chlorophyll
Amino acids

Physical Key Figures of Pumpkin Seed Oil

refractive index: n_D^{40} 1.4714–1.4740
density: d_4^{15} 0.918–0.927
saponification number: 185–197
acid number: 11–13

iodine number: 113–134
unsaponifiable: 1–1.5%
point of solidification: −15 to −16 °C

(Roth and Kormann 2000, p. 131)

Shelf Life

If refrigerated, pumpkin seed oil has a shelf life of up to 12 months (Roth and Kormann 2000, p. 131).

5 Application

In Pharmacy and Medicine

Pumpkin seed oil contains selenium and vitamin E. It therefore has antioxidant effects and protects the organism from free radicals. Due to its high content of linoleic acid and phytosterols, cholesterol-lowering properties *are ascribed to Oleum Cucurbitae*. It also has antiphlogistic properties and can be used as an additional therapy for rheumatoid arthritis. In vitro, the amino acid *cucurbitin* contained in pumpkinseed oil shows antiparasitic activity.The extent to which this is applicable to humans would have to be examined scienfically. The healing properties of pumpkin seed oil are described in several studies and books, but there are no clinical studies on this subject. The German Federal Ministry of Health (*Bundesgesundheitsamt*) certified that pumpkin seeds have a positive effect in the treatment of adenoma of the prostate glands in their initial stages, but it is not certain whether pumpkin seed oil has the same effect (Boucek 1997, p. 13).

In Cosmetics

Pumpkinseed oil contains many natural active components, such as vitamin A, vitamin E and carotenoids, which would otherwise have to be added to cosmetic products in other ways. For this reason, pumpkin seed oil is often used as a carrier oil in care cosmetics. It is effective in treating (Waniorek 1997, p. 60; Löw 2003, p. 126; Roth and Kormann 2000, p. 131):

- skin dehydration;
- wrinkles;
- skin ageing;
- flaky, chapped skin, and
- stretch marks.

In Folk Medicine

Pumpkinseed oil is used in folk medicine to treat the following complaints (Waniorek 1997, p. 41 et seq; Kircher 2002, p. 36 et seq):

- arteriosclerosis;
- difficulties in bladder emptying;
- complaints of the prostate glands;
- bladder infections;
- an irritable bladder;
- high blood pressure;
- problems with intervertebral discs;
- muscle cramps;
- kidney complaints, and
- deworming.

As Food

Like many vegetable oils, pumpkin seed oil is an excellent edible oil. It should not be heated, but this does not limit the diversity of its uses, which range from salad oil to ice cream (Hager 1978 volume 7b p. 175; Roth and Kormann 2000 p. 131; Kerschbaum and Schweiger 2001 p. 45).

Quinoa Oil

synonyms: Quinoaöl (D); *Chenopodium quinoa* seed oil (E); huile d'Quinoa (F).

1 Source Plant

Chenopodium quinoa (Chenopodiaceae), Quinoa

Habitat

Quinoa is an important provider of starch, and, together with tuberous products, the staple food of the population of the Andes from Bolivia to Chile (Lieberei and Troisdorf 2007, p. 89).

S. Krist, *Vegetable Fats and Oils*, https://doi.org/10.1007/978-3-030-30314-3_99

Q

Description

Quinoa is a herbaceous plant that reaches a height of 1–2 m. It is highly branched. The large leaves are different colours, from green to red, depending on their betacyanin content. In addition, they are polymorphic. The bottom leaves are rhomboidal, the upper leaves lanceolate. The root system is well-developed and highly branched. It extends to a depth of 1.5 m and protects the plant against dry spells. The inflorescence is a panicle that is 15–70 cm long; inflorescences are found at the top and in the axils of the leaves further down. A characteristic feature of the plant is that there are androgynous flowers as well as unisexual, female flowers. The fruit is an achene. The size and colour of the corn varies. It is usually black, sometimes red or yellow, and rarely white (Bhargava et al. 2006).

The corn is disc-shaped and has a diameter of about 2 mm and a thickness of 0.5 mm (Ando et al. 2002). It basically consists of three parenchyma. The central perisperm contains starch and is surrounded by the embryo and a narrow layer of endosperm, which contain protein and oil reserves. This parenchyma is in turn surrounded by the testa and pericarp. The pericarp contains a high amount of saponin and is removed from the corn by rubbing if it is used as food (Belton and Taylor 2002, p. 13).

2 Cultivation and Extraction

Cultivation

The Incas cultivated quinoa in the Andes region of South America, i.e. in Peru, Chile and Bolivia, from 3000 B.C. It was an important food product. After the Spanish conquest, cultivation declined (Bhargava et al. 2006) and lost its importance. From 1970 onwards, there was a rediscovery; because of its dietary composition, it forms an important part of nutrition programmes for undernourished children in South American countries. Varieties with small amounts of saponin are now available. Large corn and a low saponin content are important breeding goals (Aufhammer 1998, p. 161).

Global production of quinoa acc. to Belton and Taylor (2002), p. 94

	Area (ha)	Production (t)	Yield (kg/ha)
1970–1974	30,269	18,768	620
1975–1979	37,587	21,334	568
1980–1984	46,630	27,209	584
1985–1989	61,136	33,590	549
1990–1994	55,022	33,279	605
1995–1999	66,083	45,653	691

Extraction of the Oil

The original method of oil production is simple and uncomplicated. The plants are cooked in an iron vessel with a lid made of steatite. The oil condenses against the lid and is decanted. As the oil was primarily sold in Baltimore, it was called "Baltimore oil" (Ahamed et al. 1998).

Today, the oil can be obtained by various methods. A physical method is that of pressing in a mechanical press or a twin screw extruder. Chemical extraction is possible with organic solvents like aliphatic alkanes or alcohols. The oil can also be extracted in a supercritical medium, either with carbon dioxide alone or with a co-solvent. Prior to extraction, the saponins are removed either by rubbing off the outer layer, or by washing. After extraction, the oil is refined by chemical or physical means (Msika 2010).

3 Character

Colour

Colourless to yellowish (Ahamed et al. 1998)

Odour

Pungent, acrid, caustic, unpleasant (Ahamed et al. 1998)

Flavour

Bitter, caustic (Ahamed et al. 1998)

4 Ingredients

The total fat content of quinoa amounts to 6–8% in the corn and 11–12% in the bran (Ahamed et al. 1998).

Composition of Fatty Acids

Fatty acids	Content in % acc. to Koziol (1993)	Content in % acc. to Ando et al. (2002)	Content in % acc. to Ahamed et al. (1998)	Content in % acc. to Jahanival et al. (2000)
Myristic acid	0.2	0.1–0.2	0.1	0.32
Palmitic acid	9.9	9.5–10.8	9.9	11.4
Palmitoleic acid	0.1	–	–	0.07

Q

Fatty acids	Content in % acc. to Koziol (1993)	Content in % acc. to Ando et al. (2002)	Content in % acc. to Ahamed et al. (1998)	Content in % acc. to Jahanival et al. (2000)
Stearic acid	0.8	0.7–0.9	0.6	0.79
Oleic acid	24.5	19.7–29.5	24.5	25.6
Linoleic acid	50.2	49.0–56.4	52.3	52.8
Linolenic acid	5.4	8.7–11.7	3.8	7.00
Arachidic acid	0.7	–	0.4	0.29

The ratio of polyunsaturated to saturated fatty acids is 4:9 (Ahamed et al. 1998).

Composition of Triglycerides

Triglycerides	Content in % quinoa oil acc. to Jahanival et al. (2000)
LnLnL	1.09
LnLL	6.04
LLL	19.22
PLnL	2.86
OLL + OOLn	21.09
PLL + PLnO	11.34
OOL + PoOO	12.78
POL	9.47
PPL	1.54
OOO + MSO	1.99
OOP + PSL	3.88
POP	1.97
PPO	0.61
PPP	0.34
OOS	0.41

Composition of Sterols: (Content in %)

Sterols	Ahamed et al. (1998)	Ryan et al. (2007) in mg/100 g
Total content	–	82.5
Δ^7-Stigmasterol	43	–
Cholesterol	3.6	–
Δ^5-Campesterol	2.3	15.6
$\Delta^{5,22}$-Stigmasterol	5.5	3.2
Δ^7-Campesterol	8	–
$\Delta^{5,24}$-Avenasterol	21.7	–
β-Sitosterol	15	63.7

Composition of Tocopherols

Tocopherols	Repo-Carrasco et al. (2003) (ppm)	Koziol et al. (1993) (ppm)	Ryan et al. (2007) (mg/100 g)	Bhargava et al. (2006) (mg/100 g)
Total content	–	–	–	–
α-Tocopherol	721.4	690–740	2.1	5.3

Tocopherols	Repo-Carrasco et al. (2003) (ppm)	Koziol et al. (1993) (ppm)	Ryan et al. (2007) (mg/100 g)	Bhargava et al. (2006) (mg/100 g)
β-Tocopherol	–	–	3.1 (β + γ)	–
γ-Tocopherol	797.2	790–930	–	2.6
δ-Tocopherol	–	–	–	–

Other Ingredients: (Ryan et al. 2007)

Squalene	85.4 mg/100 g

Physical Key Figures of Quinoa Oil: (Koziol 1993)

refractive index: n_D^{20}	1.464
relative density:	0.891
saponification number:	190
iodine number:	129
unsaponifiable:	5.2%

5 Application

In Pharmacy and Medicine

Quinoa oil is a biological source of many substances used in pharmaceutics, such as phytosterols, squalene, tocopherols and polyunsaturated fatty acids (Ryan et al. 2007).

The high amount of polyunsaturated fatty acids is effective in treating cardiovascular diseases. It improves the blood lipid profile as well as insulin sensitivity, and lowers the risk of developing type 2 diabetes and antiarrhythmic events (Ryan et al. 2007).

In Cosmetics

Quinoa oil is added to skin care and hair care products. It has a moisturising effect and protects the skin and hair against drying out and environmental stress. It soothes the skin and produces a radiant complexion. Quinoa oil strengthens the hair and fingernails, which is why it forms part of nail balms (www.cosmoty.de).

Other Uses

Quinoa oil is used as a biological dye and a surfactant (Ahamed et al. 1998).

Rapeseed Oil

Oleum Rapae synonyms: Rapsöl (D); lear oil (E); huil de colza (F)

1 Source Plant

Brassica rapa subsp. oleifera (DC.) Metzg., *Brassica napus subsp. napus* L. (Brassicaceae), field mustard, bird rape

S. Krist, *Vegetable Fats and Oils*, https://doi.org/10.1007/978-3-030-30314-3_100

R

Habitat

Today, rapeseed oil is one of the most commonly produced vegetable oils worldwide. Field mustard is one of the most important oil plants of temperate climes. The origins of *Brassica napus* probably lie in the region around the Baltic Sea and Russia. A distinction is made between winter and summer rape. Winter rape grows mainly in maritime locations of Western, Eastern and Central Europe, whereas summer rape is predominant in Canada, China and India. Field mustard is an important cultivated plant, but also grows wild outside of fields, for example on roadsides, railway tracks and on the central reserve of motorways, where it is plainly visible when in bloom. *Brassica napus* needs deep, rich soil. The summer variety grows best in continental climes, whereas the winter variety, due to its susceptibility to frost, prefers maritime environments (Löw 2003, p. 12 et seq).

Description

Brassica napus ssp. napus is an upright plant 60–140 cm tall. The leaves are lyrate, bluish green and hairless. The bottom leaves are sparse, stem-clasping, and covered with stiff hair. The flowers are golden yellow and arranged in racemes. The buds surmount the already open flowers; the sepals are upright and protrude. *Brassica rapa ssp.oleifera* is also cultivated as an oil plant. It differs from *Brassica napus* inasmuch as the flowers surmount the buds, and the sepals stick out horizontally. In both cases, the fruit is a husk 5–10 cm long. Rapeseed oil is obtained from both subspecies (Roth and Kormann 2000, p. 44; Hänsel et al. 1999, p. 255).

2 Cultivation and Extraction

Cultivation

Field mustard is a plant that grows naturally in temperate climes. It is best cultivated on rich, alkaline, either sandy or loamy soil with good water-retaining capacities. The largest growing areas are in China and India. Canada and Europe produce more rapeseed oil, however, as the yield per unit area is 3–5 times higher. The following table lists the most important producing countries (Roth and Kormann 2000, p. 146):

Producing countries	Yield 2001 (mio.t) acc. to fao.org/29.04.04	Yield 2002 (mio.t) acc. to fao.org/29.04.04
EU	8.9	9.2
USA	0.9	0.7
China	11.3	10.5
India	4.2	5.0
Canada	4.9	3.6
CIS	0.3	0.3
Global	35.9	33.2

Extraction of the Oil

Rapeseed oil is obtained from the ripe seeds of *Brassica napus L.* and other varieties containing little erucic acid. It is cold pressed. The harvested seeds are small, brown or black, and contain about 40% oil. After being thoroughly cleaned, they are cracked in a roller mill. By pressing the seed twice and filtration at relatively low temperatures without adding chemical substances, the greater part of rapeseed oil can be extracted. The remains consist of a relatively compact oilcake with a rest content of oil amounting to about 4%, which is used in the production of forage. The still cloudy oil is filtrated, the suspended matter removed, and the oil subsequently refined (Hager 1978, volume 7b, p. 201; Hunnius 1998, p. 232; Aepli 2003, p. 23 et seq; Roth and Kormann 2000, p. 146).

3 Character

Colour

Refined: yellow tinge, nearly colourless, clear.
Unrefined: dark yellow.

Yellow to brownish yellow, sometimes greenish, slightly viscous (Roth and Kormann 2000, p. 146; Hunnius 1998, p. 232).

Slightly yellow (Hager 1978, volume 7b, p. 201)

Odour

Refined: fresh, slightly green, touch of citrus, slightly oily, slightly pungent.
Unrefined: roasted, nutty, slightly fruity and herbaceous.

Volatile Compounds

Aroma Analysis of the Volatile Compounds of Rapeseed Oil Using Capillary Gas Chromatography

(Snyder et al. 1985):

Ethane	1-Octen-3-ol
Propane	Pentylfurane
Propenal	t,c-2,4-Heptadienal
Pentene	Octanal
Pentane	t,t-2,4-Heptadienal
Propanal	Octenal
Hexane	Nonanal
2-Butenal	t-2-Decenal
1-Penten-3-ol	Octane

Pentanal	t-2-Hexenal
Heptane	Heptanal
Pentenal	c-2-Heptenal
Pentanol	t-2-Heptenal
Hexanal	

Aroma Analysis of the Volatile Compounds of Cold Pressed Rapeseed Oil Using h-SPME-GC-MS

R

(Jelen et al. 2000):

Pentane	Octanal
Butanal	Limonene
Acetous acid	3-Octen-2-one
2-Butenal	Hexanoic acid
2-Ethyl furane/pentanal	5-Ethyldihydrofuranone
Propanoic acid	2-Octenal
2-Pentenal	1-Octanol
Octane	2-Nonanone
Hexanal	3,5-Octadien-2-one
2,4-Dimethyl 2-pentane	Nonanal
2-Hexenal	Nonanoic acid
2-Heptanone	2-Nonenal
Heptanal	2-Decanone
2,4-Hexadienal	Decanal
2-Heptenal	2,4-Decadienal
1-Octen-3-ol	2-Undecenal
6-Methyl-5-hepten-2-one	6-Methyl-5-hepten-2-ol
2-Pentylfurane	2,4-Heptadienal

Flavour

Mild (Roth and Kormann 2000, p. 146). Pleasantly neutral (Aepli 2003, p. 23 et seq). Slightly scratchy (Hager 1978, volume 7b, p. 201).

4 Ingredients

Composition of Fatty Acids

Since rapeseed oil contains a high amount of erucic acid, it was not suitable for human consumption for a long time; pathological changes of the myocardium and fatty degeneration of the heart were possible consequences of intake. In order to solve this problem, new varieties were bred, which contained a lower amount of erucic acid and a higher amount of oleic acid. This new variety was called 00-rapeseed oil (Roth and Kormann 2000, p. 146; Fleischhacker 2002, p. 16).

Fatty acids	Content in % acc. to Roth and Kormann (2000), p. 146 (old varieties)	Content in % acc. to Roth and Kormann (2000), p. 146 (new varieties)	Content in % acc. to Hunnius (1998), p. 232	Content in % acc. to Kerschbaum and Schweiger (2001), p. 19
Palmitic acid	2.5	4	3–8	4.7
Linoleic acid	13.5	20	18–32	19.6
Palmitoleic acid	–	–	–	0.2
Oleic acid	15	63	45–65	58.8
α-Linolenic acid	8	9	6–14	9.2
Stearic acid	–	–	–	1.6
Arachidic acid	–	–	–	0.6
Behenic acid	–	–	–	0.3
Vaccenic acid	–	–	–	3.5
Myristic acid	–	–	–	0.1
Erucic acid	48	0.5	<5	0.7
Eicosenoic acid	–	–	–	1.3

Composition of Sterols

(concentration in mg/100 g and content in %)

Sterols	Rapeseed oil acc. to Firestone (1999), p. 84	Refined rapeseed oil acc. to Carstensen (2001), p. 43	Cold pressed rapeseed oil acc. to Carstensen (2001), p. 43
Total content	88.1	568.1	592.9
Cholesterol	–	–	1.7
Brassicasterol	12–13	13.3	15.1
Campesterol	30–33	29.8	32.3
Stigmasterol	0.4–0.6	0.8	1.0
β-Sitosterol	49–55	51.3	46.6
Δ5-Avenasterol	1–2	0.8	1.9
Δ7-Stigmasterol	–	2.2	0.4
Δ7-Avenasterol	–	1.4	0.6

Composition of Tocopherols

(concentration in mg/100 g and content in %)

Tocopherols	Refined rapeseed oil acc. to Carstensen (2001), p. 45	Cold pressed rapeseed oil acc. to Carstensen (2001), p. 43
Total content	82.0	89.4
α-Tocopherol	30.0	36.4
β-Tocopherol	3.8	1.1
γ-Tocopherol	66.2	62.0
δ-Tocopherol	–	0.5

R

Other Ingredients

(Kerschbaum and Schweiger 2001 p. 43; Salunkhe et al. 1992, p. 59 et seq)

Phospholipids
Sterols
Hydrocarbonate
Vitamin E
Lipochromes
Pigments
Betacarotene
Vitamin A
Vitamin K
Lecithin
Sulphur compounds
Minerals
Squalene
Provitamin A

Physical Key Figures of Rapeseed Oil

(Roth and Kormann 2000, p. 146)

refractive index: n_D^{20}	1.4720–1.4752
density: d_{15}^{15}	0.913–0.916
saponification number:	170–180
acid number:	not over 8
iodine number:	94–120
unsaponifiable:	max. 1.5%
point of solidification:	−2 to −10 °C

Nutritional Values of Rapeseed Oil per 100 g

large calories: 865 kcal
kilojoules: 3620 kJ

Shelf Life

Edible rapeseed oil has a shelf life of up to 12 months if refrigerated (Roth and Kormann 2000, p. 146).

If stored in a cool and dark place, rapeseed oil has a shelf life of up to 12 months.

5 Application

In Pharmacy and Medicine

In pharmaceutics, rapeseed oil is used in the production of ointments and liniments, as a filling material for soft gelatine capsules and as a DC reagent for the identification of fat oils (Fleischhacker 2002, p. 16; Hager 1978, volume 7b, p. 201).

In Cosmetics

Rapeseed oil is added to cosmetic formulations for dry, ageing, flaky and chapped skin as well as skin poorly supplied with blood (Kircher 2002, p. 64 et seq).

Technical Uses

As a Lubricant

Most of the rapeseed oil produced is used for technical purposes, for example as a filling material in the production of rubber goods or as a lubricant (Roth and Kormann 2000, p. 146). Rapeseed oil is especially suitable as a lubricant if it does not have to meet high demands, for example in the case of loss lubrication of chain saws and band saws, where the oil is left behind in the chips. Rapeseed oil cannot be used as a high-quality lubricant, for example as a motor oil. The main advantage of rapeseed oil as a loss lubricant is that it is 100% biodegradable and there is therefore no danger of polluting ground water or drinking water.

As an Anti-corrosion Agent

Rapeseed oil can be used as an anti-corrosion agent. Industry applies the oil to semi-finished or finished parts so that they can be stored until they are further processed, without beginning to rust.

Industrial Uses

As a Raw Material

The oil is used as a raw material in the production of insect lime, grafting wax, plasters and leather-greasing agents. The oil rich in erucic acid is a source of C20 and C22 fatty acids that are used in the polymer processing industry and by detergent manufacturers.

As a Pesticide

In biodynamic agriculture, rapeseed oil is used as a pesticide and insecticide. It forms an oil film, underneath which insect pests and their eggs suffocate. Rapeseed oil has proved of value against bloodsucking insects, but spares bees and other pollinators. In addition, it is biodegradable (Roth and Kormann 2000, p. 146).

As Fuel

The use of rapeseed oil and rapeseed oil methyl ester (RME), which is also a bio diesel, has many advantages over traditional fossil fuels. Rapeseed oil and RME are biodegradable and are much less ecologically toxic than diesel. Rapeseed oil is 100% carbon neutral. This means that on combustion, the amount of carbon dioxide produced does not exceed the amount absorbed by the plant to produce the oil. Additionally, rapeseed oil produces less soot than diesel fuel. The cold-pressed vegetable oil, in particular, has a very favourable energy balance.

Another advantage is that rapeseed oil is safe in storage and transport, due to its high flash point of more than 300 °C and its low sulphur content.

As Food

Since varieties containing only small amounts of erucic acid have been bred, rapeseed oil is also an edible oil. Of all traditional edible oils, it has the lowest content of saturated fatty acids. It can be used for roasting, frying and as a salad oil. In addition, rapeseed oil does not become bitter in products containing eggs and milk, and is thus suitable to be added to mayonnaise and marinades (Hager 1978, volume 7b, p. 201; Kerschbaum and Schweiger 2001, p. 18; Löw 2003, p. 134; Kircher 2002, p. 64 et seq).

6 Possible Unwanted Side Effects

After ingesting learge amounts of rapeseed oil rich in erucic acid, test animals developed fat infiltrations of the heart, liver and adrenal glands and sustained severe damage to the myocardium. The tissues of the adrenal glands and heart mitochondria have an especially strong affinity to erucic acid, which is stored instead of linoleic acid and as a consequence inhibits the enzymatic system responsible for the oxidation of fatty acids in the mitochondria of the heart. This effect is close to toxic for the heart (Roth and Kormann 2000, p. 188).

Rice Bran Oil

Oleum Oryzae synonyms: Reiskeimöl (D); huile de son de riz (F)

1 Source Plant

Oryza sativa L. (Poaceae), rice

S. Krist, *Vegetable Fats and Oils*, https://doi.org/10.1007/978-3-030-30314-3_101

R

Habitat

Oryza sativa is one of the world's oldest cultivated plants, dating to the Pleistocene. According to the current level of knowledge, the origins of cultivated rice lie in South-East Asia. The oldest written documents date from China in 2800 B.C. In the ninth century, the Moors brought rice cultivation to Spain. In Italy, rice has been cultivated since the fifteenth century. It was imported to North America in the seventeenth century. Rice does not make many demands on the soil apart from that it should be rich in humus, loamy and heavy, so that it can absorb sufficient amounts of water. Rice is much more sensitive to weather conditions, rain and warmth. Tropical varieties germinate at 18 °C, subtropical varieties at 10–12 °C. Rice does not tolerate frost. During bloom, it needs temperatures ranging from 25 to 35 °C (Hager 1978, volume 6a, p. 340; Roth and Kormann 2000, p. 147).

Description

Oryza sativa is an annual grass plant and, depending on the variety, grows to a height of 80–150 cm. Cultivated rice develops a blade 50–160 cm tall with a narrow, overhanging panicle consisting of spikes with one flower. On the spikes, the hard lemmas are arranged. The lemma forms the outer layer of the rice corn. Beneath, there is the pericarp and testa, which have grown together. The fruit is a caryopsis. The plant can develop up to 30 bastard branches, 80–100 corns developing on each of them. The corn itself consists mainly of pure rice starch. The embryo is located on its narrow end; it develops into a new plant and contains the most nutrients (Hager 1978, volume 6a, p. 340).

2 Cultivation and Extraction

Cultivation

More than 90% of the global yield is from Asia, especially from India, Malaysia, the Philippines, China, Korea and Japan. In Europe, rice is cultivated in Italy, namely in the Po Valley, as well as in Spain and Portugal. Additionally, rice is cultivated in tropical West Africa, the southern USA and in South America. The following table lists the most important producing countries (Salunkhe et al. 1992, p. 424 et seq):

Rice producers in the year 2000	mio.t
China	190
India	135
Indonesia	51
Vietnam	32
Thailand	23

Extraction of the Oil

Rice bran oil is extracted with hexane from the embryos or the the outer layers of the rice corn. The raw oil can be used neither for medical purposes nor as an edible oil, as it contains about 3–20% free fatty acids and 5–8% unsaponifiable portion, which have to be removed. For this reason, the oil is cleaned and neutralised after extraction (Roth and Kormann 2000, p. 147; Salunkhe et al. 1992, p. 438).

3 Character

Colour

Slightly yellowish.
Golden yellow or greenish brown (Roth and Kormann 2000, p. 147).

Odour

Slightly flowery, touch of boiled rice.

Flavour

Neutral.

4 Ingredients

Composition of Fatty Acids

Fatty acids	Content in % acc. to Roth and Kormann (2000), p. 147	Content in % acc. to Nicolosi et al. (1994)	Content in % acc. to fao.org/29.04.04
Palmitic acid	12–18	12–18	13–18
Linoleic acid	29–42	20–42	29–42
Palmitoleic acid	0.2–0.4	0.2–0	–
Oleic acid	40–50	40–50	0–50
Linolenic acid	Traces	0–1	0–1
Stearic acid	1–3	1–3	1–3
Arachidic acid	–	0–1	–
Myristic acid	0.4–1	0.1–1	0.4–1

Composition of Sterols

(concentration in mg/kg and content in %)

Sterols	Rice bran oil acc. to Firestone (1999), p. 86
Total content	10,550
Cholesterol	–
Brassicasterol	–
Campesterol	20–28
Stigmasterol	8–15
β-Sitosterol	49–54
Δ5-Avenasterol	5–11
Δ7-Stigmasterol	1–2
Δ7-Avenasterol	2–4

Composition of Tocopherols

Rice bran oil contains 32.3 mg tocopherol/100 g oil.

Other Ingredients

(Hager 1978, volume 6a p. 342; Nicolosi et al. 1994; Salunkhe et al. 1992, p. 442)

γ-Oryzanol
Tocopherol
Tocotrienol
Phytosterols
Polyphenols
Lecithin
Wax
Squalene

Physical Key Figures of Rice Bran Oil

refractive index: n_D^{25} 1.470–1.473
density: d_{25}^{25} 0.916–0.921
saponification number: 181–194
acid number: 4–120
iodine number: 89–108
unsaponifiable: 1–5%

(Roth and Kormann 2000, p. 147)

Shelf Life

If stored in a cool, dark place, the oil has a shelf life of up to 12 months. However, unrefined oil becomes rancid quickly (Roth and Kormann 2000, p. 147).

Characteristics of the Oil

Rice bran oil contains a high percentage of γ-oryzanol. This is a compound of ferulic acid and phytosterols. Ferulic acid, like caffeic acid, commonly occurs in the outer layers of plants and acts as an antioxidant, protecting the plant from free radicals. The effects of γ-oryzanol are described in more detail below (Cicero and Gaddi 2001; Nicolosi et al. 1994).

5 Application

In Pharmacy and Medicine

Rice bran oil contains various components like γ-oryzanol, tocotrienols, tocopherols and squalene, which have antioxidative properties. In this context, an anticancerogenic effect and a positive effect on the cardiovascular system are ascribed to the oil. In addition, it contains phytosterols, of which β-sitosterol has the largest share; phytosterols have been proven to competitively inhibit the synthesis of cholesterol. In Asia, rice bran oil has been applied as a cholesterol-lowering agent for some time. γ-Oryzanol, which is mainly obtained from rice bran oil, is added to many dietary supplement products in the sports sector. There is speculation that the phytosterols in γ-oryzanol may cause an increase in the testosterone level and may stimulate the release of growth hormones, and could therefore have an anabolic effect. Since less than 10% of the phytosterols is absorbed from the intestines, an enhancement of muscle formation is unlikely. It is also thought that the effect of γ-oryzanol can be explained by an increase in norepinephrine and β-endorphin, but this could not be proven in clinical studies. Nevertheless, rice bran oil continues to be advertised with slogans such as "a natural alternative to anabolic steroids", "increases muscular strength and size" and "reduces fatigue" (Cicero and Gaddi 2001; Nicolosi et al. 1994).

In Cosmetics

Rice bran oil is added to cosmetic products for all skin types, especially to babycare oil.

It also forms part of many sunscreens, as the γ-oryzanol that it contains is thought to be a natural sunblock. Wax extracted from rice bran oil is used in lipsticks (Roth and Kormann 2000, p. 147).

Technical Uses

Rice bran oil is used as a raw material in the production of soap and candles, for the treatment of leather and textiles, and is also employed as an anti-corrosion agent (Roth and Kormann 2000, p. 147; Salunkhe et al. 1992, p. 442).

As Food

In Europe, rice bran oil is hardly of importance as an edible oil, but it is popular in Asia, where it is used as a salad oil as well as for cooking and frying. It is also added to mayonnaise and salad dressings (Roth and Kormann 2000, p. 147; Salunkhe et al. 1992, p. 442).

Rose Hip Oil

Oleum rosae aff. rubiginosa synonyms: Hagebuttenöl, Wildrosenöl (D)

1 Source Plant

Rosa canina L., *Rosa rugosa Thunb.*, *Rosa pendulina* L. *(Rosaceae)*, dog rose, rugosa rose

S. Krist, *Vegetable Fats and Oils*, https://doi.org/10.1007/978-3-030-30314-3_102

Habitat

The supposed region of origin of the rose plant is Chile. The bush grows at heights of up to 3000 m and is native to Europe and the temperate zones of America, Africa and Asia.

Description

The rose hip is the fruit of the dog rose, a tall bush with overhanging branches and consistent, sickle-shaped thorns. The leaves are pinnate. The sepals only have a few, narrow leaflets, which fold back after the plant has flowered and drop off before the fruit is ripe. The flowers are about 5 cm broad and sit on hairless stems. The spurious fruit (rose hips) are the fleshy receptacles. They are bright red, hairless, oval to round, with hair on the inside. The actual fruit is the angular, light-coloured, hard nut (Schoenfelder and Schoenfelder 2001, p. 208).

2 Cultivation and Extraction

Cultivation

The natural habitat and preferred locations of the dog rose are hedges, shrubberies and sparse forests (Schoenfelder and Schoenfelder 2001, p. 208).

Extraction of the Oil

The fruit are harvested in dry weather as soon as they are fully ripe. At this point, they are already red, but not yet hard. The oil is obtained by mechanical pressing or extraction. The raw oil is then refined and in most cases stabilised by adding natural vitamin E.

3 Character

Colour

Clear, light yellow to orange-red oil.

Odour

Sweet-sour, nutty, metallic, oily, fruity, floral, of aldehyde.

Flavour

Taste characteristic of the species.
Flavourless, mild.

4 Ingredients

Composition of Fatty Acids

Fatty acids	Content in % acc. to Roth and Kormann (2000), p. 160	Content in % acc. to speiseoele.com/07.03.06
Oleic acid	29.3	15
Linoleic acid	56.7	42
α-Linolenic acid	9.4	13
Saturated fatty acids	4.5	7.5

Composition of Tocopherols

100 ml rose hip oil contain 47 mg α- and β-tocopherol (Hunnius 1998, p. 1198).

Physical Key Figures of Rose Hip Oil

refractive index: 1.478–1.4793
relative density: 0.9270
iodine number: 152–169
saponification number: 189–193
acid number: 0.5
unsaponifiable: 1.9–2.5%

(Roth and Kormann 2000, p. 160)

Shelf Life

If stored in a cool, dry place, rose hip oil has a shelf life of about 12 months.

5 Application

In Pharmacy and Medicine

In pharmacy, rose hip oil is employed for treating dry, flaky, chapped and dull skin. It is also used to treat eczema, psoriasis, strongly pigmented skin and scars, and to support the treatment of burns and blunt injuries. A treatment with rose hip oil restores the skin's elasticity and natural colour and prevents a pathological thickening of the epithelium. In several countries, for example in the USA and Japan, rose hip oil is used prior to surgical treatment. Two to three weeks before surgery, it is applied to the area on which surgery is to be performed, to guard against unaesthetic scarring. Another area of application is wounds of the gums and oral mucosa caused by prostheses and other external causes. Applying the oil directly to the sore areas is thought to speed up the healing process.

In Cosmetics

The oil is part of ointments and (night) creams. The trans-retinoic acid contained in rose hip oil speeds up the process of skin regeneration and stimulates collagen formation, helping dry skin to retain moisture and protecting it from drying out.

Sacha Inchi Oil

synonyms: Sacha Inchiöl (D); Inca peanut oil (E) huile végétale d'Inca Inchi (F)

1 Source Plant

Plukenetia volubilis (Euphorbiaceae), Sacha Inchi, Mountain Peanut, Inca Peanut

Habitat

Sacha inchi is a native plant of the rain forests of the Amazonas Region. Apart from Peru, the plant also grows in Brazil, Colombia, Ecuador and Venezuela; the Incas and their ancestors cultivated it for centuries (Pies 2010, p. 7, 10).

S. Krist, *Vegetable Fats and Oils*, https://doi.org/10.1007/978-3-030-30314-3_103

Description

The species *Plukenetia* is named after the English botanist Leonard Plukenet. The variety *volubilis* means “circling, revolving”, indicating that it is a climbing plant. Sacha inchi is a perennial, perfect plant with hairy, heart-shaped leaves. The upper side of the leaves is shiny and deep green, the leaf edge is serrated. The leaves are 10–12 cm long and 8–10 cm broad; the leaf stem is 2–6 cm long. On the base, there are frequently stipules. The male flowers develop first. They are small, white and arranged in clusters. Shortly afterwards, the female flowers appear at the base of the inflorescence. The fruit are capsules, first green, then brown or black when they are ripe. The star-shaped dehiscent fruit are 3–5 cm in diameter. They consist of 4–7 segments, each of which contains a seed. The seeds are oval, dark brown and 1.5–2 cm in diameter (Hanssen and Schmitz-Hübsch 2011; Pies 2010, p. 10–11).

2 Cultivation and Extraction

Cultivation

At the beginning of the twenty-first century, the Peruvian government declared sacha inchi an endangered species and began to support projects for the sustainable cultivation of the plant. *Plukenetia volubilis* grows best in a warm climate at altitudes of up to 1700 m. Since it grows quickly, a sufficient water supply and well-drained soil are important. The plant prefers acidic soil. Provided that there is sufficient water supply, the seed germinates 2 weeks after sowing. The plant blossoms after 3–5 months for about 7–19 days. Three to four months later, the first fruit develop. The harvest takes place by hand about 7.5–9.5 months after sowing. The fruit are then dried in the sun to obtain the seeds. The seeds used to be removed by hand, but are now removed by machines. The harvest is still, however, manual (Hanssen and Schmitz-Hübsch 2011; Pies 2010, p. 7, 10, 11, 16).

Extraction of the Oil

Sacha inchi oil is usually cold pressed. Another possibility is extraction with solvents or supercritical carbon dioxide. Pressing is preferable because it is an environment-friendly method and results in high-quality oil. When the oil is extracted with solvents, the most commonly used is hexane, which is toxic. Additionally, the extraction takes place at high temperatures, and compounds like polyunsaturated fatty acids are destroyed in the process. Off-flavours may also be produced. Carbon dioxide, on the other hand, is neither toxic nor inflammable, and can be easily separated from the extract. A study by Follegatti-Romero et al. tested the various methods of extraction of sacha inchi oil. The result was that cold pressing and subsequent extraction with carbon dioxide at 400 bar and 60 °C achieved the greatest yield of 99.1% (Follegatti-Romera et al. 2009).

3 Character

Colour

Yellow (Axtell 1992, p. 118).

Odour

Like beans (http://www.inkanatural.com, 23.01.2012).

Flavour

Like beans (http://www.inkanatural.com, 23.01.2012).

4 Ingredients

Composition of Fatty Acids

Fatty acids	Content in % acc. to Hamaker et al. (1992)	Content in % acc. to do Prado et al. (2011)	Content in % acc. to Follegatti-Romero et al. (2009)	Content in % acc. to Fanali et al. (2011)
Palmitic acid	4.5	3.90	4.24	4.3
Stearic acid	3.2	2.82	2.50	3.0
Oleic acid	9.6	9.98	8.41	9.0
Vaccenic acid	–	–	–	0.6
Linoleic acid	36.8	33.74	34.08	36.2
α-Linolenic acid	45.2	44.79	50.41	46.8
Gondoic acid	–	0.23	0.16	–

Composition of Triglycerides

Triglycerides	Amount in % acc. to Fanali et al. (2011)
LnLnLn	12.3
LnLLn	22.2
LLnL	18.2
LnOLn	7.3
PLnLn	3.2
LLL	5.3
OLLn	9.3
PLLn	4.4
SLnLn	2.1
LOL	3.0
OLnO	1.8
LLP	6.9

Triglycerides	Amount in % acc. to Fanali et al. (2011)
OLO	0.8
SLL	2.7
OOO	0.1
SLO	0.4

The study by Fanali et al. detected 21 different triglycerides in total. The table does not list SLLn, POLn, OLP, SOLn and SLP, for which no specifications of quantity are available. More than 80% of the triglycerides contain linolenic acid. This high percentage of linolenic acid is very important, as it lowers the total cholesterol and LDL levels, which in turn protects against cardiovascular diseases (Fanali et al. 2011).

Composition of Tocopherols

Tocopherols	Content in g/kg acc. to Fanali et al. (2011)	Content in g/kg acc. to Follegatti-Romero et al. (2009)	Content in g/kg acc. to Hamaker et al. (1992)
Total content	–	2.39	–
α-Tocopherol	0.004	–	0.038–0.063
γ-Tocopherol	1.257	1.14	–
δ-Tocopherol	0.869	1.25	–

Due to the high amount of unsaturated fatty acids, it could be surmised that sacha inchi oil is liable to oxidate. The unrefined oil seems, however, to be fairly stable. Hamaker et al. suggest that the reason is the oil's content of α-tocopherol and carotene. A study by Follegatti-Romero et al. could not, however, find any α-tocopherol, but a higher content of γ-tocopherol and δ-tocopherol (Follegatti-Romero et al. 2009).

Other Ingredients

(Hamaker et al. 1991)

carotene:	0.008 g/kg

Physical Key Figures

Iodine number:	193.1
Saponification number:	185.2
Refractive index: n^{25}_D	1.4791
Density at 25 °C:	0.9187
Viscosity at 20 °C:	35.4

(Gutiérrez et al. 2011)

Iodine number:	198
Saponification number:	193

(Follegatti-Romero et al. 2009)

5 Application

In Pharmacy and Medicine

In sacha inchi oil, α-linolenic acid amounts to about 50%; it is the origin of eicosapentaenoic acid, which is contained in fish oil, and docosahexaenoic acid. They have a positive influence on the cardiovascular system, inflammation, nervous disorders and malignant tumours. Sacha inchi oil is thus a good alternative to fish oil and has the advantage of a neutral taste (Pies 2010, p. 7–8).

Industrial Uses

The oil is used in the production of paint, varnish and linoleum (Axtell 1992, p. 117).

In Cosmetics

Traditionally, the Mayoruna, Chayuita, Campa, Huitoto, Shipibo, Yagua and Bora people add sacha inchi oil to cosmetic products. They mix the oil with flour made from the seeds and apply it as a skin cream. In addition, they use it to treat rheumatic complaints and muscle pain. Alternative practitioners use sacha inchi oil as a massage oil (Pies 2010, p. 17).

As Food

Since it contains a high amount of unsaturated fatty acids, sacha inchi oil is an edible oil of high quality (Pies 2010, p. 7).

Safflower Oil

Oleum Carthami synonyms: Safloröl, Distelöl (D); safflower seed oil (E); huile de carthame (F)

1 Source Plant

Carthamus tinctorius L. (Asteraceae), safflower

S. Krist, *Vegetable Fats and Oils*, https://doi.org/10.1007/978-3-030-30314-3_104

Habitat

Safflower probably originates in the region between Asia Minor and the Indian subcontinent. It has been cultivated in Iran, eastern India, China, Japan, North Africa and Southern Europe since antiquity. From what is now Turkey, the plant spread to the Mediterranean area. The Romans brought it to Central Europe. In medieval times, it was primarily used for dyeing. In the nineteenth century, safflower cultivation for the purpose of colour extraction lost its importance because of the invention of synthetic dyes (Roth and Kormann 2000, p. 50; Hager 1978, volume 3, p. 725; Löw 2003, p. 36).

Description

Safflower is an annual or, in cultivation sometimes biennial plant that reaches a height of about 10–60 cm. It has a sturdy taproot. The stem is upright, either unbranched or grooved with a few rigid branches at the top, hairless, shiny and a whitish-yellow colour. The leaves are hairless, quite soft, with denticulate edges and thorned at the apex, and clearly veined on the blade. The lower leaves are oval, elongated, gradually narrowing to a short stem, the others are elongated to oval or lanceolate, slightly heart-shaped and stem-clasping at the bottom, the uppermost gradually becoming bracts. The flowers are first a reddish yellow, then a vivid orange red. They have a tube about 20–25 mm long, which separates at the top into 5 lanceolate, about 4–6 mm long, two-nerved flaps of a light red colour. The fruit is an ivory-coloured achene. It is obovate to pear-shaped, 6–8 mm long, 3–4 mm broad and 2.5 mm thick. The seed is surrounded by a seed coat 0.5 mm thick, 5.7 mm long and yellowish grey to dirty white in colour. Besides fat oil, safflowers contain the pigment safflower red, also called carthamin (Hager 1978, volume 3, p. 725; Löw 2003, p. 36).

2 Cultivation and Extraction

Cultivation

For centuries, the safflower was cultivated for dyeing, for example in India and Germany. In Ancient Egypt, mummies were dyed with dyestuff derived from safflowers. Today, the plant is mainly cultivated for its oil. The USA, Australia, Greece, Turkey and especially India cultivate safflowers for oil production.

Safflowers grow best in climates with warm summers. They are inured to droughts and even need dry conditions to flourish, as they are susceptible to bud rot during and after flowering. *Carthamus tinctorius* grows well in loamy, sandy soil as well as neutral loam with open subsoil. The seeds are sown in spring, 2–3 cm deep and at a distance of 30–60 cm from each other. In this way, 20–30 kg of seeds can be planted per hectare. The growing season spans 4 or 5 months and results in an average crop of 0.6 t/ha (Roth and Kormann 2000, p. 50; Hager 1978, volume 3, p. 725; Löw 2003, p. 36).

Extraction of the Oil

The harvest takes place when the plants are already quite dry, but not yet dried up. This is from August to September. The corn is harvested with a combine harvester, whereas the petals for the extraction of the dye are gathered by hand.

Safflower oil is obtained from the seeds of *Carthamus tinctorius* by pressing or extraction. Cold-pressed oil is a good edible oil, whereas oil pressed by adding heat is only suitable for technical purposes. From 100 kg seeds, 25 kg of raw oil are gained by pressing, and 28 kg by extraction (Roth and Kormann 2000, p. 148). To get rid of its scratchy taste, the raw oil is at least partly refined before use (Ulmer 1996, p. 57).

3 Character

Colour

Intense, vivid light to dark yellow.
Golden yellow to reddish (Roth and Kormann 2000, p. 148).
Light yellow (Hunnius 1998, p. 283).
Golden yellow to reddish yellow (Löw 2003, p. 37).

Odour

Herbaceous, nutty, peppery.
Peculiar (Roth and Kormann 2000, p. 148; Löw 2003, p. 37).
Weak (Hunnius 1998, p. 283).

Volatile Compounds

Aroma analysis of the volatile compounds of safflower oil using the electronic nose and solid phase microextraction gas chromatography-mass spectrometry (Shin et al. 2003):

The following table lists the aroma components of safflower oil as well as of SL (Structured Lipids) safflower oil. The composition of fatty acids of these lipids has been changed by chemical or enzymatic esterification.

Aroma components – safflower oil	Aroma components – SL safflower oil
Hexanal	Hexanal
(E)-2-Heptenal	Heptanal
Hexane acid	(E)-2-Heptenal
4-Methyl-2-propyl-1-pentanol	3 Pentyl-2,4-pentadien-1-ol
2-Butyl-1-octanol	2-Hexylfuran
2-Ethyl-1-decanol	3-Nonen-1-ol
Hexyl-octyl-ether	(E)-Tridecenal
7-Tetradecene	Caprylic acid ethyleser
	5-Octen-1-ol
	3,5-Octadien-2-ol
	3-Propyl-2,4-pentadien-1-ol
	(E,E)-2,4-Nonadienal
	1,3-Tridecenal
	(E, E)-2,4- Dodecadienal
	2-Hexyl-1-octanol
	(E,E)-2,4-Decadienal

Flavour

Slightly pungent, lingering taste (Roth and Kormann 2000, p. 148). Neutral (Kircher 2002, p. 24). Nutty (Löw 2003, p. 37).

4 Ingredients

Composition of Fatty Acids

Fatty acids	Content in % acc. to Hager (1978), volume 7b, p. 204	Content in % acc. to Salunkhe et al. (1992), p. 339	Content in % acc. to Kerschbaum and Schweiger (2001), p. 12
Palmitic acid	4	2–10	6.9
Linoleic acid	63–79	55–81	79.0
Palmitoleic acid	–	0.5	0.1
Oleic acid	14–24	7–42	10.4
α-Linolenic acid	0.1–6	1	0.1
Stearic acid	1.5	1–10	2.1
Arachidic acid	0.4	0.5	0.3
Behenic acid	–	0.5	0.2
Vaccenic acid	–	–	0.8
Myristic acid	0.04	1	0.1

The composition of safflower oil varies depending on the growing area. Two kinds of oil can be distinguished: one with a higher proportion of oleic acid and one in which linoleic acid predominates. The following table illustrates the differences in the composition of the two oils:

Fatty acids	Content in %, high amount of linoleic acid acc. to Karleskind (1996), p. 149	Content in %, high amount of linoleic acid acc. to Firestone (1999), p. 88	Content in %, high amount of oleic acid acc. to Karleskind (1996), p. 149	content in %, high amount of oleic acid acc. to Firestone (1999), p. 89
Myristic acid	<0.1	0–0.2	<0.1	0–0.1
Palmitic acid	6–7	5.3–8.0	6–7	5–6
Palmitoleic acid	<0.5	0–0.2	<0.2	0–0.2
Stearic acid	2–3	1.9–2.9	<2	1.5–2.0
Oleic acid	10–20	8.4–30	73–80	74–80
Linoleic acid	68–83	67.8–83.2	12–20	13–18
α-Linolenic acid	<0.2	0–0.1	<0.2	0–0.2
Arachidic acid	<0.5	0.2–0.4	<0.3	0–0.3
Eicosanoic acid	<0.2	0.1–0.3	<0.2	0.1–0.2
Behenic acid	<0.2	0.2–0.8	<0.2	0–0.2
Cetoleic acid	-	0–1.8	-	-
Lignoceric acid	-	0–0.2	-	-
Nervonic acid	-	0–0.2	-	-

Composition of Sterols

(concentration in mg/100 g and content in %)

Sterols	Safflower oil with a high amount of linoleic acid acc. to Firestone (1999), p. 88	Safflower oil with a high amount of oleic acid acc. to Firestone (1999), p. 89	Refined safflower oil acc. to Carstensen (2001), p. 43	Cold pressed safflower oil acc. to Carstensen (2001), p. 43
Total content	209.5–265	-	334.7	377.3
Cholesterol	0–0.5	0–0.2	0.2	0.6
Brassicasterol	-	-	0.1	0.1
Campesterol	9.2–13	10–16	11.3	12.1
Stigmasterol	6.5–9.6	8–15	-	-
β-Sitosterol	40.2–49.8	52–60	53.8	57.4
Δ5-avenasterol	2.1–4.0	5–6	5.2	2.9
Δ7-stigmasterol	15.7–22.4	13–18	18.7	16.0
Δ7-avenasterol	2.9–5.3	5–6	3.2	1.5
Δ5-stigmasterol	-	-	7.3	9.4

Composition of Tocopherols

(concentration in mg/100 g and content in %)

Tocopherols	Refined safflower oil acc. to Carstensen (2001), p. 45	Cold pressed safflower oil acc. to Carstensen (2001), p. 45	Safflower oil acc. to Karleskind (1996), p. 150
Total content	46.7	152.3	30–60
α-Tocopherol	90.1	94.4	83–96
β-Tocopherol	2.2	2.5	2–3
Γ-Tocopherol	6.7	1.3	7–17
Δ-Tocopherol	1.0	1.8	1–4

Other Ingredients

(Kerschbaum and Schweiger 2001, p. 43)

Squalene: 15.7–17.2 mg/100 g
Triterpene alcohols: 60–66 mg/100 g
Carbon hydrides: 126–162 mg/100 g
Protein
Minerals
Vitamin A
Vitamin K
Glutamine
Potassium
Asparagine

Physical Key Figures of Safflower Oil

refractive index: n_D^{20} 1.4731–1.4754
density: d_{15}^{15} 0.922–0.938
saponification number: 186–203
iodine number: 140–150
unsaponifiable: 0.5–1.5%
melting point: −5 °C
point of solidification: −13 to −20 °C

(Roth and Kormann 2000, p. 148)

Shelf Life

Cold-pressed safflower oil has a shelf life of up to 12 months if it is refrigerated and kept in a dark place (Roth and Kormann 2000, p. 148).

Characteristics of the Oil

Safflower oil is one of the highest grade oils in human nutrition; compared with other known vegetable oils, it has the highest content of linoleic acid (nearly 80%). The amount of saturated fatty acids is small (under 10%). Despite its high amount of linoleic acid, the oil is very stable (Kerschbaum and Schweiger 2001, p. 12; Ulmer 1996, p. 57).

5 Application

In Pharmacy and Medicine

For Prophylaxis and for the Treatment of Hypercholesterolemia

Patients with hypercholesterolaemia have an up to ten times higher concentration of lipoproteins of low density (LDL, etc.) in the vascular walls than in the serum. These lipoproteins are involved in the pathogenesis of arteriosclerosis. Due to its high amount of unsaturated fatty acids, safflower oil as food for particular nutritional uses not only lowers the cholesterol level, but also serves as prophylactic against arteriosclerosis. This effect is only possible, however, if the patient follows a low-fat diet (Hager 1978, volume 7b, p. 204; Roth and Kormann 2000, p. 50; Sutherland et al. 2002).

To Treat Cardiovascular Diseases

Hypertriglyceridaemia increases the coagulation ability of the blood and thus the danger of thrombosis. Safflower oil has a positive effect on cardiovascular diseases because it inhibits agonists of blood clotting, for example thrombin, and therefore lowers the risk of a thrombosis. Changing from a diet of predominantly animal fats to one with an increased amount of unsaturated fatty acids also leads, however, to a lower concentration of antioxidants, for example tocopherol, in the blood, whereby the risk of cancer is increased. Safflower oil contains 45.3 mg tocopherol/100 g oil. It is questionable whether the above-mentioned risk applies to safflower oil and whether other antioxidants would have to be substituted in the case of corresponding dietary change (Watkins and Bierenbaum 2001; Kerschbaum and Schweiger 2001, p. 36).

Topically

Safflower oil is recommended as a basis for unctions to treat rheumatism, paralyses, sprains and contusions (Hager 1978, volume 3, p. 727).

As Food

Safflower oil is one of the healthiest edible oils and excellent for human nutrition because of its high proportion of unsaturated fatty acids. It is not heat-resistant, however, and therefore not suitable for frying. If it is heated, this should only be over short periods of time, for example for steaming and simmering (pflanzenoel.ch/25.05.06). The oil is especially well suited for vegetable dishes if it is stirred into the dish shortly before serving (Kircher 2002, p. 24).

In Skincare and Body Care

Safflower oil is frequently used in cosmetics (Hager 1978, volume 7b, p. 204):

- in soap production, for a soft soap;
- as a re-fattening agent in bath oils and shower gels;
- as a skin oil for oily skin with acne, normal skin with a tendency towards inflammation and combination skin with an oily tendency, since the oil has a matting effect, and
- as a lotion for the eye area.

Technical Uses

Safflower oil has many technical uses (Hager 1978, volume 7b, p. 204; Roth and Kormann 2000, p. 50):

- as a substitute for linseed oil in the production of linoleum (12 hours' cooking results in a gelatinous mass called Afridi wax, which is used instead of linseed oil);
- to produce varnish and paint (safflower oil films do not yellow with age);
- for alkyd resin;
- as a paint for houses, as well as for printing ink;
- as lamp oil;
- for margarine production, and
- as a lubricant; in India, in particular, it is used for leather, ropes, etc.

Sea Buckthorn Oil

Oleum Hippophae rhamnoides synonyms: Sanddornöl/Sanddornfruchtfleischöl (D)

1 Source Plant

Hippophae rhamnoides L. (Elaeagnaceae), sea buckthorn, sandthorn, sallowthorn, seaberry

S. Krist, *Vegetable Fats and Oils*, https://doi.org/10.1007/978-3-030-30314-3_105

Habitat

There are two subspecies of sea buckthorn: a maritime subspecies that grows near the North Sea and the Baltic Sea, and a continental subspecies, which is common in the Pyrenees, the Alps, the Carpathian Mountains, the Pamir Mountains, the Altai Mountains to Lake Baikal, and Tibet. The ecological types differ in habitus and build, and have different types of thorns, fruit and ingredients. In Central Europe, *Hippophaes rhamnoides ssp. rhismnoides* grows along the coasts on sandy soil, *ssp. fluviatilis* along alpine rivers on broken rocks, in riparian forests or in clearings of dry pine forests. Sea buckthorn generally grows best in slightly moist, sandy soil with dry and calciferous upper layers, or in loamy or clayey soil, especially on river banks, seashores and dunes (Hager 1978, volume 5, p. 84).

S

Description

Depending on its location, sea buckthorn is either a branchy, thorny bush similar to a willow, about 1–3.5 m tall, or a densely branched, upright tree that can reach a height of more than 6 m. The irregular, bulky branches have a smooth, dark, reddish-brown bark. Male shrubs are usually more branched than female ones. The annual branches have a silvery shine because of their top layer of hair, but usually carry thorns later. In their second year, the branches produce thorny bastard branches. The thorns are stiff, with a straight point. The leaves are alternate, narrow and lanceolate. The surface of the leaves is first a dull sea-green because of stellate hair, then hairless, and reddish to reddish brown in autumn. The underside is greyish white and silvery to copper red; it is shiny because it is densely covered with peltate scales. The leaves have a protruding longitudinal vein. The male flowers are thick and brown, with a protruding stigma. The female flowers are small and yellowish green. The fruit are round mockberries of a fluorescent-orange colour (Hager 1978, volume 5, p. 84; Pilaske 2002, p. 78).

2 Cultivation and Extraction

Cultivation

Until 1980, the Soviet Union was the leading country in the breeding, cultivation and processing of sea buckthorn. In the GDR, about 300 hectares of sea buckthorn plantations were cultivated until 1989; they declined after the German reunification but are now slowly being rebuilt. Today, China is the largest cultivating country of sea buckthorn, with a cultivation area of one million hectares. Sea buckthorn is popular in China, which is illustrated by the fact that it was added to the official Chinese Pharmacopoeia in 1977 (Pilaske 2002, p. 80).

Extraction of the Oil

Sea buckthorn oil comes in three different varieties: Cold pressing or centrifugation of the raw pressed juice extracts the oil from the fruit pulp. Another method to extract the fruit pulp is with supercritical CO_2. For this method, whole, dried berries are crushed in a quern. This prevents the seeds from being crushed, so that only the lipids in the fruit pulp are extracted. The yield amounts to 12.5%. Oil is obtained by separating the seeds from the fruit pulp and then cold pressing and subsequently extracting the oil from the seeds. Pomace oil is a mixture of the oil from the seeds and the oil from the fruit pulp. After the first pressing of the fruit, amounts of oil remain in the residues of the fruit pulp, skins and seeds. These are extracted and mixed with the oils from the fruit pulp and seeds. The pomace oil is the actual sea buckthorn oil (Pilaske 2002, p. 93; Luetjohann 2001, p. 121; Quirin and Gerard 1993).

3 Character

Colour

Oil from the fruit pulp: orange-red.
Heavily orange-red due to carotenoids (Löw 2003, p. 135).
Highly fluid, vividly orange-red (Pilaske 2002, p. 93).
Orange-red, clear oil, semi-solid at room temperature (Quirin and Gerard 1993).
Oil from the seeds: yellowish-red.

Odour

Oil from the fruit pulp: fruity, sweet, like dried fruit, vinegar note, touch of sea buckthorn.
Typical of sea buckthorn, at times unpleasant (Quirin and Gerard 1993).
Oil from the seeds: no smell of sea buckthorn
Oil from the fruit pulp and seeds: sour, fruity, lemony, of sea buckthorn.

Flavour

Oil from the fruit pulp: typically aromatic, fruity. Typical of sea buckthorn (Luetjohann 1999, p. 67).

4 Ingredients

Composition of Fatty Acids

Sea Buckthorn Oil from the Seeds

Fatty acids	Content in % acc. to Kerschbaum and Schweiger (2001), p. 21
Palmitic acid	19.5
Linoleic acid	11.3
Palmitoleic acid	10.5
Oleic acid	49.8
α-Linolenic acid	1.1
Stearic acid	0.7
Arachidic acid	0.1
Vaccenic acid	6.5
Myristic acid	0.1

S

Sea Buckthorn Oil from the Fruit Pulp

Fatty acids	Content in % acc. to Luetjohann (1999), p. 69	Content in % acc. to Pilaske (2002), p. 94	Content in % acc. to Xing et al. (2002)	Content in % acc. to Steen and Mörsel (2003)	Content in % acc. to Quirin and Gerard (1993)
Myristic acid	0.2–0.5	–	–	0.56	0.4
Palmitic acid	30.8–34.0	30.8	33.4	34.98	33.0
Palmitoleic acid	34.2–35.6	35.6	24.9	31.27	32.9
Stearic acid	0.5–0.9	–	1.0	0.89	0.6
Oleic acid	23.2–25.4	25.4	26.2	19.39	18.9
Linoleic acid	5.1–5.5	5.5	5.1	3.49	5.4
α-Linolenic acid	1.7	1.7	1.6	1.00	–
Eicosenoic acid	–	–	–	–	2.0

Composition of Fatty Acids Compared

(Luetjohann 2001, p. 126)

Fatty acids	Content in % in oil from the fruit pulp	Content in % in oil from the seeds	Content in % in pomace oil
Palmitic acid	30.8–34	7.5–8.3	23.5
Linoleic acid	5.1–5.5	33.0–34.6	14.8
Palmitoleic acid	34.2–35.6	0.6–1.7	20.8
Oleic acid	23.2–25.4	19.3–23.3	24.7
α-Linolenic acid	1.7	30.4–32.9	13.2
Stearic acid	0.5–0.9	2.1–3.4	1.8
Arachidic acid	–	0.3–0.6	0.3
Myristic acid	0.2–0.5	0.1	0.1

Composition of Sterols in Sea Buckthorn Oil Made from Fruit Pulp

Sterols	Content in % acc. to Steen and Mörsel (2003)	Content in % acc. to Yang et al. (2001) (free sterols/sterol esters)
Campesterol	6.7	2.4/0.7
Δ5,23-Stigmastadienol	6.8	–
β-Sitosterol	46.1	66.1/1.8
Isofucosterol	10.7	–
Cycloartenol	6.5	–
24-Methyl-cycloartenol	18.8	–
Citrostadienol	6.9	–
Stigmastanol	–	2.3/1.0
Stigmasta-5,24-dien-ol	–	4.3/3.0
Stigmasta-8,24-dien-ol	–	1.4/2.4
Stigmast-7-en-ol+cycloartenol	–	1.4/6.7
4,14-Dimethyl-9,19-cyclo-ergost-enol	–	1.9/7.8
Stigmasta-7,24-dien-ol	–	1.0/3.0
9,19-Dimethyl-9,19-cyclo-ergost-enol	–	1.0/1.3
24-Methyl-cycloart-24(24′)-en-3β-ol	–	1.0/20.5
4-Methyl-stigmasta-7,24-dien-ol	–	0.1/7.9
4,14,24-Trimethyl-9,19-cyclo-ergost-24(24′)-enol	–	0.4/4.0

Composition of Tocopherols

Sea Buckthorn Oil from the Seeds

The oil contains 58.2 mg α-tocopherol/100 g oil. The literature does not define which kind of oil was examined (Kerschbaum and Schweiger 2001, p. 36).

Sea Buckthorn Oil from the Fruit Pulp

Tocopherols	Content in mg/kg acc. to Bat and Tannert (1993)
α-Tocopherol	184.4
Δ-Tocopherol	64.8
γ-Tocopherol	55.2
Tocopherols in total	330.4

Other Ingredients

(Kerschbaum and Schweiger 2001, p. 36)

Vitamin A
Vitamin E
Vitamin K
Lecithin
β-Carotene
Carotenoids
Flavonoids
Tanning agents
Sitosterol

Physical Key Figures of Sea Buckthorn Oil

refractive index: n_D^{20} 1.465–1.477
density: 0.8900–0.9550
saponification number: 180–210
acid number: (mg KOH/g): <40
iodine number: 130–200
unsaponifiable: <1%

S

Shelf Life

Sea buckthorn oil has a shelf life of about 3 months.

Characteristics of the Oil

The three varieties of sea buckthorn oil differ considerably in their content of active components and their composition of fatty acids. The oil obtained from the fruit pulp contains a particularly high amount of palmitoleic acid, of which there are only small amounts in other vegetable oils, with the exception of macadamia oil and avocado oil. As palmitoleic acid is also part of the fat in human skin, skin diseases can be treated with sea buckthorn oil obtained from the fruit pulp. Oil extracted from the seeds of sea buckthorn contains a high amount of α-linolenic acid. This ω-3 fatty acid influences the synthesis of prostaglandins and the lipid metabolism, and prevents arteriosclerosis. In general, sea buckthorn oil has a particularly high vitamin E content, twice as high as that of wheat-germ oil, nine times higher than that of corn oil, and 35 times higher than that of soya bean oil. For this reason, sea buckthorn oil is a valuable antioxidant employed in cancer therapy (icrts.org/27.10.03 – Kerschbaum and Schweiger 2001, p. 23; Pilaske 2002, p. 93 et seq).

5 Application

In Pharmacy and Medicine

To Treat Inflammation in the Gastrointestinal Tract

Sea buckthorn oil lowers the activity level of proteolytic enzymes, the secretion of gastric acid and the motility of the stomach. If sea buckthorn oil is administered orally, it not only provides the organism with supplementary vitamins, carotenoids, unsaturated fatty acids, sterines and micronutrients, but also has a regulating effect on the mucosa of the gastrointestinal system. Sea buckthorn oil can thus be employed in this field against the following indications (Hunnius 1998, p. 664; Luetjohann 2001, p. 134 et seq):

- hyperacidity of the stomach and heartburn;
- gastritis (superficial inflammation of the gastric mucosa as well as atrophic gastritis with degeneration of the gastric mucosa);
- *Ulcus ventriculi et duodeni*;
- benign tumours of the gastrointestinal system;
- malign tumours at the entrance of the stomach, and
- gastric carcinoma in early stages.

A study on rats examined the effect of sea buckthorn oil obtained from the fruit pulp on experimentally induced stomach ulcers. The results showed the protective and healing effect of oil extracted with supercritical CO_2 on four different experimental models. The oil showed a dose-dependent inhibition of the formation of gastric ulcers. After ligation of the pylorus, inhibition amounted to 20% on average, independent of dosage. The results suggest that the positive effect of the oil not only depends on its fatty acid content, but that the high amounts of β-carotene, ***α-t***ocopherol and β-sitosterol are also contributing factors. The healing effect is probably due to an increase in the hydrophobia of the gastric mucosa, the delayed emptying of the stomach, the inhibitory effect on the proteolytic activity of the gastric acid and the stimulation of wound healing of the mucosa (Xing et al. 2002).

In Radiotherapy

Scientific research in Russia and China before and especially after the Chernobyl accident shows that sea buckthorn oil can be used in the treatment of radiation sickness, both orally and topically. It is not only employed in cases of radiation injuries, but also as an adjunctive therapy, for example in cases of oesophageal cancer as protection from radiation injuries (Hager 1978, volume 5, p. 85; Hunnius 1998, p. 664; Pilaske 2002, p. 84, 98).

In Cancer Therapy

The tumour-inhibitory effect of sea buckthorn oil cannot exclusively be traced back to the scavengers tocopherol and β-carotene, but is rather a result of the entirety of the ingredients and the resulting spectrum of activity, for example the effects of cell regeneration and of strengthening the immune defence (Pilaske 2002, p. 84).

To Treat Inflammation in the Genital Area (Vagina, Uterus, Cervix)

These can be, for example, erosions of the cervix or chronic cervicitis caused by infection, fractures or scars, accompanied by discharge. In some cases, the reason for erosions can be traced back to deficiencies of β-carotene and tocopherol, and the healing effect of sea buckthorn oil is thus based on the fact that it supplies these substances. The antiphlogistic effect of sea buckthorn oil is beneficial in cases of inflammatory redness of the mucosa and lesions, as well as in cases of oozing inflammation of the cervical mucosa, which is often accompanied by tissue damage and changes to the epithelial tissue. In cases of colpitis of various genesis, topical as well as systemic treatment with sea buckthorn oil showed a good effect (Luetjohann 2001, p. 131 et seq).

As Adjunctive Therapy in Cases of Mycosis

The administration of sea buckthorn oil in combination with evening primrose oil and tea-tree oil has proved of value as an adjunctive therapy in cases of mycosis caused predominantly by *Candida albicans*. As an antioxidant, sea buckthorn oil has an anti-inflammatory effect and intercepts free radicals released by the fungal toxins as a consequence of mycosis therapy (Pilaske 2002, p. 97; Luetjohann 2001, p. 138 et seq).

To Treat Inflammation of the Oral Mucosa

Aphtha of the oral mucosa, bleeding or inflamed gums and also inflamed tonsils and pharyngeal catarrh can be treated topically with sea buckthorn oil. Depending on the complaint, it is either applied in undiluted form to the afflicted area of the skin, or used in diluted form, and thus has an antiphlogistic effect. In Russia, there are sea buckthorn sprays to treat sore throats and hoarseness (Pilaske 2002, p. 97; Luetjohann 2001, p. 133 et seq).

In Cases of Bedsores, Burns and Frostbite

In cases of bedsores, sea buckthorn oil soothes inflammation and promotes wound healing. It shortens the healing process and supports wound healing in cases of damage to the tissues caused by burns of all kinds and by frostbite. Sea buckthorn oil is applied to the damaged skin in undiluted form. It can also be applied if transplantations of skin are necessary, to speed up the process of adhesion of the skin transplant (Pilaske 2002, p. 99 et seq; Luetjohann 2001, p. 156 et seq).

Inhibiting Blood Coagulation

A small-scale study on 12 healthy men with normal lipidemia showed that a mixture of the oils from the fruit pulp and the seeds minimised cardiovascular risk factors. The oil mixture did not have any influence on the phospholipid fatty acids of the plasma or platelets, but it had a significant influence on ADP-induced platelet aggregation. Both the percentage of aggregation/min and the maximal aggregation after 4 minutes were reduced. It can be deduced that supplementation with a mixture of fruit pulp and seed oils has a positive effect on patients with an increased risk of blood coagulation. The exact mechanism of this effect remained unclear, and further research is therefore necessary (Johansson et al. 2000).

To Treat Skin Diseases

In China and Russia, sea buckthorn oil is traditionally used in the treatment of skin diseases. A study intended to show the effects of sea buckthorn oil made from the fruit pulp on the composition of the fatty acids in the glycophospholipids of the skin of patients suffering from atopic dermatitis. Over a period of 4 months, the patients received capsules containing 5 g oil daily. At the end of the study, a slight increase in stearic acid in the phospholipids of the skin was observed. The results show that the composition of fatty acids in the glycophospholipids of the skin is well buffered against dietary changes. Further studies are necessary to verify the results (Yang et al. 2000).

In Cosmetics

All kinds of sea buckthorn oil have antibacterial, antiphlogistic and analgesic properties and promote wound healing. In addition, sea buckthorn oil provides effective protection against sunburn and actinic dermatitis. Due to the high amount of palmitoleic acid, which forms part of human skin fat, it reduces wrinkles and delays skin ageing, and is especially useful against dry and scabby skin. Sea buckthorn oil is added to ointments, creams, lotions and emulsions that serve as sunblock, as well as to products for ageing, brittle skin. In shampoos and deep conditioners, it is effective in treating itchy scalps and dandruff. In general, all sea buckthorn oil products as well as the pure oil are effective and skin-friendly products for external use. The oil's high content of the antioxidants tocopherol and β-carotene makes it especially valuable in cosmetics for the following purposes (Luetjohann 2001, p. 145 et seqq; Pilaske 2002, p. 95 et seq):

- protection against free radicals, UV rays and harmful substances in the environment;
- strengthening the protective and defence mechanisms of the skin;
- protection of the elastic and collagenous fibres of the connective tissue;
- prevention of wrinkles and smoothing of already existing wrinkles due to the so-called "skin repair effect";
- removal of dead skin cells;
- tightening of skin tissue, and
- pigment equilibrium.

In the Food Industry

Sea buckthorn oil and other preparations with sea buckthorn are used in the production of sweets, flavourings, and to flavour gravy, canned fruit and canned vegetables.

In Folk Medicine

In Tibetan medicine, burns, frostbite, female complaints, stomach complaints, tumours and respiratory problems are treated with sea buckthorn oil in the form of ointments, pastes, compresses, tinctures, pills and powder. The oil is also used for general physical and mental strengthening. In Mongolia, it is applied in cases of coughs, rheumatism, diseases of the liver and lungs, and for blood thinning. In Europe, it is mainly used to treat skin problems (Kerschbaum and Schweiger 2001, p. 47; Pilaske 2002, p. 84).

Sesame Oil/Sesame Oil from Roasted Seeds

Oleum Sesami indicae synonym: Sesamöl (D)

1 Source Plant

Sesamum indicum L. (Pedaliaceae), *syn.: Sesamum Orientale* L., sesame

S. Krist, *Vegetable Fats and Oils*, https://doi.org/10.1007/978-3-030-30314-3_106

Habitat

There are 35 different species of sesame, *Sesamum Orientale* being the most important economically (Kim 2001, p. 9). *Sesamum Orientale* is one of the oldest oil plants. It has been cultivated for centuries and originates in tropical Africa, from where it spread to the Middle East, India and China (Kim 2001, p. 9). Today, sesame is mainly cultivated in India, China, Sudan, Egypt, the former Soviet Union and Mexico. It is almost exclusively grown in developing countries (Salunkhe et al. 1992, p. 371).

Description

Sesame is an annual, herbaceous plant 10–120 cm tall. The leaves can look quite different, from elongated-ovate and entire to 3- to 5-lobed with a serrated edge. The bottom leaves are opposite, the upper ones either opposite or alternate. The short-stemmed flowers are white or reddish and usually stand alone in the upper leaf axils. The fruit is a brownish capsule with a length of 2–3 cm and a breadth of 0.5–1 cm. It usually has two compartments containing 80–100 seeds, which are yellowish-white, reddish, brown or black (Hager 1994, p. 688).

2 Cultivation and Extraction

Cultivation

Sesame is cultivated in tropical, subtropical and temperate climes throughout the world (Hager 1994, p. 690).

The plant grows in various kinds of soil, but good harvests are only possible if the soil is neither too sandy nor too heavy. Dry soil is usually not a problem, but higher yields can be achieved under irrigation. The seeds should be sown at a minimum temperature of 24 °C; as the seeds are very small, they are only covered with small amounts of earth (3–6 cm deep). The distance between rows should be 45–100 cm; the harvest begins before the seeds are fully ripe (Roth and Kormann 2000, p. 101).

Extraction of the Oil

Depending on the production method, there are two kinds of sesame oil: oil made from unroasted seeds, which is light yellow, and oil made from roasted seeds, which is dark brown (Kim 2001, p. 13).

The harvest is usually carried out by hand, as the seeds ripen at different times. The capsules are thus cut before the seeds are fully ripe, and then dried. The seeds are subsequently removed by shaking. The unpeeled sesame seeds are cleaned, dried and conditioned by applying steam, so that it is easier to separate the oil. The seeds are subsequently roasted and then rehydrated using steam. The oil is obtained by either pressing or extraction. Pressed oil is of higher quality, but extraction has the advantage that the yield is nearly 100%. Finally, the oil is filtrated (Kim 2001, 14 et seq).

3 Character

Colour

Sesame oil:
Cold-pressed oil: clear, golden yellow.
Refined oil: clear, light-yellow liquid.
Cold pressed oil is a light-yellow colour; oil pressed by adding heat is darker (Roth and Kormann 2000, p. 150).
Sesame oil from roasted seeds: dark-brown oil (Kim 2001, p. 13).

Odour

Sesame oil: cold pressed oil: nearly odourless (Roth and Kormann 2000, p. 150).
Typical sesame note, fatty, roasted, slightly pungent.
Sesame oil from roasted seeds: An important sensory characteristic of roasted sesame oil is its smell of popcorn, which is caused by acetylpyrazine and 2-acetylpyridine. Pyrroles are responsible for the burnt, earthy scent components, thiazoles for the green, nutty components smelling of vegetables, and furanes for the roasted caramel scent components (Ryu et al. 1999).

Volatile Compounds

Sesame Oil from Roasted Seeds

Flavour substances in sesame oil from roasted seeds (content in ppb):

Compounds	Olfactory impression acc. to Kim (2001), p. 70	Predominant olfactory impression acc. to Cadwallader and Heo (2001)	Heavily roasted oil (10 min at 180 °C) acc. to Shimoda et al. (1997)	Slightly roasted oil (10 min at 150 °C) acc. to Shimoda et al. (1997)
Aldehydes:				
3-methylbutanal	Like solvents, like banana	Malty, chocolate		
2-methylbutanal	Fruity, sourish	Malty, musty	31	15

S

Compounds	Olfactory impression acc. to Kim (2001), p. 70	Predominant olfactory impression acc. to Cadwallader and Heo (2001)	Heavily roasted oil (10 min at 180 °C) acc. to Shimoda et al. (1997)	Slightly roasted oil (10 min at 150 °C) acc. to Shimoda et al. (1997)
Pentanal	–	–	36	15
Hexanal	Green, like grass	Green, cut grass	263	59
Heptanal	Sweetish	Sweet, ethanolic	30	7
(E)-2-heptenal	Green, like grass	–	82	13
Octanal	–	Orange oil, sweet	26	6
(E)-2-octenal	–	–	48	16
Nonanal	Like cucumber	–	–	–
(E,E)-2,4-nonadienal	–	Fatty, fried	32	12
(E)-nonenal	Burnt, dusty like paper	Stale, like hay	–	–
(E,Z)-2,4-decadienal	Burnt, heavily fried	–	35	Traces
(E,E)-2,4-decadienal	–	Fatty, fried	154	36
(E)-2-decenal	–	Green, coriander	–	–
(E)-2-undecenal	–	Waxy, coriander	–	–
4,5-Epoxy-(E)-2-decenal	–	Metallic, fatty	–	–
Methylpropanal	–	Malty, chocolate	–	–
Acetaldehyde	–	Spicy, like yoghurt	–	–
Ketones, aliphatic alcohols and acids:				
4-octen-3-one	Fungoid, roasted, like coffee	Fungoid		
2-heptanone	–	–	31	6
2,3-butadione	–	Cream cheese	–	–
Hexanol	–	–	79	12
Octanol	–	–	26	9
Hexanoic acid	–	–	32	8
Heptanoic acid	–	–	32	5
Dodecanol	–	–	69	12
Octanoic acid	–	–	86	Traces
3-Methyl-butanoic acid	–	Dried fruit	–	–
2-Methyl-butanoic acid	–	Malty, musty	–	–
2-Nonanone	–	–	62	11
Phenols and aromates:				
Benzaldehyde	Like hay	–	240	53
Benzenmethanol	–	–	69	16
Benzenethanol	–	–	47	10
α-Ethylidenbenzenacetaldehyde	–	–	47	Traces
2-methoxy-5-(1-propenyl)phenol	–	Smoky	33	7

Compounds	Olfactory impression acc. to Kim (2001), p. 70	Predominant olfactory impression acc. to Cadwallader and Heo (2001)	Heavily roasted oil (10 min at 180 °C) acc. to Shimoda et al. (1997)	Slightly roasted oil (10 min at 150 °C) acc. to Shimoda et al. (1997)
Phenylacet-aldehyde	Flowery, sweetish, like honey	Flowery, like hay		
1-Phenylethanone	–	–	142	24
1-(3-methoxy-phenyl)ethanone	–	–	177	37
2-Methoxy-4-vinylphenol	–	Smoky, cloves	–	–
Acetophenone	Aromatic, Nutty, Sweetish			
4-Methyl-acetophenone	Roasted, like almonds	–	–	–
o-Guaiacol	Burnt, smoked	–	321	32
Ethylguaiacol	Burnt, smoked	–	–	–
4-Hydroxy-2-methyl-acetophenone	Sweetish, like vanilla	–	–	–
Pyridines and pyrroles:				
1H-pyrrole	Burnt, alcoholic	–	70	16
2-Ethyl-1H-pyrrole	–	–	19	3
2-Methylpyridine	Unpleasant, pungent	–	58	9
3-Methyl-1H-pyrrole	–	–	26	Traces
1-Methyl-1H-pyrrole-2-acetonitrile	–	–	40	7
1-Ethyl-1 H-pyrrole-2-carboxyaldehyde	–	–	29	5
1-(1-methyl-1H-pyrrole-2-yl) ethanone	–	–	22	Traces
1-(1H-pyrrole-2-yl)ethanone	–	–	71	17
1H-pyrrole-2-carboxyaldehyde	–	–	127	143
Methyl-pyrrole-2-carboxylate	–	–	33	Traces
2,5-Dimethyl-1 H-pyrrole	Burnt	–	–	–
3,4-Dimethylpyridine	Stinking	–	–	–
Methyl-4-pyridine-carboxylate	–	–	55	10
4-Pyridinylacetate	–	–	67	13
1-(2-pyridinyl)-ethanone	–	–	16	Traces
2-Pentylpyridine	Fatty, slightly burnt	Fatty, metallic	–	–
Pyrazines:				
Methylpyrazine	Roasted, nutty	–	779	195
2,5-Dimethylpyrazine	Roasted, musty	–	735	158

Compounds	Olfactory impression acc. to Kim (2001), p. 70	Predominant olfactory impression acc. to Cadwallader and Heo (2001)	Heavily roasted oil (10 min at 180 °C) acc. to Shimoda et al. (1997)	Slightly roasted oil (10 min at 150 °C) acc. to Shimoda et al. (1997)
2,6-Dimethylpyrazine	–	–	383	82
2-Ethylpyrazine	Roasted, like baked cake		153	35
2,3-Dimethylpyrazine	Roasted, sweetish	–	101	23
2-Ethyl-5(6)-methylpyrazine	Roasted	Ethanolic, sweet	212 (237)	40 (47)
Trimethyl-pyrazine	Musty, burnt	–	–	65
2-Ethenylpyrazine	–	–	46	12
2-Ethyl-3-methylpyrazine	Roasted, Aromatic, nutty	–	–	–
1-Acetylpyrazine	Aromatic, heavily roasty	–	–	–
2-Acetylpyrazine	–	Roasted, popcorn	178	35
2-Ethyl-3,5-dimethylpyrazine	Earthy, like coffee	Earthy, nutty	–	–
3-Ethyl-2,5-dimethylpyrazine	–	–	542	104
3-Ethyl-3,5-dimethylpyrazine	–	–	154	45
2-(1-methylpropyl)-pyrazine	–	–	23	5
2-Ethenyl-6-methylpyrazine	–	–	55	9
2,3-Diethyl-5-methylpyrazine	–	–	22	4
3,5-Diethyl-5-methylpyrazine	–	Nutty, potato	92	21
2-Acetyl-3-methylpyrazine	–	Popcorn, coffee	–	–
2-Methyl-6-(1-propenyl) pyrazine	–	–	47	8
2,6-Diethylpyrazine	Roasted, aromatic, sweetish			
5-Methyl-6,7-dihydro-5(H)-cyclopenta-pyrazine	Aromatic, heavily roasted	–	–	–
3(2),5-Dimethyl-6,7-dihydro-5(H)-cyclopenta-pyrazine	Aromatic, malty, rancid			
3,5(6)-Dimethyl-2isobutylpyrazine	Roasty, aromatic	–	–	–
2-Isopropenyl-pyrazine	–	–	48	8
(E)-2-methyl-6-(1-propenyl)-pyrazine	–	–	51	11
2,3-Dimethyl-5-isopentylpyrazine	–	–	28	Traces
Sulphuric compounds:				
Thiophene	Spicy	–	–	–

Compounds	Olfactory impression acc. to Kim (2001), p. 70	Predominant olfactory impression acc. to Cadwallader and Heo (2001)	Heavily roasted oil (10 min at 180 °C) acc. to Shimoda et al. (1997)	Slightly roasted oil (10 min at 150 °C) acc. to Shimoda et al. (1997)
Thiazole	Nutty, aromatic, spicy	–	–	–
Dimethyldisulfide	Roasted, aromatic	–	–	–
2-Methylthiophene	Aromatic, burnt	–	–	–
3-Methylthiophene	Spicy	Spicy, plastic	–	–
2-Methylthiazole	Of faeces	–	–	–
Methylethyl-disulfide	Malty, aromatic	–	–	–
2,4-Dimethylthiazole	Slightly burnt, like coffee		58	12
2,5-Dimethylthiazole	–	–	115	20
Furfurylthiol	Fungoid, musty, roasted, like coffee	Roasted, like coffee		
4,5-Dimethylthiazole	Burnt	–	110	22
4,5-Dimethyl-isothiazole	–	–	39	8
4-Methyl-5-ethylthiazole	–	–	23	5
2-Ethyl-5-methylthiazole	–	–	28	8
Dimethyl-trisulfide	Like vegetables, grass	–	–	–
Furfurylmethyl-sulfide	Burnt, like coffee	–	–	–
2-Acetylthiophene	Burnt	–	–	–
3-Acetylthiophene	–	–	38	6
2-Acetylthiazole	–	Roasted, popcorn	–	–
2-Acetyl-5(4)-methylthiazole	Dusty, like burnt wood			
Isopropyl-2-methylthiazole	Like burnt meat			
2-Ppropyl-4-methylthiazole	–	–	123	21
2-Butyl-5-methylthiazole	–	–	45	5
2-Methyl-3-methyl-thiopyrazine	Aromatic, heavily roasted, burnt			
Furfurylmethyl-disulfide	Roasted, meaty	–	–	–
5-Methyl-2-furfuryl-mercaptan	Nutty, roasted	–	–	–
2-Furan-methanthiol	–	–	40	6

Compounds	Olfactory impression acc. to Kim (2001), p. 70	Predominant olfactory impression acc. to Cadwallader and Heo (2001)	Heavily roasted oil (10 min at 180 °C) acc. to Shimoda et al. (1997)	Slightly roasted oil (10 min at 150 °C) acc. to Shimoda et al. (1997)
Methanthiol	–	Sulphuric, mouldy	–	–
3-(methylthio)-propanal	–	Cooked potato	–	–
2-Methyl-(3-methyldithio-furan)	–	Vitamin, meaty	–	–
4-Methyl-3-thiazoline	–	Garlic	–	–
2-Acetyl-2-thiazoline	–	Roasted, popcorn	–	–
Other compounds:				
Furfuryl alcohol	Rancid, like cooked sugar		316	49
2,4,5-Trimethyloxazole	Like hay	–	–	–
3,5,5-Trimethyl-2-cyclopenten-1-one	–	–	37	Traces
2-Acetylfuran	Burnt	–	–	–
2-Furfural	–	–	51	10
1-(5-methyl-2-furanyl)-1-propanone	–	–	70	10
α-Methyl-α-vinyl-2-furanacetaldehyde	–	–	25	8
Pentylfuran	Sweetish, fruity, flowery		50	8
2,5-Dimethyl-4-hydroxy-3(2H)-furanone	–	Burnt sugar	–	–
2-Acetyl-5-methylfuran	Sweetish, like caramel	–	75	13
5-Methyl-2-furfural	–	–	220	47
2,3-Dihydro-1H-indole	Like wood	–	37	7
3-Methylindole	–	Spicy, of faeces	–	–
Limonene	–	–	27	7
β-Damascenone	–	Tea, Apple gravy	–	–
Ethylacetate	–	–	45	12
4,5-Dihydro-2(phenylmethyl)-1H-imidazole	Malty, tart			

Sesame Oil

Using GC-MS, the following volatile compounds were identified in slightly roasted sesame oil (Shimoda et al. 1998):

2-Acetylpyrazine	2-Ethylpyrazine	1-Methyl-1H-pyrrole
4-Acetylpyrazole	4-Ethylthiazole	3-Methyl-2,4-pentandione
Benzaldehyde	Furfural	2-Methylpropanal
2-Butanal	Furfuryl alcohol	2-Methyl propanoic acid
2-Butanone	Guaiacol	2-Methylpyrazine
Dimethyldisulfide	n-Hexanal	5-Methylpyridinamine

2,5-Dimethylfuran	4-Hepten-2-ol	2-Methylthiophene
3,6-Dimethyl-2H-pyran-2-ol	1H-pyrrole	3-Methylthiophene
3,6-Dimethyl-2H-pyran-2-one	3-Hydroxy-2-butanone	Octane
3,5-Dimethylisoxazole	1-Hydroxy-2-propanone	7-Octen-2-one
2,5-Dimethylpyrazine	Isothiazole	n-Pentanal
2,6-Dimetylpyrazine	Methylacetate	4-Pentyn-2-ol
2,6-Dimethylpyrimidine	2-Methylbutanal	2-Propanol
Dimethyltrisulfide	3-Methylbutanal	2-Propanone
Acetic acid	3-Methylbutanol	Propanoic acid
2-Ethylhexanal	3-Methyl-2-butanone	Pyrazine
2-Ethyl-6-methylpyrazine	2-Methylfuran	Pyridine
2-Ethyl-5-methylpyrazine	5-Methylfurfural	Thiophene

Flavour

Sesame oil: spicy taste (Roth and Kormann 2000, p. 150). Slightly nutty taste (Ulmer 1996, p. 73).

Sesame oil from roasted seeds: characteristic, strong, nutty, pleasant roasted flavour (Kim 2001, p. 8).

4 Ingredients

Composition of Fatty Acids

Sesame Oil

Fatty acids	Content in % acc. to Firestone (1999), p. 90 et seq	Content in % acc. to Karleskind (1996), p. 152	Content in % acc. to Nature certificate of analysis/ 06.02.06
Lauric acid	–	<0.1	–
Myristic acid	0–0.1	<0.1	–
Palmitic acid	7.9–9.2	8–11	8.2
Palmitoleic acid	0.1–0.2	<0.3	0.1
Stearic acid	4.4–6.1	4–6	5.6
Oleic acid	33.5–44.1	37–42	39.0
Linoleic acid	40.3–50.8	39–47	45.7
α-Linolenic acid	0.3–0.5	<0.6	0.4
Arachidic acid	0.3–0.7	<1	0.6
Behenic acid	0–0.3	<0.5	0.1

Sesame Oil from Roasted Seeds

In comparison to unroasted sesame oil, the composition of fatty acids and tocopherols changes only at a roasting temperature between 220 and 240 °C (Kim 2001, p. 147 – content in %)

Fatty acids	140 °C	160 °C	180 °C	200 °C	Commercially roasted
Myristic acid	1	0.9	0.8	0.9	0.9
Palmitic acid	8.8	8.7	9	8.8	9.2
Stearic acid	4.9	5	4.8	5.1	5.1
Oleic acid	40	39.2	40	39.4	38.3
Linoleic acid	45.3	45.7	45	45.1	45.9
Arachidic acid	0.4	0.6	0.3	0.6	0.6

Composition of Triglycerides in Sesame Oil

Triglycerides	Content in % acc. to Firestone (1999), p. 141	Triglycerides	Content in % acc. to Firestone (1999), p. 141
PSO	0–0.6	POL	0–8
SOS	0.3–4	SOL	4–21
PPO	0–0.6	OOL	15–20
POO	0–3	PLL	0–11
SOO	2–10	SLL	3–10
PPL	0–2	OLL	18–25
PSL	0–1	LLL	5–20
SSL	0.5–5	ALO	0–0.3
OOO	4–7	LnLL	0–0.5

Composition of Sterols in Sesame Oil

(concentration in mg/100 g and content in %)

Sterols	Refined sesame oil acc. to Carstensen (2001), p. 43	Cold pressed sesame oil acc. to Carstensen (2001), p. 43	Sesame oil acc. to Firestone (1999), p. 91	Sesame oil acc. to Karleskind (1996), p. 152
Total content	521.3	626.6	450–1896	539–636
Cholesterol	0.8	0.4	0.1–0.2	<0.3
Brassicasterol	–	<0.1	0.1–0.2	<0.1
Campesterol	18.8	18.6	10.1–20	18–19
Stigmasterol	6.7	7.0	3.4–6.4	6–7
β-Sitosterol	61.4	62.6	57.7–61.9	59–62
Δ5-Avenasterol	9.4	10.5	6.2–7.8	10–11
Δ7-Stigmasterol	1.3	0.5	1.8–7.6	0.6–0.8
Δ7-Avenasterol	1.6	0.4	1.2–5.6	0.8–1.1
Fucosterol	–	–	–	<0.5

Composition of Tocopherols in Sesame Oil

(concentration in mg/100 g and content in %)

Tocopherols	Refined sesame oil acc. to Carstensen (2001), p. 45	Cold pressed sesame oil acc. to Carstensen (2001), p. 45	Sesame oil acc. to Karleskind (1996), p. 152
Total content	27.3	272	20–50
α-Tocopherol	2.2	1.3	4
β-Tocopherol	2.2	1.3	2
γ-Tocopherol	92.1	92.7	83
δ-Tocopherol	4.0	6.2	11
γ-Tocotrienol	–	–	–

Physical Key Figures

Sesame Oil

refractive index: 20 °C	1.474–1.476
40 °C	1.465–1.469
density:	0.915–0.926
iodine number:	100–120
saponfication number:	187–199
acid number:	<0.6
point of solidification:	−8 °C to −6 °C
unsaponifiable:	0–2%

(Roth and Kormann 2000, p. 150; Firestone 1999, p. 150)

Sesame Oil from Roasted Seeds

combustion point: 160 °C
(Kim 2001, p. 14)

Shelf Life

If stored appropriately, sesame oil has a shelf life of about 12 months.

5 Application

In Pharmacy and Medicine

In pharmacy, sesame oil is a solvent and carrier for medicines, especially in the production of oleaginous solutions for injection and injection suspensions for steroids. In patches and ointments, it can substitute olive oil (Hager 1993, p. 693).

The lignan sesamin has immunosuppressive proprieties. In vitro, it inhibits the proliferation of human peripheral lymphocytes of the blood, which is induced by Concavalin A (Hager 1993, p. 692).

As Food

As a result of its strong flavour, dark sesame oil from roasted seeds is traditionally used in Asian cuisine for seasoning ("food flavouring"), and as an edible oil for cooking, frying, roasting and smoking meat, fish, shrimps, vegetables and products made from soya beans (Kim 2001, p. 8).

In Japan, oriental sesame oil (from peeled, toasted seeds) is used for seasoning tempura. In Chinese and Korean cuisine, it is used in small amounts for seasoning purposes.

In Folk Medicine

In cases of constipation, sesame oil is applied rectally. Externally, it is used to remove scabs and encrustations of the skin, and as a massage oil. In East and South-East Asia, it is used as a liniment to treat rheumatism and dizziness (Hager 1993, p. 692).

In addition, sesame oil protects the skin from the sun with a sun protection factor of 2. In India, it is the skin care oil par excellence, and is also regarded as a tonic for the nerves.

Industrial Uses

Sesame oil is used in the baking industry and in margarine production (pflanzenoel.ch/06.02.06). As lignans are easily detectable as specific ingredients of sesame oil, some countries have stipulated that sesame oil must be added to margarine, so that it can be distinguished from butter (Hager 1993, p. 693).

The oil is also used in the perfume industry and the cosmetics industry. It is the base of most perfumes and fragrance oils, as it does not become rancid easily and has no unpleasant flavour or smell (Salunkhe et al. 1992, p. 395).

In Agriculture

In agriculture, sesame oil is added to insecticides as a carrier and as a synergistic component. For example, the spyrethrum can be reduced by 50% by adding 5% sesame oil (Hager 1993, p. 693).

Other Applications

Oil from the second and third pressing of sesame seeds is processed to produce soap and lubricants (Senti et al. 2000).

6 Possible Unwanted Side Effects

Allergies to sesame are uncommon but are increasing because sesame seeds and sesame oil are used more frequently. Certain proteins in the seeds are triggers of allergic reactions. Cross-reactions with hazelnuts, soya and peanuts are possible.

To date, there are only few reports on hypersensitivity after using sesame oil. In one case, a treatment of burns with an ointment containing 60% sesame oil lea to allergic contact dermatitis after 10 days. Another case of allergic contact dermatitis was caused by applying lipstick containing sesame oil. Sesame seeds are regarded as a potent allergen, but in comparison, the sensitising potency of the oil is weak. Sesamin and sesamolin have been identified as the allergens of sesame seeds (Hager 1993, p. 693).

Shea Butter

synonyms: Sheabutter, Bambukbutter, Karitébutter, Galambutter (D); beurre de karité (F)

1 Source Plant

Vitellaria paradoxa C.F. Gaertn., *syn. Butyrospermum paradoxum* (C.F. Gaertn.) Hepper *(Sapotaceae)*, shea tree

S. Krist, *Vegetable Fats and Oils*, https://doi.org/10.1007/978-3-030-30314-3_107

Habitat

Shea trees are common from West Africa to the Upper Nile; they are characteristic trees of the savannah. The tree grows naturally in the Sahel and the northern regions of the Guinea zone (Axtell and Fairman 1992, p. 838 et seqq).

Description

S

Shea trees are 6–12 m tall and gnarled like oaks. The large leaves sit in bunches at the end of the thick branches; they are elongate, rounded at the front, and have a wavy edge. The flowers are arranged in dense, head-shaped bunches of 50–100 flowers, which are creamy white and smell intensely of honey. The ovate stone fruit are 4–5 × 3 cm in size, have a thick shell and are usually one-seeded. The seeds are shaped similarly to the fruit; they are crusty, brown and shiny, resembling chestnuts. The ripe fruit pulp is edible. Shea trees reach their maximum yield when they are 40–50 years old. The harvest takes place in June and July (Krist et al. 2006a; Neuwinger 1998, p. 838).

2 Cultivation and Extraction

Cultivation

The tree grows in the savannah, where oil palms cannot thrive because there is not sufficient rainfall. The main producers are countries in West Africa: Mali, Senegal, Côte d'Ivoire, Ghana, The Gambia, Nigeria and Benin (Axtell and Fairman 1992, p. 121). Although shea butter is an important article of trade – in 2003, Africa produced 651,000 tons of shea nuts – there are no commercial shea tree plantations yet, as expenditure is not profitable (Krist et al. 2006a).

Extraction

The fruit of the shea tree are gathered manually and piled on top of each other in containers or in a pit. The fruit pulp is removed. Subsequently, the nuts are dried in the sun or in ovens built from branches, underneath which a fire burns. Another method is to cook the nuts first and then dry them in the sun. To remove the shell, the nuts are ground in oil mills and sifted in wide-meshed baskets to separate the shells from the kernels. The wind blows away small parts, the shells fall to the ground, and the kernels stay in the basket. The kernels are extracted in an oven and reduced to small pieces in a mill or mortar. The fatty mass is mixed manually with water and kneaded for a long time until it is nearly white and the impurities have been removed. A great quantity of water is necessary for this process (Karleskind

1996, p. 758 et seqq). The extract obtained thereby is cooked until all the water has been removed, then it is filtrated and finally packed (Lovette 2004).

Black shea butter is obtained by roasting the kernels for about 15–30 min in hot sand or over a fire in suitable containers. The raw butter obtained by this method is used especially for culinary purposes. White shea butter is not roasted (Krist et al. 2006a).

3 Character

Colour

White shea butter: ivory, slightly yellowish, like white chocolate (Krist et al. 2006a). Like soft wax, lard or tallow, greenish-yellow to nearly white fat (Olberg 1987).
Black shea butter: not homogenous, grey, lumpy (Krist et al. 2006a).

Odour

White butter from Chad: fatty, slightly fruity, aromatic, nutty, slightly greasy smell (Krist et al. 2006a).
White butter from Cameroon: fatty, woody, spicy, smoky, ligneous (Krist et al. 2006a).
Black shea butter from Chad: aromatic, balmy, pungent, dominantly smoky (Krist et al. 2006a).

Volatile Compounds

Examination using SPME headspace analysis with GC-MS of various samples of shea butter (Krist et al. 2006a):

Components	Content in % in white shea butter (Chad, cooking time 15 min)	Content in % in white shea butter (Cameroon, cooking time 30 min)	Content in % in black shea butter (Chad, roasting time 30 min, cooking time 15 min)
Acetic acid	9.9	10.5	2.7
2-Methylfuran	–	–	5.4
Ethylacetate	0.4	–	–
Trans-2-pentanone	–	–	0.7
2-Methylpentanal	–	2.7	–
Pentanol	–	0.3	–
Butyric acid	–	0.1	–
Butyl acetate	1.1	0.5	0.2
4-Methylpyridine	–	–	0.1
2-Methylpyrazine	–	–	2.1
Furfural	0.2	0.3	–

Components	Content in % in white shea butter (Chad, cooking time 15 min)	Content in % in white shea butter (Cameroon, cooking time 30 min)	Content in % in black shea butter (Chad, roasting time 30 min, cooking time 15 min)
2,4-Dimethyl-1-heptene	0.9	–	0.2
Trans-2-hexenal	–	0.4	–
2-Hydroxymethylfuran	–	0.5	–
4-Methyloctane	0.3	0.5	–
Ethylbenzene	0.7	0.2	–
Hexanol	–	0.2	–
1,4-Dimethylbenzene	4.0	0.4	0.4
4-Heptanone	–	–	0.5
Valeric acid	–	0.5	–
2-Heptanone	0.2	0.4	0.5
Styrene	1.4	1.7	3.3
Nonane	–	0.5	–
Heptanal	2.2	1.9	–
2-Butoxyethanol	2.2	–	–
(E,E)-2,4-Hexadienal	–	0.1	–
2-Acetylfuran	–	0.2	2.7
2-Ethylpyrazine	–	–	0.8
Methylethylbenzene	0.1	–	–
α-Pinene	1.3	0.3	–
Propylbenzene	0.9	–	–
Trans-2-heptenal	0.7	1.6	–
Benzaldehyde	–	1.0	–
Hexanoic acid	1.9	2.8	–
Heptanol	–	0.8	–
Sabinene	–	1.0	–
1-Octen-3-ol	–	0.5	–
Phenol	–	1.2	–
6-Methyl-5-hepten-2-one	–	1.5	–
Decene	–	0.2	–
β-Myrcene	0.4	1.0	–
2-Octanone	–	0.9	–
1,2,3-Trimethylbenzene	3.2	1.2	–
Decane	2.6	0.7	–
*(E, Z)-2,4-h*eptadienal	–	0.2	–
Octanal	1.8	1.8	–
(2-furanyl)-1-propanone	–	–	0.6
3-Carene	0.8	1.6	–
2-Ethyl-1-hexanol	7.2	–	–
ρ-Cymene	2.9	2.9	–
α-Terpinene	1.3	–	–
Limonene	2.1	4.5	–
Phenylmethanol	–	0.5	–
Trans-3-octen-2-one	–	0.2	–

S

Components	Content in % in white shea butter (Chad, cooking time 15 min)	Content in % in white shea butter (Cameroon, cooking time 30 min)	Content in % in black shea butter (Chad, roasting time 30 min, cooking time 15 min)
2,5-Furandion-3,4-dimethyl	–	0.4	–
1-Methyl-3-propylbenzene	0.5	–	–
Trans-2-octenal	1.5	8.3	–
2-Acetylpyrrole	–	0.3	3.4
Octanol	–	1.3	–
Acetophenone	–	0.8	0.5
m-Cresol	–	–	0.9
Heptanoic acid	–	1.9	–
2-Nonanone	–	1.7	1.4
Guaiacol	–	0.5	–
3,5-Octadien-2-one	–	0.3	–
Undecane	2.6	0.5	1.7
Nonanal	6.5	4.9	–
2-Phenylethanol	–	0.4	–
2-Ethyl hexanoic acid	14.7	–	–
Trans-2-nonenal	–	4.9	–
Nonanol	–	0.8	–
Octanoic acid	5.0	0.9	–
Furfurylpyrrole	–	–	2.0
Naphthalene	1.2	–	1.8
Decan-2-one	–	2.6	–
Dodecane	6.5	1.8	3.3
Decanal	–	1.1	–
2,6-Dimethylundecane	–	–	0.6
*(E, E)-2,4-n*onadienal	–	0.2	–
2,3-Dimethoxytoluene	–	–	0.8
3,4-Dimethoxyphenol	–	–	1.8
2-Phenylethylacetate	–	0.6	–
Nonanoic acid	4.4	0.9	–
4-Ethylguaiacol	–	–	1.9
Undecan-2-one	–	0.1	–
*(E, Z)-2,4-d*ecadienal	0.3	0.6	–
Tridecane	–	0.6	3.2
Methylnaphthalene	–	–	1.7
Trans-2-undecenal	–	9.1	–
γ-Nonalactone	–	0.4	–
Tetradecane	2.0	0.2	1.6

4 Ingredients

The unsaponifiable portion of shea butter amounts to 7.0–10.0% and consists of sterols, 90 mg/100 g tocopherols and 2–3 g/100 g karitene (Karleskind 1996, p. 211).

Composition of Fatty Acids

Fatty acids	Content in % acc. to Firestone (1999), p. 92	Content in % acc. to Karleskind (1996), p. 210	Content in % acc. to Rossell and Pritchard (1991), p. 295
Lauric acid	0.4	<1	0.4
Myristic acid	0.3	<0.7	0.3
Palmitic acid	4–8	3–5	4.6
Palmitoleic acid	–	<0.3	–
Stearic acid	36–41	28–45	40.4
Oleic acid	45–50	42–59	44.9
Linoleic acid	4–8	3–9	7.5
α-Linolenic acid	0–0.4	<1	0.4
Arachidic acid	1–2	<1.5	1.6

Composition of Triglycerides

Triglycerides	Content in % acc. to Karleskind (1996), p. 209	
	Author 1	Author 2
SLL	–	1.0
POL	0.7	0.8
OOO	7.5	–
SOL	6.0	5.5
POO	3.5	2.0
PSL	0.8	2.0
PPO + MSO	0.6	–
SOO	30	24
SSL	–	4.5
PSO	7.5	6.0
PPS	0.3	0.1
SSO	31	42
PSS + AAPP	1.5	0.1
SSS + PSAA	2.0	0.1

Composition of Sterols

(concentration in mg/kg and content in %)

Sterols	Shea butter acc. to Firestone (1999), p. 92	Shea butter acc. to Karleskind (1996), p. 211
Cholesterol	–	1–3
β-Sitosterol	–	1–4
Δ7-Stigmasterol	38	38–41
Δ7-Avenasterol	11	2–6
24- Methyl-Δ7-cholestanol	6	–
Δ7, 22-Stigmastadien-3β-ol	45	–
Total content	2470	1500–2600

Composition of Triterpene Alcohols

Triterpene alcohols	Content in % acc. to Karleskind (1996), p. 212
β-Amyrin	7–8
Butyrospermol	15–20
α-Amyrin	38–48
Lupeol	25–28

Physical Key Figures of Shea Butter

density $d^{15,5}$	0.916–0.918
refractive index: n_D^{40}	1.463–1.467
saponification number:	178–198
iodine number:	52–66
unsaponifiable:	2–11%
melting point:	32–45 °C
titre °C:	48–54

(Firestone 1999, p. 92)

5 Application

In Pharmacy and Medicine

A clinical study tested the application of shea butter to treat common colds and ascertained that the oil has better effects than common nose drops (Tella 1979).

In Dermatotherapy

A clinical study tested pure shea butter on 35 patients suffering from skin diseases and a 15% shea butter cream on 15 patients. The test persons applied the products twice daily for a period of 10 days and 5 months respectively. Results showed that quicker and better effects were achieved against scaly dermatitis and dry, ageing skin than with a fat corticosteroid cream. Pure shea butter as well as the 15% cream are an effective treatment of atrophic and degenerative symptoms caused by ageing and persistent radiation exposure of the skin (Tran 1984).

The triterpene alcohols contained in shea butter, amyrin, lupeol and butyrospermol, have an anti-inflammatory effect. They inhibit protein-disintegrating enzymes and proteases affecting the skin and disintegrating collagen and elastin, which are important structural proteins contributing to the skin's strength and resilience. The production rate of these proteins decreases with age, which in turn leads to thinner, less elastic skin. Additionally, triterpene alcohols can be used as an alternative to conventional products like hydrocortisone and indomethacin to treat rheumatoid arthritis (Alander 2004).

S

Lowering Cholesterol Levels

A controlled double-blind study examined the effect of a shea butter spread on healthy test persons with normal or slightly increased serum cholesterol levels. Fifty-three test persons ate 30 g/d of the spread for a period of six consecutive weeks. The control group (52 test persons) ate a spread made of sunflower oil, also 30 g/d. The "shea nut diet" reduced the total cholesterol level by 5%, the LDL cholesterol level by 8%, the LDL cholesterol/HDL cholesterol ratio by 9%, the apolipoprotein B level by 7% (in each case the average value in the serum), the mean diastolic blood pressure by 2% and the mean systolic blood pressure by 3%. The HDL cholesterol, triglyceride and lipoprotein α levels in the serum remained the same. The mean BMI (body mass index) value was significantly reduced by 0.6%. The results show that a diet containing shea butter reduces the total and LDL cholesterol levels of healthy adults effectively and can thus contribute to preventing coronary cardiovascular diseases (Schmidt et al. 2003).

In Cosmetics

The high unsaponifiable portion (up to 11%) has a very positive effect on the skin when shea butter is applied topically, making the skin supple, preventing it from drying out, and retaining moisture in the epidermis. It is therefore added to care products like ointments, lotions and after sun products. It is skin-friendly and has a natural sun protection factor of 3. Shea butter is added to lip balms, ointments for the nose and breast, and creams to treat stretch marks.

In the Food Industry

Stearins from shea butter are used as equivalents to cocoa butter; its oleins are used as oil, after hardening as margarine, and in shortening and other food products. A study evaluated the toxicity of oleins in shea butter. Rats were given 7% refined shea olein and hardened shea olein, as well as 15% raw shea nut oil, shea olein and hardened shea olein during their pregnancy; subsequently, their young were fed on the mixture. This had no significant negative consequences on the general condition and on hemic parameters, but the mean cholesterol levels were significantly lower. Shea oleins had neither a significant influence on the number of offspring nor on their birth weight or weight during breeding. No signs of reproduction toxicity on rats were observed, neither in unhardened nor in hardened shea oleins, at least not at a dosage of up to 7.5 g/kg body weight/d (Baldrick et al. 2001).

Other Applications

Due to its high amount of allantoin, shea butter forms a base for pharmaceutical products in West Africa, where it is also used in soap production and in the building trade as rain protection for house walls (Axtell and Fairman 1992, p. 121).

Shea butter is an edible oil in Africa. In Europe, it is used for margarine production, as an ointment base and as a substitute for cocoa butter (Hunnius 1998, p. 1466).

Sour Cherry Oil

synonym: Weichselkernöl (D)

S. Krist, *Vegetable Fats and Oils*, https://doi.org/10.1007/978-3-030-30314-3_108

1 Source Plant

Prunus cerasus L., *Cerasus vulgaris* Mill., (Rosaceae), Sour Cherry

Habitat

Prunus cerasus is a cultivar that developed from the wild cherry (*Cerasus avium*) and the dwarf cherry (*Cerasus fruticosa*). The sour cherry grows in Asia Minor and Europe. Presumably, the Romans imported and cultivated it (Lieberei and Reisdorff 2007, p. 190; Lampe and Hilsendegen 2006).

S

Description

The sour cherry is usually a shrub when growing wild, but a tree in cultivation. The tree can grow up to 10 m high and has thin branches and ovate to lanceolate leaves with stub points and a double serrated edge. The leaves' upper side is smooth and shiny. Both sides are hairless. The flowers are long-stemmed and arranged in umbels in the axils of the bracts. After autogamy or exogamy, the half-inferior ovary develops into a round, dark stone fruit with a fleshy mesocarp. The fruit are soft and taste sour. They contain twice as many fruit acids as those of the sweet cherry (Guimpel et al. 1815, p. 82; Lieberei and Reisdorff 2007, p. 190; Lampe and Hilsendegen 2006)

2 Cultivation and Extraction

Cultivation

All kinds of soil are suitable for the cultivation of sour cherries, as long as they are not too moist. Sowing takes place in the autumn. The trees blossom from the end of April to early May; the fruit can be harvested in the summer. Cultivation of *Prunus cerasus* is increasing. Since 1960, the worldwide production has doubled. In 2004, one million tons of sour cherries were produced, about 70% of them in Europe. The main producers are Russia, Poland, Ukraine, Turkey, Serbia, Montenegro, the USA, Germany and Hungary (Guimpel et al. 1815, p. 82; Lieberei and Reisdorff 2007, p. 190; Lampe and Hilsendegen 2006).

Extraction of the Oil

The dried kernels are ground and the oil is subsequently extracted with hexane. The oil can also be obtained by pressing (Chandra and Nair 1993).

3 Character

Colour

Golden yellow (Hackbarth 1944, p. 342).

Odour

Intense odour of bitter almonds.

Flavour

Bitter almonds, marzipan, fruity, sweet.

4 Ingredients

Composition of Fatty Acids

Fatty acids	Content in % acc. to Matthäus and Özcan (2009)	Content in % acc. to Bak et al. (2010)	Content in % acc. to Roth and Kormann (2005), p. 155	Content in % acc. to Chandra and Nair (1993)
Myristic acid	–	–	0.2	–
Palmitic acid	5.3	3–4	4.3	–
Stearic acid	1.5	–	2.9	–
Oleic acid	63.9	50–53	49.5	63.6
Vaccenic acid	1.0	–	–	–
Linoleic acid	27.0	35–38	42.3	31.5
Linolenic acid	0.1	–	–	
Arachidic acid	0.1	–	0.9	

Composition of Triglycerides

Triglycerides	Content in % acc. to Deineka (2003)
LnLnLn	0.64
LnLnL	11.1
LnLL	59.9
LLL	28.3

Composition of Sterols

Sterols	Content in % acc. to Bak et al. (2010)
γ-Sitosterol	4–5

Composition of Tocopherols

Tocopherols	Content in % acc. to Bak et al. (2010)	Content in mg/kg acc. to Matthäus and Özcan (2009)
Total content	–	240.2
α-Tocopherol	–	4.7
α-Tocotrienol	–	21.5
β-Tocopherol	1–1.6	0.4
γ-Tocopherol		197.2
δ-Tocopherol		15.1

S

Other Ingredients

	Content in % acc. to Bak et al. (2010)
Squalene	1–1.2
Vitamin E	<1

Physical Key Figures

refractive index: n^{25}_{D}	1.4753–1.4769
density: d^{15}_{15}	0.920–0.927
saponification number:	190–198
iodine number:	110–116
unsaponifiable:	0.5–0.9%

(Roth and Kormann 2005, p. 155)

specific weight:	0.924
saponification number:	190–198
iodine number:	110–116
unsaponifiable:	0.7%

(Löw 2003, p. 66)

Sour cherry oil is a non-drying oil (Hackbarth 1944, p. 342).

Thermal decomposition takes place at 352 °C (Chandra and Nair 1993).

Shelf Life

6 months (if stored in a cool, dark place).

5 Application

In Pharmacy and Medicine

As the oil mainly consists of unsaturated fatty acids, it is thought to protect against various chronic diseases. According to Bak et al., it can be used as a dietary supplement to prevent and alleviate chronic diseases due to the interaction of other compounds like β-tocopherol, γ-sitosterol, vitamin E and squalene (Bak et al. 2010).

In Cosmetics

Sour cherry oil forms part of body-care products. Due to its anti-wrinkle effect, it is added to facial creams; additionally, it is thought to provide protection against UV rays and dehydration. The oil is also used for producing soap and deodorants (Tősaki et al. 2010).

As Food

Thermal decomposition begins at 352 °C, which is a slightly higher temperature than that of other vegetable oils like olive oil and sunflower oil. In addition, sour cherry oil almost exclusively contains unsaturated fatty acids, and is therefore a high-quality oil. It is a good edible oil (Chandra and Nair 1993).

Soya Bean Oil

Oleum Sojae synonyms: Sojaöl (D); huile de soja (F)

1 Source Plant

Glycine max (L.) Merr. (Fabaceae), soya bean

S. Krist, *Vegetable Fats and Oils*, https://doi.org/10.1007/978-3-030-30314-3_109

S

Habitat

The origins of the soya bean presumably lie in South-East Asia, as it was already cultivated in China during the Shang dynasty about 1700–1100 B.C. Until the end of the eighteenth century, it was only grown in China, Manchuria, Korea and Japan. In 1790, it was exported to England, but did not thrive in this climate. At the beginning of the nineteenth century, the soya bean started to be grown in the USA, but it took another 50 years until cultivation in the USA increased as a result of a trade agreement with Japan. Today, soya is cultivated in the USA on an industrial scale. Soya beans need subtropical or tropical climes to thrive, with temperatures between 25 °C and 32 °C, soil temperatures that are not too low and a sufficient water supply (Hänsel et al. 1999, p. 262; Hager 1978, volume 4, p. 1154).

Description

Glycine soja is regarded as the wild form of *Glycine max*. *Glycine max.* is an annual plant with an upright, fairly branched, hairy stem that can grow 40–90 cm tall. The leaves are trifoliate, large, entire and densely covered with hair. The flowers are inconspicuous, violet to white, and short-stemmed. *Soja max.* Produces 1–4 seeds, which are ovate, 4–10 mm long, 3–7 mm broad and 2–7 mm thick, smooth, with a whitish to brownish-yellow, greenish, reddish or black tint (Hänsel et al. 1999, p. 262; Hager 1978, volume 4, p. 1154).

2 Cultivation and Extraction

Cultivation

Soya beans dominate oilseed production. According to the United States Department of Agriculture, over 58% of the global oil seed production consisted of soya beans in the business year of 2003/04, with an estimated harvest of 201 mio.t. The USA is the main producer, followed by Brazil and Argentina. In Central Europe, there is little cultivation of soya beans as a result of the climate. Only about 1% of the global production is located in the EU. Important European producers are Italy and France. The following table illustrates the enormous extent of soya bean cultivation and the most important producing countries (Hänsel et al. 1999, p. 262; Roth and Kormann 2000, p. 151 – Hager 1978, volume 4, p. 1154):

Producing countries	2001 (mio.t) acc. to fao.org/29.04.04	2002 (mio.t) acc. to fao.org/29.04.04
EU	1.2	0.8
USA	78.7	74.3
Brazil	37.9	41.9
Argentina	26.9	30.0
China	15.4	16.9
India	5.9	4.3
Paraguay	3.5	3.3
Canada	1.6	2.3

Producing countries	2001 (mio.t) acc. to fao.org/29.04.04	2002 (mio.t) acc. to fao.org/29.04.04
CIS	0.5	0.5
Nigeria	0.4	0.4
global	176.7	179.9

Extraction of the Oil

Oleum Sojae is obtained from soya beans by pressing or extraction (Roth and Kormann 2000, p. 151; Hager 1978, volume 7b, p. 209).

3 Character

Colour

Dark yellow.

Pressed oil: light yellow; extracted oil: brownish yellow (Roth and Kormann 2000, p. 151).

Yellow (Hager 1978, volume 7b, p. 209).

Yellow to brownish yellow (Hunnius 1998, p. 608).

Odour

Pungent, greenish-mouldy, nutty. Pleasant (Hager 1978, volume 7b, p. 209).

Volatile Compounds

Aroma analysis of the volatile compounds of soya bean oil using capillary gas chromatography (Snyder et al. 1985):

Ethane	1-Octen-3-ol
Propane	Pentylfuran
Propenal	t,c-2,4-Heptadienal
Pentene	Octanal
Pentane	t,t-2,4-Heptadienal
Propanal	Octenal
Hexane	Nonanal
2-Butenal	t-2-Decenal
1-Penten-3-ol	Hexanal
Pentanal	Octane
Heptane	t-2-Hexenal
Pentenal	Heptanal
Pentanol	c-2-Heptenal
Octene	t-2-Heptenal

Flavour

Mild (Roth and Kormann 2000, p. 151). Pleasant (Hager 1978, volume 7b, p. 209).

4 Ingredients

Composition of Fatty Acids

Fatty acids	Content in % acc. to Hager (1978), volume 7b, p. 209	Content in % acc. to Nature certificate of analysis 28.08.03	Content in % acc. to Kerschbaum and Schweiger (2001)p.16
Palmitic acid	2–11	10.9	11.7
Linoleic acid	49–51	53.9	55.2
Palmitoleic acid	–	0.1	0.1
Oleic acid	23–31	23.6	20.0
α-Linolenic acid	2–11	6.4	6.2
Stearic acid	2–7	3.6	3.9
Arachidic acid	0.9–2.4	0.4	0.3
Behenic acid	–	0.5	0.3
Vaccenic acid	–	–	1.6
Myristic acid	0.1–0.4	0.1	0.1

Composition of Triglycerides

(Karleskind 1996, p. 138)

Triglycerides	Content in %	Triglycerides	Content in %
LLnLn	0.8	OOL	7.6
LLLn	7.5	SLL	4.2
OLnLn	0.3	POL	7.8
LLL	19.3	PPL	1.5
OLLn	5.0	OOO	1.9
PLLn	2.8	SOL	3.0
OLL	20.4	POO	1.3
OOLn	0.8	PPO	1.0
PLL	13.1	PSO	0.7
POLn	0.7		

Composition of Sterols

(concentration in mg/100 g and content in %)

Sterols	Soya bean oil acc. to Firestone (1999), p. 94	Soya bean oil acc. to Karleskind (1996), p. 139	Refined soya bean oil acc. to Carstensen (2001), p. 43
Total content	184–409	250–418	474.6
Cholesterol	0.6–1.4	<1	–
Brassicasterol	0–0.3	–	<0.1

Sterols	Soya bean oil acc. to Firestone (1999), p. 94	Soya bean oil acc. to Karleskind (1996), p. 139	Refined soya bean oil acc. to Carstensen (2001), p. 43
Campesterol	15.8–24.2	19–23	22.5
Stigmasterol	15.9–19.1	17–19	22.4
β-Sitosterol	51.7–57.6	47–59	50.4
Δ5-Avenasterol	1.9–3.7	2–4	2.2
Δ7-Stigmasterol	1.4–5.2	1–3	1.0
Δ7-Avenasterol	1.0–4.6	1–2	0.9
Others	0–1.8	<3	–

Composition of Tocopherols

(concentration in mg/100 g and content in %)

Tocopherols	Soya bean oil acc. to Karleskind (1996), p. 139	Refined soya bean oil acc. to Carstensen (2001), p. 45
Total content	80–167	77.7
α-Tocopherol	5–10	13.3
β-Tocopherol	2–3	1.2
γ-Tocopherol	44–60	53.9
Δ-Tocopherol	30–43	31.6

Other Ingredients

(Kerschbaum and Schweiger 2001, p. 17, p. 43; Ulmer 1996, p. 75; Löw 2003, p. 137)

Lecithin
Mineral compounds
Isoflavones
Carotenoids
Vitamin A
Vitamin B
Vitamin E

Physical Key Figures of Soya Bean Oil

refractive index: n_D^{20} 1.4747–1.4765
density: d_{25}^{25} 0.916–0.922
saponification number: 189–195
acid number: 0.3–3.0
iodine number: fluctuates, usually 124–136
unsaponifiable: 0.5–2%
flash point: 282 °C
point of solidification: −8 to −16 °C

(Roth and Kormann 2000, p. 151)

Shelf Life

Cold-pressed edible soya bean oil has a shelf life of up to 9 months at temperatures from 5 to 10 ° C (Roth and Kormann 2000, p. 151).

5 Application

In Pharmacy and Medicine

Natural soya bean oil has a high lecithin content of about 4%. It is used as a pharmaceutical auxiliary material in the production of parenteral nutrient solutions with lipids. In addition, lecithin is part of several commercially available dietary and roborant preparations. There are also attempts to use lecithin in choline substitution therapy for various diseases of the central nervous system, such as *Dyskinesia tarda*. Lecithin also plays an important role as a surfactant in cases of respiratory distress syndrome (RDS) of premature babies. *Oleum Sojae* is one of few vegetable oils that contain high amounts of α-linolenic acid. Also as a result of the sterin content, it lowers the risk of developing arteriosclerosis, as well as lowering trigylceride and cholesterol levels. Soya bean oil also contains isoflavones, the so-called phytoestrogens. They are related to oestrogen and have numerous indications. For example, they significantly lengthen the follicular phase of the menstrual cycle. As the cell division rate of the mammary gland is four times lower in the follicular phase than in other phases, women with a longer menstrual cycle have a lower risk of developing breast cancer. Epidemiological studies suggest that soya products in general, and soya bean oil in particular, have a protective function, particularly against carcinoma of the colon, lungs, stomach and prostate glands. Other possible indications for phytoestrogens are cardiovascular risks and menopausal complaints. The substances have not yet been sufficiently examined toxicologically, however, so they should be taken with care (Fleischhacker 2002, p. 111, 202 et seq; Kerschbaum and Schweiger 2001, p. 17; Hänsel et al. 1999, p. 246; Hager 1978, volume 7b, p. 209; Hunnius 1998, p. 608).

As Food

For nutritional purposes, soya bean oil is the most frequently produced vegetable oil worldwide. It is not only a valuable salad oil, but can also be used for stewing, cooking and frying. The food industry uses soya bean oil, especially because of its lecithin content, in the production of margarine and mayonnaise; it also improves immediate solubility of instant products, as it serves as an emulsifier, and the flow characteristics of chocolate, as it prevents sugar from crystallising. It also improves the volume of bakery products and delays their becoming stale (Kerschbaum and Schweiger 2001, p. 17; Roth and Kormann 2000, p. 152; Fleischhacker 2002, p. 111; Hunnius 1998, p. 608; Ulmer 1996, p. 75; Löw 2003, p. 104).

Technical Uses

The paint industry uses soya bean oil in the production of paint, varnish and print colours. The oil crisis of 1970 can serve as an example. It caused the American newspaper industry to search for alternative print colours. Soya bean oil proved to be an equivalent carrier for rotary colours. Since 1987, the American newspaper industry has used print colours on a soya bean oil basis. 50% of American newspapers and 75% of the daily papers are now printed with soya bean oil colours. In Europe, they only amount to 15%, but it is increasing. Furthermore, biodiesel can be produced from soya bean oil, just like from rapeseed oil; usually the mixture used consists of 20% soya biodiesel and 80% diesel, as pure bio diesel would be too expensive. In the USA and several European countries, bio diesel is already a success, and France is currently the largest producer (Roth and Kormann 2000, p. 152).

In Naturopathy

Soya bean oil is well known in folk medicine and applied in the following areas (Kerschbaum and Schweiger 2001, p. 47):

- the stomach/intestines;
- the heart/circulation; and
- the immune system.

In Cosmetics

In cosmetics, soya bean oil is mainly used as a vehicle for liposoluble phytonutrients and vitamins. It also serves as a base for ointments and bath oils, as it penetrates the skin quickly, keeps it moisturised and does not leave a fatty film. It is especially important in care cosmetics (Roth and Kormann 2000, p. 152).

Stillinga Tallow/Stillingia Oil

Oleum Stillingiae synonyms: Stillingiatalg, Stillingiaöl (D); huile de stillingia (F)

1 Source Plant

Sapium sebiferum (L.) Roxb. (Euphorbiaceae), Chinese tallow tree

Habitat

The origins of the Chinese tallow tree lie, as the name suggests, in China, especially in the provinces of Central China in the region of the Yangtze, but also in Japan. The tree is also cultivated on the island Hainan, in Hong Kong, Taiwan and Korea. It was imported to Sri Lanka, Indo China, Bengal, India, Sudan, Martinique and the USA,

S. Krist, *Vegetable Fats and Oils*, https://doi.org/10.1007/978-3-030-30314-3_110

where it is found in California, Arizona, Texas, Florida, North Carolina and South Carolina. It is also cultivated in the south of France and in Algeria.

Description

The Chinese tallow tree is a fast-growing, monoecious, summergreen tree about 10–13 m tall. It has a gnarled trunk with a grey or greyish-white bark. Like all *Euphorbiaceae*, the tree contains a white, poisonous sap. The leaves are alternate, rhombic-ovate to heart-shaped in their form, and 3.5–8.5 cm long and 4–9 cm broad. The leaf stem is about 1.5–7 cm long and has two conspicuous glands at the end, as well as a bract on both sides. The flowers are greenish-yellow, terminate, 5–10 cm long, and arranged in spikes. The fruit are round capsules measuring 0.95–1.7 cm in diameter (hort.purdue.edu/11.08.05). They discard their septi and thus expose their three seeds coated in wax, which remain on the tree. The colour of the capsule is first green, then nearly black when it is ripe.

2 Cultivation and Extraction

Cultivation

The Chinese tallow tree prefers subtropical, warm temperatures. It is found on steep embankments, stony slopes, but also on sandy soil. The tree grows on alkaline, salty and acidic soil; moist soil rich in humus or loamy, boggy soil is best. Potting soil to which sand, perlite or vermiculite have been added to ensure water permeability is suitable. Although the tree needs temperatures of 12–30 °C, it survives several days of frost. The best time for planting is in late autumn or in spring. The tree grows relatively fast, and the first harvest is possible after 3–8 years. It then continues to bear fruit for up to 100 years. The tree blooms from January to February; the fruit are ripe in November. They are harvested by shaking the capsules from the trees, and then collecting the capsules by hand. The capsules are opened with special knives; the seeds are subsequently freed from the capsule shell by sifting.

Extraction

The tallow and the oil are obtained from the seeds of *Sapium sebiferum Roxb. by applying various methods.*

Extraction of Stillingia Tallow

Stillingia tallow is contained in the fruit pulp of *Sapium sebiferum*. In China, the fruit are freed from their hard outer shell by pounding, and are then steamed until they form a tallowy mass. The liquefied tallow flows off and is filtrated through primitive filters in the form of layers of straw. The melting out with steam can also be carried out in finely perforated metal cylinders. The fat obtained by this method is considered to be the best and is called *Pi-ieou* or *Pi-yu* (Axtell 1994, p. 48).

Extraction of Stillingia Oil

The seeds left over after the extraction of stillingia tallow are heated in water and pressed strongly. Stillingia oil can also be extracted with hexane. The oil is known as *ting-yu* (Axtell 1994, p. 48).

A Mixture of Both Methods Is Also Possible

A method exists of obtaining the fat of the fruit pulp and the oil of the seeds together. The seeds with their outer layer of tallow are first reduced to small pieces, then treated with hot steam, and finally pressed slightly. The oil is finally warmed over hot ash and filtered through straw. It is called *mou-ieou* (Axtell 1994, p. 48).

3 Character

Colour

Stillingia oil: liquid, brownish-yellow to brown oil (Roth and Kormann 2000, p. 153).
Yellowish brown to brown (Bauer 1928, p. 227).
Stillingia tallow: hard, brittle fat of a white to greenish colour (Roth and Kormann 2000, p. 154).
White to greenish yellow (Bauer 1928, p. 18).
Tallow melted out with steam: white to greenish (Von Wiesner 1927, p. 723).
Hot-pressed tallow: greenish yellow (Von Wiesner 1927, p. 723).

Odour

Stillingia oil: smells like China wood oil (tung oil), which in turn smells like lard (Bauer 1928, p. 227).
Stillingia tallow: odourless (Roth and Kormann 2000, p. 154).

Flavour

Stillingia oil: For a long time, it was believed that stillingia oil was poisonous, which is why no descriptions of its taste could be found. Stillingia tallow: without flavour (Roth and Kormann 2000, p. 154).

S

4 Ingredients

Composition of Fatty Acids

Stillingia Oil

Fatty acids	Content in % acc. to Roth and Kormann (2000), p. 154	Content in % acc. to Firestone (1999), p. 98	Content in % acc. to Axtell (1994), p. 49	Content in % acc. to bagkf. de/10.07.06
Palmitic acid	4–6	6–9	7–9	9
Stearic acid	1–3	3–5	3–5	5
Oleic acid	7–11	7–10	20	10
Linoleic acid	50–63	24–30	40–54	30
α-Linolenic acid	22–26	41–54	24–30	54
Arachidic acid	Traces	–	–	–

Stillingia Tallow

Fatty acids	Content in % acc. to Roth and Kormann (2000), p. 154	Content in % acc. to Firestone (1999), p. 28	Content in % acc. to Axtell (1994), p. 49
Lauric acid	2–2.5	0–2.5	0–2.5
Myristic acid	3.6	0.5–3.7	0.5–3.7
Palmitic acid	57–67	58–72	58–72
Stearic acid	1.2–1.8	1–8	1.2–7.6
Oleic acid	27–35	20–35	20–35
Linoleic acid	–	0–2	0–1.6

Composition of Sterols

Sterols	Content in % acc. to Jeffrey and Padley (1991)
Cholesterol	0.8
Campesterol	8.7
Stigmasterol	4.1
β-Sitosterol	77.8
Δ5-Avenasterol	7.2

Physical Key Figures of Stillingia Tallow and Stillingia Oil

(Roth and Kormann 2000, p. 153)

Stillingia Oil

refractive index: n_D^{25} 1.4817–1.4830
density: d_{15}^{15}, 0.939–0.946
saponification number: 202–212
iodine number: 169–178
unsaponifiable: 0.5–1%

Stillingia Tallow
refractive index: n_D^{40} 1.4560–1.4574
density: d_{15}^{15} 0.918
melting point: 45–53 °C
point of solidification: 35–40 °C
saponification number: 199–210
iodine number: 27–33
unsaponifiable: 0.4–1.3%

5 Application

In Pharmacy and Medicine

Stillingia Tallow
As stillingia tallow has similar properties to cocoa butter and also has a similar melting point, it is a suitable ground mass for suppositories (Shao 1958).

In Folk Medicine

Stillingia Oil
In folk medicine, stillingia oil is used to treat hookworm disease, maggot-ridden wounds and inflamed eyes. It is also applied as a laxative, emetic and externally as a vulnerary drug and in cases of oedema and skin diseases. These applications have not yet been clinically examined (List and Hörhammer 1978, 4th edition, 6th volume, p. 282).

In Cosmetics

Stillingia Tallow
Stillingia tallow can be used in soap production, but only in small quantities, as soap made purely from this kind of tallow produces hardly any foam, is greasy at first and then becomes brittle (Raie et al. 1983).

As Food

Stillingia Tallow
In China, stillingia tallow is used as an edible oil (Khan et al. 1977).

Other Uses

Stillingia Tallow

In Candle Production
Stillingia tallow is well suited to produce candles. The candles are coated with a mixture of vegetable wax and stillingia tallow to prevent them from burning down too quickly. They have good combustion properties and burn with a bright, odourless flame. The candles are used especially for cultic purposes (Raie et al. 1983; Von Wiesner 1927, 4th edition, 1st volume, p. 723).

S

As Lamp Oil
In China, stillingia tallow is used as lamp oil (Khan et al. 1973).

In the Technical Industry
Stillingia tallow is used in the production of printing ink, canvas, linoleum and as a textile additive (Raie et al. 1983).

Stillingia Oil
Stillingia oil has extremely quick-drying properties. It is used in the paint industry to produce paint, varnish and translucent coatings. Its drying properties are comparable to linseed oil (Raie et al. 1983; Khan et al. 1977).

6 Possible Unwanted Side Effects

The assumption that stillingia oil is poisonous was proven wrong by a feeding study on rabbits. This – already very old – study did not detect any toxicity (Bailey 1920).

Nevertheless, it is reported that workers who came into contact with the leaves of the plant suffered from dermatitis and allergic skin reactions. Animals, for example sheep and goats, which ate leaves of the Chinese tallowtree developed symptoms of poisoning, but these are a result of the chyle contained in the leaves.

St. John's Wort Oil

Oleum Hyperici synonym: Johanniskrautöl (D)

St. John's wort oil is not a vegetable oil in the true sense of the word, but rather macerated plant parts of St. John's wort in a non-drying oil.

1 Source Plant

Hypericum perforatum L. (Clusiaceae), St. John's wort, perforate St. John's wort, common St. John's wort

S. Krist, *Vegetable Fats and Oils*, https://doi.org/10.1007/978-3-030-30314-3_111

Habitat

St. John's wort is a native plant of Europe, West Asia, the Canary Islands and North Africa. It was introduced into many other regions (Hänsel et al. 1999, p. 924). It is often found by the wayside and in dry, sunny locations. Generally, St. John's wort is a plant of open areas and does not grow in deep shade.

Description

St. John's wort is a perennial, herbaceous plant. The plant parts above the ground die back in winter, and the plant reappears from its rhizome the following year. St. John's wort can grow up to 1 m tall; its root is highly branched. The upright stem is branched at the top, with many longitudinal edges; it is frosted and covered with glands all the way to the top. The leaves are alternate, oval to lanceolate, entire and hairless; a typical feature is that they appear perforated when held against the light, due to their many oil and resin repositories. The bright yellow flowers are arranged in cymes on stems with black glands. They have five sepals and five petals each, covered in many black glands in the form of stripes or dots; if they are rubbed, a red dyestuff emerges. The numerous stamina (50–60) have grown together, forming bunches. The fruit is a capsule covered in pointed or linear glands. The plant blooms from June to August (Wenigmann 1999, p. 147).

2 Cultivation and Extraction

Cultivation

St. John's wort is cultivated because of its uses as a medical plant. Species selected from wild plants are liable to fungal infestation, anthracnose of St. John's wort. The plant needs poor soil for optimal growth; sandy and gravelly soil are also a good basis.

Extraction

St. John's wort oil is made from the fresh flowering twig tips or, better, from fresh flowers. The plant parts are squashed, mixed with olive oil, peanut oil (Hänsel et al. 1999, p. 156) or sunflower oil (Wenigmann 1999, p. 149), and macerated for 6 weeks. After pressing, the oil is dehydrated with sodium sulphate to delay rancidity (Hänsel et al. 1999, p. 156).

3 Character

Macerated St. John's wort with peanut oil:

Colour

Clear to slightly cloudy, reddish.

Odour

Weak.

Flavour

Neutral.

Ingredients

The exact composition of St. John's wort oil is not fully known. Characteristic ingredients of St. John's wort are naphthodianthrones (especially the red dyestuff hypericin and pseudohypericin, about 0.1–0.15%), phloroglucinol derivates like hyperforin and adhyperforin (2–4%, depending on the age of the plant), flavonoids (especially the quercetin glycosides hyperoside, rutoside, quercitrin and isoquercitrin, as well as flavonoid glycans and biflavones), procyanidines and catechin tannins (6–15%). In addition, St. John's wort contains xanthones (1,3,6,7-tetrahydroxyxanthone), sterols, phenolcarboxylic acids (Hänsel et al. 1999, p. 927), pectin, choline, fat and essential oil with monoterpenes (for example α-pinene, myrcene) and sesquiterpenes in schizogenous, roundish oil repositories (Hunnius 1998, p. 691). In the essential oil, 2-methyl-3-butenol was detected, which is known as a degradation product of hop acids. It is a possible explanation for the sedative effect of hop (Wenigmann 1999, p. 147).

As St. John's wort oil can be produced by maceration of twigs or flowers with different kinds of oil, its ingredients depend on the oil used. The effectiveness seems to be te result, however, of a coaction of all ingredients (Klotz 2002). The bright-red colour is caused by degradation products of hypericin (Hänsel et al. 1999, p. 927).

Physical Key Figures of St. John's Wort Oil

relative density: 0.9150
refractive index: 1.470

acid number: 0.68
iodine number: 92
hypericin content: 2.9 mg/100 g

(macerated St. John's wort with peanut oil): (nature.de/15.03.06)

4 Application

In Pharmacy and Medicine

External Application

Since St. John's wort oil has cooling, disinfectant and pain-soothing properties, it is an effective treatment for open wounds, inflammation and sporting injuries like bruises, strains, contusions, sprains, blisters, abrasions and burns. It also alleviates painful scars, trigeminal neuralgia, sciatica and gout.

Internal Application

The main areas of use are psychovegetative syndromes, depressive disorders, anxiety and nervous restlessness. St. John's wort oil is a household remedy for gastroenteritis, nervous gastrointenstinal disorders, mood swings, sleeping disorders, migraines, meteorosensitivity and enuresis of children.

Effectiveness

The effect of preparations with St. John's wort is partly due to its content of hypericin and hyperforin, but the synergistic effect – the interaction of all the ingredients – plays an important role. Hypericin and hyperforin cause a weak to medium cerebral inhibition of the reuptake of serotonin, noradrenaline and dopamine. This is the mode of action of synthetic antidepressants. The doses of St. John's wort preparationsmust be high, however, to achieve the desired effect (Hänsel et al. 1999, p. 926).

In Folk Medicine

The oil is used as a liniment to treat lumbago, gout and rheumatism, to alleviate pain and to promote wound healing after dislocations and sprains, bruising and shingles. The flowers of St. John's wort are placed in cold-pressed olive oil or sunflower oil, and the jar is left standing on the windowsill in the sun for 2 months.

In Cosmetics

St. John's wort oil has a soothing effect and heals inflamed, dry and chapped skin. It supports skin regeneration and stimulates the metabolism of the skin.

Possible Unwanted Side Effects

Interactions

In the past, repeated cases of possible interactions of St. John's wort with other drugs have been reported, for example the lowering of the plasma level in combination with cyclosporine, indinavir, oral contraceptives and anticoagulants of the coumarin type (Klotz 2002). In combination with inhibitors of serotonin reuptake like fluoxetine, paroxetine, citalopram, etc., serotonergic side effects like nausea, diarrhoea, fluctuations of blood pressure and agitation may intensify. For this reason, undeclared oily preparations were taken off the shelves (Klotz 2002).

Photosensitisation

St. John's wort oil may cause gastrointestinal disorders, allergic reactions, fatigue or restlessness, although these symptoms rarely appear. Hypericin has photosensitising properties (light disease of grazing animals after ingesting hay containing St. John's wort), but photosensitisation has not been observed with any certainty in humans who have taken St. John's wort for medical reasons. The reference to the assertion that light-skinned individuals may suffer from inflammation similar to sunburn when they have been taking the medication for a long time and expose themselves to high solar radiation only seems to apply to doses that are much higher than recommended (Frohne 2002, p. 312). St. John's wort oil is not suitable for persons suffering from severe depressive moods or for persons with actinic dermatitis.

Sunflower Oil

Oleum Helianthi synonyms: Sonnenblumenöl (D); sunflower seed oil (E); huile de tournesol (F)

1 Source Plant

Helianthus annuus L. (Asteraceae), sunflower

S. Krist, *Vegetable Fats and Oils*, https://doi.org/10.1007/978-3-030-30314-3_112

Habitat

Sunflowers originally came from North America, where they were already used for oil production three or four millennia ago. The Native Americans cultivated it long before the arrival of the Europeans. It is thought that Spanish discoverers imported the plant to Europe, where it spread to Italy, Germany and France. Today, it is one of the most important oil plants in Russia (Hamilton 1995, p. 132), and is cultivated in North America, South America and Europe. As an oil plant, the sunflower ranks third after the soya bean and the oil palm (Roth and Kormann 2000, p. 69).

Description

The sunflower reaches a height of 1.5–2.5 m. It has a sturdy, branched, leafed and rough-haired stem (Roth and Kormann 2000, p. 69). The plant is an excellent example of phototropism. This means that the plant bends towards the light; the bending is enabled by growth. The leaves are heart-shaped, alternate or opposite, and covered with short rough hair on their upper as well as their lower sides. They are long-stemmed, and their edges are dentate. The flower heads are large (10–40 cm broad). The ray florets are 5–10 cm long and yellow or orange. The disc florets are numerous, brown, and are visited by many insects. The fruit from which the oil is obtained are 0.8–1.7 cm long and 0.4–0.9 cm broad (Roth and Kormann 2000, p. 69). After the flower has withered, the head often contains about 2000 seeds.

The colour of the seeds varies from black to whitish; they sometimes also have black and white longitudinal stripes. The shell is leathery and ligneous and can easily be separated from the seed. The size and weight of the seeds vary. In addition to fat oil (40–50%), they contain protein (approx. 24%), saccharose, choline, betaine and tanning agents. If mould fungi attack the seeds, mycotoxin rubratoxin B is produced (Roth and Kormann 2000, p. 69).

2 Cultivation and Extraction

Cultivation

The development of the plant depends on the temperature. In typical cultivation areas, the summers are short and hot. The depth and distance of sowing varies, depending on the variety. Varieties with small seeds are sown about 3 cm deep, with a distance of 30–45 cm between rows and 15–20 cm within a row. Large-sized varieties are sown at greater distances from each other (Roth and Kormann 2000, p. 69).

Extraction of the Oil

The ripe seeds are shelled, ground and then hot or cold pressed to obtain the oil (Roth and Kormann 2000, p. 153). The oil can also be extracted (Hänsel et al. 1999, p. 263) or obtained by fluid extraction with supercritical CO_2 (Andrich et al. 2001). The raw oil can be refined if required (Hager 1993, p. 413).

3 Character

Colour

Cold pressed sunflower oil is light yellow; oil pressed by adding heat is reddish yellow, but can easily be decoloured (Roth and Kormann 2000, p. 153).

Extracted and refined oil is clear and a light-yellow colour (Hager 1993, p. 414)

Odour

Aromatically fatty, nutty, roasted smell, woody, green smell.

Cold- and hot-pressed as well as extracted oil is nearly odourless (Hager 1993, p. 413 et seq).

Volatile Compounds

Via HS-SPME and subsequent GC-MS (Jelen et al. 2000), as well as using head-space GC-MS (after storage for 8 days at 60 °C) (Snyder et al. 1985), the following volatile compounds were identified:

Volatile compounds	Identified via HS-SPME/GC-MS	Identified via headspace GC-MS
Pentane	+	+
2-Butanone	+	
Butanol	+	
Pentanal	+	+
Pentanol	+	+
Hexanal	+	+
2-Heptanone	+	
Heptanal	+	+
2-Heptenal	+	+
1-Octen-3-ol	+	+
Octanal	+	+
2-Nonanone	+	
Nonanal	+	+
Decadienal	+	
Ethane		+
Propane		+
Propenal		+
Hexane		+

Volatile compounds	Identified via HS-SPME/GC-MS	Identified via headspace GC-MS
2-Butenal		+
1-Pentan-3-ol		+
Heptane		+
Octane		+
t-2-Hexenal		+
Octenal		+
t-2-Decenal		+
Decenol		+
t,c-2,4- Decadienal		+
t,t-2,4- Decadienal		+
Undecenal		+

S

Flavour

Mild, pleasant taste (Roth and Kormann 2000, p. 153).

4 Ingredients

Composition of Fatty Acids

(Firestone 1999/p. 98 et seqq)

Sunflower oil can be differentiated into the following categories:

- sunflower oil (SO);
- sunflower oil with a high amount of linoleic acid (HL);
- sunflower oil with a high amount of oleic acid (HO);
- sunflower oil with high amounts of palmitic acid and linoleic acid (HP/HL);
- sunflower oil with high amounts of palmitic acid and oleic acid (HP/HO), and
- sunflower oil with high amounts of stearic acid and oleic acid (HS/HO).

The different compositions of fatty acids are illustrated in the following table:

Fatty acids	Content in % in SO	Content in % in HL	Content in % in HO	Content in % in HP/HL	Content in % in HP/HO	Content in % in HS/HO
12:0	0–0.1	–	–	–	–	–
14:0	0–0.2	0.1	–	0.1	–	–
16:0	5–8	7.5	3–5	27.3	24.6	4.6
16:1	0–0.3	0.1	0.1–0.2	4.4	6.1	0.1
18:0	2.5–7	1.9	3–5	2.7	2.9	11
18:1	13–40	13.3	70–92	17.1	59.8	79.1
18:2	48–74	76	2–20	46.8	3.5	2
18:3	0–0.3	0.1	–	0.1	0.1	0.1
20:0	0.2–0.5	0.1	0.3	0.3	0.4	0.9
20:1	0–0.5	0.2	0.2	0.1	0.2	0.2
22:0	0.5–1.3	0.4	1	0.8	1.8	1.8
22:1	0–0.5	–	0.1	0.1	0.1	–
24:0	0–0.4	0.2	0.4	0.3	0.6	0.3

Composition of Triglycerides

(Firestone 1999, p. 144 et seq)

A different composition of fatty acids in different sunflower oils results in a different composition of triglycerides:

Triglycerides	Content in % in SO	Content in % in HL	Content in % in HO	Content in % in HP/HL	Content in % in HP/HO	Content in % in HS/HO
PSO	0.6	0.6	1.1	1.4	3.7	3.7
SOS	–	–	–	–	2.3	3.6
PPO	0.5	0.7	0.2	3.5	13.8	–
POO	1.6	0.7	10	3	31.2	10.5
SOO	1.2	–	4.3–11	–	5	25.1
PPL	0.8	1.1	–	11.9	–	–
PSL	1.2	0.8	–	2.7	–	–
SSL	0.7	0.7	–	1.2	–	–
OOO	2.5	0.7	73.9	1.3	12.9	41.9
POL	6.4	3.2	1.6	6.6	6.6	–
SOL	4.2	1	10	1.9	–	–
OOL	11.8	2.5	6.7	1.8	4.4	3.2
PLL	8.9	12.9	1	17.4	–	–
SLL	6.4	2.9	–	5.1	3.5	–
OLL	25.9	19.1	2.3	4.9	–	–
LLL	20.6	37.5	1.5	7.6	–	–
AOO	0.5	0.7	1.2	–	2.6	3.5
PBL	–	–	–	1.4	–	–
PoLL	–	–	–	2.6	–	–
PoPL	–	–	–	3.5	–	–
PoPO	–	–	–	1.4	–	–
BOO	–	–	–	–	2.9	3.1
PoOO	–	–	–	–	2.7	–

Composition of Triglycerides of Sunflower Oil from Different Countries of Origin

(Krist et al. 2006c)

Triglycerides	Content in % in sunflower oil from Germany	Content in % in sunflower oil from France	Content in % in sunflower oil from Austria
LnPP	0.2 ± 0.2	0.6 ± 0.6	0.1 ± 0.2
LPP	1.4 ± 0.1	1.5 ± 0.3	1.9 ± 0.1
OPP	0.1 ± 0.1	0.2 ± 0.2	–
LnLP	0.1 ± 0.2	0.2 ± 0.1	0.3 ± 0.0
LLP	13.1 ± 0.6	14.2 ± 1.2	13.9 ± 0.2
LOP	6.2 ± 0.2	5.4 ± 0.4	5.7 ± 0.2
LSP	3.0 ± 0.3	2.1 ± 0.6	2.6 ± 0.5
OOP	–	–	–
SSP	0.4 ± 0.2	0.2 ± 0.1	1.2 ± 0.7
LnLnLn	0.5 ± 0.0	0.3 ± 0.1	0.5 ± 0.2
LnLnL	0.3 ± 0.3	–	0.2 ± 0.0
LLL	24.9 ± 0.6	29.8 ± 1.0	27.5 ± 1.9
LLO	23.4 ± 0.6	24.3 ± 2.7	23.6 ± 1.2
LOO	14.4 ± 0.9	13.0 ± 0.9	14.0 ± 1.4

Triglycerides	Content in % in sunflower oil from Germany	Content in % in sunflower oil from France	Content in % in sunflower oil from Austria
LOS	7.2 ± 0.8	4.1 ± 1.0	3.7 ± 0.7
LSS	0.6 ± 0.3	0.3 ± 0.4	0.6 ± 0.4
LLG	0.2 ± 0.2	0.2 ± 0.2	0.2 ± 0.2
LLA	0.8 ± 0.2	0.7 ± 0.1	0.7 ± 0.2
LOA	0.3 ± 0.1	0.2 ± 0.2	0.2 ± 0.1
OOA	0.3 ± 0.0	0.2 ± 0.1	0.3 ± 0.0
OSA	0.1 ± 0.1	0.4 ± 0.7	0.1 ± 0.1
LLB	0.5 ± 0.0	0.4 ± 0.3	0.6 ± 0.0
LOB	0.4 ± 0.1	0.2 ± 0.2	0.4 ± 0.1
OOB	0.2 ± 0.0	0.1 ± 0.0	0.1 ± 0.0

S

Composition of Sterols

(concentration in mg/100 g and content in %)

Sterols	SO acc. to Firestone (1999), p. 99	HO acc. to Firestone (1999), p. 99 et seq	Cold-pressed SO acc. to Carstensen (2001), p. 43	Refined SO acc. to Carstensen (2001), p. 43	SO acc. to Karleskind (1996), p. 123
Total content	244–455	–	447.1	449.7	325–515
Cholesterol	0–0.7	0–0.5	0.6	–	<0.4
Brassicasterol	0–0.2	0–0.1	0.1	–	0
Campesterol	7–13	7–12	8.4	8.6	8–11
Stigmasterol	7–12	8–13	11.5	10.1	7–10
β-Sitosterol	56–65	53–61	64	63.1	58–64
Δ5-avenasterol	1.5–7	1.5–5	2.4	2.4	2–7
Δ7-Stigmasterol	7–24	7–21	3.7	6.1	9–14
Δ7-Avenasterol	3–6.5	3–6	6.6	6.2	4–6
Δ7-Campesterol	2–3	1–3	0.6	0.5	–
Clerosterol	0.7–1	–	–	–	–
Sitostanol	–	0.3–1.5	–	–	–
Isofucosterol	–	–	–	–	0.4–1
Fucosterol	–	–	–	–	2–3

Composition of Tocopherols

Tocopherols	Concentration and composition (mg/100 g) in SO acc. to Firestone (1999), p. 99	Concentration and composition (mg/100 g) in HO acc. to Firestone (1999), p. 100	Concentration (mg/100 g) and composition (%) in refined SO acc. to Carstensen (2001), p. 45	Concentration (mg/100 g) and composition (%) in cold-pressed SO acc. to Carstensen (2001), p. 45
Total content	–	45	110.8	126.6
α-Tocopherol	40–95	9.4–43	95.1	96
β-Tocopherol	0–0.5	2	3.9	2.4
γ-Tocopherol	0–0.5	1	0.8	1.0
Δ-Tocopherol	0–0.1	–	0.2	0.5

Other Ingredients

content of squalene:

- refined SO 36.4 mg/100 g
- cold-pressed SO 38.2 mg/100 g
- aliphatic alcohols: 100 mg/100 g

(Carstensen 2001, p. 43; Karleskind 1996, p. 123)

Physical Key Figures of Sunflower Oil

refractive index: 25 °C 1472–1476
    40 °C 1467–1469
density: 20 °C 0.918–0.923
    15 °C 0.922–0.926
iodine number: 118–145
    115–136
saponification number: 186–194
acid number: 0.6
point of solidification: −16 to −18 °C
unsaponifiable: 0–1.5%

(Firestone 1999, p. 98; Roth and Kormann 2000, p. 153)

Shelf Life

If stored in a cool, dark place, sunflower oil has a shelf life of about 9 months. As the seeds are processed with their shells when they are cold pressed, sunflower oil also contains wax, which may cause the oil to become cloudy when it is stored in a cool place. This does not mean that it is contaminated or ruined.

5 Application

In Pharmacy and Medicine

The pharmaceutical industry uses sunflower oil as a neutral filling material for soft gelatine capsules, in the production of ointments and creams and as a substitute for olive oil and peanut oil in pharmaceutical preparations (Hager 1993, p. 31).

As Food

Cold pressed sunflower oil is an excellent and valuable edible oil and salad oil, and, like wheat-germ oil, is also used as dietary food (Roth and Kormann 2000, p. 153). It can be added to baby food. The food industry uses it in the production of mayonnaise, salad dressings and margarine (Carstensen 2001, p. 4).

Industrial Uses

S

Sunflower oil is used in the production of soap, varnish (Roth and Kormann 2000, p. 153), oil paint and artists' paints (Carstensen 2001, p. 3). It is also used in leatherwork and as a preservative for draperies (Hager 1993, p. 415). Hot-pressed oil is only suitable for industrial purposes. Since it is liable to oxidation, it is seldom used in cosmetics (Roth and Kormann 2000, p. 153).

In Folk Medicine

In experience-based medicine, sunflower oil is applied internally to treat constipation and externally in cases of bad wound healing, skin lesions, psoriasis and rheumatism. There are, however, no pharmacological studies or clinical tests proving the oil's effectiveness.

Tamanu Oil

synonyms: Domba fat (E); Calophyllumöl (D)

1 Source Plant

Calophyllum inophyllum L. (Clusiaceae), Alexandrian laurel, ball tree

S. Krist, *Vegetable Fats and Oils*, https://doi.org/10.1007/978-3-030-30314-3_113

Habitat

Hundreds of different species of *Calophyllum* are common in the Pacific region. The tree is widespread in Africa, eastern India and Polynesia (Hunnius 1998, p. 260).

Description

The Alexandrian laurel is 10–15 m tall; its large, crooked branches are 2–3 m long. The fruit are 3–4 cm in size; they grow in small bundles, their colour changes from green to yellow, and they taste like apples. The fruit contains a smooth, round nut with a thin shell that protects a light-yellow pip.

2 Cultivation and Extraction

Cultivation

The evergreen tree, called *undi* in India, grows in sandy locations on the beach, near water and in gardens (Salunkhe et al. 2003, p. 520).

Extraction of the Oil

After the harvest, the fruit are crushed slightly to remove the nuts. The nuts are cut into thin layers and sun-dried for about 8 weeks; during this process, they turn brown and lose about one third of their weight. It is unique that the oil is only generated during this drying process. The dried nuts are cold pressed, and the oil is filtered. 100 kg of fruit result in about 5 kg of cold-pressed oil.

3 Character

Colour

Bluish-yellow.
Dark, green, thick oil (Dweck and Meadows 2002).

Odour

Sweet, smells of honey, slightly of hay, slightly acetous.

Flavour

Flavourless (Dweck and Meadows 2002).

4 Ingredients

Composition of Fatty Acids

Fatty acids	Content in % acc. to Kalayasiri et al. (1996)	Content in % acc. to Salunkhe et al. (2003)	Content in % acc. to Firestone (1999), p. 38
Myristic acid	–	–	–
Palmitic acid	12.6	5.7–18.5	15–19
Palmitoleic acid	0.2	–	–
Stearic acid	19.8	6.1–12.2	6–12
Oleic acid	45.2	48.5–53.1	48–53
Linoleic acid	20.3	24.1–58.8	16–24
α-Linolenic acid	0.1	–	–
γ-Linolenic acid	0.7	–	–
Arachidic acid	0.3	–	–
In total	40.5	37	–

Composition of Tocopherols and Tocotrienols

Tocopherols/tocotrienols	Content in mg/kg acc. to Matthaus et al. (2003)
α-Tocopherol	58
β-Tocopherol	36
γ-Tocopherol	42
α-Tocotrienol	49
γ-Tocotrienol	57
δ-Tocotrienol	94

Physical Key Figures of Tamanu Oil

density: 0.9415–0.9452 mg/ml (at 15 °C)
refractive index: 1.4699–1.4772 (at 15 °C)
saponification number: 192–202
iodine number: 82–98
unsaponifiable (%): 0.25–1.4
acid number: 27–78
melting point: 8 °C

(Firestone 1999, p. 38; Salunkhe et al. 2003, p. 504)

5 Application

In Pharmacy and Medicine

Effects of the Coumarins and Neoflavonoids Contained in Tamanu Oil

The seeds of *Calophyllum inophyllum* and other species of *Calophyllum* contain several well-known coumarins, among them costanolid and inophyllum P. They are effective reverse-transcriptase inhibitors for treating HIV and can therefore be used as a renewable source for these agents (Spino et al. 1998, p. 3477).

Tamanu oil contains three important natural neoflavonoids: calophyllolide, which has antibacterial, anti-inflammatory and anticoagulant properties, inophyllolide, which has antiviral properties, and calophyllic acid, which has mollusc-repellent and healing properties. The ED_{50} of calophyllolide is 140 mg/kg body weight, administered orally. It was proven to have anti-inflammatory and anti-arthritic effects on rats with arthritis induced by formaldehyde and adjuvant arthritis. The LD_{50} is 2.5 g/kg body weight, administered orally. Even the double ED_{50} does not trigger cancer (Dweck and Meadows 2002).

In Cases of Acne Vulgaris

Many adolescents suffer from common acne. It is a trifold syndrome, which is characterised by inflammation, infection and suppurative foci. Since it has anti-inflammatory, anti-infective and healing properties, tamanu oil from *Calophyllum inophyllum* plays an important role in the treatment of this physiological syndrome (Martini 2001).

Other Uses

In India, tamanu oil is used externally as a massage oil to treat rheumatism. It is added to ointments and is used for the production of soaps and as a lubricant (Salunkhe et al. 2003, p. 520).

In the Pacific region, tamanu oil is applied to deep wounds to support wound healing, and to treat excoriations in cases of diabetes, psoriasis, herpes and haemorrhoids. It is also applied to insect bites, acne and sunburn, and against lice.

As Fuel

A study examined the potential suitability of tamanu oil as a substitute for diesel fuel. The oil was esterified with ethanol in the presence of KOH. The burning of glyceride (77.46% in 60 min) reached a maximum at a temperature of 80 °C and a KOH concentration of 0.304 N. The ester produced in this way has a higher degree of viscosity than diesel oil, but it is comparable to diesel in all other respects. The fuel value of this product amounts to 39,182 J/g (Warnigati et al. 1992, p. 1479 et seqq).

Taramira Seed Oil

synonyms: Öl der Ölrauke (D); jamba oil (E); Eruca sativa seed oil (E)

S. Krist, *Vegetable Fats and Oils*, https://doi.org/10.1007/978-3-030-30314-3_114

1 Source Plant

Eruca sativa (Brassicaceae), Rucola, Colewort, Salad Rocket

Habitat

Eruca sativa presumably originates in the Mediterranean area. It is also found in the Middle East (Schuster 1992, p. 48–49) and grows wild in North Africa in many varieties (Hackbarth 1944, p. 170).

Description

T

Rucola is an annual plant, which may winter in mild climates. The root is spindle-shaped, 10–15 cm long and not very branched. The shoot grows to a height of 90 cm and has numerous side shoots. The lower part is sparsely covered with hair, whereas the upper part is hairless. The leaves are lyrate; there are only few. The flower has 4 sepals. The petals are twice as long as the sepals and yellow with dark-violet veins. The plant blossoms from early May to early June. The fruit are husks, 22–34 mm long and 4–7 mm broad, hairless and beaked. Within, there are about 20–34 seeds, which are 2 mm long, 1.6 mm broad and 1.1 mm thick. The thousand-seed weight is 3.6–3.2 g. The surface of the seed is yellow to yellowish-brown with greenish dots (Hackbarth 1944, p. 169–170).

Rucola is pollinated via exogamy, although autogamy is also possible (Schuster 1992, p. 48).

2 Cultivation and Extraction

Cultivation

Eruca sativa has long been cultivated in Iran, Afghanistan and India. Originally, it was as widespread as a weed as false flax, but was then cultivated separately. It only gained importance as an oil plant after Arnold Scheibe (University of Gießen) brought the seeds with him from an expedition to the Hindu Kush in 1936. He experimented with cultivating the plant and began breeding it systematically (Schuster 1992, p. 49).

Rucola is an undemanding plant, which grows even with a poor water supply and poor soil (Lieberei and Reisdorff 2007, p. 249).

As a result of its resistance to droughts, it is ideal for regions with low amounts of precipitation (Kanya and Urs 1989).

Sowing the plant in spring is preferable to sowing in autumn, to prevent it from dying in the winter. The harvest takes place in early July if the plant was sown in autumn, and in early- to mid-August if it was sown in spring. Although *Eruca sativa* survives dry conditions, moist conditions are advantageous (Hackbarth 1944, p. 170–172).

Extraction of the Oil

To obtain the oil, the seeds are dried and ground. The oil is extracted in a Soxhlet extractor (Gulfraz et al. 2011).

3 Character

Colour

The oil is a golden-yellow to golden-brown colour (Hackbarth 1944, p. 172).

Flavour

The oil tastes bitingly hot. This is the result of glucosinolates in the seeds, which are transformed into isothiocyanates by the enzyme myrosinase during pressing (Kanya and Urs 1989). The sharp taste can be removed by distillation (Löw 2003, p. 39).

4 Ingredients

Composition of Fatty Acids

Fatty acids	Content in % acc. to Schuster (1992), p. 10	Content in % acc. to Hackbarth (1944), p. 172	Content in % acc. to Kanya and Urs (1989)	Content in % acc. to Gulfraz et al. (2011)
Capric acid	–	–	–	0.008
Lauric acid	–	–	0.03	0.017
Myristic acid	–	–	0.11	–
Palmitic acid	2–6	–	3.76	2.378
Palmitoleic acid	–	–	–	0.136
Stearic acid	2–4	–	0.93	–
Oleic acid	15–20	5.4	19.88	15.1
Linoleic acid	12–28	28.5	9.28	6.938
α-Linolenic acid	2–10	1.5	11.70	0.005
Eicosenoic acid	8–10	–	12.61	12.514
Behenic acid	3–5	–	–	1.261
Erucic acid	35–58	58.5	42.40	51.212
Lignoceric acid	1–3	–	–	0.148
Nervonic acid	–	–	–	0.218

Physical Key Figures

The iodine number of taramira seed oil is about 105, which makes it a slow drying oil (Hackbarth 1944, p. 172).

The density at 27 °C is 0.8990 g/ml, and the kinematic viscosity is 27.73 mm^2/s (Le et al. 2009).

5 Application

In Pharmacy and Medicine

A current study by Khoobchandani et al. detected the oil's antimicrobial activity against antibiotic-resistant gram-negative and gram-positive bacteria. It is therefore used in traditional medicine in cases of skin infections, fever, urinary tract infections and diarrhoea. The antimicrobial properties can presumably be traced back to the high amount of erucic acid (Khoobchandani et al. 2010).

T

A study by Gulfraz et al. tested various extracts from *Eruca* seeds for their antimicrobial properties. The result showed that the seed oil and methanol extracts have a higher antimicrobial potential against gram-positive and gram-negative bacteria than other extracts or the standard antibiotic gentamicin. The minimal inhibitory concentration of the seed oil against *Staphylococcus aureus*, *Pseudomonas aeruginosa*, *Escherichia coli*, *Salmonella typhi*, *Klebsiella pneumoniae*, *Staphylococcus epidermidis* and *Pasteurella* lies between 50 and 72 μg/ml, whereas in the case of gentamicin, it is 56–70 μg/ml (Gulfraz et al. 2011).

Industrial Uses

Taramira seed oil is used as a lamp oil and a hair oil in India. It is also a lubricant and a motor oil (Löw 2003, p. 39).

As Food

The oil obtained from the seeds by pressing is also called jamba oil and is an edible oil (Lieberei and Reisdorff 2007, p. 249).

The amount of erucic acid is, howsever, noteworthy. According to the decree on erucic acid (BGBl. No. 468/1994), it is prohibited to market edible oil, edible fat and food produced by adding edible oils or fats with a total fat content of more than 5%, if they are intended for human consumption and contain more than 5% erucic acid – based on the total content of fatty acids in the fat phase (http://www.ris.bka.gv.at/04.07.2011).

Tea Seed Oil

Camellia Oleum synonym: Teesamenöl, Kamelliasamenöl (D)

S. Krist, *Vegetable Fats and Oils*, https://doi.org/10.1007/978-3-030-30314-3_115

1 Source Plant

Camellia sasanqua Thunb. (Theaceae), tea sasanqua

Habitat

Camellia sasanqua grows in the mountains of South East Asia, and in the evergreen forests along the coasts of southern Shikoku, Kyushu and many other smaller Japanese islands. In its wild form, it grows to a height of 900 m above sea level. The Japanese never cultivated it as an ornamental plant, like other species of *Camellia*, but only as a useful plant. They called the plant *sasanqua*, which means "plum-blossomed tea". The plant is undemanding; it thrives in the cold, dry Japanese winters and even blossoms beneath the snow in January and February.

Description

Camellia sasanqua is an evergreen bush or tree, which can reach a height of 6 m. The symmetrical crown has a diameter of 4–8 m. The leaves are green and covered with silky hair while they are young. Later, they are thick and hairless. In late winter, the single white or pink flowers appear, from which the fruit develop. The fruit are capsules, about 2 cm in size, and contain the fatty seeds.

2 Cultivation and Extraction

Cultivation

In Japan, *Camellia sasanqua* has been cultivated for a long time. The leaves have been used for centuries to brew tea, and the seeds for oil extraction. This tea plant was not known in Europe until 1869, when Dutch merchants imported it.

Even though cold winters are predominant in Europe, the cultivation of *Camellia sasanqua* has not been as successful as in Australia or in the southern USA. Other cultivation areas are located in India, Sri Lanka and China. The cultivation of *Camellia sasanqua* gives the farmers of southern China the opportunity to keep their traditional cultivable land, which would otherwise be destroyed by the cultivation of sugar cane and maize.

Extraction of the Oil

All *Theaceae* contain 17–45% fat oil, but the tea seed oil on the market comes exclusively from *Camellia sasanqua. Camellia sinensis* is cultivated for its leaves, and the seeds are not allowed to ripen. *Camellia sasanqua*, on the other hand, is cultivated for its oil. The seeds contain 58–60% oil, which is obtained by pressing or liquid extraction. The raw oil must not be used as food and has to be refined before use (Hager 1978, volume 7b, p. 210).

3 Character

Colour

Colourless to pale yellow.

Odour

Green, fatty, like aldehyde, delicate, paint, medical, slightly sweetish.

Flavour

Characteristic.

4 Ingredients

Composition of Fatty Acids

Fatty acids	Content in % acc. to Hager (1978), volume 7b, p. 210	Content in % acc. to Roth and Kormann (2000), p. 164	Content in % acc. to greencottage.com/04.11.03
Palmitic acid	7.6	7.6	9
Stearic acid	0.8	<1	1
Oleic acid	83.3	83	80
Linoleic acid	7.4	7.4	9
Myristic acid	0.3	<1	–
Arachidic acid	0.6	<1	1

Other Ingredients

terpenes
(Akihisa et al. 2004)

Physical Key Figures of Tea Seed Oil

refractive index: n_D^{40} 1.462
density: d_{40}^{40} 0.899–0.903
saponification number: 190–195

iodine number: 80–87
point of solidification: −5 to −10 °C

(Roth and Kormann 2000, p. 164)

5 Application

Tea seed oil resembles olive oil chemically and physically and can be used accordingly. The extraction of *Camellia sasanqua* oil is more time-consuming than that of sunflower oil, wheat-germ oil and many other oils. For this reason, it is not important technically. It is mainly used in Japan in the home industry to produce novelties for souvenir shops, for example as a basis for paint for dolls' eyes.

In Pharmacy and Medicine

Due to its composition of fatty acids and other ingredients, *Oleum Camellia* is promising as an oil for therapeutical, medical purposes. It can be applied in the following cases (Akihisa et al. 1997, 2004):

- highly sensitive skin with a tendency towards allergies;
- prevention of chronic eczema;
- prophylaxis in cases of high blood pressure and coronary diseases;
- cancer prevention (because the oil contains antioxidants);
- influencing blood coagulation, and
- anti-inflammatory properties (as a result of triterpenes).
- inhibitory effect on the Epstein-Barr virus (which is based on the inhibition of the EBV-EA/Epstein-Barr virus early antigen by the triterpenes contained in tea seed oil and is currently being examined in clinical studies of tea seed oil in Japan).

In Cosmetics

Oleum Camellia forms part of the following cosmetic products:

- shampoo and lotions for dry, dull hair;
- after-sun products for thin and normal hair;
- moisturising after-sun body lotions;
- preparations to protect against the cold;
- anti-wrinkle creams for the eyes, and
- nail hardeners and massage oils.

As Food

In South-East Asia, teaseed oil is used traditionally as an edible oil (Hager 1978, volume 7b, p. 210).

Technical Uses

Teaseed oil is used for the following technical purposes (Hager 1978, volume 7b, p. 210):

- as fuel oil;
- in soap production;
- as a lubricant for precision instruments;
- as a textile oil, and
- for lighting.

Tiger Nut Oil

Oleum Cyperi esculenti – synonyms: Chufaöl, Erdmandelöl (D); Cyperus esculentus root oil, chufa oil (E)

1 Source Plant

Cyperus esculentus L. (Cyperaceae), *chufa sedge*

S. Krist, *Vegetable Fats and Oils*, https://doi.org/10.1007/978-3-030-30314-3_116

Habitat

Chufa sedge is a native plant of the tropics of East Africa. It was already used in Ancient Egypt and is still cultivated in the Mediterranean area today, as well as in eastern India, West Africa and Brazil. The cultivation of *chufa sedge* is of local importance only (Franke 1989, p. 151).

Description

The part above ground is a dense group of many leaf blades, which are covered with coarse hair. Similar to the potato plant, *chufa sedge* produces thin underground runners, which stop their linear growth after some time and swell in thickness at the tip instead, forming tubers about the size of an acorn. As the plant produces hardly any seeds, the tubers serve the purpose of vegetative propagation (Franke 1989, p. 151).

2 Cultivation and Extraction

Cultivation

C*hufa sedge* does not make high demands on the soil. It prefers light to medium soil, especially boggy, wet ground. The yield is greatest in warm and humid weather, but the plant is extremely frost-susceptible. It is sown by distributing the tubers on the field. Since the soil must be warm for the tubers to germinate, the sowing must not take place too early in the year, but only from the second half of May onwards. The harvest takes place 3–5 months later; it is toilsome, as the tiger nuts must be dug out of the earth and collected (Hackbarth 1944, p. 167).

Extraction of the Oil

The oil is extracted from the dried and chopped-up tubers of *Cyperus esculentus* by pressing them twice and adding heat. The first pressing takes place at low temperatures, the second pressing at higher temperatures. Extraction methods can also be applied (Hackbarth 1944, p. 168).

3 Character

Colour

A pale golden yellow (Franke 1989, p. 151).
Golden-yellow to brown oil (Roth and Kormann 2000, p. 122).
Golden yellow (Hackbarth 1944, p. 168).

Odour

Nutty-fatty, rounded, sweet, of roasted almonds.

Volatile Compounds

SPME-GC-MS analysis (Bail et al. 2009b):

Acetic acid	Heptanal
1-Hydroxy-2-propanone	Gamma-butyrolactone
2,3-Butandiol	Trans-2-heptenal
Hexanal	Benzaldehyde
Furfural	Octanal

Flavour

Pleasant nutty taste (Roth and Kormann 2000, p. 122).

4 Ingredients

Composition of Fatty Acids

Fatty acids	Content in % acc. to Roth and Kormann (2000), p. 122	Content in % acc. to tigernuts.com/09.08.05
Oleic acid	67–73	68.83
Linoleic acid	6–15	11.70
Linolenic acid	–	0.15
Palmitic acid	12	13.19
Stearic acid	5	4.32
Arachidic acid	<1	0.64
Myristic acid	<1	0.08
Lignoceric acid	<1	0.19

Composition of Sterols

Sterols	Content in % acc. to tigernuts.com/09.08.05
Cholesterol	0.27
Brassicasterol	0.17
Campesterol	14.83
Stigmasterol	15.48
β-Sitosterol	54.67
Sitostanol	0.62
Δ7-Campesterol	0.84
Δ5-Avenasterol	2.95
Δ7-Avenasterol	2.55
24-Methylencholesterol	0.06

Physical Key Figures of Tiger Nut Oil

refractive index: n_D^{25} 1.4680
density: d_{15}^{15} 0.917–0.924
point of solidification: below 3 °C
saponification number: 190–194
iodine number: 74–89
unsaponifiable: approx. 0.6%

(Roth and Kormann 2000, p. 122)

Shelf Life

Since tiger nut oil has a high content of vitamin E and monounsaturated oleic acid, it has a very long shelf life.

5 Application

In Pharmacy and Medicine

Since tiger nut oil contains high amounts of oleic acid, it lowers high cholesterol levels (Hitzenberger 2005) and prevents the formation of thrombi, and is therefore used to treat arteriosclerosis and circulatory disorders.

Tiger nut oil has a positive effect on the metabolism as well as on digestion, and also has anti-inflammatory properties. For this reason, it is applied internally to treat gastrointestinal diseases. The fatty acids in tiger nut oil also have positive effects on osteoporosis: More calcium is absorbed in the bones, thereby counteracting bone loss (Hitzenberger 2005).

In Folk Medicine

Tiger nut oil is applied in cases of fertility disorders of males and females. It is also thought to be an aphrodisiac.

In Cosmetics

Since tiger nut oil has a high content of vitamin E, it has a positive effect on cells and stimulates microcirculation in the skin. It is used to treat eczema and other skin alterations. It is also added to anti-wrinkle creams, as its high amount of vitamin E delays cellular ageing, improves skin elasticity and therefore reduces wrinkles (Hitzenberger 2005).

As Food

Since tiger nut oil has a fine aroma, it is used as an edible oil. It is a good accompaniment to sweet dishes such as parfaits, fruit salads, dessert creams, cakes and pastries, but can also be used to season salads, sauces and pasta. As tiger nut oil is heat-resistant, it is better suited for frying than many other oils. In the process of frying, a crust forms on the surface of the food, which prevents the oil from penetrating the inside (Hitzenberger 2005).

In Industry

In the pharmaceutical and chemical industries, tiger nut oil is a useful source of tocopherol, which can only be synthesised chemically with considerable effort (Hitzenberger 2005).

In the textile industry, textile fibres are impregnated with tiger nut oil.

In addition, tiger nut oil is used in the production of soap and lubricants, and in precision mechanics (Roth and Kormann 2000, p. 122).

6 Possible Unwanted Side Effects

In very rare cases, tiger nuts and tiger nut oil cause allergic reactions such as swelling of the tongue and lips, itching and urticaria, but also bronchospasms and anaphylactic shock (Bonilla et al. 2002).

Tobacco Seed Oil

synonyms: Tabaksamenöl (D)

S. Krist, *Vegetable Fats and Oils*, https://doi.org/10.1007/978-3-030-30314-3_117

1 Source Plant

Nicotiana tabacum L. *(Solanaceae),* Tobacco

Habitat

Nicotiana tabacum originates in America and arrived in Spain in about 1560 as an ornamental plant. The French ambassador in Lisbon, Jean Nicot, imported the tobacco plant to France, which is why the plant was later named after him (Lieberei and Reisdorff 2007).

Description

Tobacco is an annual herb that can reach a height of 2–3 m. *Nicotiana tabacum* is presumably a hybrid of *Nicotiana sylvestris* and *Nicotiana tomentosiformis*. Nicotine is produced in the root, from where it is transported to the leaves. The leaves are alternate and have a broad lanceolate form. The flowers are funnel-shaped, and either red or yellow (Lieberei and Reisdorff 2007).

The main root is sturdy, spiral and of medium length, whereas the lateral roots are smaller. The stem is upright and only branches at the top. The flowers sit on the top of the stem and have five petals and stamina each (Barth and Jehle 2003).

The tobacco fruit is a capsule with 2–4 compartments. It is oval and dark to light brown. When it is ripe, it opens to show numerous seeds. The seed is also oval and has a conspicuous stigma. It is 0.65 mm long and 0.50–0.56 mm thick, and consists, to two thirds, of endosperm. The embryo makes up about one third of the content. The seeds are small, but the yield is nevertheless high, as each plant produces about 400,000 seeds (Hackbarth 1944).

2 Cultivation and Extraction

Cultivation

Tobacco is mainly cultivated for its leaves, which contain nicotine. From the sixteenth century onwards, the leaves were rolled and smoked. Soon afterwards, their use was prohibited, but this had hardly any effect. As a result, the smoking of tobacco leaves was permitted again, but taxed. In general, the cultivation of tobacco is possible in tropical as well as temperate climates, due to the plant's numerous varieties adapted to many different conditions. Tobacco does not tolerate salty soil or stagnant moisture; it prefers warmth and precipitation rates between 250 and 1000 mm. Tropical conditions and sandy, loamy soil are therefore necessary to obtain tobacco of the best quality (Lieberei and Reisdorff 2007).

As an oil plant, tobacco was first used in 1915 in Hungary, then from 1930 onwards in Russia, and later in Greece. The use of seeds for oil production met with interest, since as a by-product, they are a cheap raw material, and it is possible to gather the seeds without disturbing the leaf harvest. If the seeds are to be gathered, the inflorescences must not be removed. This has the advantage that infertile bastard branches do not grow back and thus do not have to be cut off. On the other hand, another step is added, as the plants have to be shaken repeatedly. This is necessary to prevent rotten spots, which emerge on the leaves in moist weather conditions if petals fall on them. If a rotten spot has emerged despite the shaking, the leaves concerned have to be harvested separately and to undergo a more intense drying process. In Central European climates, the seeds only ripen at the end of September or the beginning of October, and their harvest therefore takes place much later than the leaf harvest. The plants with the leaves removed are left standing until half of the capsules have turned brown. The inflorescences are then cut and hung up in bunches for post-ripening. After completion of the drying process, the seeds are removed from the capsules and cleaned (Hackbarth 1944).

Extraction of the Oil

The oil can be extracted by pressing or by extraction with solvents in a Soxhlet extractor or a Soxtec apparatus. A current study by Stanisavljevic et al. tested ultrasonic extraction. It became apparent that the yield is better at higher temperatures and with a lower ratio of seeds and solvent. In summary, ultrasonic extraction proved more efficient than "classic" extraction via batch-mode. The best method, however, is extraction with a Soxhlet extractor. Preparatory to all extraction methods, the seeds are dried in shady, well-aired locations directly after the harvest. Afterwards, they are put in a paper bag and stored in a dry, dark place. Before the oil is extracted, the seeds are ground. Grounding ensures a higher yield because there is no oil in the aril; the larger part is found in the thin-walled cells of the endosperm and a small amount also in the embryo (Stanisavljević et al. 2007).

3 Character

Colour

Depending on the degree of heating before pressing, tobacco seed oil is light yellow to brown (Hackbarth 1944).

Flavour

Tobacco seed oil is a very bitter oil. This problem can be alleviated by refining it (Hackbarth 1944).

4 Ingredients

Composition of Fatty Acids

Fatty acids	Content in % acc. to Frega et al. (1991)	Content in % acc. to Giannelos et al. (2002)	Content in % acc. to Mukhtar et al. (2007)	Content in g/kg acc. to Zlatanov et al. (2007)	Content in % acc. to Maestri and Guzmán (1993)
Caprylic acid	–	<0.1	–	–	–
Capric acid	–	<0.1	–	–	–
Lauric acid	–	<0.1	–	1–102	–
Myristic acid	–	0.09	1.13	2–76	–
Palmitic acid	8.9–9.5	10.96	8.72	126–367	8.16–10.3
Palmitoleic acid	0.1	0.2	1.3	2–127	–
Margaric acid	0.1	0.1	–	1–17	–
Stearic acid	2.5–2.8	3.34	2.64	16–87	1.42–3.63
Oleic acid	9.5–11.1	14.54	13.46	152–263	10.4–13.2
Vaccenic acid	0.6–0.8	–	–	–	–
Linoleic acid	74.9–76.1	69.49	71.63	176–627	72.4–78.5
Linolenic acid	1.1–1.4	0.69	0.93	1–8	0.46–1.04
Elaidic acid	–	0.02	–	–	–
Arachidic acid	0.2	0.25	–	–	–
Behenic acid	0.1	0.12	–	–	0.11–0.30
Erucic acid	–	<0.01	–	–	–
Lignoceric acid	–	0.04	–	–	–

Composition of Triglycerides

Triglycerides	Content in % acc. to Frega et al. (1991)
PPS	0.1
PPO	0.4
PLP	2.9–3.2
PPoL	0.2–0.6
POO	0.7–1.1
PLS	1.2–1.3
PLO	6.6–7.2
PLL	22.9–25.0
SOO	0.2–0.5
OOO	0.9–1.4
SLO	2.4–3.7
OLO	5.2–5.3
OLL	15.0–16.1
LLL	36.0–40.1

Composition of Sterols

Sterols	Content in % acc. to Frega et al. (1991)	Content in g/kg acc. to Zlatanov et al. (2007)	Content in % acc. to Maestri and Guzmán (1993)
Cholesterol	7.3–9.8	56–112	4.01–6.91
Cholestanol	0.2–0.4	–	–
Brassicasterol	0.1–1.1	2–18	–
24-Methylencholesterol	0.4–0.9	–	–
Campesterol	13.0–13.6	93–149	10.7–13.0
Campestanol	0.3	–	–
Stigmasterol	9.3–10.2	102–188	11.3–15.9
Stigmastenol	0.6	–	–
Δ7-Campestanol	0.9–1.9	–	–
Clerosterol	0.7–1.3	–	–
β-Sitosterol	39.5–43.8	433–682	41.2–47.2
Δ5-Avenasterol	19.2–22.7	7–144	16.8–19.2
Δ3,2-Stigmastadienol	0.8–1.0	–	–
Δ7-Stigmastenol	0.1–0.2	3–14	3.86–6.32
Δ7-Avenasterol	0.6–0.7	1–14	1.0–2.76

The sterol content increases continually during germination, the greatest increase being noticeable in the levels of stigmasterol and campesterol. However, sitosterol remains the main sterol at all times (Bush and Grunwald 1972).

Composition of Tocopherols

Tocopherols	Content in $g.kg^{-1}$ acc. to Zlatanov et al. (2007)
Total content	–
α-Tocopherol	169–981
α-Tocotrienol	30–199
β-Tocopherol	8–93
γ-Tocopherol	9–351
δ-Tocopherol	2–283

Other Ingredients

Phospholipids	Content in $g.kg^{-1}$ acc. to Zlatanov et al. (2007)
Phosphatidylcholine	247–405
Phosphatidylinositol	215–276
Phosphatidylethanolamine	143–320
Phosphatidic acid	13–57
Lysophosphatidylcholine	42–103
Lysophosphatidylethanolamine	21–71
Sphingomyeline	22–69
Phosphatidylserine	18–43

Physical Key Figures

refractive index: n_D^{25}	1.4739–1.4770
density: d_{15}^{15}	0.922–0.925
saponification number:	189–198
iodine number:	136–147
unsaponifiable:	3.0%

(Roth and Kormann 2005, p.161)

viscosity at 40 °C:	27.7 mm²/s
cetane number:	38.7
energy content:	39.4 MJ/kg
cloud point:	−7.8 °C
flow point:	−14 °C
flash point:	220 °C

(Giannelos et al. 2002)

5 Application

Technical Uses

Tobacco seed oil can be used in bio diesel production. Biodiesel is produced from renewable raw materials. In order to reduce the costs, cheap raw materials are advantageous. As tobacco seeds are a by-product (since the plant is cultivated for its leaves), they are among the most cost-efficient raw materials for biodiesel.

Bio diesel is made from alkyl esters of fatty acids. The production process of methyl esters of fatty acids from tobacco seed oil consists of two steps. First, free fatty acids are reduced by esterification catalysed by acids. The second step is a methanalysis catalysed by alkali, which is responsible for the transfer of the first product into fatty acid methyl esters and glycerol. A current study by Veljkovic et al. reduced free fatty acids from 35% to 2% by means of catalysed esterification. Bio diesel produced in this way met all the prescribed standards with regard to fuel characteristics and threshold values. Only the acid number was relatively high. Tobacco seed oil can thus be used as an economic, renewable raw material for producing bio diesel (Veljković et al. 2006).

Tobacco seed oil can also be employed in the production of alkyd resins, which are products consisting of a higher alcohol and a bivalent acid with the modified oil. As tobacco seed oil contains a high amount of unsaturated fatty acids, the result is an alkyd resin that dries well. Other characteristics are good adhesion, resistance to water and acids, durability, shine and a light colour. These characteristics make it suitable for the production of architectural paint, road marking paint and high-gloss printing ink (Mukhtar et al. 2007).

As Food

Tobacco seed oil is not actually an edible oil, but can be used as such if it has been refined (Hackbarth 1944).

In some European countries, refined tobacco seed oil is used as a salad oil (Eshetu 2000).

Tomato Seed Oil

Oleum Lycopersici synonyms: Tomatenkernöl (D); huile de pépins de tomate (F)

1 Source Plant

Lycopersicon esculentum Mill. (Solanaceae), tomato

S. Krist, *Vegetable Fats and Oils*, https://doi.org/10.1007/978-3-030-30314-3_118

Habitat

The tomato originates in the Andes of Peru and other Central and South American countries, where it grew as a tropical wild plant (Dassler and Heitmann 1991, p. 236), as well as in the Caribbean (Schoenfelder and Schoenfelder 2001, p. 160). In 1498, Columbus brought it to Europe on his second voyage. The first attempt at cultivation is reported to have taken place in Italy in 1554. At first, the tomato was regarded as a rare ornamental plant in Europe, a view that is still reflected in its names in some countries, which translate as "apple of gold" or "apple of paradise" (note, for example, the Italian name *pomodoro* or the Austrian *Paradeiser*). The history of the tomato as a culinary plant began in the mid-eighteenth century. It was only after the Second World War that it became a basic food. The main producers in the annual global production of nearly 70 million t are the USA (14%), the former USSR (10%), China (8%), Italy (8%), Turkey (8%), Egypt (6%), Spain (4%), Romania (3%), Greece (3%), Brazil (3%) and Mexico (2%) (Dassler and Heitmann 1991, p. 236 et seqq).

Description

Tomatoes are annual, herbaceous plants. The colour of their fruit is usually red, which is caused predominantly by the colourant lycopene and, to a lesser extent, by carotene and xanthophyll. Some tomatoes are yellow or violet. With regard to their outer form, there are five main types: round tomatoes, beef tomatoes, plum tomatoes, Roma tomatoes and cherry tomatoes (Dassler and Heitmann 1991, p. 236 et seqq). The solanaceous plant is glandular, bristly, with odd-pinnate leaves. The yellow flower has a short tube and 5–6 pointed lappets (Schoenfelder and Schoenfelder 2001, p. 160). The seeds contain 12–18% fat oil (Roth and Kormann 2000, p. 165).

2 Cultivation and Extraction

Cultivation

Tomatoes can be grown either in the open air or in conservatories. They need warmth and are very frost-susceptible, which is why they are usually cultivated in conservatories in Western Europe. Early tomatoes from Southern Europe are usually from a hothouse; they are generally varieties of stake tomatoes that are cultivated under protection. After the fruit have been plucked, they are best stored at temperatures of 8–10 °C if ripe. They have a shelf life of about 2 weeks (Dassler and Heitmann 1991, p. 236 et seqq).

Extraction of the Oil

Tomato seed oil is obtained from the seeds of tomatoes. The fruit are first dried (Lazos et al. 1998) and then either cold pressed or extracted with hot petroleum ether (Lazos et al. 1998). For cold pressing, the yield amounts to 17%, while the yield for extraction is 33.1% of the seed weight (Calvacanti 1941). Today, the oil can also be extracted with supercritical CO_2, which results in a yield of 80% of oil from the seeds (Sun and Ding 1998). The extracted oil is subsequently neutralised and bleached (Lazos et al. 1998).

3 Character

Colour

Golden yellow (Calvacanti 1941).

4 Ingredients

Composition of Fatty Acids

Fatty acids	Content in % acc. to Rossell and Pritchard (1991), p. 295	Content in % acc. to Firestone (1999), p. 104
Lauric acid	0.5	–
Myristic acid	0.5	0–0.2
Palmitic acid	15	12–16
Palmitoleic acid	0.5	0–0.6
Stearic acid	6.5	4–7
Oleic acid	23	16–25
Linoleic acid	50.5	50–60
α-Linolenic acid	2	1.5–2.5
Arachidic acid	0.5	0–0.6
Eicosanoic acid	–	0–0.2
Behenic acid	–	0–0.2

Composition of Sterols

Sterols	Content in % acc. to Malecka (2002)	Content in % acc. to Karleskind (1996), p. 182	Content in % acc. to Firestone (1999), p. 104 et seq
Cholesterol	–	18–30	7–27
Brassicasterol	9.8	1–4	–
Campesterol	29.5	1–2	4–7
Stigmasterol	4.5	14–15	7–17
β-Sitosterol	43.2	48–64	54–73
Δ5-Avenasterol	10.8	<2	0–8

Sterols	Content in % acc. to Malecka (2002)	Content in % acc. to Karleskind (1996), p. 182	Content in % acc. to Firestone (1999), p. 104 et seq
Δ7-Stigmasterol	–	–	1
Δ7-Cholesterol	–	–	1.5–4

Tomato seed oil also contains small amounts of 24-methylencholesterol, clerosterol, Δ7, 24-stigmastadienol, Δ7-stigmastanol, Δ7-avenasterol, Δ7-campesterol and erythrodiol (Lazos et al. 1998).

Composition of Tocopherols

Tocopherols	Content (ppm) and amount (%) acc. to the 21st World Congress of the ISF (1995), p. 197	Content (mg/kg) acc. to Lazos et al. (1998)
Tocopherol content	950	–
α-Tocopherol	14	202
β-Tocopherol	1	–
γ-Tocopherol	85	–
Δ-Tocopherol	–	1059

Physical Key Figures of Tomato Seed Oil

refractive index: 25 °C 1.4715–1.4728
density: 15 °C 0.919–0.924
iodine number: 107–125
saponification number: 183–196
unsaponifiable: 2.6%

(Roth and Kormann 2000, p. 165)

5 Application

In Pharmacy and Medicine

During a study, rats suffering from arteriosclerosis were given 1 ml tomato seed oil daily in oral form over a period of 30 days. As a consequence, the cholesterol level decreased and the phospholipid level in the blood serum increased in comparison to arteriosclerotic rats that were fed sunflower oil. The reason could be the high amount of linoleic acid in sunflower oil (Babadzhanov and Khadzhiew 1982).

As Food

Tomato seed oil can be used as an edible oil for salads and to produce margarine (Calvacanti 1941).

In Cosmetics

On an industrial scale, the oil is used in cosmetic products (Nisshin Flour Milling Co. 1981) and in soap production (Calvacanti 1941).

Tung Oil

Oleum Dryandrae synonyms: Tungöl, Chinesisches Holzöl, Lackbaumöl, Ölfirnisbaumöl (D); China wood oil, lumbang oil, wood oil (E)

1 Source Plant

Aleurites fordii Hemsl. *(Euphorbiaceae), Aleurites montana* (Lour.) E.H. Wilson *syn. Vernicia montana*, tung tree

Habitat

Aleurites fordii is a native plant of Central and Western China. *Aleurites montana* originates in south-eastern China, Cambodia, Laos and Vietnam (Hager 1998, p. 54, 62).

Description

Aleurites fordii

The tung tree is a 2–12 m tall, deciduous tree that branches at a low height like a bush and has long, thin, hairy, pendant branches. The leaves are dark green, 7.5–20 cm in size, and ovate with pointed ends. The tree blossoms before the leaves appear. The flowers are white with pale red spots. The oval fruit are 4–8 cm in diameter and, depending on their state of ripeness, olive green to dark brown. The pericarp is 3–4 mm thick and contains 3–4 longish-oval seeds. They have a brown testa with a yellowish-white, noticeably oily pulp (Hager 1998, p. 54 et seqq).

S. Krist, *Vegetable Fats and Oils*, https://doi.org/10.1007/978-3-030-30314-3_119

Aleurites montana

The tung tree appears, according to its habitus, in 2 types: The A-type reaches a height of 15 m and has a narrow crown; the branches are arranged in levels. The B-type has a low, bushy crown at only two levels. The leaves are reddish, and either entire or 3- to 5-lobed. The flowers are a pure white colour with a red base, and appear before the leaves. The greenish fruit are oval and 4–5 cm broad, with prominent longitudinal ribs; they contain slightly warty seeds (Hager 1998, p. 62).

2 Cultivation and Extraction

Cultivation

Aleurites fordii is mainly cultivated in China, but also in Paraguay, Argentina, Brazil and the southern USA, and, on a smaller scale, in some African and Asian countries. The tree needs periods of cold in order to bloom.

Aleurites montana is cultivated in Australia, Brazil, Myanmar, China, Guinea, India, Indonesia, Cameroon, Cambodia, Kenya, Laos, Madagascar, Malawi, Morocco, New Caledonia, Zambia, in the region of the Black Sea in the CIS, Sri Lanka, South Africa, the southern USA, Tanzania, Uganda, Vietnam, the Democratic Republic of the Congo and Cyprus. Altitudes of about 1000 m in tropical regions provide the best conditions (Hager 1998, p. 54 et seqq).

Tung trees grow best on loose, stony ground. The harvest is profitable in the 5th year after sowing.

Extraction of the Oil

In a first step, the shell of the seeds must be removed. In China, the fruit either remain lying on the ground until the shell can be removed easily, or the ripe fruit are shaken from the trees, stacked on top of each other and covered to start a process of fermentation that decomposes the shell. The deshelled kernels are either cold or hot pressed (Hager 1998, p. 58).

3 Character

Colour

Light-yellow to dark-brown, clear oil.
Light-yellow oil (Roth and Kormann 2000, p. 155).
Viscous, yellowish to brown oil (Hager 1998, p. 58).

Odour

Like lard, very characteristic.
Characteristic, unpleasantly reminiscent of lard oil and plant bugs (Roth and Kormann 2000, p. 155).

Flavour

Peculiar, slightly obnoxious taste (Hager 1998, p. 59).

4 Ingredients

Kernels contain 50–60% oil (Karleskind 1996, p. 222); the main ingredients of *Aleurites fordii oil* are elaeostearic acid (79.7%), oleic acid (14.9%), palmitic acid (4.1%) and stearic acid (1.3%) (Roth and Kormann 2000, p. 155).

Composition of Fatty Acids in *Aleurites fordii*

Fatty acids	Content in % acc. to Firestone (1999), p. 106	Content in % acc. to Karleskind (1996), p. 222
Palmitic acid	2	3–5
Stearic acid	3	1–2
Oleic acid	4–10	4–9
Linoleic acid	8–15	4–10
α-Linolenic acid	2	1–3
Elaeostearic acid	71–82	77–88

Composition of Tocopherols and Tocotrienols in *Aleurites montana*

Tocopherols/tocotrienols	Content in mg/kg acc. to Matthaus et al. (2003)
α-Tocopherol	255
γ-Tocopherol	1206
Δ-Tocopherol	44
γ-Tocotrienol	34

Other Ingredients

(Schneebeli-Graf 1995, p. 123)

Tannins
Phytosterols
Saponin

T

Physical Key Figures of Tung Oil

density: d^{25} 0.913–0.917
refractive index: n_D^{25} 1.514–1.520
saponification number: 189–195
iodine number: 160–175
unsaponifiable: 0–1%
point of solidification: −17 to −21 °C
solubility: in chloroform, diethyl ether, carbon disulphide, oil
acid number: around or lower than 5

(Firestone 1999, p. 106 – kremer-pigmente.de/13.07.06; Roth and Kormann 2000, p. 155)

Characteristics of the Oil

The properties of tung oil when heated are important for technical, but also for analytical purpose. If the oil is heated to high temperatures, it becomes more viscous, which, depending on the temperature, leads to gelation to a greater or lesser extent. The changes that tung oil undergoes in this process of thickening are:

- increase in mass per unit area;
- decrease in iodine number;
- decrease in refractive index, and
- the acid number and saponification number hardly change.

5 Application

In Pharmacy and Medicine

Tung oil contains 80% α-eleostearic acid (9,11,13-cis,trans,trans-octadecatrienoic acid), which is an intrinsic chromophore with strong UV-absorbing properties and, according to a newly developed assay, can be used to measure lipase activity (Pencreach et al. 2002).

In Folk Medicine

In Chinese folk medicine, the oil is applied to treat skin diseases, wounds, burns, ulcers and contusions, and is thought to be part of nearly every Chinese patch. Its effects have not been scientifically proven (Hager 1998, p. 59).

Technical Uses

Like linseed oil, tung oil is used in the technical industry to produce quick-drying wood varnish, linoleum, insulating material, waterproof textiles and paper. In China, it is used in the production of China ink and as a base for binding agents for antique lacquer works of the Far East (Roth and Kormann 2000, p. 155).

6 Possible Unwanted Side Effects

A study examined the potential of tung oil to activate the Epstein-Barr virus (EBV). The results show that tung oil has a similar, albeit lower potential on human lymphoblastoid cells carrying the EBV genome than croton oil, which was used as a positive comparison. Whether these results are correlated to the increased appearance of nasopharyngeal carcinomas in regions where tung trees grow will be the subject of further examinations (Ito et al. 1983).

If ingested orally, the oil results in nausea; applied topically, it leads to slow-healing wounds. It is a strong purgative (Roth and Kormann 2000, p. 155). The oil is classified as poisonous (Hager 1998, p. 60). A study by Igarashi and Miyazawa examined the cytotoxic effect of a conjugated, threefold-unsaturated fatty acid (α-eleostearic acid) extracted from tung oil. The result showed that the highest cytotoxicity against DLD-1- (colorectal adenocarcinoma), HepG2- (hepatoma), A549- (pulmonary adenocarcinoma), MCF-7- (breast adenocarcinoma) and MKN-7- (tubular adenocarcinoma, belly) cells was found at concentrations of more than 25/zM. The exact mode of action is still unknown (Igarashi and Miyazawa 2000).

Walnut Oil

Oleum Juglandis synonyms: Walnussöl (D); huile de noix (F)

1 Source Plant

Juglans regia L. (Juglandaceae), walnut tree

S. Krist, *Vegetable Fats and Oils*, https://doi.org/10.1007/978-3-030-30314-3_120

Habitat

The walnut tree originates in Persia, from where the Romans imported it to Asia and to Central, Western and Southern Europe. Today, if the climate allows, it is an avenue tree, grows in parks or is a solitary tree near farms and in gardens. It grows best in sunny, warm areas, alluvial forests, river valleys and deep, rich, loamy soil. Even if the climate is favourable, it takes about 15 years until a tree bears fruit abundantly, and fifty years until it reaches its peak of productivity (Kircher 2002, p. 27 et seq; Fischer and Krug 1984, p. 54; Hager 1978, volume 5, p. 328).

Description

Juglans regia grows to a height of 10–25 m. The crown is rather broad and not very dense. The bark is at first ash-coloured and smooth, and later becomes dark, greyish black, with deep cracks. The leaves are unevenly pinnate, first reddish, then green. The male flowers form long, pendant, green catkins; the female flowers sit in twos or threes at the ends of the branches. The fruit are one-seeded, round stone fruit surrounded by a smooth, fleshy outer shell that is first green and then changes to brown with white dots, and a hard, ligneous, wrinkly inner shell. The seed is 2- to 4-lobed, and its cotyledons are rich in oil (Hager 1978, volume 5, p. 328).

W

2 Cultivation and Extraction

Cultivation

Walnut trees are mainly cultivated in regions with a wine climate in Central and Southern Europe, Central Asia, northern India, China, the Mediterranean area, California and Mexico. They are grown in large, garden-like plantations. The first harvest usually takes place 6–8 years after planting. The trees then bear up to 4000 fruit annually for nearly a century. The main producers of walnuts sold in Europe are France, Italy, Turkey, Greece and Iran (Hager 1978, volume 5, p. 328).

Extraction of the Oil

Walnut oil is cold pressed or extracted from walnut kernels, which are the cotyledons of the embryo, and subsequently usually refined. Before being processed, the nuts have to be stored for 2–3 months. One ton of walnuts yields 500 litres of walnut oil (Hänsel et al. 1999, p. 263; Roth and Kormann 2000, p. 155; Hunnius 1998, p. 741).

3 Character

Colour

Dark yellow to light brown.

Light yellow to greenish yellow (Roth and Kormann 2000, p. 156). Yellow (Hager 1978, volume 7b, p. 210).

Odour

Light, touch of walnuts.

Pleasant (Hänsel et al. 1999, p. 263; Roth and Kormann 2000, p. 156; Hager 1978, volume 7b, p. 210).

Volatile Compounds

Analysis of volatile compounds of walnut oil by means of liquid extraction, liquid and gas chromatography (Bail et al. 2009a; Torres et al. 2005; Del Mar Caja et al. 2000):

Hexanal	2-Hexanal
1-Acetylcyclohexene	Heptanal
Hydrogen carbonate	Octanal
Furancarboxaldehyde	2-Nonenal
2,4-Heptadienal	Decanal
Furfuryl alcohol	2-Decenal
Nonanal	2-Hexanone
2,4-Decadienal	2-Pentylfuran
(E,E)-2,4-Decadienal	2-Octylfuran
n-Pentane	Pentanal
n-Octane	1-Octanol
n-Nonane	1-Heptanol
Ethanol	1-Hexanol
Cyclobutanol	1-Pentanol

Flavour

Nutty (Roth and Kormann 2000, p. 156). Pleasant (Hager 1978, volume 7b, p. 210).

4 Ingredients

Composition of Fatty Acids

Fatty acids	Content in % acc. to Roth and Kormann (2000), p. 156	Content in % acc. to Nature certificate of analysis 28.8.03	Content in % acc. to Kerschbaum and Schweiger (2001), p. 12
Palmitic acid	4–5	7.0	7.8
Linoleic acid	47–83	61.3	62.0
Palmitoleic acid	–	–	0.1
Oleic acid	14–30	16.1	14.9
α-Linolenic acid	3–16	13.2	10.6
Stearic acid	1–2.5	2.0	2.5
Arachidic acid	0.01	–	0.1
Eicosenoic acid	–	–	0.3
Vaccenic acid	–	–	1.1
Myristic acid	0.01	–	–

Composition of Fatty Acids in Walnut Oil from Different Countries of Origin

(Concentration in % acc. to Crews et al. 2005)

Fatty acids	China	France	Hungary	Italy
Palmitic acid	5.1–5.4	6.5–7.3	5.8–7.7	7.3–8.1
Palmitoleic acid	0.1	0.1	0.1	< 0.1
Stearic acid	2.7–2.9	1.7–2.9	2.1–2.2	2.2–2.9
Oleic acid	16.9–21.0	15.1–18.9	17.4–22.2	14.5–15.3
Linoleic acid	60.1–64.1	57.4–64.3	58.3–60.8	60.2–63.1
α-Linolenic acid	10.3–10.4	11.3–15.4	10.8–11.6	11.8–14.3
Arachidic acid	0.1	<0.1	<0.1	0.1
Eicosenoic acid	0.1	0.2–0.3	0.2	0.2

Composition of Fatty Acids at Position 2 of the Triglycerides

(Concentration in % acc. to Crews et al. 2005)

Fatty acids	China	France	Hungary	Italy
Palmitic acid	0.4–0.5	0.4–0.8	0.7–1.0	0.7–0.9
Palmitoleic acid	0.1	<0.1	0.1	< 0.1
Stearic acid	0.2	0.1–0.3	0.2–0.3	0.2–0.3
Oleic acid	18.9–22.8	15.8–23.8	19.0–30.9	16.2–18.8
Linoleic acid	67.3–71.5	65.8–73.5	61.9–69.4	69.9–71.5
α-Linolenic acid	8.9–9.0	9.4–13.1	5.7–10.4	8.6–12.8

Composition of Sterols

(Concentration in mg/100 g and content in %)

Sterols	Walnut oil acc. to Firestone (1999), p. 107	Walnut oil acc. to Karleskind (1996), p. 167
Total content	176	170–245
Cholesterol	–	< 0.5

Sterols	Walnut oil acc. to Firestone (1999), p. 107	Walnut oil acc. to Karleskind (1996), p. 167
Brassicasterol	–	< 0.4
Campesterol	5	4–7
Stigmasterol	–	< 2
β-Sitosterol	89	82–90
Δ5-Avenasterol	5	4–7
Δ7-Stigmasterol	–	<4
Δ7-Avenasterol	–	< 1
Ergosterol	–	< 0.3
Fucosterol	–	< 1

Composition of Sterols in Walnut Oil from Different Countries of Origin

(Concentration in % acc. to Crews et al. 2005)

Sterols	China	France	Hungary	Italy
Cholesterol	0.2–0.3	0.0–0.6	0.0–0.3	0.2–1.2
Brassicasterol	–	0.0–0.1	–	–
24-Methylencholesterol	0.3	0.0–0.2	–	–
Campesterol	4.0–4.5	4.3–5.0	4.3–4.6	4.4–5.3
Stigmasterol	0.4–0.5	0.0–0.4	0.0–0.2	0.0–0.3
Clerosterol	0.8	0.7–1.5	1.4–2.2	1.2–3.1
β-Sitosterol	83.6–85.4	85.2–89.6	85.0–86.4	83.9–86.7
Δ5-Avenasterol	8.2–9.2	4.4–7.7	6.7–8.4	5.1–7.1
Δ7-Stigmastenol	0.0–0.2	0.0–0.2	0.0–0.6	0.0–0.7
Δ7-Avenasterol	0.0–0.1	0.0–0.2	–	–

Composition of Tocopherols

Tocopherols	Content in mg/100 g acc. to Kerschbaum and Schweiger (2001), p. 36
Total content	36.5
α-Tocopherol	10.5
β-Tocopherol	–
γ-Tocopherol	18.5
Δ-Tocopherol	7.4

Composition of Tocopherols in Walnut Oil from Different Countries of Origin

(Concentration in mg/kg acc. to Crews et al. 2005)

Tocopherols	China	France	Hungary	Italy
Total content	309–366	383–507	460–569	358–460
α-Tocopherol	10	0.0–16	0.0–19	0.0–10
β-Tocopherol	–	–	–	–
γ-Tocopherol	276–329	308–447	424–525	318–445
Δ-Tocopherol	23–27	27–63	20–34	12–30

Other Ingredients

(Kerschbaum and Schweiger 2001, p. 14, 43)

Vitamin A
Vitamin B
Vitamin E
Vitamin K
Lecithin
Phosphorus

Physical Key Figures of Walnut Oil

refractive index: n_D^{40} 1.4690–1.4710
density: d_4^{20} 0.920–0.924
saponification number: 188–196
acid number: not over 8
iodine number: 141–153
unsaponifiable: 0.2–0.4%
point of solidification: −28 to −29 °C

(Roth and Kormann 2000, p. 118)

W

Shelf Life

If stored in a cool, dark place, the oil has a shelf life of up to 12 months (Roth and Kormann 2000, p. 156).

5 Application

In Pharmacy and Medicine

With regard to the composition of fatty acids of *Oleum Juglandis*, its high content of mono- and polyunsaturated fatty acids is conspicuous. In addition to oleic and linolenic acid, there is a high amount of linoleic acid. Walnut oil can thus be regarded as a dietary edible oil. It lowers the triglyceride level and significantly increases the HDL cholesterol level in the blood. It lowers the risk of arteriosclerosis and has a positive influence on lipid metabolism. It also has pronounced fungicidal and disinfectant properties, which makes it an ideal ointment base. It supports the healing process of the actual active components in cases of mycosis, decubitus and other open wounds (Hänsel et al. 1999, p. 263; Zibaeenezhad et al. 2003).

In Cosmetics

Although walnut oil is a very fat oil, it penetrates the skin quickly and is therefore used in care cosmetics for the following purposes (Roth and Kormann 2000, p. 156; Hager 1978, volume 7b, p. 210; Kircher 2002, p. 77 et seq):

- as a base for ointments and creams;
- in shampoo, and
- in lipid-replenishing shower gels.

Juglandis nucum infusum is an oily extract of green, unripe fruit shells of the walnut, which contains hydrojuglone. The extract is made by infusion. In cosmetics, it is a tanning lotion and a sunscreen (Hänsel et al. 1999, p. 263; Roth and Kormann 2000, p. 156; Hager 1978, volume 7b, p. 210; Hunnius 1998, p. 741).

Technical Uses

Although walnut oil is primarily used for culinary purposes, it has technical uses as well, for example in soap production, as a dispersant for artists' paints and to impregnate wood (Roth and Kormann 2000, p. 156; Hager 1978, volume 7b, p. 210; Kerschbaum and Schweiger 2001, p. 14).

As Food

Walnut oil should not be heated because of its high amount of unsaturated fatty acids, and is thus mainly used as a salad oil (Hänsel et al. 1999, p. 263; Roth and Kormann 2000, p. 156; Hager 1978, volume 7b, p. 210; Hunnius 1998, p. 741).

In Folk Medicine

Probably due to its fungicidal, disinfectant and anti-arteriosclerotic properties that *Oleum juglandis* is used to treat (Kircher 2002, p. 77 et seq; Kerschbaum and Schweiger 2001, p. 47):

- liver and gall bladder complaints;
- metabolic disorders;
- pruritus ani;
- bee stings;
- athlete's foot;
- skin irritations;
- clavi;
- oral herpes;

- lacerated corners of the mouth;
- warts, and
- respiratory diseases.

6 Possible Unwanted Side Effects

For a long time, the walnut and its oil were regarded as food allergens, to which patients with a pollinosis caused by early-flowering trees are especially sensitive. In rare cases, patients suffering from a walnut allergy are also sensitised to birch pollen. Correspondingly, no allergens similar to Bet-v-1 could be found in walnuts (Senti et al. 2000).

Wheat-Germ Oil

Oleum Tritici synonyms: Weizenkeimöl (D); huile de germe de ble (F)

1 Source Plant

Triticum aestivum L. (Poaceae), wheat

S. Krist, *Vegetable Fats and Oils*, https://doi.org/10.1007/978-3-030-30314-3_121

Habitat

Wheat is one of the oldest cultivated plants of humanity. It is only known in its cultivated form. Its origins probably lie in northern Persia (Roth and Kormann 2000, p. 107). The various species of wheat are the second most often cultivated crop worldwide (after maize, but before rice). It is cultivated on all continents. Wheat prefers moderate climes; it is also cultivated, on a smaller scale, in subtropical zones, but not in the tropics. There is spring wheat and winter wheat; the latter tolerates a minimum temperature of −22 °C. In Central Europe, winter wheat is the variety usually cultivated (Roth and Kormann 2000, p. 107).

Description

Wheat is a grass that is 60–120 cm tall and has parallel-veined leaves, which are about 30 cm long and pointed. Its spikes usually consist of four flowers; they have four sides and may be awned. Wheat blooms from June to July (Roth and Kormann 2000, p. 107). It can be distinguished from barley by its short, fringed auricles that are not stem-clasping. The fruit are one-seeded achenes (caryopses).

These fruit are golden-yellow to brownish, ovate grains, which have a deep longitudinal furrow and a bunch of hair at the crest. Within lies the embryo, which contains 7–12% fat oil. Good wheat consists of fully ripe, dry, light-coloured grains of similar size. Besides 2% fat oil, the grains contain about 70% carbohydrates (starch, sugar) and about 12% protein (Roth and Kormann 2000, p. 107 et seq).

2 Cultivation and Extraction

Cultivation

Wheat makes higher demands on the climate, soil and water supply than other crops. It needs heavy, rich soil (loam or black earth) with high field capacity. Winter wheat is sown in late autumn and harvested in July. Spring wheat, on the other hand, is sown in early spring, and the harvest takes place in August. Breeding has created productive, high-quality varieties, so that today, wheat can be cultivated in nearly all climes (Roth and Kormann 2000, p. 107). The following table gives an overview of the 15 most successful producers worldwide in 2005:

The 15 most important producing countries in 2005 (faostat.fao.org/01.02.07)	
2005	Production (tons)
Global production	629,566
China	96,340
India	72,000
USA	57,106
Russia	47,608
France	36,922

The 15 most important producing countries in 2005 (faostat.fao.org/01.02.07)	
2005	Production (tons)
Canada	25,547
Australia	24,067
Germany	23,578
Pakistan	21,591
Turkey	21,000
Ukraine	18,700
Argentina	16,000
Great Britain	14,950
Iran	14,500
Kazakhstan	11,070

Extraction of the Oil

Wheat-germ oil is a by-product of the production of wheat flour. Extraction of the oil only became profitable when methods were found to separate the embryo, which is relatively rich in fats, by mechanical means. Modern methods of grinding are designed in such a way that not only flour and bran, but also the embryos accumulate as a separate fraction (Hänsel et al. 1999, p. 264). The embryos are pressed in hydraulic presses to obtain the oil or the oil is extracted with solvents. The oil may be refined subsequently. Another method is extraction without solvents, using supercritical CO_2; in this case, the oil does not have to be refined afterwards (Panfili et al. 2003).

3 Character

Colour

Raw wheat-germ oil is golden-yellow to brown in colour.
Refined oil is yellow (Roth and Kormann 2000).

Odour

Like grain, dried herbs, slightly peppery.

Flavour

Typical, distinct taste of wheat.

4 Ingredients

Composition of Fatty Acids

Fatty acids	Content in % acc. to Panfili et al. (2003)	Content in % acc. to Firestone (1999), p. 108	Content in % acc. to Rossell and Pritchard (1991), p. 295	Content in % acc. to Belitz and Grosch (1987), p. 611
Myristic acid	–	0–0.2	–	–
Palmitic acid	18.0	12–20	19	17
Palmitoleic acid	1.3	0.2–0.5	–	–
Stearic acid	1.3	0.3–3	1	1
Oleic acid	21.6	13–23	17	20
Linoleic acid	51.3	50–59	56	52
α-Linolenic acid	6.4	2–9	6	10
Arachidic acid	–	0.3	0.3	–
Eicosanoic acid	–	0.3	0.3	–
Behenic acid	–	0–0.1	–	–
Cetoleic acid	–	0.3	0.3	–
Lignoceric acid	–	0–1	–	–

Composition of Sterols

W

(Concentration in mg/100 g and content in %)

Sterols	Wheat-germ oil acc. to Malecka (2002)	Wheat-germ oil acc. to Lercker et al. (1977)	Refined wheat-germ oil acc. to Carstensen (2001), p. 43	Raw wheat-germ oil acc. to Carstensen (2001), p. 43	Wheat-germ oil acc. to Firestone (1999), p. 109
Total content	–	–	1484.6	2121.5	550
Cholesterol	–	–	0.3	–	–
Brassicasterol	–	0.4	–	–	0–0.4
Campesterol	22.2	29.0	25.4	29.5	19–29
Stigmasterol	2.0	1.0	1.5	1.6	0.3–4
β-Sitosterol	67.1	63.8	62.8	62.1	56–67
Δ5-Avenasterol	6.2	2.0	1.9	4.1	2–6
Δ7-Stigmasterol	–	1.1	1.5	0.7	1–4
Δ7-Avenasterol	–	2.0	1.9	1.0	2
Cholestanol	–	–	–	–	2
Campestanol	–	–	–	–	1–2

Comparison of the Composition of Sterols in Oils made from Spring and Winter Wheat

(Homberg and Bielefeld 1990)

Sterols	Content in % in spring wheat-germ oil	Content in % in winter wheat-germ oil
Cholesterol	0.3	0.3
Campesterol	30.1	27.2
Stigmasterol	0.7	0.7

Sterols	Content in % in spring wheat-germ oil	Content in % in winter wheat-germ oil
β-Sitosterol	58.8	63.5
Δ5-Avenasterol	7.0	4.6
Δ7-Stigmasterol	0.9	1.1
Δ7-Avenasterol	1.8	2.0

Composition of Tocopherols and Tocotrienols

(Concentration in mg/100 g and content in %)

Tocopherols/ tocotrienols	Wheat-germ oil acc. to Panfili et al. (2003)	Refined wheat-germ oil acc. to Carstensen (2001), p. 45	Raw wheat-germ oil acc. to Carstensen (2001), p. 45	Wheat-germ oil acc. to Firestone (1999), p. 109
Content of tocopherols	217.3	241.3	591.5	135–310
α-Tocopherol	163.0	59.3	64.3	45–310
β-Tocopherol	54.3	20.3	26.6	20–115
γ-Tocopherol	–	0.9	0.2	1.8–95
Δ-Tocopherol	–	0.4	0.3	2–10
Content of tocotrienols	11.4	57.2	55.3	2–40
α-Tocotrienol	2.9	–	3.2	1–20
β-Tocotrienol	8.5	19.2	5.4	1–20

Other Ingredients

(Carstensen 2001, p. 43)

Squalene
Sterols
Plant hormones
Lecithin
Amino acids
Provitamin A

Physical Key Figures of Wheat-Germ Oil

refractive index: 25 °C 1474–1483
40 °C 1469–1478
density: 15.5 °C 0.928–0.938
25 °C 0.925–0.933
iodine number: 115–128
saponification number: 179–190
190–196

unsaponifiable: 2–5%
peroxide number: 7.0
(Firestone 1999, p. 108; Karleskind 1996, p. 182)

Shelf Life

Wheat-germ oil has a shelf life of about half a year if stored in a cool, dark place.

If the oil is extracted with supercritical gas, it keeps for longer. CO_2 and high pressure lead to a decrease in activity of the enzyme peroxidase or lipoxygenase, as well as inhibiting microorganisms and delaying rancidity by diminishing contact with oxygen (Panfili et al. 2003).

5 Application

In Pharmacy and Medicine

As a result of its natural active ingredients, wheat-germ oil is used to fill soft gelatine capsules and to produce ointments and powders (Carstensen 2001, p. 3).

In addition, feeding animals with wheat-germ oil resulted in lower cholesterol levels of the plasma and liver (Panfili et al. 2003).

In Cosmetics

In Pregnancy

Percutaneously applied preparations containing vitamin E improve blood circulation in peripheral vessels and are recommendable in cases of skin badly supplied with blood. Wheat-germ oil is suitable as an ingredient of the oil used for perineum massage of pregnant women, which is intended to improve blood circulation and tissue elasticity. It can also be used for belly care and to prevent stretch marks, which may form because of the extreme stretching of the connective tissue during pregnancy. It helps to strengthen and stabilise the connective tissue.

Skincare and Hair-Care

Wheat-germ oil is a good skin care oil to treat symptoms of ageing skin, as it moisturises the skin and regulates the metabolism. It has a positive influence on the production of new cells for dry and ageing skin prone to wrinkles. Products containing wheat-germ oil are also suitable to keep hair healthy; they are ideal for scalps poorly supplied with blood.

The Cosmetics Industry

In the cosmetics industry, wheat-germ oil is useful for the production of skin oils, powder, soap, face packs (Roth and Kormann 2000, p. 156) and other care cosmetics products (Carstensen 2001, p. 3). In addition, it is a good base for essential oils (Ulmer 1996, p. 82).

As Food

Wheat-germ oil is recommendable as part of a healthy diet as a result of its good effects. It can be used as a salad oil and for other cold dishes. Due to its high tocopherol content, it is a popular dietary oil. It is also a source of natural vitamin E and unsaturated fatty acids (Roth and Kormann 2000, p. 156).

White Lupin Seed Oil

synonyms: Öl der Öllupine (D); huile de lupin (F), olio di semi di lupino bianco (I)

1 Source Plant

Lupinus albus L. (Fabaceae), white lupin, field lupine

S. Krist, *Vegetable Fats and Oils*, https://doi.org/10.1007/978-3-030-30314-3_122

Habitat

The white lupin is a native plant of the Mediterranean area. It was domesticated in antiquity (Schuster 1992, p. 90).

Description

In comparison to other lupins, *Lupinus albus* has a high oil content. It has a strong taproot with many lateral roots, which are covered with nodules. The shoot can grow to a height of one metre; before blossoming, there is only one shoot, but afterwards, the plant produces 3–4 side shoots, which in turn also produce shoots after blooming. In moist conditions, the plant can therefore grow in 3–5 levels. The leaves are large, 5- to 9-pinnate and hairy. Due to their size, the upper leaves cast a shadow, which is why the bottom leaves often die back. The flowers appear in whorls at the end of the shoots. The petals are either white or white with blue ends. The pollen is usually released straight away when the flowers open. Autogamy is thus prevalent, but exogamy by insects like bees or bumblebees is also possible. The fruit is a legume about 6–11 cm long. It is yellow and hairy; cultivated varieties have legumes that do not burst open. Each legume contains 3–6 white or yellow seeds. Depending on the variety, the thousand-seed weight is between 250 and 800 g. Small seeds tend to be roundish, whereas larger seeds have a squarish form (Hackbarth 1944, p. 281–281; Schuster 1992, p. 89–90).

W

2 Cultivation and Extraction

Cultivation

Lupinus albus was first cultivated in the sixteenth and seventeenth centuries as a medical plant. In the nineteenth century, it was used for green manuring on loamy, sandy soil in Eastern Europe. It was only in the 1st and 2nd World Wars that it gained importance as a source of protein and oil. Cultivation stopped after the 2nd World War because the harvest was endangered by plant diseases. However, interest in the white lupin continued. In 1984, 3 million hectares were cultivated worldwide, 0.96 million of which were grown for their corn (Schuster 1992, p. 90–91).

The white lupin needs warm, dry conditions to thrive. If the conditions are too wet, it never fully ripens, as it keeps on producing new shoots. It does not make many demands on the soil; loamy, sandy soil is well suited. The seeds should be sown as early in the year as possible, between early March and early April, so that the seeds are ripe in the summer. The white lupin tolerates night frosts down to −6 °C. 200–250 kg of seeds are required per hectare; they should be sown at a distance of about 25 cm from each other, to limit the number of side shoots. *Lupinus albus* tends to become infested with the fungus *Ceratophorum setosum*, especially in wet conditions. The time of harvest depends on the plants' development. If there are not many side shoots, the harvest takes place when all the legumes have turned brown. The harvest should not be delayed to wait for more side shoots, because it might be too late by then to harvest. Usually, the harvest takes place from late August to early September. The yield amounts to between 2500 and 4000 kg/ha (Hackbarth 1944, p. 283–284).

White lupins are harvested with a combine harvester. It is important to ensure gentle threshing, for example by reducing the rotation speed of the drum and leaving the threshing concave open. The harvest must be dried immediately to a moisture content of 12–14% (Löw 2003, p. 55).

In order to use the white lupin as an oil plant efficiently, several breeding goals still have to be achieved. By making the plant more adaptable to environmental conditions, better yields would be possible. Higher potential yields, i.e. more legumes per plant, would be desirable as well. The quality of the seeds could also be improved, i.e. seeds with a low content of alkaloids (below 0.01%) that could be used as forage, and with a high oil content of more than 20% in order to improve oil yield (Schuster 1992, p. 92).

The main alkaloids of the white lupin are the quinolizidine alkaloids sparteine and lupanine. They are the reason for the bitter taste of the seeds and cause respiratory problems and liver damage (Erbas et al. 2005).

Older varieties contain bitter alkaloids, which is why the seeds could not be used as forage. New varieties contain only 0.01–0.04% of these alkaloids (Löw 2003, p. 54).

Extraction of the Oil

To obtain white lupin seed oil, the seeds are deshelled and ground. The oil is then extracted with hexane in a Soxhlet extractor (Boschin et al. 2008). The seeds can also be pressed (http://www.cosmeticanalysis.com 26.3.2018).

3 Character

Colour

Brown (Hackbarth 1944, p. 284).

Odour

No characteristic smell detectable (Hackbarth 1944, p. 284).

4 Ingredients

Composition of Fatty Acids

Fatty acids	Content in % acc. to Schuster (1992), p. 10	Content in % acc. to Uzun et al. (2007)	Content in % acc. to Boschin et al. (2007)	Content in % acc. to Erbas et al. (2005)	Content in % acc. to Boschin et al. (2008)
Myristic acid	–	–	–	–	–
Palmitic acid	6–10	7.6	14.44–21.57	11.6	15.17–19.85
Palmitoleic acid	–	–	0.36–1.03	–	0.29–0.80
Stearic acid	6–9	1.5	2.48–3.91	1.9	1.34–3.56
Oleic acid	50–70	47.6	42.78–52.87	55.4	40.76–50.46
Linoleic acid	18–22	20.3	9.20–17.23	22.4	7.79–15.81
Linolenic acid	<2.5	9.2	4.81–9.02	8.7	5.31–10.36
Arachidic acid	–	3.5	1.61–2.30	–	1.15–1.99
Eicosenoic acid	–	–	3.86–5.30	–	4.57–6.93
Behenic acid	–	–	4.75–5.99	–	4.81–6.56
Erucic acid	–	–	0.51–1.47	–	0.78–3.52

Composition of Sterols

Sterols	Content in % acc. to Hamama and Bhardwaj (2004)	Content in % acc. to Blaschek et al. (1998), p. 109
Total content	199–287 g/kg	–
Fucosterol	–	43.0
Campesterol	22.5–29.0	22.0
Stigmasterol	7.1–14.0	0–3
β-Sitosterol	50.0–60.0	31.0
Δ5-Avenasterol	0.8–3.0	–
Δ7-Stigmasterol	0.9–2.7	–

Composition of Tocopherols

Tocopherols	Content in % acc. to Lampart-Szczapa et al. (2003)
Total content	1.86 g/kg
α-Tocopherol	2.8
β-Tocopherol	–
γ-Tocopherol	86.1
δ-Tocopherol	11.1

Other Ingredients

Phospholipids	Content in % acc. to Hamama and Bhardwaj (2004)
Total content	26–28 g/kg
Lysophosphatidylcholine	1.9–5.1
Phosphatidylcholine	32.6–46.3
Phosphatidylserine	11.5–17.8
Phosphatidylinositol	1.8–6.5
Phosphatidylethanolamine	21.4–37.5
Phosphatidic acid	2.8–4.6
Phosphatidylglycerol	1.9–3.7
Diphosphatidylglycerol	0.9–2.3

Physical Key Figures

specific weight (20 °C):	0.920
point of solidification:	–9 °C
refractive index: n^{20}_{D}	1.4742
saponification number:	185.0–192.8
iodine number:	61.6–83.2
unsaponifiable:	0.98–1.14%

(Hackbarth 1944, p. 284)

5 Application

In Pharmacy and Medicine

The phytosterols contained in white lupin seed oil have positive health effects. β-Sitosterol, in particular, is thought to lower blood pressure, the cholesterol levels of the serum and the risk of coronary heart diseases (Hamama and Bhardwaj 2004).

Industrial Uses

The oil is used for margarine and soap production (Hackbarth 1994, p. 284).

In Cosmetics

The fat oil of the white lupin can be used for various cosmetic purposes; it is added to conditioners and has good skin care qualities. As an emollient, it keeps the skin soft and supple (Bährle-Rapp 2007, p. 331).

Abbrevations

AA	Arachidic acid
AEDA	Aroma extract dilution analysis
CM	Extraction with chloroform/methanol (50:50)
CNSL	Cashew nut shell liquid
CP	Cold pressed
DHA	Docosahexaenoic acid
GC	Gas chromatography
GC-MS	Gas chromatography-mass spectrometry
GCOH	Gas chromatography-olfactometry of headspace samples
H	Extraction with hexane
HPLC	High-performance liquid chromatography
HS-SPME	Headspace solid-phase microextraction
I.U.	International units
KOH	Potassium hydroxide
L	Linoleic acid
Ln	Linolenic acid
M	Myristic acid
MID	Maximal individual dose
14-MHD	14-methylhexadecenoic acid
MS	Mass spectroscopy
MDD	Maximal daily dose
O	Oleic acid
P	Palmitic acid
ref.	Refined
S	Stearic acid
SFE	Supercritical fluid extraction
SPME	Solid-phase microextraction

S. Krist, *Vegetable Fats and Oils*, https://doi.org/10.1007/978-3-030-30314-3

Picture Credits

Dr. Werner Huber
Univ.-Prof. Dr. Gerhard Gottsberger
Franz Krist (†)
Univ.-Prof. Mag.pharm. DDr. Sabine Krist
Univ.-Doz. Dr. Reinhard Länger
Dr. Wolfram Lobin
Eliot Masters
Andrea Müller
Univ.-Prof. Dr. Anton Weber
Herbert Zimmermann

S. Krist, *Vegetable Fats and Oils*, https://doi.org/10.1007/978-3-030-30314-3

About the Author

Univ.-Prof. Mag.pharm. Dr.rer.nat. Dr.phil. Sabine Krist studied pharmacy at the University of Vienna. After her doctoral studies in the natural sciences and in philosophy, she qualified as a professor in the subject of pharmaceutical chemistry at the University of Vienna in 2010. Since 2016, she holds a chair at the Faculty of Medical Chemistry, Sigmund Freud University Vienna.

Her main research areas include volatile compounds in pharmaceutical/medical chemistry and analytical chemistry (device-related, sensory). She has received several science awards, namely, Kardinal-Innitzer-Förderungspreis (2009), Czedik-Eysenberg-Preis of the Gesellschaft Österreichischer Chemiker (2010), Young Investigator Award of the Faculty of Life Sciences/University of Vienna (2010), Dr. Maria Schaumayer Stiftung Würdigungspreis (2012) and Science Award 2015 of the Country Lower Austria.

She has written numerous professional articles and books such as *Die Erforschung der chemischen Sinne* (Lang, Vienna, 2006) and *Volatile Compounds: The Utilisation of Volatile Compounds in the Characterisation of Vegetable Oils and Fats and in Reducing the Bacterial Count of Ambient Air* (Lang, Vienna 2011). Lexikon der pflanzlichen Fette und Öle (Springer, Vienna 2013).

S. Krist, *Vegetable Fats and Oils*, https://doi.org/10.1007/978-3-030-30314-3

Bibliography

Abd Elreheim F.; *A method für preparation of pharmaceutical products from cashew nutshell oil as a treatment for pilosebaceous follices disorders e.g. acne vulgari, dandruff and psoriasis*, Patent No. EG 21971, Egypt, **2002**

Abdurrasheed, K. M.; Janardanan, C.; *Chemical Composition of Nutmeg and Mace (Myristica fragrans Houtt.) from Tellicherry and Kannur Regions, Kerala*, Journal of Spices and Aromatic Crops, **18**(2), 108–110, **2009**

Abreu F. R.; Lima D. G.; Hamu E. H.; Wolf, C.; Suarez Paulo A. Z.; *Utilization ofmetal complexes as catalysts in the transesterification of Brazilian vegetable oils with different alcohols*, Journal of molecular catalysis, **209**, 29–33, **2004**

Adikari, S.; Huq, F.; Begum, M.; Saha, G.; *Studies on Coriandrum sativum Linn. Part II-Chemical investigation on corianderseed oil*$_L$ Inst. Food Sei. Technol., Bangladesh Journal of Scientific and Industrial Research, **36**(1–4), 33–40, **1991**

Aepli, B.; *Rapsöl – gesund geniessen*, Fona-Verlag, Lenzburg, **2003**

Afolabi, I. S.; Akuiyibo, S. M.; Rotimi, S. O.; Adeyemi, A. O.; *In Vivo Evaluation of Lipid and Antioxidans Qualities of Carica Papaya Seed Oil*, **4**, 125–135, **2011**

Ahamed, N. T.; Singhal, R. S.; Kulkarni P. R.; Pal, M.; *A lesser-known grain, Chenopodium quinoa: Review of the chemical composition of its edible parts*, Food and Nutrition Bulletin, **19**(1), 61–70, **1998**

Aitzetmüller, K.; *Chilean hazelnut (Gevuina avellana) seed oil. Comments*, Journal of the American Oil Chemists Society, **81**(7), 721–723, **2004**

Aitzetmüller, K.; Xin, Y.; Werner, G.; Groenheim, M.; *High-performance liquid chromatographic investigations of stillingia oil*, Journal of Chromatography, **603**(1–2), 165–173, **1992**

Ajiwe, V. I. E.; Okeke, C. A.; Agbo, H. U.; Ogunleye, G. A.; Ekwuozor, S. C.; *Extraction, Characterization and Industrial Uses of Velvet-tamarind, Physic-nut and Nickernut Seed Oils*, Bioresource Technology, **57**, 297–299, **1996**

Akihisa, T.; Tokuda, H.; Ukiya, M.; Suzuki, T.; Enjo, F.; Koike, K.; Nikaido, T.; Nishino, H.; *3-Epicabraleahydroxylactone and other triterpenoids from camellia oil and their inhibitory effects on epstein-barr virus activation*, College of Science and Technology, Nihon Univerity, Chemical & Pharmaceutical Bulletin, **52**(1), 153–156, Tokyo, Japan, **2004**

Akihisa, T.; Yasukawa, K.; Kimura, Y.; Takase, S.; Yamanouchi, S.; Tamura, T.; *Triterpene alcohols from camellia and sasanqua oils and their anti-inflammatory effects*, College of Science and Technology, Nikon Univerity, Chemical & Pharmaceutical Bulletin, **45**(12), 2016–2023, Tokyo, Japan, **1997**

Ako, H.; Okuda, D.; Gray, D.; *Healthful new oil from macadamia nuts*, Department of Environmental Biochemistry, University of Hawaii, Nutrition, **11(3)**, 268–268, **1995**

S. Krist, *Vegetable Fats and Oils*, https://doi.org/10.1007/978-3-030-30314-3

Al-Jabre S.; Al-Akloby O. M.; Al-Qurashi A. R.; Akhtar N.; Al-Dossary A.; Randhawa M. A.; Thymoquinone, an active principle of Nigella sativa, inhibited Aspergillus niger. Pakistan Journal of Medical Research, **42**(3), 102–104, **2003**

Alander, J.; *Shea butter-a multifunctional ingredient for food and cosmetics,* Lipid Technology, **16**(9), 202–205, **2004**

Alarcon De la Lastra, C.; Barranco, M. D.; Motilva, V.; Herrerias, J. M.; *Mediterranean diet and health: biological importance of olive oil,* Departamento de Farmacologia, Facultad de Farmacia, Seville, Spain, Current Pharmaceutical Design **7**(10), 933–950, **2001**

Ali, A. R.; Moi, L. M.; Fisal, A.; Nazaruddin, R.; Sabariah, S.; *The Application of Borneo Tallow in Dark Chocolate Shell,* Journal of the Science of Food and Agriculture, **76**, 285–288, **1998**

Alvarez, A. M.; Rodriguez, M. L.; *Lipids in pharmaceutical and cosmetic preparations*, Universität von Sevilla, Grasas y Aceites, Sevilla, **51** (1–2), 74–96, **2000**

Andersen, F. A.; Bergfeld, W. F.; Belsito, D. V.; Hill, R. A.; Klaassen, C. D.; Liebler, D. C:; Marks, J. G.; Shank, R. C.; Slaga, T. J.; Snyder, P. W.; *Final Report of the Cosmetic Ingredient Review Expert Panel Amended Safety Assessment of Calendula officinalis – Derived Cosemtic Igredients*, International Journal of Toxicology, **29**(4), 221–243, **2010**

Ando, H.; Chen, Y.; Tang, H.; Shimizu, M.; Watanabe, K.; Mitsunga, T.; *Food Components in Fractions of Quinoa Seed,* Food Science & Technology Research, **8**(1), 80–84, **2002**

Andrews, H. J.; *Uses of oiticica oil,* Paint Manufacture, **42**(7), 21,**1972**

Andrianaivo-Rafehivola A. A.; Gaydou, E. M.; *Effects offresh and heated Baobab seed oil feeding on growth, food consumption and weight ofsome organs in rats,* Revue Francaise des Corps Gras, **41**(3–4), 53–59, **1994**

Andrianaivo-Rafehivola, A. A.; Siess, M. H.; Gaydou, E. M.; *Modification of hepatic drug metabolizing enzyme activities in rats fed baobab seed oil containing cyclopropenoid fatty acids,* Food Chem.Toxicol., **33**(5), 377–82, **1995**

Andrich, G.; Balzini, S.; Zinnai A.; De Vitis, V.; Silvestris, S.; Venturi, F.; Fiorentini, R.; *Supercritical fluid extraction in sunflower seed technology,* European Journal of lipid science and technology, **103,** 151–157, **2001**

Angelini, L. G.; Moscheni, E.; Colonna, G.; Belloni, P.; Bonari, E.; *Variation in Agronomic Characteristics and Seed Oil Composition of New Oilseed Crops in Central Italy*, Industrial Crops and Products, **6**, 313–323, **1997**

Angerosa, F.; *Influence of volatile compounds on virgin olive oil quality evaluated by analytical approaches and sensor panels,* Instituto Sperimentale per la Elaiotecnica, Italy, European Journal of Lipid Science and Technology, **104** (9–10), 639–660, **2002**

Angers, P.; Morales, M. R.; Simon, J. E.; *Basil Seed Oils*, In: J. Janick, Progress in new Crops, ASHS Press, 598-601, **1996a**

Angers, P.; Morales, M. R.; Simon, J. E.; *Fatty Acid Variation in Seed Oil Among Ocimum Species,* Journal of the American Oil Chemists' Society, **73** (3), 393–395, **1996b**

Anonym; *Madia sativa seed from South Africa,* Bulletin of the Imperial Institute, London, **13,** 344–346, **1915**

Anonym; *Myrtle wax of South America,* Bulletin of the Imperial Institute, London, **16**(3), 287–289, **1918**

Anonym; *Color additives. Carrot oil,* Federal Register 32 10980, **1967**

Anwar, F.; Bhanger, M. I.; *Analytical Characterization of Moringa oleifera Seed Oil Grown in Temperate Regions of Pakistan,* Journal of Agricultural and Food Chemistry, **51,** 6558–6563, **2003**

Appleqvist, L.-A.; *Lipids in Cruciferae VI: The Fatty Acid Composition of Seeds of Some Cultivated Brassica Species and of Sinapis alba L.*; Fette Seifen Anstrichmittel, **72**(9), 783–792, **1970**

Appleqvist, L.-A.; *Composition of Seeds of Cruciferous Oil Crops*, Journal of the American Oil Chemists' Society, **48**, 851–859, **1971**

Aquino, L. P.; Borges, S. V.; Queiroz, F.; Antoniassi, R.; Cirillo, M. A.; *Extraction of Oil from Pequi Fruit (Caryocar Brasiliense, Camb.) Using Several Solvents and Their Mixtures*, Grasas y Aceites, **62**(3), 245–252, **2011**

Arterburn, L. M.; Boswell, K. D.; Koskelo, E.-K.; Kassner, S. L.; Kelly, C.; Kyle, D. J.; *A Combined Subchronic (90-Day) Toxicity and Neurotoxicity Study of a Single-cell Source of Docosahexaenoic Acid Triglyceride (DHASCO®-Oil),* Food and Chemical Toxicology, **38,** 35–49, **2000**

Asset, G.; Bauge, E.; Wolff, R. L.; Fruchart, J. C.; Dallongeville, J.; *Pinus pinaster Oil Affects Lipoprotein Metabolism in Apolipoprotein E-Deficient Mice,* Journal of Nutrition, **129,** 1972–1978, **1999**

Athar, M.; Nasir, S. M.; *Taxonomic perspective of plant species yielding vegetable oils used in cosmetic and skin care products*, African Journal of Biotechnology, **4** (1), 36–44, **2005**

Aufhammer, W.; *Getreide- und andere Körnerfruchtarten*, Eugen Ulmer GmbH & Co, Stuttgart, **1998**

Aufhammer, W.; *Rohstoff Getreide*, Eugen Ulmer GmbH & Co, Stuttgart, **2003**

Austria Codex, *Fachinformation,* Österreichische Apotheker-Verlagsgesellschaft M.B.H. Österreich, **2004**

Averina, S.; Seewald, G.; Müller, R. H.; Radaneva, L. D.; Popov, D. V.; *Nanostructured Lipid Carriers (NLC) on the Basis of Siberian Pine (Pinus Sibirica) Seed Oil*, Pharmazie, **65**, 25–31, **2010**

Awasthi, Y. C.; Bhatnagar, S. C.; Mitra, C. R.; *Chemurgy of Sapotaceous plants: Madhuca species of India,* Util.Res.Lab., Natl. Bot. Gard., Lucknow, India, Economic Botany, **29**(4), 380–389, **1975**

Axtell, B. L.; *Minor Oil Crops*, Food and Agriculture Organization of the United Nations, 94, **1992**

Axtell, B. L.; *Minor oil crops,* FAO Agricultural Service Bulletin, **1994**

Axtell, B. L.; Fairman, R. M.; *Minor oil crops Part I-Edible oils, Part II-Non-edible oils, Part III-Essential oils,* FAO Agricultural Services Bulletin, **94**, Rom,**1992**

Ayerza, R.; *Oil Content and Fatty Acid Composition of Chia (Salvia hispanica L.) from Five Northwestern Locations in Argentinia*, Journal of the American Oil Chemists' Society, **72** (9), 1079–1081, **1995**

Ayerza, R.; Coates, W.; *An ω-3 Fatty Acid Enriched Chia Diet: Influence on Egg Fatty Acid Composition, Cholesterol and Oil Content*, Canadian Journal of Animal Science, **79**, 53–58, **1999**

Ayerza, R.; Coates, W.; *Ground Chia Seed and Chia Oil Effects on Plasma Lipids and Fatty Acids in the Rat,* Nutrition Research, **25**, 995–1003, **2005**

Ayerza, R.; Coates, W.; Lauria, M.; *Chia Seed (Salvia hispanica L.) as an ω-3 Fatty Acid Source for Broilers: Inflluence on Fatty Acid Composition, Cholesterol and Fat Content of White and Dark Meats, Growth Performance, and Sensory Characteristics,* Poultry Science, **81**, 826–837, **2002**

Babadzhanov, S. N.; Khadzhiew, A. K.; *Study of the effect of tomato oil on the cholesterol and phospholipid content of blood during experimental atherosclerosis,* Meditsinskii Zhurnal Uzbekistana, **5,** 54–56, **1982**

Babayan, V. K.; *Medium Chain Triglycerides and Structured Lipids*, Lipids, **22**(6), 417–420, **1987**

Badami, R. C.; Alagadawi, K. R.; *Minor seed oils XVIII: Physicochemical characteristics and component acids ofeight seedoils,* Dep. Chem., Karnatak, Univ. Dharwad, India, Fette, Seifen, Anstrichmittel, **85**(5), 197–198, **1983**

Badami, R. C.; Daulatabad, C. D.; *The Component Acids of Carica Papaya (Caricaceae) Seed Oil,* Journal of the Science of Food and Agriculture, **18**, 360–361, **1967**

Badami, R. C.; Morris, L. J.; *The Oxygenated Fatty Acid of Calendula Seed Oil*, The Journal of the American Oil Chemists' Society, **42**, 1119–1121, **1965**

Badifu, G. I. O.; *Food potentials of some unconventional oilseeds grown in Nigeria-A brief review,* Plant Foods for Human Nutrition, **43,** 211–224, **1993**

Bährle-Rapp, M.; *Springer Lexikon Kosmetik und Körperpflege*, Springer Verlag, Berlin Heidelberg, 3.Auflage, **2007**

Bährle-Rapp, M.; *Springer Lexikon Kosmetik und Körperpflege*, Springer Verlag, Berlin Heidelberg, 4.Auflage, **2012**

Bärtels, A.; *Farbatlas Tropenpflanzen Zier- und Nutzpflanzen*, Ulmer Verlag, Stuttgart, 4.Auflage, **1996**

Baier, C.; Biologische und pflanzenbauliche Untersuchungen an der Arzneipflanze Artischocke (Cynara spec. L.), Cuvillier Verlag, Göttingen, **2003**

Bail, S.; Stübiger, G.; Krist, S.; Unterweger, U.; Buchbauer, G.; *Characterization of various grape seed oils byvolatile compounds, triacylglycerol composition, total phenols and antioxidant capacity,* Food Chemistry, **108,** 1122–1132, **2008**

Bail, S.; Krist, S.; Masters, E.; Unterweger, U.; Buchbauer, G.; *Volatile compounds of shea butter samples made under different production conditions in western, central and eastern Africa,* Journal of Food Composition and Analysis, **22,** 738–744, **2009a**

Bail, S.; Stübiger, G.; Unterweger, U.; Buchbauer, G.; Krist, S.; *Characterisation of volatile compounds and triacylglycerol profiles of nut oils using SPME-GC- and MALDI-TOF- mass spectrometry,* European Journal of Lipid Science and Technology, **111**, 170–182, **2009b**

Bailey, H. S.; *Vegetable tallow and Stillingia oil,* Cotton Oil Press, **4**(7), 50, **1920**

Bak, I.; Lekli, I.; Juhasz, B.; Varga, E.; Varga, B.; Gesztelyi, R.; Szendrei, L.; Tosaki, A.; *Isolation and Analysis of Bioactive Constituents of Sour Cherry (Prunus cerasus) Seed Kernel: An Emerging Functional Food,* Journal of Medicinal Food, **13**(4), 905–910, **2010**

Baldrick, P.; Robinson, J. A.; Hepburn, P. A.; *Reproduction studies in the rat with shea oleine and hardened shea oleine,* Food and Chemical Toxicology, **39,** 923–930, **2001**

Banerjee S.; Kaseb A. O.; Wang Z., Kong D.; Mohammad M.; Padhye S.; Sarkar F. H.; Mohammad R. M.; *Antitumor Activity of Gemcitabine and Oxaliplatin Is Augmented by Thymoquinone in Pancreatic Cancer,* Cancer Research, **69,** 5575–5583, **2009**

Barre, D. E.; *Potential ofevening primrose, borage, black currant, and fungal oils in human health,* Annals of Nutrition and Metabolism, **45**(2), 47–57, **2001**

Barth, M.; Jehle, D.; *Tabakanbau und Tabakverarbeitung leicht gemacht*, Lumica Verlag, Berlin, **2003**

Basuki, Y.; *Bedeutung und Wirkung von Weintraubenkernöl,* Diplomarbeit, TU-Wien, Wien, **2003**

Baswa, M.; Rath, C. C.; Dash, S. K.; Mishra, R. K.; *Antibacterialactivity of Karanj (Pongamiapinnata) and Neem (Azadirachta indica) seed oil. A preliminary report,* Microbios, **105**(412), 183–189, **2001**

Bat, S.; Tannert, U.; *Oils rom Hippophae rhamnoides. A novel lipid or cosmetics,* Seifen, Öle, Fette, Wachse, **119**(1), 29–31, **1993**

Bauer, K. H.; *Fette und Öle,* Verlagsbuchhandlung Paul Parey, Verlag für Landwirtschaft, Gartenbau und Forstwesen, **1928**

Bauer, K. H.; Frömming, K.-H.; Führer, C.; *Lehrbuch der pharmazeutischen Technologie,* Wissenschaftliche Verlagsgesellschaft mbH, Stuttgart, 8.Auflage, **2005**

Behrens, R.; *Cashew as an agro orestry crop, prospects and potentials,* Margraf Verlag, Weikersheim, **1996**

Belarbi, E. H.; Molina, E.; Chisti, Y.; *A process for high yield and scaleable recovery of high purity eicosapentaenoic acid esters from microalgae and fish oil,* Enzyme and Microbial Technology, **26,** 516–529, **2000**

Belitz, H. D.; Grosch, W.; *Food Chemistry,* 2nd edition, Springer Verlag, Berlin, **1987, 1999**

Belton, P. S.; Taylor, J. R. N.; *Pseudocereals and Less Common Cereals*, Springer Verlag, Berlin Heidelberg, **2002**

Benmiloud M.; Chaouki, M. L.; Gutekunst, R.; Teichert, H. M.; Wood, W. G.; Dunn, J. T.; *Oraliodised oil for correcting iodine deficiency: optimal dosing and outcome indicator selection,* Journal of Endocrinal Metabolism, **79,** 20–24, **1994**

Bernardo, A.; Howard-Hildige, R.; O'Conell, A.; Nichol, R.; Ryan, J.; Rice, B.; Roche, E.; Leahy, J. J.; *Camelina oil as a fuel fordiesel transport engines,* Industrial Crops and Products, **17**(3), 191–197, **2003**

Bernardo-Gil, G.; Oneto, C.; Antunes, P.; Rodrigues, M. F.; Empis, J. M.; *Extraction of Lipids from Cherry Seed Oil Using Supercritical Carbon Dioxide*, European Food Research Technology, **212**, 170–174, **2001**

Bernardo-Gil, G.; Grenha, J.; Santos, J.; Cardoso, P.; *Supercritical fluid extraction and characterisation of oil from hazelnut,* European Journal of Lipid science and Technology, **104** (7), 402–409, **2002**

Bernäth, J.; *Poppy. The Genus Papaver,* Harwood Academic Publishers, Amsterdam, **1998**

Berrougui, H.; Alvarez de Sotomayor, M.; Perrez-Guerrero, C.; Ettaib, A.; Hmamouchi, M.; Marhuenda, E.; Herrera, M. D.; *Argan (Argania spinosa) oillowers bloodpressure andimproves endothelial dysfunction in spontaneously hypertensive rats,* British Journal of Nutrition, **92**(6), 921–929, **2004**

Bertoli, C.; Fay, L. B.; Stancanelli, M.; Gumy, D.; Lambelet, P.; *Characterization of Chilean Hazelnut (Gevuina avellana Mol.) Seed oil,* Journal of the American Oil Chemists Society, **75**(8), 1037–1040, **1998**

Bertomeu, M. C.; Crozier, G. L.; Haas, T.; Fleith, M.; Buchanan, M. R.; *Selectiveeffectsofdietaryfats on vascular 13-HODE synthesis and platelet/vessel wall interactions,* Thrombosis Research, **59**(5), 819–30, **1990**

Bhattacharya, S.; Novell, J. R.; Winslett, M. C.; Hobbs K. E.; *Iodized oil in the treatment of hepatocellular carcinoma,* British Journal of Surgery, **81,** 1563–1571, **1994**

Bhargava, A.; Shukala, S.; Ohri, D.; *Chenopodium quinoa – An Indian perspective,* Industrial Crops and Products, **23**, 73–87, **2006**

Blicher-Mathiesen, U.; *Borneo Illipe, a Fat Product from different Shorea spp. (Dipterocarpaceae),* Economic Botany, **48** (3), 231–242, **1994**

Biesalski, H. K.; Fürst, P.; Kasper, H.; Kluthe, R.; Pölert, W.; Puchstein, C.; Stähelin, H. B.; *Ernährungsmedizin.* Thieme Verlag, Stuttgart **1999**

Blaschek, W.; Hänsel, R.; Keller, K.; Reichling, J.; Rimpler, H.; *Hagers Handbuch der pharmazeutischen Praxis, Folgeband 3, Drogen L-Z,* Springer Verlag, Berlin Heidelberg, 5.Auflage, **1998**

Blum, C.; *Analytik und Sensorik von Gewürzextrakten und Gewürzölen,* Dissertation, Hamburg,**1999**

Bockisch, M.; *Fats and oils handbook,* AOCS Press, Champaign, **1998**

Boeke, S. J.; Boersma, M. G.; Alink, G. M.; Van Loon, J. J. A.; Van Huis, A.; Dicke, M.; Rietjens, I. M. C. M.; *Safety evaluation of neem (Azadirachta indica) derived pesticides,* Journal of Ethnopharmacology, **94,** 25–41, **2004**

Bonilla, P. A. G.; Torrijos, E. G.; Segade, J. B.; Brito, F. F.; Vega, M. L.; *Allergy to tiger nut,* Allergy, **57**(11), 1083, **2002**

Boucek, B.; *Parametrisierung von Kürbiskernöl anhand olfaktorisch relevanter Daten,* Diplomarbeit, Universität Wien, Wien, **1997**

Boulanger, R.; Crouzet, J.; *Free and bound flavour components of amazonian fruits: 2. Cupuacu volatile compounds,* Flavour and Fragrance Journal, **15**(4), 251–257, **2000**

Boschin, G.; D'Agostina, A.; Annicchiarico, P.; Arnoldi, A.; *The Fatty Acid Composition of the Oil from Lupinus albus cv. Luxe as Affected by Environmental and Agricultural Factors,* European Food Research and Technology, **225**, 769–776, **2007**

Boschin, G.; D'Agostina, A.; Annicchiarico, P.; Arnoldi, A.; *Effect of Genotype and Environment on Fatty Acid Composition of Lupinus albus L. Seed,* Food Chemistry, **108**, 600–606, **2008**

Bozan, B.; Temelli, F.; *Extraction of poppy seed oil using supercritical CO_2,* Journal of Food Science, **68**(2), 422–426, **2003**

Bressmer, Fa.; *Firmenangabe zu geliefertem Lorbeeröl,* 18.09.2003

Brod, J.; Traitler, H.; Studer, A.; De Lacharriere, O.; *Evolution of lipid composition in skin treated with blackcurrant seed oil,* International Journal od Cosmetic Science, **10,** 149–159, **1988**

Buch, J. G.; Dikshit, R. K.; Mansuri, S. M.; *Effect of certain volatile oils on ejaculated human spermatozoa,* Indian Journal Med.Res., **87,** 361–363, **1988**

Buchbauer, G.; Boucek, G.; Nikiforov, A.; *On the Aroma of Austrian Pumkin Seed Oil: Correlation of Analytical Data with Olfactoric Characteristics,* Ernährung/Nutrition **22,** 246–249, **1998**

Budin, J. T.; Breene, W. M.; Putnam, D. H.; *Some compositional properties ofcamelina (Camelina sativa L. Crantz) seeds and oils,* Journal of the American Oil Chemists Society, **72**(3), 309–315, **1995**

Burger, A.; Wachter, H.; *Hunnius Pharmazeutisches Wörterbuch,* 7.Auflage, Verlag Walter de Gruyter, Berlin und New York, **1993**

Burger, A. E. C.; De Villiers, J. B. M.; Du Plessis, L. M.; *Composition ofthe kernel oil and protein of the marula seed,* South African Journal of Science, **83**(11), 733–5, **1987**

Burns, R. A.; Wibert, G. J.; Diersen-Schade, D. A.; Kelly, C. M.; *Evaluation of Single cell Sources of Docosahexaenic Acid and Arachidonic Acid: 3-Month Rat Oral Safety Studywith an In-Utero-Phase,* Food and Chemical Toxikology, **37,** 23–36, **1999**

Bush, P. B.; Grunwald, C.; *Sterol Changes during Germination of Nicotiana tabacum Seeds,* Plant Physiology, **50**, 69–72, **1972**

Bushman, B. S.; Phillips, B.; Isbell, T.; Ou, B.; Crane, J.; Knapp, S.; *Chemical Composition of Caneberry (Rubus spp.) Seeds and Oils and Their Antioxidant Potential,* Journal of Agricultural and Food Chemistry, **52**, 7982–7987, **2004**

Cadwallader, K. R.; Heo, J.; *Aroma of roasted sesame oil: characterization by direct desorption-gas chromatography-olfactometry and sample dilution analysis,* ACS Symposium Series, **782** (Gas Chromatography-Olfactometry), 187–202, **2001**

Cahill, J. P.; *Ethnobotany of Chia, Salvia Hispanica L. (Lamiaceae)*, Economic Botany, **57** (4), 604–618, **2003**

Calvacanti, M. P. B.; *Revista de Quimica industrial* **10**(114), 21–3, 345–347, **1941**

Cao, J.-M.; Gresti, J.; Blond, J. P.; Bezard, J.; *Effects of cyclopropenoid fatty acids (Baobab seed oil) on the fatty acid profile of lipids from different tissues in the rat,* Journal of Food Lipids, **3**(1), 73–86, **1996**

Carstensen B.; *Schnellmethode zur Isolierung und Bestimmung von Minorkomponenten in pflanzlichen Ölen und Fetten,* Gustav Fischer Verlag, Jena, **2001**

Cavalli, J. F.; Fernandez, X.; Lizzani-Cuvelier, L.; Loiseau, A. M.; *Comparison ofstatic headspace, headspace solid phase microextraction, headspace sorptive extraction, and direct thermal desorption techniques on chemical composition of French olive oils,* Laboratoire „Aromes, Syntheses and Interactions", Journal of Agriculture and Food Chemistry **51**(26), 7709–16, **2003**

Chaieb K.; Kouidhi B.; Jrah H.; Mahdouani K.; Bakhrouf A.; *Antibacterial activity of Thymoquinone, an active principle of Nigella sativa and its potency to prevent bacterial biofilm formation.* BMC Complementary and Alternative Medicine **11**, 29–34, **2011**

Chan, H. T.; Heu, R. A.; *Composition of Papaya Seeds*, Journal of Food Science, **43**(1), 255–256, **1978**

Chandra, M.; Nair, M. G.; *Characterization of Pit Oil from Montmorency Cherry (Prunus cerasus L.),* Journal of Agricultural and Chemistry, **41**, 879–881, **1993**

Charrouf, Z.; Guillaume, D.; *Ethnoeconomical, ethnomedical and phytochemical study of Argania spinosa (L.) Skeels,* Journal of Ethnopharmakology, **67,** 7–14, **1999**

Chehl N.; Chipitsyna G.; Gong Q.; Yeo C. J.; Arafat H. A.; *Anti-inflammatory effects of the Nigella sativa seed extract, thymoquinone, in pancreatic cancer cells.* International Hepato-Pancreato-Biliary Association, **11**, 373–381, **2009**

Chen, F.; Jiang, Y. (Editors); *Algae and their Biotechnological Potential,* Kluwer Academic Publishers, Dordrecht, Boston, London, **2001**

Chen, Y.-G.; Lu, Y.; Liu, Y.; Hai, L.-N.; *Study on fatty acids extracted from pomegranate seed oil,* Shipin Kexue (Beijing, China), **24**(11), 111–112, **2003**

Cheng, Y. C.; Lee, M. H.; Li, C. F.; *Flavour components ofpeanut oil,* Zhongguo Nongye Huaxue huizhi, **27**(3), 336–349, **1989**

Chevallier, A.; *Das große Lexikon der Heilpflanzen*, Dorling Kindersley Verlag GmbH, Starnberg, **2001**

Child, R.; *Ceylon candlenut oil. Aleurites moluccana (linn) Willd.,* Oil and Soap **18, 224–6, 1941**

Chisholm, M. J.; Hopkins, C. Y.; *Conjugated Fatty Acids of Tragopogon and Calendula Seed Oils,* Canadian Journal of Chemistry, **38**, 2500–2507, **1960**

Choi, B. I., Park, J. H.; Kim, B. H.; Kim, S. H.; Han, M. C.; Kim, C. W.; *Smallhepatocellular carcinoma: detection with sonography, computeredtomography, angiography and Lipidol-CT,* British Journal of Radiology, **62,** 897–903, **1989**

Chung, T. Y.; Eiserich, J. P.; Shibamoto, T.; *Volatile compounds identified in headspace samples ofpeanut oil heated under temperatures ranging from 50 to 200°C,* Journal of agricultural and food chemistry **41**(9), 1467–1470, **1993**

Cicero, A. F. G.; Gaddi, A.; *Rice bran oil and j-oryzanol in the treatment of hyperlipoproteinaemias and other conditions,* University of Bologna, Italy, Phytotherapy Research, **15**(4), 277–289, **2001**

Collin, G.; Hilditch, T. P.; *CXLI. Regularities in the Glyceride Structure of Vegetable Seed-Fats,* Biochemical Journal, **23**(6), 1273–1289, **1929**

Collin, G.; *The Kernel-Fats of some Members of the Palmae: Acrocomia Sclerocarpa Mart. (Gru-Gru Palm), Manicaria Saccifera Gaertn., Astrocaryum Tucuma Mart., Maximiliana Caribaea Griseb., Attalea Excelsa Mart. Pallia Palm), and Cocos Nucifera Linn. (Coconut),* Biochemical Journal, **27**(5), 1366–1372, **1933**

Comes, F.; Farines, M.; Aumelas, A.; Soulier, J.; *Fatty Acid and Triglycerols of Cherry Seed Oil,* Journal of the American Oil Chemists' Society, **69**(12), 1224–1227, **1992**

Compton, Senja V.; Stout P.; *Application of headspace GC-FTIR. Analysis of flavor oils,* Cambridge, MA, USA, American Laboratory (Shelton, CT, United States), United States, **1992**

Conrado, F. A.; Goyco, A. J.; *Puerto Rican Fatty Oils IV. Expressed Tropical Almond (Talisay) oil,* Department of Chemistry, School of Tropical MEdicine of the University of Puerto Rico, Journal of the American Chemical Society, **65,** 1417–1418, **1943**

Coskuner, Y.; Tekin, A.; *Monitoring of seed composition of prickly pear (Opunita ficus-indica L) fruits during maturation period,* Journal of the Science of Food and Agriculture, **83**, 846–849, **2003**

Cosmetic, Toiletry and Fragrance Assoc., N.N.; *Final Report on the Safety Assessment of Candelilla wax, Carnauba wax, Japan, wax and Beeswax,* Washington DC, USA, Journal of the American College of Toxicology, **3**(3), 1–41, **1984**

Courtin, M.; *Antiaging cosmetics compositions containing metal ion binding agents and free radical scavengers,* Demand Fr., FRXXBL FR 2706294 A3 19941223 (Patent), **1994**

Crews, C.; Hough, P.; Godward, J.; Bereton, P.; Lees, M.; Guiet, S.; Winkelmann, W.; *Study of the Main Constituents of Some Authentic Hazelnut Oils,* Journal of Agricultural and Food Chemistry, **53,** 4843–4852, **2005**

Cromack, H. T. H.; Smith, J. M.; *Calendula Officinalis- Production Potential and Crop Agronomy in Southern England,* Industrial Crops and Products, **7**, 223–229, **1998**

Cserhalmi, Z.; Márkus, Z.; Czukor, B.; Baráth, A.; Tóth, M.; *Physico-Chemical Properties and Food Utilization Possibilities of RF-Treated Mustard Seed,* Innovative Food Science & Emerging Technologies, **1**, 251–254, **2001**

Cuperus, F. P.; Boswinkel, G.: Derksen, T. P.; *The Processing of New Oilseed Crops – An Economic Evaluation,* Journal of the American Oil Chemists' Society, **73**(12), 1635–1640, **1996**

Curt, M. D.; Sánchez, G.; Fernández, J.; *The potential of Cynara cardunculus L. for seed oil production in perennial cultivation system,* Biomass and Bioenergy, **23**, 33–46, **2002**

DAB; Deutsches Arzneibuch, Band 6, **1938**

Dachler, M.; Pelzmann, H.; *Arznei- und Gewürzpflanzen,* Österreichischer Agrarverlag, Klosterneuburg, 2. Auflage, **1999**

Dagne, K.; Jonsson, A.*; Oil Content and Fatty Acid Compostition of Seeds of Guizotia Cass (Compositae),* Journal of the Science of Food and Agriculture, **73**, 274–278, **1997**

Danert, S., *Urania Pflanzenreich Blütenpflanzen 1,* Urania-Verlag, Leipzig, 1.Auflage, **1993**

Dantas de Araujo, F.; *A Review of Caryocar Brasiliense (Caryocaraceae) – An Economically Valuable Species of the Central Brasilian Cerrados,* Economic Botany, **49**(1), 40–48, **1995**

Das, M.; Das, S. K.; Suthar, S. H.; *Composition of seed and characteristics of oil from karingda (Citrullus lanatus (Thumb) Mansf),* International Jornal of Food Science and Technology, **37**(8), 893–896, **2002**

Dasgupta, A.; Datta, P. C.; *Medicinal species of Piper, pharmacognostic delimitation, University* of Calcutta, Quarterly Journal of Crude Drug Research, **18**(1), 17–25, India, **1980**

Dassler, E.; Heitmann, G.; *Obst & Gemüse,* Paul Parey Verlag, Berlin, **1991**

Daulatabad, C. D.; Mulla, G. M.; Mirajkar, A. M.; *Vernolic and cyclopropenoic fatty acids in Piper nigrum seed oil,* Karnatak University, Fett Wissenschaft Technologie **97** (12), 453–4, India, **1995**

Davis, P.; Iwahashi, C.; *Whole almonds and almond fractions reduce aberrant crypt foci in a rat model ofcolon carcinogenesis,* Department of Nutrition, University of California, Cancer letters **165**(1), 27–33, **2001**

De Souza, A. H.; *Chemical and toxicological studyofthe seeds residues ofthe oiticica, Licania rigida,* Revista Brasileira de Farmacia, **27,** 245–247, **1946**

Deferne, J.-L.; Leeds, A. R.; *Resting blood pressure and cardiovascular reactivity to mental arithmetic in mild hypertensive males supplemented with blackcurrant seed oil,* Journal of Human Huypertension, **10,** 531–537, **1996**

Dehelean, C.; Dragomiresu, A.; Heghes, A.; *Vegetable oils as new sources of surfactans and their role in cream properties,* Fac. of Pharmacy, Rom, Communicaciones presentadas a las Jornadas del Comite Espanol de la Detergencia, **33,** 301–307, Rom, **2003**

Deiler, A. C.; Fraps, G. S.; *Pecan Oil,* American Chemical Journal, **43,** 90–91, **1910**

Deineka, V. I.; *Triglyceride Composition of Seed Oils from Certain Plants*, Chemistry of Natural Compounds, **39**(6), 523–527, **2003**

Deineka, V. I.; Deineka, L. A.; *Type Composition of Triglycerides from Seed Oils. II. Triglycerides from Certain Cultivated Plants of the Rosaceae Family*, Chemistry of Natural Compounds, **40**(3), 293–294, **2004**

Deineka, V. I.; Gabruk, N. G.; Deineka, L. A.; Manokhina, L. A.; *Triglyceride Composition of Oil from Stones of Nine Rosaceae Plants*, Chemistry of Natural Compounds, **28**(5), 410–412, **2002**

Del Mar Caja, M.; Del Castillo, M.; Alvarez, R.; Herraiz, M.; Blach, C.; *Analysis of volatile compounds in edible oils using simultaneous distillation – solvent extraction and direct coupling of liquid chromatography with gas chromatography,* Instituto de Fermentaciones Industriales, Madrid, European Food Research and Technology **211**(1), 45–51, Spain, **2000**

Denke Margo, A.; *Effects ofcacao butteron serum lipids in humans: historical highlights,* American Journal of Clinical Nutrition, **60,** 1014–1016, **1994**

Derksen, J. T. P.; Cuperus, F. P.; Kolster, P.; *Paints and Coatings from Renewable Resources*, Industrial Crops and Products, 3, 225–236, **1995**

Destaillats, F.; Cruz-Hernandez, C.; Giuffrida, F.; Dionisi, F.; *Identification of the Botanical Origin of Pine Nuts Found in Food Products by Gas-Liquid Chromatography Analysis of Fatty Acid Profile*, Journal of Agricultural and Food Chemistry, **58**, 2082–2087, **2010**

Deutsche Gesellschaft für Fettwissenschaft (DGF), **2003**, http://www.dgfett.de/

DGE (Deutsche Gesellschaft für Ernährung), Österreichische Gesellschaft für Ernährung, Schweizerische Gesellschaft für Ernährungsforschung, Schweizerische Vereinigung für Ernährung. *Referenzwerte für die Nährstoffzufuhr.* Umschau. Baus Verlag, Frankfurt am Main, **2000**

Dhar, K. L.; Chauhan, R. N. S.; *D.S.B. Govt. Coll,* Naini Tal, India, Agra Univ. J. Res **12**(1), 1,**1963**

Diederichsen, A.; *Promoting the conservation and use of underutilized and neglected crops,* International plant genetic resources institut, **1996**

Diener, H.; *Arzneipflanzen und Drogen,* Fachbuchverlag Thun, 1.Auflage, **1987**

Diersen-Schade, D. A.; Hansen, J. W.; Harris, C. L.; Merkel, K. L. Wisont, K. D.; Boettcher, J. A.; *Docosahexaenoic acid plus arachidonic acid enhance preterm infant growth,* Essential Fatty Acids and Eicosanoids, Invited Papersfrom the International Congress, 4th, Edingburgh, July 20–24, 1997 (1998), 123–127, **1998**

Dietrichs, A.; Knorr, L.; *„Babassonnuts and theiroil,* Neuss. Z. Nahr. Genussm., **40,**152–153, **1920**

Do Padro, I. M.; Giufrida, W. M.; Alvarez, V. H.; Cabral, V. F.; Quispe-Condori, S.; Saldana, M. D. A.; Cardozo-Filho, L.; *Phase Equilibrium Measurements of Sacha Inchi Oil (Plukenetia volubilis) and CO_2 at High Pressures*, Journal of the American Oil Chemists' Society, **88**, 1263–1269, **2011**

Domokos, J.; Perédi, J.; *Studies on the seed oils of basil (Ocimum basilicum L.) and summer savory (Satureja hortensis L.)*, Acta Horticulture, **344**, 312–314, **1993**

Downey, R. K.; *Agricultural and Genetic Potentials of Cruciferous Oilseed Crops*, Journal of the American Oil Chemists' Society, **48**, 718–722, **1971**

Dörfler, H. P.; Roselt, G.; *Heilpflanzen,* Urania Verlag, Leipzig, Jena, Berlin, **1990**

Drissi, A.; Girona, J.; Cherki, M.; Godas, G.; Derouiche, A.; El Essal, M.; Saile, R.; Kettani, A.; Sola, R.; Masana, L.; Adlouni, A.; *Evidence of hypolipemiant and antioxidant properties ofargan oil derived from the argan tree (Argania spinosa),* Clinical Nutrition, **23**(5), 1159–1166, **2004**

Drogen E-O, Springer Verlag, Berlin, Heidelberg 1993

Drogen P-Z, Springer Verlag, Berlin, Heidelberg 1994

Drogen L-Z, Springer Verlag, Berlin, Heidelberg 1998

Dutta, P. C.; Helmersson, S.; Kebedu, E.; Alemaw, G.; Appelqvist, L.; *Variation in Lipid Composition of Niger Seed (Guizotia abyssinica Cass.) Samples Collected from Different Regions in Ethiopia,* Journal oft the American Oil Chemist's Society, **71**(8), 839–843, **1994**

Dweck, A. C.; Meadows, T.; *Tamanu (Calophyllum inophyllum) – the African, Asian, Polynesian and Pazific Panacea,* International Journal of Cosmetic Science, **24**,1–8, **2002**

Earle, F. R.; McGuire, T. A.; Mallan, J.; Bagby, M. O.; Wolff, I. A.; *Search for New Industrials Oils. II. Oils with high iodine values,* Journal of the Oil Chemists' Society, **37**, 48–50, **1960**

Earle, F. R.; Mikolajczak, K. L.; Wolff, I. A.; *Search for New Industrial Oils. X. Seed Oils of the Calendulae,* Journal of the American Oil Chemists' Society, **41**(5), 345–347, **1964**

Earle, F. R.; Wolff, I. A.; Glass, C. A.; *Search for New Industrial Oils. VII.,* Journal of the American Oil Chemists' Society, **39**(9), 381–383, **1962**

Ebermann, R.; Elmadfa, I.; *Lehrbuch Lebensmittelchemie und Ernährung,* Springer Verlag, Wien, **2008**

EFSA Panel on Contaminants in the Food Chain (CONTAM): Risks for human health related to the presence of 3- and 2-monochloropropanediol (MCPD), and their fatty acid esters, and glycidyl fatty acid esters in food. EFSA Journal, **14**(5), e04426, 1–159, **2016**

Eidhin, D. Ni.; Burke, J.; Lynch, B.; O'Beirne, D.; *Effects ofdietarysupplementation with camelina oil on porcine blood lipids,* Journal of Food Science, **68**(2), 671–679, **2003a**

Eidhin, D. Ni.; Burke, J.; O'Beirne, D.; *Oxidative stability of* ***oj3****-rich camelina oil and camelina oil-based spread compared with plant and fish oils and sunflower spread,* Journal of Food Science, **68**(1), 345–353, **2003b**

Eklund, A.; Agren, G.; *Nutritive value of poppy seed protein.* Journal of the American Oil Chem. Society, **52**(6), 188–190, **1975**

El-Aal, M.; Khalil, M.; Rahma, E.; *Apricot kernel oil: characterization, chemical composition and utilization in some baked products,* Fac. Agric., Alexandria Univ., Alexandria, Egypt, Food Chemistry **19**(4), 287–98, **1986**

El-Adawy, T. A.; Taha, K. M.; *Characteristics and Composition o Watermelon, Pumpkin, and Paprika Seed Oils and Flours,* Journal of Agricultural and Food Chemistry, **49,** 1253–1259, **2001**

Elias Luiz, G.; Bressani, R.; *Nutritive Value of Brazilnut oil,* Inst. Nutrition Central Am. and Panama, Guatemala, Journal of the American Oil Chemists Society, **38,** 450–452, **1961**

Elmadfa, I.; Freisling, H.; König, J.; Blachfelner, J.; Cvitkovich-Steiner, H.; Genser, D.; Grossgut, R.; Hassan-Hauser, C.; Kichler, R.; Kunze, M.; Majchrzak, D.; Manafi, M.; Rust, P.; Schindler, K.; Vojir, F.; Wallner, S.; Zilberszac, A.; *Österreichischer Ernährungsbericht 2003.* Bundesministerium für Gesundheit und Frauen, Wien, **2004.**

Ennouri, M.; Bourret, E.; Mondolot, L.; Attia, H.; *Fatty acid composition and rheological behaviour of prickly pear seed oils,* Food Chemistry, **93**, 431–437, **2005**

Ennouri, M.; Fetoui, H.; Bourret, E.; Zeghal, N.; Attia, H.; *Evaluation of some biological parameters of Opunita ficus indica. 1 Influence of a seed oil supplemented diet on rats,* Bioresource Technology, **97**, 1382–1386, **2006**

Erbas, M.; Certel, M.; Uslu, M. K.; *Some Chemical Properties of White Lupin Seeds (Lupinus albus L.),* Food Chemistry, **89**, 341–345, **2005**

Esdorn, I.; *Die Nutzpflanzen der Tropen und Subtropen der Weltwirtschaft,* Gustav Fischer Verlag Stuttgart, **1981**

Eshetu, B.; Nicotiana Tabacum L. Seed Oil, 2000, http://ipp.boku.ac.at/pz/oilseeds/eshetu.html, 09.06.2011

Espada, C. E.; Berra, M. A.; Martinez, M. J.; Eynard, A. R.; Pasqualini, M. E.; *Effect of Chia Oil (Salvia Hispanica) rich in ω-3 Fatty Acid on the Eicosanoid Release, Apoptosis and T-Lymphocyte Tumor Infiltration in a Murine Mammary Gland Adenocarcinoma,* Prostaglandins, Leukotriens and Essential Fatty Acids, **77**, 21–28. **2007**

Eteshola, E.; Oraedu, A. C. I.; *Fatty acid compositions of tigernut tubers (Cyperus esculentus L.), baobab seed (Adansonia digitata L.), and their mixture,* Journal of the American Oil Chemists Society, **73**(2), 255-7, **1996**

Europäisches Arzneibuch, **1976, 2002**

Ezaki, O.; Takahashi, M.; Shigematsu, T.; Shimamura, K.; Kimura, J.; Ezaki, H.; Gotoh, T.; *Long-Term Effects of Dietary α-Linolenic Acid from Perilla Oil on Serum Fatty Acids Composition and on the Risk Factors of Coronary Heart Disease in Japanese Elderly Subjects,* Journal of nutritional science and vitaminology, **45**(6), 759–772, **1999**

Ezeagu, I. E.; Petzke, K. J.; Lange, E.; Metges, C.; *Fat Content and Fatty Acid Composition of Oils Extracted from Selected Wild-Gathered Tropical Plant Seedes from Nigeria,* Journal of the American Oil Chemists Society, **75**(8), 1031–1035, **1998**

Falasca, S. L.; Flores, N.; Lamas, M. C.; Carballo, S. M.; Anschau, A.; *Crambe abyssinica: An Almost Unknown Crop with a Promissory Future to Produce Biodiesel in Argentinia,* International Journal of Hydrogen Energy, **35**, 5808–5812, **2010**

Fallico, B.; Arena, E.; Zappala, M.; *Roasting ofhazelnuts. Role ofoil in colour development and hydroxymethylfurfural formation,* Food Chemistry **81**, 569–573, **2003**

Fanali, C.; Dugo, L.; Cacciola, F.; Beccaria, M.; Grasso, S.; Dachá, M.; Dugo, P.; Mondello, L.; *Chemical Characterization of Sacha Inchi (Plukenetia volubilis L.) Oil,* Journal of Agricultural and Food Chemistry, **59**, 13043–13049, **2011**

Feder, W. J.; Holtgrefe, R.; Butte, W.; Biermann, U.; *Calendulaöl als Lackrohstoff für Naturfarben,* Abschlußbericht, Jever und Oldenburg, **2009**

Feldl, F.; *Einfluß von Formelnahrungen mit Borretschöl oder mit Borretschöl und Fischöl auf die Fettsäuren in den Phospholipiden von Plasma und Erythrozyten sowie auf den Vitamin E-Status bei Frühgeborenen,* Dissertation an der TU München, München, **2000**

Fernández, J.; Curt, M. D.; *Low-Cost Biodiesel from Cynara Oil,* 2nd World Conference and Exhibition on Biomass for Energy, Industry and Climate Protection, Rome, Italy, 10–14 May 2004

Ferree, D. C.; Warrington, I. J.; *Apples Botany, Production and Uses,* CABI Publishing, Cambridge, **2003**

Filsoof, M.; Mehran, M.; Farrohi, F.; *Determination and comparison ofoil characteristics in Iranian almond, apricot and peach nuts,* Coll.Pharm, Teheran Univ., Teheran, Iran, Fette, Seifen, Anstrichmittel, **78**(4), 150–1, **1976**

Firestone, D.; *Physical and chemical characteristics of oils, fats and waxes,* AOCS Press, Champaign, **1999**

Fischer, G.; Krug, E.; *Heilkräuter und Arzneipflanzen,* Karl F. Haug Verlag, Heidelberg, 7. Auflage, **1984**

Flamini, G.; Cioni, P. L., Morelli, I.; *Volatiles form Leaves, Fruits, and Virgin Oil from Olea europaea cv. Olivastra Seggianese from Italy,* Dipartimento di Chimica Bioorganica e Biofarmacia, Pisa, Italy, Journal of Agricultural and Food Chemistry, **51**(5), 1382–1386, **2003**

Fleischhacker, W.; *Skriptum zur Vorlesung: Pharmazeutische Chemie I,* Universität Wien, Wien, **2002**

Follegatti-Romero, L.; Piantino, C. R.; Grimaldi, R.; Cabral, F. A.; *Supercritical CO_2 Extraction of Omega-3 Rich Oil from Sacha inchi (Plukenetia volubilis L.) Seeds,* The Journal of Supercritical Fluids, **49**, 323–329, **2009**

Franke, W.; *Nutzpflanzenkunde,* 4.Auflage, Georg Thieme Verlag, **1989**

Frankel, E. N.; Satue-Gracia, T.; Meyer, A. S.; German, J. B.; *Oxidative Stability of Fish and Algae Oils Containing Long-Chain Polyunsaturated Fatty Acids in Bulkand in Oil-in-Water Emulsions,* Journal of Agricultural and Food Chemistry, **50,** 2094–2099, **2002**

Frega, N.; Bocci, F.; Conte, L. S.; Testa, F.; *Chemical Composition of Tobacco Seeds (Nicotiana tabacum L.),* Journal of the American Oil Chemists' Society, **68**(1), 29–33, **1991**

Fritsche, S.; Rijken, P.; *The potential health benefits of the Mediterranean diet, olive oil and olive oil phenols,* Unilever Research + Development Vlaardingen, Vlaardingen, Netherlands, Lipid Technology, **15**(1), 9–13, **2003**

Frohn, B.; *Natürlich Heilen mit Olivenöl, Mit der Heilkraft von Olivenöl Erkrankungen und Beschwerden behandeln,* Midena Verlag, München, **1998**

Frohne, D.; *Heilpflanzenlexikon, ein Leitfaden auf wissenschaftlicher Grundlage,* 7.Auflage, Wiss. Verl.Ges., Stuttgart, **2002**

Frohne, D.; Pfänder, H. J.; *Giftpflanzen,* Wissenschaftliche Verlagsgesellschaft, Stuttgart, **1997**

Furnee, C. A.; West, C. A.; van der Haar, F.; Hautvast, J. G.; *Effect of intestinal parasite treatment on the efficacy of oral iodized oil for correcting iodine deficiency in schoolchildren,* American Journal of Clinical Nutrition, **66,** 1422–1427, **1998**

Gafur, M. A.; Toregard, B.; *Investigation on fat from mango (Mangifera indica) kernel. Part I Fatty acid composition of lipid classes,* BCSIR Rajshahi, Bangladesh, Bangladesh Journal of Scientific and Industrial Research, **19**(1–4), 184–192, **1984**

Gallina Toschi, T.; Caboni, M. F.; Penazzi, G.; Lercker, G.; Capella, P.; *A Study on Cashew Nut Oil Composition,* Journal of the American Oil Chemists Society, **70**(10), 1017–20, **1993**

Gamel, T. H.; Mesallam, A. S.; Damir, A. A.; Shekib, L. A.; Linssen, J. P.; *Characterization of Amaranth Seed Oil,* Journal of Food Lipids, **14**, 323–334, **2007**

Garber, K.; *Über die Preßrückstände bei der Ölgewinnung aus Lallemantia iberica*, Zeitschrift für Lebensmitteluntersuchung und –forschung, **89**(7), 587–594, **1949**

Garcia, C. C.; Franco, P. I. B. M.; Zuppa, T. O.; Antoniosi Filho, N. R.; Leles, M. I. G.; *Thermal Stability Studies of Some Cerrado Plants*, Journal of Thermal Analysis and Calorimetry, **87**, 645–648, **2007**

Gaydou, E. M.; Bianchini, J. P.; Ralaimanarivo, A.; *Oil of the African baobab: Fatty acid and sterol composition od Adansonia digitata,* Univ. Madagascar, Antananarivo, Madagascar, Revue Francaise des Corps Gras, **26**(11), 447–448, **1979**

Gebauer, J.; El-Siddig, K.; Ebert, G.; *Baobab (Adansonia digitata L.): A review on a multipurpose tree with promising future in the Sudan,* Gartenbauwissenschaft, **67**(4), 155–160, **2002**

Geitel, A. C.: *Über Minjak Tengkawang, ein aus Borneo stammendes Pflanzenfett*, Journal für Praktische Chemie, **36** (1), 515–518, **1887**

Gheita T. A.; Kenawy S. A.; *Effectiveness of Nigella sativa Oil in the Management of Rheumatoid Arthritis Patients: A Placebo Controlled Study,* Phytotherapy Research **26**(8), 1246–1248, **2011**

Giannelos, P. N.; Zannikos, F.; Stournas, S.; Lois, E.; Anastopoulos, G.; *Tobacco Seed Oil as an Alternative Diesel Fuel: Physical and Chemical Properties,* Industrial Crops and Products, **16**, 1–9, **2002**

Gilad, R.; Hochner, H.; Savitsky, B.; Porat, S.; Hochner-Celnikier, D.; *Castor oil for induction of labor in post-date pregnancies: A randomized controlled trial*, Women Birth, **31**(1), 26–31, **2018**

Goffman, F. D.; Galletti, S.; *Gamma-Linolenic Acid and Tocopherol Contents in the Seed Oil of 47 Accessions from Several Ribes Species,* Journal of Agricultural and Food Chemistry, **49,** 349–354, **2001**

Goffman, F. D., Pinson, S.; Bergman, C.; *Genetic diversity for lipid content and fatty acid profile in rice bran,* Rice Research Unit, USA, Journal of the American Oil Chemists' Society, **80**(5), 485–490, **2003**

Gomes Da Silva, W.; Cortesi, N.; Novellini, P.; *The Brazil nut (Bertholletia excelsa H.B.K.-Family: Lecythidaceae) Note II. Lipids: Study of chemical composition,* Revista Italiana delle Sostanze Grasse, **74**(7), 311–314, **1997**

Gresti, J.; Mignerot, C.; Bezard, J.; Wolff, R. L.; *Distribution of Δ5-Olefinic Acid in the Triacylglycerols from Pinus korainensis and Pinus pinaster Seed Oils,* Journal of the American Oil Chemists Society, **73**(9), 1539–1547, **1996**

Grossgut, R.; *Österreichische trans-Fettsäuren Verordnung.* ÖGE-Jahrestagung, Wien, **2009**

Grynberg, H.; Szczepanska, H.; *The Structure of Triglycerides in Selected Oils Containing Erucic Acid*, The Journal of the American Oil Chemists' Society, **43**(3), 151–152, **1966**

Gübitz, G. M.; Mittelbach, M.; Trabi, M.; *Exploitation ofthe tropical seed plant Jatropha curcas L.,* Bioresource Technology, **67,** 73–82, **1998**

Guimpel, F.; Wildenow, K. L.; Hayne, F. G.; *Abbildungen der deutschen Holzarten für Forstmänner und Liebhaber der Botanik*, In der Schüppischen Bibliothek, Berlin, **1815**

Gulfraz, M.; Sadio, A.; Tario, H.; Imran, M.; Qureshi, R.; Zeenat, A.; *Phytochemical Analysis and Antibacterial Activity of Eruca Sativa Seed,* Pakistan Journal of Botany, **43** (2), 1351–1359, **2011**

Gunstone, F. D.; Hilditch, T. P.; *The component acids and glycerids of Australian lumbang oil,* Journal of the Society of Chemical Industry, London, Transactions and communications, **66,** 205–8, **1947**

Gurr, M. I.; Blades, J.; Appleby, R. S.; Smith, C. G.; Robinson, M. P.; Nichols, B. W.; *Studies on Seed-Oil Triglycerides*, European Journal of Biochemistry, **43**, 281–290, **1974**

Gutiérrez, L.; Rosada, L.; Jiménez, A.; *Chemical Composition of Sacha Inchi (Plukenetia volubilis L.) Seeds and Characteristics of their Lipid Fraction*, Grasas y Aceites, **62**(1), 76–83, **2011**

Hachfeld, R.; Reisner, S.; Reschke, K.; *Apfel Kultur.Mythos.Gesundheit.Rezepte,* Umschau/Braus Verlag, **1999**

Hackbarth, J.; *Die Ölpflanzen Mitteleuropas,* Wissenschaftliche Verlagsgesellschaft m.b.H., Stuttgart, Band 15, **1944**

Hagemann, J. M.; Earle, F. R.; Wolff, I. A.; *Search for New Industrial Oils. XIV. Seed Oils of Labiatae, Lipids*, **2**(5), 371–380, **1967**

Hager; *Handbuch derpharmazeutischen Praxis,* Springer Verlag, Berlin, Heidelberg, **1978**

Hager; *Handbuch der pharmazeutischen Praxis,* **(1993)**

Hager; *Handbuch der pharmazeutischen Praxis,* **(1994)**

Hager; *Handbuch der pharmazeutischen Praxis,* **(1998)**

Halawani E.; *Antibacterial Activity of Thymoquinone and Thymohydroquinone of Nigella sativa L. and Their Interaction with Some Antibiotics;* Advances in Biological Research, **3,** 148–152, **2009**

Halter, L.; Systemabhängige Qualität von Artischocken (Cynara scolymus L.), Der Andere Verlag, Osnabrück, **2003**

Hamaker, B. R.; Valles, C.; Gilman, R.; Hardmeier, R. M.; Clark, D.; Garcia, H. H.; Gonzales, A. E.; Kohlstad, I.; Castro, M.; Valdivia, R.; Rodriguez, T.; Lescano, M.; *Amino Acid and Fatty Acid Profiles of the Inca Peanut (Plukenetia volubilis)*, Cereal Chemistry, **69**(4), 461–463, **1992**

Hamama, A. A; Bhardwaj, H. L.; *Phytosterols, Triterpene Alcohols, and Phospholipids in Seed Oil from White Lupin,* Journal oft the American Oil Chemists' Society, **81**(11), 1039–1044, **2004**

Hamilton, R. J.; *Developments in Oils and Fats,* Blackie Academic & Professional, Chapham & Hall, London, **1995**

Hammer, M. L. A.; Johns, E. A.; *Tapping an Amazonian plethora: four medicinal plants of Marajo Island, Para (Brazil),* Journal of Ethnopharmacology, **40,** 53–75, **1993**

Hannon, J.; *Pistachio nut oil: a natural emollient for the cosmetic formulator,* Cedar Grove, NJ, USA, Drug & Cosmetic Industry **160**(2), 30–34,36,80, USA, **1997**

Hänsel, R.; Keller, K.; Rimpler, H.; Schneider, G. (Hrsg.); *Hagers Handbuch der Pharmazeutischen Praxis*, Springer Berlin-Heidelberg, **1994**

Hänsel, R.; Sticher, O.; Steinegger, E.; *Pharmakognosie – Phytopharmazie,* Springer Verlag, Berlin-Heidelberg, **1994, 1999**

Hänsel, R.; Sticher, O.; *Pharmakognosie Phytopharmazie*, Springer Medizin Verlag, Heidelberg, 4th edition, **1992**

Hänsel, R.; Sticher, O.; *Pharmakognosie Phytopharmazie*, Springer Medizin Verlag, Heidelberg, 8.Auflage, **2007**

Hanssen, H.; Schmitz-Hübsch, M.; *Sacha Inchi (Plukenetia volubilis L.) Nut Oil and Its Therapeutic and Nutritional Uses*, Nuts & Seeds in Health Disease Prevention, **2011**

Hao, Y.-Y.; Huang, R.-H.; Deng, X.; Wang, P.-Y.; *Microwave extraction ofseed ofpassionflower,* Huangdong Ligong Daxue Xuebao, **27**(2), 117–120, **2001**

Harlow, R. D.; Litchfield, C.; Fu, H.-C.; Reiser, R.; *The triglyceride composition of Myrica pensylvanica fruit coat fat (bayberrytallow),* Texas Agr. Expt. Sta., College Station, Journal of the American Oil Chemists Society, **42**(9), 747–750, **1965**

Harwood, J. L.; Yaqoob, P.; *Nutritional and health aspects of olive oil,* School of Biosciences, Cardiff University, Cardiff, U.K., European Journal of Lipid Science and Technology **104**(9–10), 685–697, **2002**

Hassanein, M. M. M.; El-Shami, S. M.; El.Mallah, M. H.; *Investigation of Lipids Profiles of Nigella, Lupin and Artichoke Seed Oils to be Used as Healthy Oils*, Journal of Oleo Science, **60** (3), 99–107, **2011**

Hata; Chuta Inst. of Research of Chem. Industry, Taiwan. Kogyo Kagaku Zasshi, 46Suppl. Binding 41–2 CODEN: KGKZA7 ISSN: 0368–5462, **1993**

He, H. P.; Cai, Y. Z.; Sun, M.; Corke, H.; *Extraction and purification of squalene from Amaranthus grain,* J. Agric. Food Chem., **50**, 368–372, **2002**

He, H. P.; Corke, H.; *Oil and Squalene in Amaranthus Grain and Leaf,* J. Agric. Food Chem., **51**, 7913–7920, **2003**

Heftmann, E.; *Antioxidant properties of carrot oil,* Journal of American oil Chemists society **24,** 404–409, **1947**

Heilscher, K.; Morsel, J.-T.; Westphal, G.; *Physiological efficacyand stability ofsea buckthorn oil (Hippophae rhamnoides),* Parfümerie und Kosmetik, **80**(9), 10–12, **1999**

Heller, V. G.; Heston, E. S.; *A chemical and nutritive study of pecan oil,* Proceedings of the Oklahoma Academy of Science, **12,** 38–42, **1932**

Helv. V; Pharmacopöa Helvetica, 5.Ausgabe, **1953**

Henon, G.; Vigneron, P.; Stoclin, B.; Calgniez, J.; *Rapeseed oil deodorization study using the response surface methodology,* European Journal of Lipid Science and Technologie **103** (7), 467–477, **2001**

Herr, M.; *Einfluss von Melanoidinen auf die oxidative Beständigkeit von Speiseölen und Speisefetten,* Diplomarbeit, Universität Wien, **2003**

Higashi, S.; Setoguchi, T.; *Hepatic arteriol injection chemotherapy for hepatocellular carcinoma with epirubicin aqueous solution as numerous vesicles in iodinated poppy seed oil microdroplets: Clinical application of water-in-oil emulsion prepared using a membrane emulsification technique,* Department of Surgery I, Miyazaki Medical College, Miyazaki, Kiyotake-cho, Japan, Advanced Drug Delivery Reviews, **45**(1), 57–64, **2000**

Hilditch, T. P.; Murti, K. S.; *Fat acids and glycerides of solid seed fats. X. Seed fats of Garcinia Morella and Garcinia Indica,* Journal of the Society of Chemical Industry, London, **60,** 16–18, **1941**

Hinojosa, M.; *Stipatosis or hypersensitivity pneumonitis caused by esparto (Stipa tenacissima) fibers,* J. Investig. Allergol. Clin. Immunolog., **11**(2), 67–72, **2001**

Hirota, M.; Suttajit, M.; Suguri, H.; Endo, Y.; Shudo, K.; Wongchai, V.; Hecker, E.; Fujiki, H.; *A New Tumor Promoter from the Seed oil of Jatropha curcas L., an Intramolecular Diester of 12-Deoxy-16-hudroxyphorbol,* Cancer Research, **48,** 5800–5804, **1988**

Hitzenberger, V.; *Erdmandelöl,* Art of Beauty, Herbst/Winter, 9, **2005**

Holmes, A. D.; Deuel, H. J.; *Digestibility of Certain Miscellaneous Vegetable Fats,* Journal of Biological Chemistry, **41**, 227–235, **1920**

Homberg, E.; Bielefeld, B.; *Sterin- und Fettsäurezusammensetzung in Keimölen,* Fat Science Technology, **92**(3), **1990**

Hourihane, J. O.; Bedwani, S. J.; Dean, T. P.; Warner, J. O.; *Randomised, double blind, crossover challenge study of allergenicity of peanut oils in subjects allergic to peanuts*, British Medical Journal, **314**(7087), 1084–1088, **1997**

Hornok, L.; *Cultivation and Processing of Medicinal Plants,* John Wiley & Sons, **1992**

Höggerl, M.; *Phänometrische Untersuchungen an Purgiernuss (Jatropha curcas L.) im Rahmen eines Ölpflanzenprojekts in Nicaragua,* Diplomarbeit, Wien, Universität für Bodenkultur, **1992**

http://eur.lex.europa.eu/legal-content/DE/TXT/? uri=CELEX:32018R0290, consulted 14.4.2019

http://www.efsa.eurpoa.eu/de/press/news/160503-0, consulted 14.4.2019

http://www.ages.at/themen/rueckstaende-kontaminanten/mcpds-glycidyl-ester/, consulted 14.4.2019

http://www.verbrauchergesundheit.gv.at/Lebensmittel/neuartige_lm/neuartigelm.html, consulted 12.4.2019

http://www.bvl.bund.de/DE/01_Lebensmittel/04_AntragstellerUnternehmen/05_NovelFood/lm_novelFood_node.html, consulted 12.4.2019

Huang, K. C.; *The Pharmacology of Chinese Herbs,* Second Edition, CRC Press LLC, Boca Raton, **1999**

Huang, W.; Huang, Q.; Luo, R.-N.; Wu, J.-Z.; Chen, Y. Q.; *Studies on seasonal variability ofaroma components and principle sugar acids contents in passion fruit,* Ziran kexueban, **24**(4), 84–87, **2003**

Hui, Y. H.; *Baileys industrial oil & fat products Vol. 1,* Wiley Interscience Public, New York, **1996**

Hunnius, *Pharmazeutisches Wörterbuch,* de Gruyter Verlag, Berlin – New York, 5th Auflage, **1978**

Hunnius, Pharmazeutisches Wörterbuch, de Gruyter Verlag, Berlin – New York, 8. Auflage, **1998**

Hunnius, Pharmazeutisches Wörterbuch, de Gruyter Verlag, Berlin – New York, 9.Auflage, **2004**

Hussain, M. G.; Haque, M. E.; Gafur, M. A.; Ali, M. H.; Ali, M. M.; *Studies on the kernel fatofmango of Rajshahi region,* BCSIR Lab., Rajshahi, Bangladesh, Bangladesh Journal of Scientific and Industrial Research, **18**(1–4), 146–149, **1983**

Hussain, S. A.; Dollear, F. G.; O'Connor, R. T.; *Oil from the kernels of Lalob fruit, Balanites aegyptiaca,* Journal of the American Oil Chemists Society, **26,** 730–732, **1949**

Idiem'Opute, F.; *The Seed Lipids of the Palm Familiy*, Journal oft he American Oil Chemists' Society, **56**, 528–530, **1979**

Igarashi, M.; Miyazawa, T.; *Newly recognized cytotoxic effect of conjugated trienoic fatty acids on cultured human tumor cells,* Cancer Letters, **148**,173–179, **2000**

Ihara-Watanabe, M.; Umekawa, H.; Takahashi, T.; Furuichi, Y.; *Comparative effects of safflower oil and perilla oil on serum and hepatic lipid levels, fatty acids compositions of serum and hepatic phospholipids, and hepatic mRNA expressions of 3-hydroxy-3-methylglutaryl CoA reductase, LDL receptor, and cholesterol 7-alpha-hydroxylase in young and adult rats,* Food Research International, **33,** 893–900, **2000**

Illés, V.; Daood, H. G.; Perneczki, S.; Szokonya, L.; Then, M.; *Extraction of Coriander Seed Oil by CO_2 and Propane at Super- and Subcritical Conditions*, Journal of Supercritical Fluids, **17**, 177–186, **2000**

Imbs, A. B.; Pham, L. Q.; *Fatty Acids and Triacylglycerols in Seeds of Pinaceae Species,* Phytochemistry, **42**(4), 1051–1053, **1996**

Innis, S. M.; Hansen, W.; *Plasma fatty acids responses, metabolic effects and safety of microalgal and fungal oil rich in arachidonic and docosahexaenoic acids in healthy adults,* American Journal of Clinical Nutrition, **64,** 159–167, **1996**

Inove, H.; Asaka, T.; Nagata, N.; Yasuko, K.; *Mechanism ofmustard oil-induced skininflammation in mice,* European Journal of Pharmacology, **333**(2–3), 231–240, **1997**

ISF; *Proceedings of the 21st World congress of the ISF; Oil-fats-lipids,* P.J. Barnes & Associates, Bridgewater **1995**

Ito, Y.; Yanase, S.; Tokuda, H.; Kishishita, M.; Ohigashi, H.; Hirota, M.; Koshimizu, K.; *Epstein-Barr virus activation by tung-oil, extracts of Aleurites fordii and its diterpene ester 12-O-hexadecanoyl-16-hydroxyphorbol-13-acetate,* Cancer Letters (Shannon, Ireland), **18**(1), 87–95, **1983**

Ivie, G. W.; *Linear Furocoumarins (Psoralens) from the seed of Texas Ammi majus L. (Bishops Weed),* Journal of Agricultural and Food Chemistry, **26**(6), 1394–1402, **1978**

Ixtaina, V. Y.; Martinez, M. L.; Spotorno, V.; Mateo, C. M.; Maestri, D. M.; Diehl, B. W. K.; Nolasco, S. M.; Tomás, M. C.; *Characterization of Chia Seed Oils Obtained by Pressing and Solvent Extraction*, Journal of Food Composition and Analysis, **24**, 166–174, **2011a**

Ixtaina, V. Y.; Mattea, F.; Cardarelli, D. A.; Mattea, M. A.; Nolasco, S. M.; Tomás, M. C.; *Supercritical Carbon Dioxide Extraction and Characterization of Argentinean Chia Seed Oil*, Journal of the American Oil Chemists' Society, **88**, 289–298, **2011b**

Ixtaina, V. Y.; Nolasco, S. M.; Tomás, M. C.; *Physical Properties of Chia (Salvia hispanica L.),* Seeds, Industrial Crops and Products, **28**, 286–293, **2008**

Jackson, J. E.; *Biology of Apple and Pears,* Cambridge University Press, Cambridge, **2003**

Jackson, F. L.; Longenecker, H. E.; *Fat acids and glycerides of Babassu oil,* Oil and Soap (Chicago), **21,** 73–75, **1944**

Jafri S. H.; Glass J.; Shi R.; Zhang S.; Prince M; Kleiner-Hancock H.; *Thymoquinone and cisplatin as a therapeutic combination in lung cancer: In vitro and in vivo.* Journal of Experimental and Clinical Cancer Research, **29,** 87–97, **2010**

Jagella, T.; *Untersuchung über das Aroma und Fehlaroma von schwarzem und weißem Pfeffer,* Dissertation an der TU München, München **1999**

Jahanival, F.; Kakuda, Y.; Marcone, M. F.; *Fatty Acid and Triacylglycerol Compositions of Seed Oils of Five Amaranthus Accessions and Their Comparison to Other Oils*, Journal of the American Oil Chemist's Society, **77**(8), 847–852, **2000**

Jain, M.; Banerjee, A. K.; *Studies on Balanities aegyptiaca seed oil,* Journal of the American Oil Chemists Society, **65**(6), 994, **1988**

Janystin, H.; *Handbuch der Kosmetika und Riechstoffe,* Huettig Verlag, Heidelberg, **1978 (1)**

Jeffrey, B. S.; Padley, F. B.; *Chinese vegetable tallow – characterization and contamination by stillingia oil,* Journal of the American Oil Chemists Society, **68**(2), 123–127, **1991**

Jelen, H. H.; Obuchowska, M.; Zawirska-Wojtasiak, R.; Wasowicz, E.; *Headspace Solid-Phase Microextraction Use for the Characterization of Volatile Compounds in Vegetable Oils of Different Sensory Quality,* Institute of Food Technology, Agricultural University Poznan, Journal of Agricultural and Food Chemistry, **48**(6), 2360–2367, **2000**

Jellinek, G.; *Sensorische Lebensmittelprüfung,* D&PS-Verlag, Pattensen, **1979**

Johannson, A.; *The content and composition ofsterols and sterol esters in sunflower and poppy seed oils,* Dep. Food Hyg., Swedish Univ. Agric. Science, Stockholm, Sweden, Lipids, **14**(3), 285–291, **1979**

Johansson, A. K.; Korte, H.; Yang, B.; Stanley, J. C.; Kallio, H. P.; *Sea buckthorn berryoil inhibits platelet aggregation,* Journal of Nutrition and Biochemistry, **11,** 491–495, **2000**

John, W. D.; *Esparto wax – its use in the manufacture of modern polishes and shoe-finishing requisites,* Oil and Colour Trades Journal, **91,** 1140–2, **1937**

Jung, M. Y.; Bock, J. Y.; Baik, S. O.; Lee, J. H.; Lee, T. K.; *Effects of Roasting on Pyrazine Contents and Oxidative Stability of Red Pepper Seed Oil Prior to Its Extraction,* Journal of Agricultural and Food Chemistry, **47**(4), 1700–1704, **1999**

Kalayasiri, P.; Jeyashoke, N.; Krisnangkura, K.; *Survey of Seed Oils for Use as Diesel Fuels,* Journal of the American Oil Chemists Society, **73**(4), 471–474, **1996**

Kamal-Eldin, A.; Appelqvist, L. A; *The chemistry and antioxidant properties of tocopherols and tocotrienols.* Lipids, **31,** 671–701, **1996**

Kamel, B. S.; Kakuda, Y.; *Characterization of the Seed Oil and Meal from Apricot, Cherry, Nectarine, Peach and Plum*, Journal of the American Oil Chemists' Society, **69**(5), 492–494, **1992**

Kandpal, J. B.; Madan, M.; *Jatropha curcas: a renewable source of energy for meeting future energyneeds,* Renewable Energy, **6**(2), 159–160, **1995**

Kang, Han-Chul; Park, Won-Jong; Kim, Si-Dong; Park, Jong-Cheon; *Charakterization of grape seedo/l*, Grape Experiment Center, Agricultural Chemistry and Biotechnology **41**(8), 578–582, Korea, **1998**

Kanya, T. C. S.; Urs, M. K.; *Studies on Taramira (Eruca sativa) Seed Oil and Meal,* Journal of the American Oil Chemists' Society, **66** (1), **1989**

Karleskind A.; *Oilsand fatsmanual Volume 1/2,* Lavoisier Publishing, Springer Verlag, Berlin, **1996**

Karvonen, H. M.; Aro, A.; Tapola, N. S.; Salminen, I.; Uusitupa, M. I. J.; Sarkkinen, E. S.; *Effect of α-Linolenic Acid-Rich Camelina sativa Oil on Serum Fatty Acid Composition and Serum Lipids in Hypercholesterolemic Subjects,* Metabolism, **51**(10), 1253–1260, **2002**

Kasbekar, M. G.; Talekar, R. R.; Bringi, N. V.; *Major Cyanolipid of Kusum Oil (Schleicheria trijuga),* Indian Journal of Chemistry, **10**(2), 244–6, **1971**

Kaßner, G.; *Über die Verwertung der Samenkerne von Prunus domestica L.*, Archiv der Pharmazie, **256**(1–4), 106–112, **1918**

Kaßner, G.; Eckelmann, K.; *Über den Öl- und Amygdalingehalt der Samenkerne von Prunus domestica L.*, Archiv der Pharmazie, **252**(6–8), 402–408, **1914**

Kato, T.; Hancock, R. L.; Mohammadpour, H.; McGregor, B.; Manalo, P.; Khaiboullina, S.; Hall, M. R.; Pardini, L.; Pardini, R. S.; *Influence of Omega-3 fatty acids on the growthof human colon carcinoma in nude mice,* Cancer Letters, **187,** 169–177, **2002**

Kaushik, N.; Vir, S.; *Variations in fatty acid composition ofneem seedscollected from the Fajasthan state of India,* Biochemical Society Transactions, **28**(6), 880–882, **2000**

Kehren, L.; *Candlenut oil and its drying proeperties,* Oleagineux **5,** 359–63, **1950**

Kellard, B.; Busfield, D.; Kinderlerer, J.; *Volatile Off-flavour Compounds in Desiccated Coconut,* Journal of Food and Agriculture, **36**(5), 415–20, **1984**

Kelly, D.; Bessiere, J.; Crimmins, J.; Renard, S.; *Anti-inflammatory properties of Amazonian oils,* SOFW Journal, **129**(4), 16–17, **2003**

Kerschbaum, S.; Schweiger, P.; *Untersuchungen über die Fettsäure- und Tocopherolgehalte von Pflanzenölen, Abschlußbericht über das Arbeitsprojekt „Pflanzenöle"*, Information für die Pflanzenproduktion, Sonderheft 1/2001, Hrsg: Landesanstalt für Pflanzenbau Forchheim, Rheinstetten, **2001**

Keskin, O.; Sekerel, B.E.; *Poppy seed allergy: a case report and review of the literature*, Allergy and Asthmatic Proceedings, **27**(4), 396–398, **2006**

Kershaw, S. J.; *Technical note: Heterogeneity in commercial contractual samples of illipe nut (Shorea spp.)*, International Journal of Food Science an Technology, **22**, 67–72, **1987**

Khallouki, F.; Younos, C.; Soulimani, R.; Oster, T.; Charrouf, Z.; Spiegelhalder, B.; Bartsch, H.; Owen, R. W.; *Consumption of argan oil (Morocco) with its unique profile of fatty acids, tocopherols, squalene, sterols and phenolic compounds should confer valuable cancer chemo-preventive effects,* European Journal of cancer prevention, **12**(1), 67–75, **2003**

Khan, F. W.; Khan, K.; Malik, M. N.; *Vegetable tallow and stillingia oil from the fruits of Sapium sebiferum,* Pakistan Journal of Forestry, **23**(3), 257–266, **1973**

Khan, M.; Arsala, A. J.; Nowshervi, A. R.; *Comparative study of vegetable tallow stillingia oil from the seeds of Sapium sebiferum Roxb.*, Journal of Science and Technology, **1**(2), 138–140, **1977**

Khanturgaev, A. G.; Shiretorova, V. G.; Radnaeva, L. D.; Khanturgaeva, G. I.; Averina, E. S.; Bodoev, N. V.; *Investigation of the Composition of Lipids of Siberian Pine Seeds*, Chemistry for Sustainable Development, **11**, 589–593, **2003**

Khoddami A.; Ghazali H.M.; Yassoralipour A.; Ramakrishnan Y.; Ganjloo A.: Physicochemical Characteristics of Nigella Seed (Nigella sativa L.) Oil as Affected by Different Extraction Methods. Journal of the American Oil Chemists' Society 88, 533–540, **2011**.

Khoobchandani, M.; Ojeswi, B. K.; Ganesh, N.; Srivastava, M. M.; Gabbanini, S.; Matera, R.; Iori, R.; Valgimigli, L.; *Antimicrobial Properties and Analytic Profile of Traditional Eruca Sativa Seed Oil: Comparison with Various Aerial and Root Plant Extracts*, Food Chemistry, **120**, 217–224, **2010**

Kilibardia, V.; Ivanic, R.; Savin, K.; Miric, M.; *Fatty oil from fruit of wild and cultivates carrot,* Pharmazie **44**(2), 166–167, **1989**

Kim, J. S.; *Einfluss der Temperatur beim Rösten von Sesam auf Aroma und antioxidative Eigenschaften des Öls,* Dissertation, TU Berlin, **2001**

Kim, S. J.; Yoon, H. N.; Rhee, J. S.; *The Effects of Roasting Temperatures on the Formation of Headspace Volatile Compounds in Perilla Seed Oil,* Journal of the American Oil Chemists Society, **77**(49), 451–456, **2000**

Kircher, N.; *Heilen, pflegen, kochen mit Speiseölen,* Oeschverlag, **2002**

Kiritsakis, A. K.; *Olive Oil,* American Oil Chemists' Society, USA, **1990**

Kleiman, R.; Koritala, S.; Hill, J. C.; James, H.; *Moringa esters and cosmetic and pharmaceutical preparations and methods of manufacture thereof,* Patent No. US 2004096417, USA, **2004**

Klotz, M.; *Nachweis von Johanniskrautwirkstoffen in öligen Zubereitungen,* Diplomarbeit, Universität Wien, **2002**

Körber-Grohe, U.; *Nutzpflanzen in Deutschland.* Theiss Verlag Stuttgart, **1995**

Koizumi, Y.; Shimomura, K.; *Dentifrices containing cashew nut oils or their extracts,* Jpn. Kokai Tokkyo Koho, Patent No. JP 03240721, **1991**

Koziol, M. J.; *Quinoa: A Potential New Oil Crop*, In: Janick, J.; Simon, J. E. (Eds.), New Crops. Wiley, New York, **1993**

Kratochvil, H.; *Lexikon exotischer Früchte*, Verlag Brüder Hollinek, Wien, **1995**

Kraus, R.; Spiteller, G.; Bartsch, W.; *(10E,12Z)-9-Hydroxy-10,12-octadecadiensäure, ein Aromatase-Hemmstoff aus dem Wurzelextrakt von Urtica dioica*, Liebigs Annalen der Chemie, **4**, 335–339, **1991**

Krenn, L.; *Passion flower (Passiflora incarnata L.), a reliable herbal sedative*, Wiener medizinische Wochenschrift, **152**, 404–406, **2002**

Kris-Etherton, P. M.; Hecker, K. D.; Bonanome, A.; Coval, S. M.; Binkoski, A. E.; Hilpert, K. F.; Griel, A. E.; Etherton, T. D.; *Bioactive compounds in foods: Their role in the prevention of cardiovascular disease and cancer*, Pennsylvania State University, USA, American Journal of Medicine, 113(9B), 71–88, **2002**

Krishna De, A.; *Capsicum: the genus Capsicum*, Taylor and Francis, London, **2003**

Krishnani, K. K.; Gupta, B. P.; Joseph, K. O.; Muralidhar, M.; Nagavel, A.; *Studies on the useofneem products for removal ofammonia from brackishwater*, Journal of Environmental Science and Health, Part A: Toxic/Hazardous Substances & Environmental Engineering, **A37**(5), 893–904, **2002**

Krisnawati, H.; Kallio, M.; Kanninen, m.; *Aleurites moluccana* (L.) *Willd.* Ecology, silviculture and productivity. Center for International Forestry Research, **2011**

Krist, S.; *Untersuchungen zum Aroma von Mohnölen und Samen diverser Mohnsorten*, Dissertation an der Universität Wien, **2002**

Krist, S.; *Volatile Compounds*, Peter Lang internationaler Verlag der Wissenschaften, Wien, 2011

Krist, S.; Stuebiger, G.; Unterweger, H.; Bandion, F.; Buchbauer, G.; *Analysis of Volatile Compounds and Triglycerides of Seed Oils extractes from different Poppy Varieties*, Journal of Agricultural and Food Chemistry, **53**, 8310–8316, **2005**

Krist, S.; Bail, S.; Unterweger, H.; Ngassoum, M. B.; Mohagir, A. M.; Buchbauer, G.; *Volatile compounds of original African black and white shea butter from Tchad and Cameroon*, European Journal of Lipid Science and Technology, **108**, 583–588, **2006a**

Krist, S.; Stuebiger, G.; Bail, S.; Unterweger, H.; *Analysis of volatile compoundsand triacylglycerol composition of fatty seed oil gained from flaxand false flax*, European Journal of Lipid Science and Technology, **108**, 48–60, **2006b**

Krist, S.; Stuebiger, G.; Bail, S.; Unterweger, H.; *Detection of Adulteration of Poppy Seed Oil with Sunflower Oil Based on Volatiles and Triacylglycerol Composition*, Journal of Agricultural and Food Chemistry, **54**, 6385–6389, **2006c**

Krist, S.; Stuebiger, G.; Bail, S.; Unterweger, H.; *Volatile compounds and triaclyglycerid composition of original Indian fatty plant oils*, European Journal of Lipid Science and Technology, **110**, 127–140, **2008**

Kritchevsky, D.; Tepper S. A.; Scott, D. A.; Klurfeld D. M.; Vesselinovitch, D.; Wissler R. W.; *Cholesterol vehicle in experimental atherosclerosis. Part 18. Comparison of North American, African and South American peanut oils*, Atherosclerosis, 38(3–4), 291–9, **1981**

Krzizan, R.; *Zur Kenntnis des Brombeerkernöles*, Chemische Revue über die Fett- und Harzindustrie, **15** (1), 7–9, **1908**

Krstiae, B.; Merkulov, L. J.; Gvozdenoviae, D.; Pajeviae, S.; *Anatomical and physiological characteristics ofseedinpepper(Capsicum annuum L.) varieties*, Acta Agronomica Hungaria, **49**(3), 221–229, **2001**

Kubelka, W.; Länger R.; *Phytokodex 2001/2002*, Verlag für Medizin und Wirtschaft, **2001**

Kulnik, D.; *Olivenöl, Stellenwert und ernährungsphysiologische Bedeutung in traditionellen und neuen Konsumländern*, Diplomarbeit, Wien **2003**

Kushi, L.; Giovannucci, E.; *Dietary fat and cancer*, Kaiser Permanente, Oakland, USA, American Journal of Medicine, **113** (9B), 63–70, **2002**

Kusmenoglu, S.; *Fatty acid composition of oils in Cucurbitaceae seeds*, Journal of Faculty of Pharmacy of Gazi University, **13**(2), 167–170, **1996**

Küssmann, G.; *Die Nadelgehölze*, Verlag Paul Parey, Berlin und Hamburg, 3. Auflage, **1979**

Labuschagne, M. T.; Hugo, A.; *Oil contant and fatty acid composition of cactus pear seed compared with cotton and grape seed*, Journal of Food Biochemistry, **34**, 93–100, **2010**

Lakshminarayana, G.; Rao, K. V. S. A.; Sita Devi, K.; Kaimal, T. N. B.; *Changes in Fatty Acids During Maturation*, Journal of the American Oil Chemists Society, **58** (8), 839–839, **1981**

Lalas, S.; Tsaknis, J.; *Extraction and identification of natural antioxidant from the seeds of the Moringa oleifera tree variety of Malawi,* Journal of the American Oil Chemists Society, **79**(7), 677–683, **2002**

Lalas, S.; Tsaknis, J.; Sflomos, K.; *Characterisation of Moringa stenopetala seed oil variety MM-arigatffrom island Kokwa,* European Journal of Lipid Science and Technology, **105,** 23–31, **2003**

Lampe, I.; Hilsendegen, P.; Die Inhaltsstoffe der Sauerkirschen.pdf-Obstbau-Rheinland-Pfalz, **2006**

Lampart-Szczapa, E.; Korczak, J.; Nogala-Kalucka, M.; Zawirska-Wojtasiak, R.; *Antioxidant Properties of Lupin Seed Products*, Food Chemistry, **83**, 279–285, **2003**

Lansky, E. P.; Newman, R. A.; *Punica granatum (pomegranate) and its potential forprevention and treatment of inflammation and cancer,* Journal of Ethnopharmacology, **109**(2), 177–206, **2007**

Lannes, S. C. S.; Medeiros, M. L.; Gioielli, L. A.; *Physicalinteractions between cupuassu and cocoa fats,* Grasas y Aceites, **54**(3), 243–258, **2003**

Lautenschläger, H.; *Lipophil – Öle und Fette in der Kosmetik,* Kosmetik International, **11**,46–48, **2004**

Lawson, H.; *Food, Oils and Fats – Technology, Utilization and Nutrition,* Chapman & Hall, New York, **1995**

Lazos, E. S.; Tsaknis, J.; Lalas, S.; *Characteristics and composition of tomato seed oil,* Grasas y Aceites **49**(5–6), 440–445, **1998**

Leclere, J.; *Chaulmoogra oil for treatment of hypopigmentation oftheskin,* Patent Nr. DE 4420690, Ger. Offen., **1994**

Lee, J.-W.; Lee, K. W.; Lee, S.-W.; Kim, I.-H.; Rhee, C.; *Selective Increasein Pinolenic Acid(all-cis-5,9,12-18:3) in Korean Pine Nut Oil by Cristallization and Its Effect on LDL-Receptor Activity,* Lipids, **39**(4), 383–387, **2004a**

Lee, W-J.; Lee, M-H.; Su, N-W.; *Characteristics of Papaya Seed Oils Obtained by Extrusion-Expelling Processes*, Journal of Science of Food and Agriculture, **91**(13), 2348–2354, **2011**

Lee, K.-W.; Yu, K.-W.; Kim, K. M.; Suh, H. J.; Lee, S.-W.; Rhee, C.; *Effects of Pine (Pinus koraiensis)-seed Oil Supplementation on Serum Lipid Composition in Rats and Immune Responses in Mice,* Food Science and Biotechnology, **13**(3), 358–361, **2004b**

Leichtfried, D.; *Begleitende chemische Untersuchungen zur Fermentierung und thermischen Exposition der Samen von Papaver somniferum und Corylus avellana,* Dissertation, Technische Universität Wien, **2002**

Lehari, G.; *Beeren*, Verlag Eugen Ulmer GmbH & Co, Stuttgart, **2001**

Le Poole, H. A. C.; *Behen oil: A classic oil for modern cosmetics,* Cosmetics & Toiletries, **111,** 77–80, **1996**

Leon-Camacho, M.; Garcia-Gonzalez, D. L.; Aparicio, R.; *A detailed and comprehensive study of amaranth (Amaranthus cruentus L.) oil fatty profile,* Eur. Food Res. Technol., **213**, 349–355, **2001**

Lercker, G.; Capella, P.; Conte, L. S.; Folli, B.; *Composition of wheat germ oil,* **54**(4), 177–182, **1977**

Levy, L.; *The activity ofchaulmoogra acids against Mycobacterium leprae,* American Review of Respiratory Disease, **111**(5), 703–705, **1975**

Le, S.; Wang, Y.; Dong, S.; Chen,Y.; Cao, F.; Wang, X.; *Biodiesel production from Eruca Sativa Gars vegetable oil and motor, emission properties*, Renewable Energy, **34**(7), 1871–1876, **2009**

Lewerenz, H-J.; Bleyl, D. W. R.; Plass, R.; *Subacute oral toxicity study of benzyl isothiocyanate in rats*, Food/Nahrung, **36**(2), 190–198, **1992**

Lieberei, R.; Reisdorff, C.; *Nutzpflanzenkunde*, Georg Thieme Verlag KG, Erscheinungsort, 7.Auflage, **2007**

Lipp, M.; Anklam, E.; *Review of cocoa butter and alternative fats for use in chocolate – Part A. Compositional data,* Food Chemistry, **62** (1), 73–97, **1998**

List, P. H.; Hörhammer, L.; *Hagers Handbuch der pharmazeutischen Praxis,* Springer Verlag, Berlin-Heidelberg-New York, **1973**

List, P. H.; Hörhammer, L.; *Hagers Handbuch der pharmazeutischen Praxis,* Springer Verlag, Berlin-Heidelberg-New York, **1976**

List, P. H.; Hörhammer, L.; *Hagers Handbuch der pharmazeutischen Praxis,* Springer Verlag, Berlin-Heidelberg-New York, **1978**

Litvinova, L. B.; *The effect of peach kernel oil on the sexual maturation of female rats,* Eksperimental'naia i Klinicheskaia farmakologiia, **61**(3), 43–5, **1998**

Lockett, C. T.; Calvert, C. C.; Grivetti, L. E.; *Energyand micronutrient composition ofdietaryand medicinal wild plants consumed during drought,* Study of rural Fulani, Northeastern Nigeria, International Journal of Food Science and Nutrition, **51,** 195–208, **2000**

Löffler, G.; Petrides, P. E.; Heinrich, P. C.; *Biochemie und Pathobiochemie.* Springer Verlag, Heidelberg, **2007**

Löw, H.; *Pflanzenöle – Anbau und Verarbeitung der gängigen Ölpflanzen. Herstellung von Spezial- und Gewürzölen.,* Leopold Stocker Verlag, Graz, **2003**

Longvah, T.; Deosthale, Y. G.; Uday Kumar, P.; *Nutritional and short term toxicological evaluation of Perilla seed oil,* Food Chemistry, **70,** 13–16, **1999**

Lopez, A. S.; *Lipids from the seeds of passion fruit (Passiflora edulis),* Revista Theobroma, **10**(1), 47–50, **1980**

Lovette, P.; *Shea processing technical inventory and recommended improved processing methodology,* World Agroforestry Centre Report, 09, **2004**

Lu, Y.; Foo, Y.; *Constitution of some chemical components of apple seed.,* Food Chemistry, **61**, 29–33, **1998**

Lude, R.; *The Brazilian babassupalm and the babassu-kerneloil,* Fettchem. Umschau, **41**, 51–53, **1934**

Luetjohann, S.; *Sanddorn-Starke Frucht und heilsames Öl,* Windpferd Verlag, Aitrang, **1999**

Luetjohann, S.; *Sanddorn-Starke Frucht und heilsames Öl,* Windpferd Verlag, Aitrang, **2001**

Lyon, C. K.; Becker, R.; *Extraction and refining of oil from Amaranth seed,* J. Am. Oil Chem. Soc., **64**, 233–236, **1987**

Maccarone, E.; Fallico, B.; Fanella, F.; Mauromicale, G.; Raccuia, S. A.; Foti, S.; *Possible alternative utilization of Cynara spp.II. Chemical characterization of their grain oil,* Industrial Crops and Products, **10**, 229–237, **1999**

Maestri, D. M.; Guzmán, C. A.; *Chemical Composition of Tobacco Seeds (Nicotiana tabacum L) from Argentinia,* Journal of the Science of Food and Agriculture, **61**(2), 227–230, **1993**

Maghsood, Y.; *Jojoba – Eine Literaturübersicht,* Diplomarbeit, Universität Wien, **1991**

Maguire, L. S.; O'Sullivan, S. M.; Galvin, K.; O'Connor, T. P.; O'Brien, N. M.; *Fatty acid profile, tocopherol, squalene and phytosterol content of walnuts, almonds, peanuts, hazelnuts and the macadamia nut,* International Journal of Food Sciences and Nutrition, **55**(3), 171–178, **2004**

Maheshwari, B.; Reddy, S. Y.; *Application of Kokum (Garcinia indica) fat as a cocoa improver in chocolate,* Lipid Science and Traditional Foods Department, Central Fod Technological Research Institute, Mysore, India, Journal of the Science of Food and Agriculture, **85**(1), 135–140, **2005**

Malecka, M.; *Antioxidant properties of the unsaponifiable matter isolated from tomato seeds, oat grains and wheat germ oil,* Food Chemistry, **79,** 327–330, **2002**

Marini, F.; Magrí, A. L.; Marini, D.; Balestrieri, F.; *Characterization of the Lipid Fraction of Niger Seeds (Guizotia Abyssinica Cass.) from Different Regions of Ethiopia and India and Chemometric Authentication of their Geographical Origin*, European Journal of Lipid Science and Technology, **105**, 697–704, **2003**

Marion, J. E.; Dempsey, A. H.; *Fatty acids ofpimiento pepperseed oil,* Journal of the American Oil Chemists Society, **41**(8), 548–549, **1964**

Markson, L. S.; *Dermatitis from seed and oil of Bertholletia excelsa (Brazil nut),* Archives of Dermatology and Syphilology, **46,** 831–832, **1942**

Martindale *The Extra Pharmacopoeia,* the Pharmaceutical Press, London, **1993**

Martini, L.; *Rebel acne isn't as terrible thanks to tamanu oil,* Cosmetic News, **24**(138), 176–177, **2001**

Martirosyan, D. M.; Miroshnichenko, L. A.; Kulakova, S. N.; Pogojeva, A. V.; Zoloedov, V. I.; *Amaranth oil application for coronary heart disease and hypertension*, Lipids in Health and Disease, **6**(1), **2007**

Marzouki, H.; Piras, A.; Marongiu, B.; Rosa, A.; Dessi, M. A.; *Extraction and seperation of volatile and fixed oils from berries of Laurus nobilis* L. *by supercritical Co2*, Molecules. **13**(8), 1702–1711, **2008**

Massaro, M.; De Caterina, R.; *Effects ofoleic acid on endothelial activation: possible mechanism for prevention of atherosclerosis,* CNR Institute of Clinical Physiology, Pisa, Italy, Recent Research Developments in Lipids, **5**(Pt.1), 85–96, **2001**

Matthäus, B.; Münch, E. W.; *Warenkunde Ölpflanzen/Pflanzenöle,* Agrimedia Verlag, Clenze, 2009

Matthaeus, B.; Bruhl, L.; *Quality of cold-pressed edible rapeseed oil in Germany,* Institut for Lipid Research, Germany, Nahrung, **47** (6), 413–419, **2003**

Matthaeus, B.; Öczan, M. M.; *Fatty Acid and Tocopherol Contents of Some Prunus spp. Kernel Oils*, Journal of Food Lipids, **16**, 187–199, **2009**

Matthaus, B.; Vosmann, K.; Pham, L. Q.; Aitzetmüller, K.; *FA and Tocopherol Composition of Vietnamese Oilseeds,* Journal of the American Oil Chemists Society, **80**(10), 1013–1020, **2003**

Matsui, T.; Guth, H.; Grosch, W.; *A comparative study of potent odorants in peanut, hazelnut, and pumkin seed oils on the basis of aroma extract dilution analysis (AEDA) and gas chromatography-olfactometry of headspace samples (GCOH),* Department of Food Science Technologie, Faculty Agriculture, Kyushu University, Japan, Fett/Lipid **100**(2), 51–56, **1998**

Mayer, J.; Nerger, J.; *Essbare Wildkräuter und –früchte*, Urania Verlag, Berlin, **2000**

McKillop, H.; *Prehistoric Maya Use of Native Palms: Archaeobotanical and Ethnobotanical Evidence*, University of Utah Press, Salt Lake City, **1996**

McSweeney, K.; *The Cohune Palm (Orbignya Cohune, Arecaceae) in Belize: A Survey of Uses,* Economic Botany, **49**(2), 162–171, **1995**

Meier Zu Beerentrup, H.; Röbbelen, G.; *Calendula and Coriandrum – New Potential Oilcrops for Industrial Uses*, Lipid/Fett, **89**(6), 227–230, **1987**

Micelli, A.; De Leo, P.; *Extraction, Chracterization and Utilization of Artichoke-Seed Oil,* Bioresource Technology, **57**, 301–302, **1996**

Mills, S.; Bone, K.; *Principl. and Pract. of Phytotherapy,* Churchill Livingstone, Edinburgh, **2001**

Miot, H. A.; Batistella, R. F.; Batista, K. D. A.; Volpato, D. E. C. A.; Augusto, L. S. T.; Madeira, N. G.; Haddad, V. Jr.; Miot, L. D. B.; *Comparative study of the topical effectiveness of the Andiroba oil (Carapa guianensis) and DEET 50% as repellent for Aedes sp.,* Revista do Instituto de Medicina Tropical de Sao Paulo, **46**(5), 253–256, **2004**

Miranda-Vilela, A. L.; Akimoto, A. K.; Alves, P. C. Z.; Pereira, L. C. S.; Goncalves, C. A.; Klautau-Guimaraes, M. N.; Grisolia, C. K.; *Dietary Carotenois-Rich Pequi Oil Reduces Plasma Lipid Peroxidation and DNA Damage in Runners and Evidence for an Association with MnSOD Genetic Variant – Val9Ala,* Genetics and Molecular Research **8**(4), 1481–1495, **2009a**

Miranda-Vilela, A. L.; Pereira, L. C. S.; Goncalves, C. A.; Grisolia, C. K.; *Pequi Fruit (Caryocar brasiliense Camb.) Pulp Oil Reduces Exercise-Induced Inflammatory Markers and Blood Pressure of Male and Female Runners*, Nutrition Research, **29**, 850–858, **2009b**

Mohamed, A. M.; *Untersuchungen zur mechanischen Abtrennung von Protein und Öl aus dem Samen von Balanities aegyptiaca Del. (Lalobe),* Dissertation, Universität Hohenheim, **1998**

Morice, I. M.; Shorland, F. B.; Williams, E.; Seed *Oils of Apples (Malus pumila),* Journal of the Science of Food and Agriculture, **22**, 186–188, **1971**

Moser, B. R.; Vaughn, S. F.; *Coriander Seed Oil Methylesters as Biodiesel Fuel: Unique Fatty Acid Composition and Excellent Oxidative Stability*, Biomass and Bioenergy, **34**, 550–558, **2010**

Msaada, K.; Hosni, K.; Taarit, M. B.; Chahed, T.; Hammami, M.; Marzouk, B.; *Changes in Fatty Acid Composition of Coriander (Coriandrum Sativum L.) Fruit During Maturation*, Industrial Crops and Products, **29**, 269–274, **2009**

Msika, P.; *Composition Containing a Quinoa Extract for Dermatological Use*, Patent No. US 2010/0136144 A1, Washington, **2010**

Mueller, R.; Seidel, K.; Kaczich, A.; Hollenberg, D.; Matzik, I. (Henkel K.-G.a.A., Germany); Ger. Offen., 10pp.Coden: GWXXBX DE 4405127 A1 19950831, **1995**

Mukhtar, A.; Ullah, H.; Mukhtar, H.; *Fatty Acid Composition of Tobacco Seed Oil and Synthesis of Alkyd Resin*, Chinese Journal of Chemistry, **25**, 705–708, **2007**

Murkovic, M.; Piironen, V.; Lampi, A.; Kraushofer, T.; Sontag, G.; *Changes in chemical composition of pumkin seeds during the roasting process for production of pumkin seed oil (Part1: non-volatile compounds),* Institute for Food Chemistry and Food Technology, Technical University Graz, Austria, Food Chemistry (2003), Volume Date 2004, **84**(3), 359–365, **2004**

Mustakas, G. C.; Kopas, G.; Robinson, N.; *Prepress-Solvent Extraction of Crambe: First Commercial Trial Run of New Oilseed*, Journal of the American Oil Chemists' Society, **42**(10), 550A-554A, **1965**

Mutschler, E.; Geisslinger, G.; Kroemer, H. K.; Schäfer-Korting, M.; *Mutschler Arzneimittelwirkungen*, Wissenschaftliche Verlagsgesellschaft mbH, Stuttgart, 8.Auflage, **2001**

Nagalakshmi, S.; Shankaracharya, N. B.; Naik, J. P.; Rao, L. J. M.; *Studies on chemical and technological aspects ofajowan (Trachyspermum ammisyn. Carum copticum)seeds,* Journal of Food Science and Technology, **37**(3), 277–281, **2000**

Nagao, K.; Yanagita, T.; *Conjugated Fatty Acids in Food and Their Health Benefits*, Journal of Bioscience and Bioengineering, **100**(2), 152–157, **2005**

Nagle, N.; Lemke, P.; *Production of Methyl Ester Fuel from Microalgae,* Applied Biochemistry and Biotechnology, **24-25,** 355–361, **1990**

Nakamura, Y.; Tonogai, Y.; *Effects of grape seed polyphenols on serum and hepatic lipid contents and fecal steroid excretion in normal and hypercholesterolemic rats,* Division of Food Chemistry, National Institute of Health Sciences Osaka Branch, Japan, Journal of Health Science **48**(6), 570–578, Japan, **2002**

Narayana, C.; Somayajulu, B. A. R.; Rao, S. D. T.; *Recovery of fatty oil from spent seeds of Ajowan (Trachyspermum ammi),* Indian Journal of Technology, **5**(8), 268–9, **1967**

Nature in a Bottle; Technical datasheet Chaulmoogra oil; 2019 accessed: 04122019

Neetu, J.; Meenakshi, S.; *Broad spectrum antimycotic drug for the treatment ofringworm infection in human beings,* Current Science, **85**(1), 30–34, **2003**

Nergiz, C.; Dönmez, I.; *Chemical composition and nutritive value of Pinus pinea L. seeds,* Food Chemistry, **86,** 365–368, **2004**

Neuwinger, M. D.; *Afrikanische Arzneipflanzen und Jagdgifte,* Wissenschaftliche Verlags Gesellschaft, Stuttgart, **1998**

Nikakhlagh S.; Rahim F.; Aryani F. H. N.; Syahpoush A.; Brougerdnya M. G.; Saki N.; *Herbal treatment of allergic rhinitis: the use of Nigella sativa.* American Journal of Otolaryngology-Head and Neck Medicine and Surgery, **32**, 402–407, **2011**

Nicolosi, R. J.; Rogers, E. J.; Ausman, L. M.; Orthoefer, F. T.; *Rice bran oil and its health benefits,* Univerity Massachusetts, USA, Food Science and Technology, **59,** 421–37, **1994**

Nisshin Flour Milling Co.; Ltd, Japan; Jpn. Kokai tokyo Koho, 3pp. Coden: JKXXAF JP 56099405 19810810, **1981**

Nguyen, H.; Tarandjiiska, R.; *Lipid Classes, Fatty Acids and Triglycerides in Papaya Seed Oil,* European Journal of Lipid Science and Technology, **97**(1), 20–23, **1995**

Noldt, G.; *Charakterisierung der Feinwurzeln von Swietania macrophylla King und Carapa guianensis Aubl. (Meliaceae) unter verschiedenen Standortbedingungen,* Mitteilung der Bundesforschungsanstalt für Forst- und Holzwirtschaft, Hamburg, 201, **2000**

Norten, E.; *Wunderbaum Niem, Medizin, Kosmetik, Pflanzenschutz aus der Natur,* vgs Verlagsgesellschaft, Köln, **1996**

Niyas, Z.; Variyar, P. S.; Gholap, A. S.; Sharma, A.; *Effect of γ-Irritation on the Lipid Profile of Nutmeg (Myristica fragrans Houtt.)*, Journal of Agricultural and Food Chemistry, **51**, 6502–6504, **2003**

Nowak, B.; Schulz, B.; *Tropische Früchte*, BLV Verlagsgesellschaft mbH, München, **1998**

Oderinde, R. A.; Ajayi, I. A.; *Metal and oil characteristics of Terminalia catappa,* Rivista Italiana delle Sostanze Grase, **75**(7), 361–362, **1998**

Odetokun, S. M.; *The nutritive value of baobab fruit (Adansonia digitata),* Rivista Italiana delle Sostanze Grasse, **73**(8), 371–373, **1996**

Ogbobe, O.; *Physico-chemical composition and characterisation of the seed and seed oil of Sclerocarya birrea,* Plant Foods for Human Nutrition, **42**(3), 201–201, **1992**

Oh, H. H.; Hwang, K. T.; Shin, M. K.; Lee, H. K.; Kim, S. Z.; *Oils in the Seeds of Caneberries Produced in Korea.,* Journal of the American Oil Chemists' Society, **84**, 549–555, **2007**

Okuno, M.; Kajiwara, K.; Imai, S.; Kobayashi, T.; Honma, N.; Maki, T.; Suruga, K.; Goda, T.; Takase, S.; Muto, Y.; Moriwaki, H.; *Perilla Oil Prevents the Excessive Growhtof Visceral Adipose Tissue in Rats by Down-Regulating Adipocyte Differentation,* Journal of Nutrition, **127,** 1752–1757, **1997**

Olberg, H.; *Sheabutter,* Seifen, Öle, Fette, Wachse, **113**(10), 333–334, **1987**

Oliveira, J. T. A.; Vascincelos, I. M.; Bezerra, L. C. N. M.; Silveira, S. B.; Monteiro, A. C. O.; Moreira, R. A.; *Composition and nutritional properties ofseeds from Pachira aquatica Aubl., Sterculia striata St. Hil et Naud and Terminalia catappa Linn,* Food Chemistry, **70**(2), 185–191, **2000**

Onyeike, E. N.; Acheru, G. N.; *Chemical composition of selected Nigerian oil seeds and physicochemical properties ofthe oil extracts,* Food Chemistry, **77**(4), 431–437, **2002**

Oommen, S. T.; *The effect of oil of hydnocarpus on excision wounds,* International Journal of leprosy and other mycobacterial diseases: official organ of the International Leprosy Association, **68**(1), 69–70, **2000**

Oppel, T.; Thomas, P.; Wollenberg, A.; *Cross-sensitization between poppy seed and buckwheat in a food-allergic patient with poppy seed anaphylaxis*, International Archives of Allergyand Immunology, **140**(2), 170–173, **2006**

Ortiz Moreno, A.; Dorantes, L.; Galindes, J.; Guzman, R. I.; *Effect of Different Extraction Methods on Fatty Acids, Volatile Compounds, and Physical and Chemical Properties of Avocado (Persea americana Mill.) Oil,* Instituto Politechnico Nacional, Department of Biochemical Engineering, Mexico City, Mexico, Journal of Agricultural and Food Chemistry, **51**(8), 2216–2221, **2003**

Overeem, A.; Buisman, G. J. H.; Derksen, J. T. P.; Cuperus, F. P.; Molhoek, L.; Grisnich, W.; Goemans, C.; *Seed Oils Rich in Linolenic Acid as Renewable Feedstock for Environment-Friendly Crosslonkers in Powder Coatings*, Industrial Crops and Products, **10**, 157–165, **1999**

Oyedeji, F. O.; Oderinde, R. A.; *Stability of cosmetic emulsions from water melon and paraffin oils,* International Journal of Chemistry, **14**(3), 149–154, **2004**

Özgül-Yücel, S.; *Determination of Conjugated Linolenic Acid Content of Selected Oil Seeds Grown in Turkey*, Journal of the American Oil Chemists' Society, **82**(12), 893–897, **2005**

Pahlow, M.; *Das große Buch der Heilpflanzen,* Gräfe und Unzer Verlag, München, **1985**

Palma, M. B.; Pinto, A. L.; Gombert, A. K.; Seits, K. H.; Kivatinitz, S. C.; Castilho, L. R.; Freire, D. M.; *Lipase production by Penicillium restrictum using solide waste of industrial babassu oil production as Substrate,* Department of Chemical Engineering, Universityof Blumenau, Brazil, Applied Biochemistry and Biotechnology, **84-86,** 1137–45, **2000**

Palma, F.; Donde, M.; Lloyd, W. R.; *Fixed Oils of Mexico I. Oil of Chia-Salvia Hispanica*, The Journal of the American Oil Chemists' Society, 27–28, **1947**

Panfili, G.; Cinquanta, L.; Fratianni, A.; Cubadda, R.; *Extraction of Wheat Germ Oil by Supercritical CO_2: Oil and Defatted Cake Characterization,* Journal of the American Oil Chemists' Society, **80**(2), 157–161, **2003**

Pantzaris, T. P.; Basiron, Y.; *The lauric oils; Vegetable oils in food technologie,* 157-202, **2002**

Parcerisa, J.; Richardson, D.; Rafecas, M.; Codony, R.; Boatella, J.; *Fatty acid, tocopherol and sterol content ofsome hazelnut varieties harvested in Oregon,* Journal of chromatography **A 805**(1+2), 259–268, **1998**

Park, H. S.; Kim, C. J.; Lim, Y. Y.; *Perilla oil has a protective effect against colon cancer and cardiovascular disease, but induces a high degree of lipid peroxidation in rats,* Proceedings of the International Symposium on Natural Antioxydants: Molecular Mechanisms and Health Effecta, Beijing, June 20–24, Meeting Date **1995**

Parker, T. D.; Adams, D. A.; Zhou, K.; Harris, M.; Yu, L.; *Fatty Acid Composition and Oxidative Stability of Cold-Pressed Edible Seed Oils,* Journal of Food Science, **68**(4), 1240–1243, **2003**

Pasquier, E.; Ratnayake, W. M. N.; Wolff, R. L.; *Effects of Δ5-Polyunsaturated Fatty Acids of Maritime Pine (Pinus pinaster) Seed Oil on the Fatty Acid Profile of the Developing Brain of Rats,* Lipids, **36**(6), 567–574, **2001**

Passera, C.; Spettoli, P.; *Chemical Composition of Papaya Seeds*, Plant Foods for Human Nutrition, **31**(1), 77–83, **1981**

Pencreach, G.; Graille, J.; Pina, M.; Verger, R.; *An Ultraviolet Spectrophotometric Assay for Measuring Lipase Activity Using Long-Chain Triacylglycerols from Aleurites fordii Seeds,* Analytical Biochemistry, **303,** 17–24, **2002**

Peiretti, P. G.; Gai, F.; *Fatty Acid and Nutritive Quality of Chia (Salvia hispanica L.) Seeds and Plant During Growth,* Animal Feed Science and Technology, **148**, 267–275, **2009**

Peiretti, P. G.; Meineri, G.; *Effects on Growth Performance, Carcass Characteristics, and the Fat and Meat Fatty Acid Profile of Rabbits Fed Diets with Chia (Salvia hispanica L.) Seed Supplements*, Meat Science, **80**, 1116–1121, **2008**

Pfendtner, I.; Knochenbauer, S.; *Vital und aktiv mit Kürbiskernöl,* Urania Verlag, Wien, **1998**

Pharmacopoeia Europaea, Stuttgart, **1976**

Pies, J.; *Sacha-Inchi*, VAK Verlags GmbH, Kirchzarten bei Freiburg, **2010**

Pilaske, R.; *Natürliche Hausapotheke – Weißdorn, Sanddorn,* Fachverlag Fraund, Mainz, **2002**

Pina-Rodriguez, A. M.; Akoh, C. C.; *Synthesis and Characterization of a Strucutred Lipid from Amaranth Oil as a Partial Fat Substitute in Milk-Based Infant Formula*, J. Agric. Food Chem., **57**, 6748–6756, **2009**

Piroutz, J.; *Persönliche Information des Leindotterherstellers Josef Piroutz, Millnern,* **2003**

Pitroipa, X.; Sankara, D.; Konan, L.; Sylla, M.; Doannio, J. M.; Traore, S.; *Evaluation of cocoa oil for individual protection against Simulium damnosum s.i.,* Med. Trop., **62**(5), 511–516, **2002**

Polasa, K.; Rukmini, C.; *Mutagenicity tests of cashewnut shell liquid, rice-bran oil and other vegetable oils using the Salmonella typhimurium Microsome System,* Food and Chemical Toxixology, **25**(10), 763–766, **1986**

Popov, A.; Stefanov, K.; *Über das Fenchelölwachs,* Fette, Seifen, Anstrichmittel **71**(11), 952–953, **1969**

Prentner, A.; *Bewusstseinsverändernde Pflanzen von A-Z,* Springer, Wien, 2. Auflage, **2010**

Princen, L. H.; Rothfus, J. A.; *Development of New Crops for Industrial Raw Materials*, Journal of the American Oil Chemists' Society, **61**(2), 281–289, **1984**

Pruthi, J. S.; *Chemical composition and utilization of passion fruit seed, seed oil and seed meal,* Indian Oil and Soap Journal, **28,** 55–62, **1962**

Pruthi, J. S.; *Studies on the nutritive value ofrawandblendedpassion fruit seedoil,* Indian Oilseeds Journal, **7**(1), 60–65, **1963**

Pschyrembel – Medizinisches Wörterbuch, 257. Auflage, Verlag Walter de Gruyter, Nikol Verlagsgesellschaft mbH, Hamburg, **1994**

Puangsri, T.; Abdulkarim, S. M.; Ghazali, H. M.; *Properties of Carica Papaya L. (Papaya) Seed Oil Following Extractions Using Solvent and Aqueous Enzymatic Methods*, Journal of Food Lipids, **12**, 62–76, **2005**

Puri, H. S.; *Neem, The Divine Tree, Azadirachta indica,* Overseas Publishers Association, Amsterdam, **1999**

Quirin, K. W.; *Specialty Fatty Oils for Health Skin,* Cosmetic Science Technology, **2009**

Quirin, K. W.; Gerard, D.; *Sea buckthorn lipids. Promising cosmetic ingredients,* Parfümerie und Kosmetik, **74**(10), 618–625, **1993**

Quiroga, O. E.; Bou, S.; Vigo, M. S.; Nolasco, S. M.; *Chemical Characteristics of Passiflora Caerula Seed Oil and Residual Seed Meal,* Molecules, **5**(3), 376–378, **2000**

Raab, W.; Kindl, U.: *Pflegekosmetik,* 4. Auflage, Wissenschaftliche Verlagsgesellschaft mbH, Stuttgart, **2004**

Raccuia, S. A.; Melilli, M. G.; *Biomass and grain oil yields in Cynara cardunculus L. genotypes grown in a Mediterranean environment,* Field Crops Research, **101**, 187–197, **2007**

Raie, M. Y.; Zakas Iqbal, S.; Sabir, A. W.; Khan, S. A.; *Sapium sebiferum oil and fat analysis by chromatography,* Fette, Seifen, Anstrichmittel, **85**(9), 359–362, **1983**

Rajendran, R.; Anil, S.; Vijayakumar, T.; *A Rare Human Model for Oncogenesis,* Singapore Dental Journal, **13**(1), 49–52, **1988**

Ramadan, M. F.; Amer, M. M. A.; Awad, A. E.; *Coriander (Coriandrum Sativum L.) Seed Oil Improves Plasma Lipid Profile in Rats Fed a Diet Containing Cholesterol*, European Food Research and Technology, **227**, 1173–1182, **2008**

Ramadan, M.; Kroh, L.; Mörsel, J.; *Radical Scavenging Activity of Black Cumin (Nigella sativa L.), Coriander (Coriandrum sativum L.), and Niger (Guizotia abyssinica Cass.) Crude Seed Oils and Oil Fractions$_L$* Institut für Lebensmittelchemie, TU Berlin, Journal of Agricultural and Food Chemistry, **51**(24), 6961–6969, Berlin, Germany, **2003**

Ramadan, M. F.; Mörsel, J.; *Oil Composition of Coriander (Coriandrum Sativum L.) Fruit-Seeds,* European Food Research and Technology, **215**, 204–209, **2002a**

Ramadan, M.; Mörsel, J.; *Direct isocratic normal-phase HPLC assayoffat-soluble vitamins and β-carotene in oilseeds,* European Food Research and Technology, **214**(6), 521–527,632–527, **2002b**

Ramadan, M. F.; Mörsel, J.; Oil *cactus pear (Opuntia ficus-indica L.),* Food Chemistry, **82**, 339–345, **2003a**

Ramadan, M.; Mörsel, J.; *Determination of the Lipid Classes and Fatty Acid Profile of Niger (Guizotica abyssinica Cass.) Seed Oil,* Phytochemical Analysis, **14**(6), 366–370, **2003b**

Ramadan, M. F.; Mörsel, J.; *Oxidative Stability of Black Cumin (Nigella Sativa L.), Coriander (Coriandrum Sativum L.) and Niger (Guizotia Abyssinica Cass.). Crude Seed Oils Upon Stripping*, European Journal of Lipid Science and Technology, **106**, 35–43, **2004**

Raman, N.; Nagencran, S.; Sundaramahalingam, A.; *Extraction, characterization andantimicrobial activities afmango seed oil,* Dep. Chem., VHNSN College, Virudhunagar, India, Asian Journal of Chemistry, **9**(2), 321–323, **1997**

Reddy, S. Y.; Prabhakar, J. V.; *Confectionery fat from phulwara (Madhuca butyracea) butter,* Fett, Wissenschaft, Technologie, **96**(10), 387–390, **1994**

Reinberg, V.; Geissler, S.; Klug, S.; Miehle, H.; *Äpfel & Birnen Cascade,* Berichte aus Energie- und Umweltforschung, **27**, **2009** www.bmvit.at

Reiners, J.; *Objektivierung des Aromas von nativen Olivenölen unterschiedlicher Provenienz durch chemisch-instrumentelle und sensorische Untersuchungen,* Dissertation TU-München, München, **1997**

Reiter, B.; Lechner, M.; Lorbeer, E.; *The fatty acid profiles of different Umbelliferae seed oils including petroselinic an cis-vaccenic acid.,* Fett/Lipid **100**(11), 498–502, Wien, **1998**

Repo-Carrasco, R.; Espinoza, C.; Jacobsen, S. E.; *Nutritional Value and Use of the Andean Crops Quinoa (Chenopodium Quinoa) and Kaniwa (Chenopodium pallidicaule),* Food Reviews International, **19**(1,2), 179–189, **2003**

Riar, S.; Devakumar, C.; Ilavazhagan, G.; Bardhan, J.; Kain, A. K.; Thomas, P.; Singh, R.; Singh, B.; *Volatile fraction ofneem oil as a spermicide,* Contraception, **42**(4), 479–487, **1990**

Roche Lexikon Medizin, 3.Auflage, Verlag Urban & Schwarzenberg, **1993**

Rode, J.; *Studyof Autochthon Camelina sativa (L.) Crantz in Slovenia,* Journal of Herbs, Spices and Medicinal Plants, **9**(2/3 and 4), 313–318, **2002**

Rodriguez, B.; Rodriguez, A.; DeBarrio, M.; Tornero, P.; Baeza, M. L.; *Asthma inducedbycanary food mix,* Allergy Asthma Proc., **24**(4), 265–8, **2003**

Röbbelen, G.; Downey, R. K.; Ashri, A.; *Oil Crops of the World,* Mc Graw-Hill, **1989**

Rossell, J. B.; Pritchard, J.; *Analysis of oilseeds, fats and fatty foods,* Elsevier Science Publishers LTD, Essex, England, **1991**

Roth, L.; Kormann, K.; *Ölpflanzen-Pflanzenöle,* Ecomed-Verlag, Landsberg am Lech, **2000**

Roth, L.; Kormann, K.; *Atlas of Oil Plants and Vegetable Oils*, Agrimedia GmbH, Bergen/Dumme Deutschland, **2005**

Rukmini, C.; *Reproductive toxicology and nutritional studies on mahua oil (Madhuca latifolia)*, Nat. Inst. of Nutrition, Ind. Council of Med. Res., Hyderabad, India, Food Chem. Toxicol., **28**(9), 601–605, **1990**

Rukmini, C.; Vijayaraghavan, M.; *Nutritional and toxicological evaluation ofmango kernel oil*, Nat. Inst. Nutr., Indian Counc. Med. Res., Hyderabad, India, Journal of the American Oil Chemists Society, **61**(4), 789–92, **1984**

Saad, I.; Ben Marzoug, I.; Sakli, F.; *Effects of solvent extraction on Tunisian esparto wax composition*, Algerian Journal of natural Products **4**(2), 308–315, **2016**

Ryan, E.; Galvin, K.; O'Connor, T. P.; Maguire, A. R.; O'Brien, N. M.; *Phytosterol, Squalene, Tocopherol Content and Fatty Acid Profile of Selected Seeds, Grains, and Legumes*, Plant Foods for Human Nutrition, **62**, 85–91, **2007**

Ryu, S.; Kim, S. M.; Xi, J.; Ho, C.-T.; *Influence ofseedroastingprocess on the changes in volatile compounds of the sesame (Sesamum indicum) oil*, Flavour and Chemistry of Ethnic Foods, Proceedings of a Meeting held during the 5th Chemical Congress of North America, Cancun, 229-237, **1999**

SaiRam, M.; Ilavazhagan, G.; Sharma, S. K.; Dhanraj, S. A.; Suresh, B.; Parida, M. M.; Jana, A. M.; Devendra, K.; Selvamurthy, W.; *Anti-microbial activity ofa new vaginal contraceptive NIM-76 from neem oil (Azadirachta indica)*, Journal of Ethnopharmacology, **71**, 377–382, **2000**

SaiRam, M.; Sharma, S. K.; Ilavazhagan, G.; Kumar, D.; Selvamurthy, W.; *Immunomodulatory effects of NIM-76, a volatile Fraction from Neem Oil*, Journal of Ethnopharmacology, **55**(2), 133–139, **1997**

Saleem, M.; Zaka, S.; Shakir, N.; Khan, S. A.; *Studies on Marigold Seed Oil and Seed Meal*, Fette, Seifen, Anstrichmittel, **88**(5), 178–180, **1986**

Salunkhe, D. K.; Chavan, J. K.; Adsule, R. N.; Kadam, S. S.; *World Oilseeds, Chemistry, Technology and Utilization*, AVI-Verlag, New York, **1992**

Salunkhe, D. K.; Chavan, J. K.; Adsule, R. N.: Kadam, S. S.; *World Oilseeds, Chemistry, Technology and Utilization*, Van Nostrand Reinhold, New York, **2003**

Salunkhe, D. K.; Desai, B. B.; *Postharvest Biotechnology of oilseeds*, CRC Press, Boca Raton, **1986**

Samadi, S.; Khaiyamiand, M.; Hasanzadeh, A.; *A Comparison of Important Physical and Chemical Characteristics of Six Lallemantia Iberica (Bieb.) Fisch. and Mey. Varieties*, Pakistan Journal of Nutrition, **6**(4), 387–390, **2007**

Santamaria, R. I.; Soto, C.; Zúniga, M. E.; Chamy, R.; Lopez-Munguia, A.; *Enzymatic Extraction of Oil from Gevuina avellana, the Chilean Hazelnut*, Journal of the American Oil Chemists Society, **80**(1), 33–36, **2003**

Sarin, R.; Sharma, M.; Khan, A. A.; *Studies on Guizotia Abyssinica L. Oil: Biodiesel Synthesis and Process Optimization*, Bioresource Technology, **100**, 4187–4192. **2009**

Schleicher, P.; Saleh, M.; *Natürlich heilen mit Schwarzkümmel*, Südwest Verlag, München, **2007**

Schmeda-Hirschmann, G.; *Madia Sativa, a Potential Oil Crop of Central Chile*, Economic Botany, **49**(3), 257–259, **1995**

Schmidt, E. B.; Aardestrup, I.; Mellerup, J.; Christensen, J. H.; *The effect ofsheanut oil on serum lipids and lipoproteins in normocholesterolemic and mildly hypercholesterolemic humans*, NutraCos, **2**(3), 14–18, **2003**

Schneebeli-Graf, R.; *Blütenland China, Botanische Berichte und Bilder*, Birkhäuser Verlag, Basel, **1995**

Schneidermaier, B.; roemp.thieme.de/lexicon/RD-05-01740, retrieved 02 02 2020

Schoenfelder, I.; Schoenfelder, P.; *Der neue Kosmos Heilpflanzenführer*, Franckh-Kosmos Verlags-GmbH&Co., Stuttgart, **2001**

Schubert, S. Y.; Lansky, E. P.; Neeman, I.; *Antioxodant and eicosanoid enzyme inhibition properties of pomegranate seed oil and fermented juice flavonoids*, Journal of Ethnopharmacology, **66**(1), 11–17, **1999**

Schuette, H. A.; Enz, W. W. F.; *Expressed Brazil-nut-oil,* Journal of the American Chemical Society, **53,** 2756–8, **1931**

Schuette, H. A.; Thomas Ralph, W.; Duthey, M.; *Brazil-nut-oil,* Journal of the American Chemical Society, **52,** 4114–7, **1930**

Schuster, W. H.; *Ölpflanzen in Europa,* DLG-Verlag, Frankfurt, **1992**

Schuster-Gajzágó, I.; Kiszter, A. K.; Tóth-Márkus, M.; Baráth, A.; Márkus-Bednarik, Z.; Czukor, B.; *The Effect of Radio Frequency Heat Treatment on Nutritional and Colloid-Chemical Properties of Different White Mustard (Sinapis alba L.) Varieties,* Innovative Food Science and Emerging Technologies, **7**, 74–79, **2006**

Schwartz, H.; Ollilainen, V.; Piironen, V.; Lampi, A. M.; *Tocopherol, tocotrienol and plant sterol contents of vegetable oils and industrial fat S.* Journal of Food Composition and Analysis **21,** 152–161, **2008**

Schwedes, U.; Köhler, S.; Niestroj, M.; *Dietätische Intervention von Fettstoffwechselstörungen bei Typ-2-Diabetikern mit Perillaöl und den antioxidativen Vitaminen C und E als ergänzende bilanzierte Diät,* Ernährung & Medizin, **19,** 73–78, **2004**

Schwingshackl, L.; Hoffmann, G.; *Einfach ungesättigte Fettsäuren in der Primärpräventation von Herz-Kreislauf-Erkrankungen bei Übergewichtigen/Adipösen und Diabetikern,* Aktuelle Ernährungsmedizin, 36, 44–49, **2011a**

Schwingshackl, L.; Hoffmann, G.; *Empfehlungen für die Fettzufuhr im Wandel?,* Aktuelle Ernährungsmedizin, 36, 167–168, **2011b**

Seehuber, R.; *Versuche mit zur Produktion von Erucasäure geeigneten Kulturarten,* Fett Wissenschaft Technologie, **89**(7), 263–268, **1987**

Segall, S. D.; Artz, W. E.; Raslan, D. S.; Ferraz, V. P.; Takahashi, J. A.; *Triacylglycerol Analysis of Pequi (Caryocar brasiliense Camb.) Oil by Electrospray and Tandem Mass Spectrometry,* Journal of the Science of Food and Agriculture, **86**, 445–452, **2006**

Seibel, W.; *Warenkunde Getreide*, Agrimedia, Bergen/Dumme, **2005**

Seibert, B.; Food from Dipterocarps: *Utilization of the Tengkawank Species Group for Nut and Fat Production*, in: Schulte, A. und Schöne, D.; *Diptocarp forest ecosystems: Towards ustainable Management,* World Scientific Books, Singapure, **1996**

Seidemann, J.; *Falsification of seed oil of Argania spinosa,* Deutsche Lebensmittel-Rundschau, **94**(1), 26–27, **1998**

Sempore, G.; Bezard, J.; *Triacylglycerol structure of an African peanut oil,* University Dijon, American Oil Chemists' Society, **59**(3), 124–9, **1982**

Sengupta, A.; Sengupta, C.; Das, P. K.; *Interesterfication of moringa oil for utilization in pharmazy,* Indian Journal of Technology, **12**(8), 362–5, **1974**

Sen Gupta, S.; Chakrabarty, M. M.; Bhattacharyya, D. K.; *Variation in fatty acid composition of mowrah fat,* Dep. Appl. Chem., Univ. Coll. Sci. Technol., Calcutta, India, Fette, Seifen, Anstrichmittel, **84**(6), 226–228, **1982**

Senti, G.; Ballmer-Weber, B. K.; Wüthrich, B.; *Nüsse, Samen und Kerne aus allergologischer Sicht,* Schweizer Medizinische Wochenschrift, **130,** 1795–1804, **2000**

Seri, S.; D'Alessandro, A.; *Blackcurrant seed oil, zinc, and fetal alcohol syndrom,* Journal of Biological Research (Naples), **73**(1–2), 15–21, **1997**

Servili, M.; Selvaggini, R.; Taicchi, A.; Montedoro, G. F.; *Headspace composition ofvirgin olive oil evaluated by solid phase microextraction: relationships with the oil sensory characteristics,* University of Molise, Italy, Special Publication – Royal Society of Chemistry, Food Flavors and Chemistry, **274,** 236–247, **2001**

Shahidi, F.; *Canola and Rapeseed- Production, Chemistry, Nutrition and Processing Technology$_L$* Van Nostrand Reinhold, New York, **1990**

Shao, H.-K.; *Use of Chinese Vegetable tallow as suppository base,* Yauxue Xuebao, **6,** 130–134, **1958**

Sharaf-Eldin, M. A. A.; *Studies on the Effect of Some Agricultural Treatments on Growth and Productivity of Artichoke (Cynara cardunculus var. scolymus (L.) Fiori) and their Relation to Earliness and Physical and Chemical Characters of Heads,* Dissertation an der technischen Universität München, **2002**

Shikov, A. N.; Pozharitskaya, O. N.; Makarov, V. G.; Makarova, M. N.; *Anti-Inflammatory Effect of Pinus Sibirica Oil Extract in Animal Models*, Journal of Natural Medicines, **62**, 436–440, **2008**

Shimoda, M.; Nakada, Y.; Nakashima, M.; Osajima, Y.; *Headspace gas analysis of volatile compounds of light and deep roasted sesame seed oil,* Food Science and Technology International, **4**(1), 14–17, **1998**

Shimoda, M.; Nakada, Y.; Nakashima, M.; Osajima, Y.; *Quantitative Comparison of Volatile Falvor Compounds in Deep-Roasted and Light-Roasted Sesame Seed Oil,* Journal of Agricultural and Food Chemistry, **45**(8), 3193–3196, **1997**

Shimomura, K.; Koizumi, Y.; *Topicalpreparations containing cashew nutshell oils or their extracts,* Patent No. JP 03240718, Jpn. Kokai Tokkyo Koho, **1991**

Shimomura, K.; Nakamura, M.; *Skin-lightening cosmetics containing cardolor cashew nut kernel oil (extracts),* Patent No. JP 04089419, Jpn. Kokai Tokkyo Koho, **1991**

Shin, Jung-Ah; Lee, Jong-Ho; Lee, Ki-Teak; *Identification ofvolatile compounds in structured lipid of safflower oil using electronic nose and solid phase microextraction gas chromatography-mass spectrometry,* Department of Food Science and Technology, Chungham National Univerity, Korea, Agricultural Chemistry and Biotechnology, **46**(4), 152–155, **2003**

Shukla, V. K. S.; Dutta, P. C.; Artz, W. E.; *Camelina Oil and Its Unusual Cholesterol Content,* Journal of the American Oil Chemists Society, **79**(10), 965–969, **2002**

Shukla, Y. N.; Dubey, S.; Jain, S. P.; Kumar, S.; *Chemistry, biology and uses of Adansonia digitata-areview*, Journal of Medicinal and Aromatic Plant Sciences, **23,** 429–434, **2001**

Shwaireb, M. H.; *Inflammatory effects of the tumor promoter croton oil on BALB/c mice skin,* Oncology Reports, **2**(1), 133–135, **1995**

Siegmund, B.; Murkovic, M.*; Changes in chemical composition of pumkin seeds during the roasting process for production of pumkin seed oil (Part 2: volatile compounds),* Institute for Food Chemistry and Food Technology, Technical University Graz, Austria, Food Chemistry (2003), Volume Date 2004, **84**(3), 367–374, **2004**

Siew, W. L.; *Cristallisation and melting behaviour of palm kernel oil and related products by differential scanning calorimetry,* European Journal of Lipid Science and Technology, **103,** 729–734, **2001**

Simopoulos, A. P. (Editor); *Plants in Human Nutrition,* World Review of Nutrition and Dietetics, Volume 77, Karger, Basel, **1995**

Singh, K.; Gupta, I. S.; Bansal, P. C.; Jindal, K. S.; *Extraction of Coffee oil,* Science, **21**(Pt.1–2), 49–54, **1970**

Singh, S. P.; Khanna, K. R.; Dixit, B. S.; Srivastava, S. N.; *Fatty acid composition ofopium poppy (Papaversomniferum) seedoil,* Indian Journal of Agricultural Science, **60**(5), 358–359, **1990**

Singhal, R. S.; Kulkarni, P. R.; *Effect of Puffing on Oil Characteristics of Amaranth (Rajgeera) Seeds,* JAOCS, **67**(12), 952–954, **1990**

Skakovskii, E. D.; Tychinskaya, L. Y.; Gaidukevich, O. A.; Klyuev, A. Y.; Kulakova, A. N.; Petlitskaya, N. M.; Rykov, S. V.; *NMR Analyis of Oils from Pine Nuts (Pinus Sibirica) and Seeds of Common Pine (Pinus Silvestris L.),* Journal of Applied Spectroscopy, **47**(4), 584–588, **2007**

Small, E.; *Culinary Herbs*, NRC Research Press, Ottawa, 2. Auflage, **2006**

Smith, R. L. Jr.; Malaluan, R. M.; Setianto, W. B.; Inomata, H.; Arai, K.; *Separation of cashew (Anacardium occidentale L.) nut shell liquid with supercritical carbon dioxide,* Bioresource Technology, **88**,1–7, **2003**

Snyder, J. M.; Frankel, E. N.; Selke, E.; *Capillary gas chromatographic analyses of headspace volatiles from vegetable oils,* American Oil Chemists Society, **62**(12) 1675–9, **1985**

Sotheeswaran S.; Sharif, M. R.; Moreau, R. A.; Piazza, G. J.; *Lipids from the seeds ofseven Fijian plant species,* Food Chemistry, **49,** 11–13, **1994**

Soulier, P.; Lecerf, J. C.; Farines, M.; Soulier, J.; *Chemical composition of sal and illipe butters,* Revue Francaise des Corps Gras, **36**(9–10), 361–365, **1989**

Spino, C.; Dodier, M.; Sotheeswaran, S.; *Anti-HIV Coumarins from Calophyllum Seed Oil,* Bioorganic and Medicinal Chemistry Letters, **8,** 3475–3478, **1998**

Squibb, R. L.; Love, H. T.; Wyld, M. K.; *The Digestibility of Six Tropical Fats Determined on Rats,* The Journal of Nutrition, **44**(4), 547–552. **1951**

Stadlober, H.; T*ailands Problem mit dem Palmöl,* Die Presse, 28. 6. 2012

Stanisavljević, I. T.; Lazić, M. L.; Veljković, V. B.; *Ultrasonic Extraction of Oil from Tobacco (Nicotiana tabacum L.) Seeds,* Ultrasonics Sonochemistry, **14**, 646–652, **2007**

Steen, S.; Mörsel, J.-T.; *Innovative raw materials already known in ancient Tibet. Sea buckthorn oils.* CLB Chemie in Labor und Biotechnik, **54**(4), 128–135, **2003**

Steinke, G.; Weitkamp, P.; Klein, E.; Mookherjee, K. D.; *High-Yield Preparation of Wax Esters via Lipase-Catalyzed Esterification Using Fatty Acids and Alcohols from Crambe and Camelina Oils,* Journal of Agricultural and Food Chemistry, **49,** 647–651, **2001**

Strik, B.; Finn, C.; Clark, J.; Pilar Banados, M.; *Worldwide Production of Blackberries*, **2006,** https://berrygrape.org/worldwide-production-of-blackberries, 16.05.2011

Stübiger, G.; unpublishes data, **2019**

Subrahamnyam, V. V. R.; Achaya, K. T.; *Lesser-Knwon Indian Vegetable Fats. I. Oleic-Rich Fats,* Journal of the Science of Food and Agriculture, **8**(11), 657–662, **1957**

Sugano, M.; Ikeda, I.; Wakamatsu, K.; Oka, T.; *Influence of Koreanpine (Pinus koraiensis)-seedoil containing cis-5,cis-9,cis-12-octadecatrienoic acid on polyunsaturated fatty acid metabolism, eicosanoid production and blood pressure of rats,* British Journal of Nutrition, **72**(5), 775–783, **1994**

Sun, Q.; Ding, X.; *Supercritical CO_2 extraction oftomato seed oil from tomato seeds,* Zhongguo Youzhi, **23**(3), 12–14, **1998**

Sun, H.; Wiesenborn, D.; Rayas-Duarte, P.; Mohamed, A.; Hagen, K; *Bench-scale porcessing of amaranth seed for oil,* J. Am. Oil Chem Soc., **72**, 1551–1555, **1995**

Sun, L.; Tamaki, H.; Ishimaru, T.; Teruya, T.; Ohta, Y.; Katsuyama, N.; Chinen, I.; *Inhibition of Osteoporosis Due to Restricted Food Intake by the Fish Oils DHA and EPA and PErilla Oil in the Rat,* Biosciences, Biotechnology and Biochemistry, **68**(12), 2613–2615, **2004**

Sunday, S. A.; *Physical, Chemical and Functional Properties of Nigerian Mango (Mangifera indica) Kernel and its Processed Flour,* Journal of Science and Food Agriculture, **73**(3), 321–328,**1997**

Sutherland, W.; De Jong, S.; Walker R.; Williams, S.; *Release of cholesterol from cell membranes to postprandial plasma from mildly hypercholesterolemic subjects: the effect of meals rich in olive and safflower oils$_L$* University of Otago Medical School, New Zealand, Metabolism: clinical and experimental, **51**(10), **2002**

Taga, S. M.; Miller, E. E.; Pratt, D. E.; *Chia Seeds as a Source of Natural Lipid Antioxidants,* Journal of the American Oil Chemists' Society, **61**, 928–931, **1984**

Tarandzhiiska, R.; Nguyen, H.; Lichev, V.; *Triglyceride composition oftomato, pepperand grape seed oils by argentation thin-layer chromatography,* Rivista Italiana delle Sostanze Grasse, **68**(6), 309–312, **1991**

Tatematsu, K.; Hirose, N.; Ichikawa, Y.; Fujii, Y.; Takami, A.; Okuyama, H.; *Nutritional Evaluation of an Inter-Esterified Perilla Oil and Lard in Comparison with Butter and Margarine Based on the Survival of Stroke-Prone Spontaneously Hypertensive (SHRSP) Rats,* Journal of Health Science, **50**(1), 108–111, **2003**

Tella, A.; *Preliminary studies on nasal decongestant activity from the see of the shea butter tree, Butyrospermum parkii,* British Journal of Clinical Pharmacology, **7**(5), 495–497, **1979**

Terpstra, A. H.; Katan, M. B.; Weusten Van Der Wouw, M. P.; De Roos, B.; Beynen, A. C.; *The hypercholesterolemic effect of cafestol in coffe oil in gerbils and rats,* Journal of Nutrition and Biochemistry, **11**(6), 311–317, **2000**

Teubner, C.; *Paprika: Gewürz und Gemüse,* Gräfe u.Unzer, München, **1993**

Thippeswamy, H. T.; Raina, P. L.; *Lipid composition ofkokum (Garcinia indica) and dhupa (Vateria indica),* Journal of Food Science and Technology, **26**(6), 322–326, **1989**

Toro-Vazquez, J. F.; Perez-Briceno, F.; *Chemical and physicochemical characteristics of pecan (Carya illinoinensis) oil native of the central region of Mexico,* Journal of Food Lipids, **5**(3), 211–231, **1998**

Torres, M. M.; Martinez, M. L.; Maestri, D. M.; *A Multivariate Study of the Relationship Between Fatty Acids and Volatile Flavour Components in Olive and Walnut Oils,* Journal of the American Oil Chemists' Society, **82**(2), 105–109, **2005**

Tósaki, A.; Vecsernyés, M.; Fésüs, L.; Bak, I.; Juhász, B.; Papp, L.; Toth, S.; *Application of the Oil Fraction Obtained from Sour Cherry (Prunus cerasus) Seed Kernel,* Patent No. 20100008886, Ungarn, **2010**

Traitler, H.; Wille, H. J.; Studer, A.; *Fractionation ofblack currant seed oil,* Journal of the American Oil Chemists Society, **65**(5), 755–760, **1988**

Tran, T.; *Shea butter: more than just a cosmetic ingredient,* Drug & Cosmetic Industry, **134**(2), 54,58,60-61,114,116, **1984**

Tsaknis, J.; Lalas, S.; Gergis, V.; Dourtoglou, V.; Spiliotis, V.; *Characterization of Moringa oleifera Variety Mbololo Seed Oil of Kenya,* Journal of Agricultural and Food Chemistry, **47**(11), 4495–4499, **1999**

Tsaknis, J.; Lalas, S.; Lazox, E.; *Characterization forcrude and purifiedpumpkin seed oil, Faculty of Food Technology,* Athen, Grasas y Aceites (Sevilla) **48**(5), 267–272, **1997**

Tseng, Y. C.; Lee, M. H; Li, C. F.; *Effect ofpeanut kernels roasting time onpeanut oilaroma,* Shipin Kexue (Taipei) **20**(2), 136–148, **1993**

Tsuzuki, T.; Kawakami, Y.; Abe, R.; Nakagawa, K.; Koba, K.; Imamura, J.; Iwata, T.; Ikeda, I.; Miyazawa, T.; *Conjugated Linoleic Acid Is Slowly Absorbed in Rat Intestine, but Quickly Converted to Conjugated Linoleic Acid,* Journal of Nutrition, **136**(8), 2153–59, **2006**

Tuck, K. L.; Hayball, P. J.; *Major phenolic compounds in olive oil: metabolism and health effects,* Centre for Pharmaceutical Research, Univerity of South Australia, Adelaide, Australia, Journal of Nutritional Biochemistry, **13**(11), 636–644, **2002**

Tunaru, A.; Althoff, T. F.; Nüsing, R. M.; Diener, M.; Offermanns, S.; *Castor oil induces laxation and uterus contraction via ricinoleic acid activating prostaglandin EP3 receptors*, PNAS, **109**(23), 9179–9184, **2012**

Tung, C.-Y.; Hsu, L.-C.; Chu, H.-W.; Chou, Y.; *Studies on the antifertility constituents in carrot seeds,* Zhongcaoyao, **12**(2), 13, **1981**

Tura, D.; Prenzler, P. D.; Bedgood, D. R.; Antolovich, M.; Robards, K.; *Varietal and processing effects on the volatile profile of Australian olive oils,* Charles Sturt University, Wagga Wagga, Australia, Food Chemistry, Volume Date 2004, **84**(3), 341-349, **2003**

Ulmer, Günter A.; *Heilende Öle,* Günter Albert Ulmer Verlag, Tuningen, **1996**

Umanskii, Z. M.; Krutik, Y. D.; *Making medicinals from ajowan fruits,* Farmatsiya, **8**(6), 19–24, **1945**

Umarov, A. U.; Markman, A. L.; Domanskaya, K. K.; Gusakova, S. D.; *Pomegranate seed oil,* Maslozhirovaya Promyshlennost, **35**(5), 10–12, **1969**

uses.plantnet-project.org/en/Aleurites_moluccana 2016 accessed 08012020

Uzun, B.; Arslan, C.; Karhan, M.; Toker, C.; *Fat and Fatty Acids of White Lupin (Lupinus albus L.) in Comparison to Sesame (Sesamum indicum L.)*; Food Chemistry, **102**, 45–49, **2007**

Valenta, C.; *Skriptum zu Vorlesung: Grundlagen der magistralen Arzneimittelherstellung,* Universität Wien, Wien, **2005**

Valenta, C.; *Skriptum zu Vorlesung: Grundlagen der magistralen Arzneimittelherstellung,* Universität Wien, Wien, **2008**

Valentino, A.; Fimiani, M.; Baiocchi, R.; Bilenchi, R.; Perotti, R.; Castelli, A.; Mancianti, M. L.; Raffaeli, M.; *Cosmetic acne and a test of comedogenicity,* Boll. Soc. Ital. Biol. Sper., **60**(10), 1845–8, **1984**

Valle, P. W. P. A.; Rezende, T. F.; Souza, R. A.; Fortes, I. C. P.; Pasa, V. M. D.; *Combination of Fractional Factorial and Doehlert Experimental Designs in Biodiesel Production: Ethanplysis of Raphanus sativus L. var. Oleiferus Stokes Oil Catalyzed by Sodium Ethoxide*, Energy Fuels, **23**, 5219–5227, **2009** van Hoed, V.; De Clercq, N.; Echim, C.; Andjelkovic, M.; Leber, E.; Dewettinck, K.; Verhé, R.; *Berry Seeds: A Source of Specialty Oils with High Content of Bioactives and Nutritional Value*, Journal of Food Lipids, **16**, 33–49, **2009**

Vargas-Lopez, J. M.; Wiesnborn, D.; Tostenson, K.; Cihacek, L.; *Processing of Crambe for Oil and Isolation of Erucic Acid*, Journal of the American Oil Chemists' Society, **76**(7), 801–809, **1999**

Vaughan, J. G.; Hemingway, J. S.; *The Utilization of Mustards, Economic Botany*, **13**(3), 196–204, **1959**

Vecera, R.; Skottovä, N.; Väöa, P.; Kazdovä, L.; Chmela, Z.; Svagera, Z.; Walterovä, D.; Ulrichovä, J.; Simänek, V.; *Antioxidant Status, Lipoprotein Profile and Liver Lipids in Rats Fed on High-Cholesterol Diet Containing Currant Oil Rich in n-3 and n-6 Polyusaturated Fatty Acids,* Physiological Research, **52,** 177–187, **2002**

Veljković, V. B.; Lakićević, S. H.; Stamenković, O. S.; Todorović, Z. B.; Lazić, M. L.; *Biodiesel Production from Tobacco (Nicotiana tabacum L.) Seed Oil with a High Content of Free Fatty Acids*, Fuel, **85**, 2671–2675, **2006**

Vichi, S.; Castellote, A. I.; Pizzale, L.; Conte, L. S.; Buxaderas, S.; Lopez-Tamames, E.; *Analysis of virgin olive oil volatile compounds by headspace solid-phase microextraction coupled to gas chromatography with mass spectrometric anfflame ionization detection,* Universita di Udine, Udine, Italy, Journal of Chromatography, **983** (1–2), 19–33, **2003**

Viernstein, H.; *Skriptum zur Vorlesung: Pharmazeutische Technologie I und II*, Universität Wien, Wien, **2007**

Visioli, F.; Galli, C.; *In vitro and in vivo studies ofthe effects of olive oil phenolics on oxidative stress,* Department of Pharmacological Sciences, University of Milan, Italy, Abstracts of Papers, *224th* ACS National Meeting, Boston, United States, August 18–22,2002, American Chemical Society, USA, **2002**

Von Koenen, E.; *Heil-, Gift- und essbare Pflanzen in Namibia,* Klaus Hess Verlag, Göttingen/Windhoek, **2001**

Von Schoenebeck, O.; *Chemical and physiological investigations ofmustard plants, mustard seed and mustard oils,* Biedermanns Zentr. B. Tierernahr., **8**(373–96), **1936**

Von Wiesner, J.; *Die Rohstoffe des Pflanzenreichs,* 1.Band, 4.Auflage, Verlag von Wilhelm Engelmann, Leipzig, **1927**

Wagner, K-H.; Auer, E.; Elmadfa, I.; *Content of trans fatty acids in margarines, plant oils, fried products and chocolate spreads in Austria*, European Food Research and Technology, **210**(4), 237–241, **2000**

Wagner, K-H.; Plasser E., Proell C; *Trans-Fettsäuren eine weite Observation ist notwendig*, Ernährung aktuell, 3, **2006**

Wagner, K-H; Plasser E., Proell C; *Trans-Fettsäuren in Lebensmitteln des Österreichischen Marktes*, Ernährung aktuell, 1, **2007**

Wagner, K-H.; Plasser, E.; Proell, C.; Kanzler, S.; *Comprehensive studies on the trans fatty acid content of Austrian foods: Convenience products, fast food and fats,* Food Chemistry, **108**(3), 1054–1060, **2008**

Wagner, K.-H.; Tomasch, R.; Elmadfa, I.; *Impact of diets containing corn oil or olive/sunflower oil mixture on the human plasma and lipoprotein lipid metabolism*, European Journal of Nutrition, **40**(4), 161–167, **2001**

Wang, W.; Sui, Z.; Wang, F.; *Black currant seed oil mutagenicity test,* Weisheng Dulixue Zazhi, **12**(2), 105, **1998 –** zitiert aus Abstract

Wang, X.; Lan, M.; Wu, H.-P.; Shi, Y.-Q.; Lu, J.; Wu, K.-C.; Jin, J.-P.; Fan, D.-M.; *Direct effect of croton oil on intestinal epithelial cells and colonic smooth muscle cells,* World Journal of Gastroenterology, **8**(1), 103–107, **2002**

Waniorek, A.; *Kürbis und Kürbiskernöl* – das praktische Handbuch zur Anwendung bei Infektionen, Entzündungen und Stoffwechselstörungen, mvg-Verlag, Landsberg am Lech, **1997**

Warburg, O.; Van Someren Brand, J. E.; *Kulturpflanzen der Weltwirtschaft,* R. Voigtländers Verlag, Leipzig, **1909**

Warnigati, S.; Agra, I. B.; Prasetyanigsih, E.; *Alcoholysis of Nyamplung (Calophyllum inophyllum Linn) seeds oil to diesel oil-like fuel,* Renewable Energy: Technol.Environ., Proc. World Renewable Energy Congress, **2nd, 3,** 1479–83, **1992**

Watkins, T. R.; Bierenbaum, M. L.; *Dietary polyunsaturated fat in cardiovascular disease: boon or bane?,* Jordan Heart Research Foundation, Montclair, NJ, USA, ACS Symposium Series, **788** (Omega-3 Fatty Acids), 66–78, **2001**

Wenigmann, M.; *Phytotherapie,* Urban & Fischer Verlag, München, **1999**

Westerman, D.; Santos, R. C. D.; Bosley, J. A.; Rogers, J. S.; Al-Duri, B.; *Extraction of Amranth seed oil by supercritical carbon dioxide,* Journal of Supercritical Fluids, **37**, 38–52, **2006**

Whiley, A. W.; Schaeffer, B.; Wolstenholme, B. N.; *Avocado – Botany, Production and* Uses, CABI Publishing, Wallingford, **2002**

Wiesnborn, D.; Doddapaneni, R.; Tostenson, K.; Kangas, N.; *Cooking Indices to Predict Screw-Press Performance for Crambe Seed,* Journal of the American Oil Chemists' Society, **78**(5), 467–471, **2001**

Wilen, R. W.; Barl, B.; Slinkard, A. E.; *Feasibility of Cultivation Calendula as a Dual Purpose Industrial Oilseed and Medicinal Crop,* Acta Horticulture, **629**, 199–206, **2004**

Wimmer, E.; Mackwitz, H., Schemitz, S.; Burner, U.; Stadlbauer, W.; *NaWaRo-Cascading für die Wellness-Regio,* Berichte aus eergie- und Umweltforschung, **18**, 1–222, **2003**

Witzel, D. A.; Dollahite, J. W.; Jones, L. P.; *Photosensitization in sheep fed Ammi majus (Bishop's weed) seed,* American Journal of Veterinary Research, **39**(2), 319–320, **1978**

Wolff, R.; Bayard, C. C.; *Fatty acid composition of some pine seed oils,* Journal of the American Oil Chemists Society, **72**(9), 1043–1046, **1995**

Wolff, R.; Christie, W. W.; Coakley, D.; *The Unusual Occurrence of 14-Methylhexadecanoic Acid in Pinaceae Seed Oils Among Plants,* Lipids, **32**(9), 971–973, **1997**

Wolff, R. L.; Pédrono, F.; Pasquier, E.; Marqeau, A. M.; *General Characteristics of Pinus spp. Seed Fatty Acid Compositions, and Importance of Δ5-Olefinic Acids in the Taxonomy and Phylogeny of the Genus,* Lipids, **35**(1), 1–22, **2000**

Woo C. C.; Loo S. Y.; Gee V.; Yap C. W.; Sethi G.; Kumar A. P.; Benny Tan K. H.; *Anticancer activity of thymoquinone in breast cancer cells: Possible involvement of PPAR-γ pathway,* Biochemical Pharmacology, **82,** 464–475, **2011**

Woo C. C.; Kumar A. P.; Sethi G.; Benny Tan K. H.; *Thymoquinone: Potential cure for inflammatory disorders and cancer.* Biochemical Pharmacology, **83**, 443–451, **2012**

Wu, C. M.; Chen, S. Y.; *Volatile compounds in oils after deep frying or stir frying and subsequent storage,* Journal of American Oil Chemists Society, **69**(9), 858–65, **1992**

Wu, D.; Meydani, M.; Leka, L. S.; Nightingale, Z.; Handelman, G. J.; Blumberg, J. B.; Meydani, S. N.; *Effect of dietary supplementation with black currant seed oil on the immune response of healthy elderly subjects,* American Journal for Clinical Nutrition, **70**,536–543, **1999**

Wurm, L.; *Marillen, Anbau – Pflege – Verarbeitung,* Österreichischer Agrarverlag, Leopoldsdorf, **2002**

www.umweltbundesamt.de/sites/default/files/medien/publikation/long/3540.pdf retrieved 20.2.19

www.test.de/ retrieved 20.2.19

www.test.de/Gourmet-Oele-Fast-jedes-zweite-ist-mangelhaft-4901060-0/ retrieved 20.2.19

Xing, J.; Yang, B.; Dong, Y.; Wang, B.; Wang, J.; Kallio, H. P.; *Effects of sea buckthorn (Hippophae rhamnoides L.) seed and pulp oils on experimental models ofgastric ulcer in rats,* Fitoterapia, **73,** 644–650, **2002**

Yaghmur, A.; Aserin, A.; Garti, N.; *Evaluation of Argan Oil for Deep-Fat Frying,* Lebensmittel-Wissenschaft und Technologie, **34,** 124–130, **2001**

Yan, S.; Li, L.; Gao, M.; Yu, S.; Zhao, S.; Xue, L.; Quan, L.; *Regulation ofblood lipids by fatty oil extracted from common perilla (Perilla frutescens),* Zhongcaoyao, **24**(4), 193–194, **1993**

Yang, B.; Kalimo, K. O.; Tahvonen, R. L.; Mattila, L. M.; Katajisto, J. K.; Kallio, H. P.; *Effect of dietary supplementation with sea buckthorn (Hippophae rhamnoides) seed and pulp oils on the fatty acid composition of skin glycerophospholipids of patients with atopic dermatitis,* Journal of Nutrition and Biochemistry, **11,** 338–340, **2000**

Yang, B.; Kallio, H.; Koponen, J.; Tahvonen, R.; *Free and esterified sterols in seed oil and pulp/peel oil of sea buckthorn (Hippophae rhamnoides L.),* Special Publication – Royal Society of Chemistry, **269** (Biologically-activa Phytochemicals in Food), 24–27, **2001**

Yao, M.; Li, J.; Zhang, Y.; *Investigation on development and use of kiwi seed oil,* Shipin yu Fajiao Gongye, **27**(12), 28–30, **2001**

Yaqoob, P.; *Monounsaturated fatty acids and immune function,* The Hugh Sinclair Unit of Human Nutrition, University of Reading, U.K., European Journal of clinical nutrition, **56** Suppl.3, S.9–13, **2002**

Yildiz, M.; Turcan Gurcan, S.; Ozdemir, M.; *Oil composition of pistachio nuts (Pistacia vera) from Turkey,* TUBITAK Marmara Research Center, Gebeze, Turkey, Fett/Lipid **100**(3), 84–86, **1998**

Young-Hoi, K.; Kun-Soo, K.; Chong-Ki, H.; *Volatile components of parsley leaf and seed (Petroselinum crispum),* Hanguk Nonghwahak Hoechi, **33**(1), 62–7,**1990**

Yu, H. C.; *Perilla, The Genus Perilla,* Harwood Academic Publishers, Amsterdam, **1997**

Yu, X.; Voort, F. R.; Li, Z.; Yue, T.; *Proximate Composition of the Apple Seed and Characterization of its Oil,* International Journal of Food Engineering, **3**, 5, 12, **2007**

Yukui, R.; Wenya, W.; Rashid, F.; Qing, L.; *Fatty Acids Compsition of Apple and Pear Seed Oils,* International Journal of Food Properties, **12**, 774–779, **2009**

Zachariah, T. J.; Shajiprabha, S. P.; Krishnamoorthy, B.; *Major Chemical Constituents of Nut, Mace and Leaf of Myristica fragrans*, Centennial Conference on Spices and Aromatic Plants, **20-23**, 297–300, **2000**

Zardini, E.; *Madia Sativa Mol. (Asteraceae-Heliantheae-Madiinae): An Ethnobotanical and Geographical Disjunct*, Economic Botany, **46**(1), 33–43, **1992**

Zhang, J.; Zhan, L.; He, Y.; *Composition, determination and application of passion flower seed oil,* Zhonggou Youzhi, **25**(6), 116–118, **2000**

Zibaeenezhad, M.; Rezaiezadeh, M.; Mowla, A.; Ayatollahi, S.; Panjehshahin M. R.; *Antihypertriglyceridemic effect of walnut oil,* Cardiovascular Research Center, University of Medical Science, Iran, Angiology **54**(4), 411–4, **2003**

Zlatanov, M.; Angelova, M.; Antova, G.; *Lipid Composition of Tobacco Seeds*, Bulgarian Journal of Agricultural Science, **13**, 539–544, **2007**

Zlatanov, M.; Ivanov, S.; Antonova, G.; Kouleva, I.; *Study of Phospholipid Composition of Rosaceae Seed Oils*, Rivista Italiana delle Sostanze Grasse, **74**(9), 409–4010, **1997**

Zubr, J.; *Dietary Fatty Acids and Amino Acids of Camelina Sativa Seed,* Journal of Food Quality, **26,** 451–462, **2002**

MIX
Papier aus verantwortungsvollen Quellen
Paper from responsible sources
FSC® C105338

If you have any concerns about our products,
you can contact us on
ProductSafety@springernature.com

In case Publisher is established outside the EU,
the EU authorized representative is:
Springer Nature Customer Service Center GmbH
Europaplatz 3, 69115 Heidelberg, Germany

Printed by Libri Plureos GmbH
in Hamburg, Germany